Essentials of Chemistry

Essentials of Chemistry

Edited by **Gerald Cole**

NYRESEARCH PRESS

New York

Published by NY Research Press,
23 West, 55th Street, Suite 816,
New York, NY 10019, USA
www.nyresearchpress.com

Essentials of Chemistry
Edited by Gerald Cole

International Standard Book Number: 978-1-63238-486-7 (Hardback)

The publisher's policy is to use permanent paper from mills that operate a sustainable forestry policy. Furthermore, the publisher ensures that the text paper and cover boards used have met acceptable environmental accreditation standards.

Trademark Notice: Registered trademark of products or corporate names are used only for explanation and identification without intent to infringe.

Printed in the United States of America.

Contents

Preface

Chemistry is a branch of science which studies the composition, structure, and properties of all the matter around us. Chemistry is an umbrella discipline which comprises of organic chemistry, inorganic chemistry, physical chemistry, biochemistry and many other subfields. Modern day chemistry is much more advanced; it finds its applications in many areas such as, manufacturing of dyes, medicines, fertilizers, detergents, preservatives, etc. This book traces the progress of this field and highlights some of its key concepts and applications. For all those who are interested in learning the concepts of chemistry, it will prove to be an essential guide. This book aims to serve as a valuable source of reference for engineers, scholars, researchers and students across the globe.

After months of intensive research and writing, this book is the end result of all who devoted their time and efforts in the initiation and progress of this book. It will surely be a source of reference in enhancing the required knowledge of the new developments in the area. During the course of developing this book, certain measures such as accuracy, authenticity and research focused analytical studies were given preference in order to produce a comprehensive book in the area of study.

This book would not have been possible without the efforts of the authors and the publisher. I extend my sincere thanks to them. Secondly, I express my gratitude to my family and well-wishers. And most importantly, I thank my students for constantly expressing their willingness and curiosity in enhancing their knowledge in the field, which encourages me to take up further research projects for the advancement of the area.

Editor

Dual Wavelength Spectrophotometric Method for Simultaneous Estimation of Atorvastatin Calcium and Felodipine from Tablet Dosage Form

Namdeo R. Jadhav, Ramesh S. Kambar, and Sameer J. Nadaf

Department of Pharmaceutics, Bharati Vidyapeeth College of Pharmacy, Kolhapur, Maharashtra 416013, India

Correspondence should be addressed to Namdeo R. Jadhav; nrjadhav18@rediffmail.com

Academic Editor: Yun Wei

Atorvastatin calcium (ATR) and felodipine (FEL) are beneficial in combination for elderly people in management of hypertension and atherosclerosis. Aim of present study is to develop simple, accurate, and precise method for simultaneous quantitative estimation of ATR and FEL from combined tablet dosage form. Method involves simultaneous equation, using acetonitrile—double distilled water (70 : 30)—common solvent showing absorption maxima at 245 and 268 nm. Calibration curves determination for both drugs has been carried out in 0.1 N HCl, phosphate buffer pH 6.8, and acetonitrile (ACN)—water (70 : 30 V/V). Linearity range was observed in the concentration range of 2 to 12 μg/mL for FEL and 20 to 100 μg/mL for ATR. Percent concentration estimated for ATR and FEL was 100.12 ± 1.03 and 99.98 ± 0.98, respectively. The method was found to be simple, economical, accurate and precise and can be used for quantitative estimation of ATR and FEL.

1. Introduction

Atorvastatin (ATR) is chemically described as [R-(R*, R*)]-2-(4-fluorophenyl)-dihydroxy-5-(1-methylethyl)-3-phenyl-4-[(phenylamino) carbonyl]-1H-pyrrole-1-heptanoic acid (Figure 1). It is a member of the drug class known as statins, used for lowering blood cholesterol [1]. It also stabilizes plaque and prevents strokes through anti-inflammation and other mechanisms. It inhibits HMG-CoA (3-hydroxy-3-methylglutaryl-coenzyme A) reductase, an enzyme found in liver tissue that plays a key role in production of cholesterol in the body. Inhibition of this enzyme stops the reduction of HMG-CoA to mevalonate, which is the rate-limiting step in hepatic cholesterol biosynthesis. Inhibition of the enzyme decreases cholesterol synthesis and ultimately increases expression of low-density lipoprotein receptors (LDL receptors) on hepatocytes [2, 3].

Felodipine (FEL) is a 1, 4 dihydropyridine derivative, that is, chemically described as ethyl methyl-1,4-dihydro-2,6-dimethyl-4-(2,3 dichlorophenyl)-3,5-pyridinedicarboxylate. It is a dihydropyridine calcium channel blocker used mainly for the management of hypertension and angina pectoris like the other calcium channel blockers [4].

Literature survey reveals that spectrophotometric and chromatographic methods, and a stability-indicating LC method, have been reported for determination of ATR in pharmaceutical preparations in combination with other drugs [5–13]. Several chromatographic and spectrophotometric methods have been reported for felodipine assay [14–18]. However, most of the analytical methods developed for the quantization of ATR and FEL involve analysis of single component or combination with other drugs. No effective method has been reported for quantitative estimation of ATR and FEL from combined dosage form.

2. Material and Methods

Atorvastatin calcium and felodipine were kindly gifted by Cipla Ltd., Goa, India. Acetonitrile (Loba Chemie Pvt. Ltd., Mumbai, India) and other chemicals used are of analytical grade. Distilled water was prepared in laboratory.

TABLE 1: Absorbance values for calibration curves of FEL and ATR at 268 and 245 nm.

FEL			ATR		
Concentration (μg/mL)	Absorbance		Concentration (μg/mL)	Absorbance	
	268 nm	245 nm		268 nm	245 nm
20	0.17069	0.05085	2	0.12878	0.02782
40	0.33971	0.10072	4	0.24756	0.05264
60	0.52023	0.15256	6	0.37096	0.08296
80	0.68275	0.21401	8	0.50512	0.11161
100	0.84833	0.25425	10	0.62368	0.13909
—	—	—	12	0.76792	0.16704

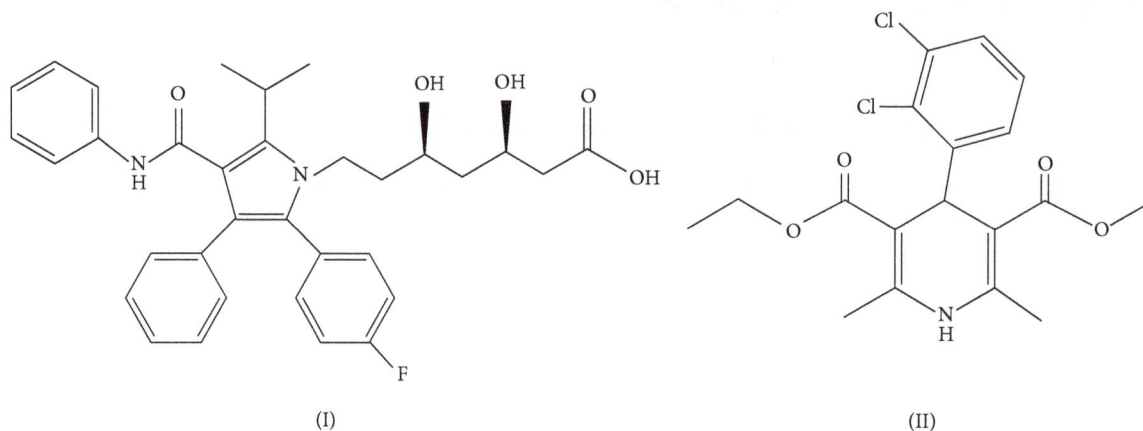

FIGURE 1: Structure of (I) atorvastatin, (II) felodipine.

2.1. Preparation of Bilayered Tablet.

Bilayered tablets of total weight of 300 mg each, containing 150 mg immediate release layer of ATR (10 mg API) and 150 mg of sustained release layer of FEL (10 mg API), were prepared by initially adding FEL granules to die of RIMEK minipress (Karnavati engineering, Gujarat, India) and compressed, above which ATR blend was poured and allowed to undergo for the final compression to prepare the bilayered tablet using 8 mm flat faced punches.

2.2. Standard Stock Solution

2.2.1. In 0.1 N HCl.

Standard stock solution was prepared by dissolving 10 mg of ATR and 10 mg of FEL separately in 100 mL of volumetric flask containing 10 mL of 0.1 N HCl. Then, the final volume of the solution was made up to 100 mL with 0.1 N HCl to get stock solution of 100 μg/mL. Adequate quantities were sampled out from the standard stock solution in 10 mL volumetric flask to get concentration of 10, 20, 30, 40, 50, and 60 μg/mL ATR and 2, 4, 6, 8, 10, and 12 μg/mL FEL. Then, the absorbances of the solution was measured at 268 nm (λ max of FEL) and 245 nm (λ max of ATR) using double beam UV visible spectrophotometer against 0.1 N HCl as blank.

2.2.2. In Phosphate Buffer pH 6.8.

All the above procedure was repeated using phosphate buffer pH 6.8 instead of 0.1 N HCl. Calibration curve of ATR and FEL in 0.1 N HCl and phosphate buffer (PB) pH 6.8 are shown in Figures 2 and 3, respectively. Absorbance values are shown in Table 1.

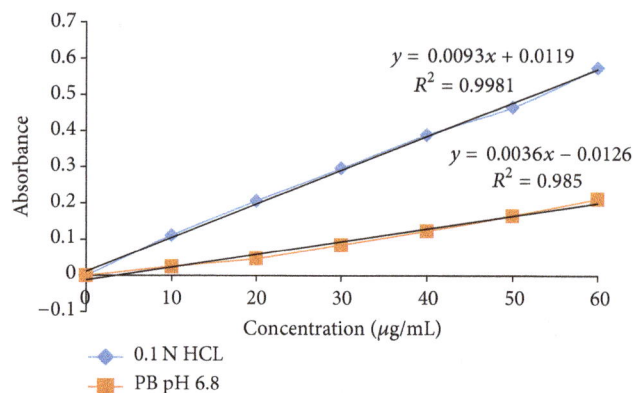

FIGURE 2: Calibration curve of ATR in 0.1 N HCl and phosphate buffer (PB) pH 6.8.

2.3. UV Method Development and Optimization

2.3.1. Selection of Common Solvent.

ACN—double distilled water (70 : 30% V/V)—was selected as common solvent for developing spectral characteristics of drugs. The selection was made after assessing the solubility of both drugs in different solvents.

2.3.2. Preparation of Standard Drug Solution

(1) ATR Standard Stock Solution (100 μg/mL). Accurately weighed quantity of Atorvastatin calcium (10 mg) was transferred into 100 mL volumetric flask dissolved in 60 mL of

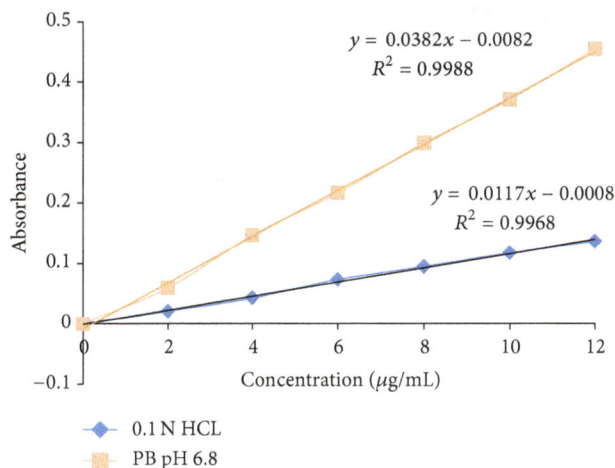

FIGURE 3: Calibration curve of FEL in 0.1 N HCl and phosphate buffer (PB) pH 6.8.

ACN—water (70 : 30% V/V)—and diluted up to mark with same solvent. This will give a stock solution having strength of 100 μg/mL.

(2) FEL Standard Stock Solution (100 μg/mL). Accurately weighed quantity of felodipine (10 mg) was transferred into 100 mL volumetric flask dissolved in 60 mL of ACN—water (70 : 30% V/V)—and diluted up to mark with same solvent. This will give a stock solution having strength of 100 μg/mL.

2.3.3. Construction of Calibration Curve

(1) Calibration Curve for ATR. Different aliquots were withdrawn from the standard stock solution and diluted with appropriate quantity of ACN—water (70 : 30% V/V)—to get a series of concentration ranging from 20 to 100 μg/mL. Absorbance was measured at different concentrations and the calibration curve was prepared by plotting absorbance versus concentration.

(2) Calibration Curve for FEL. Different aliquots were withdrawn from the standard stock solution and diluted with appropriate quantity of ACN—water (70 : 30% V/V)—to get a series of concentration ranging from 2 to 12 μg/mL. Absorbance was measured at different concentrations and the calibration curve was prepared by plotting absorbance versus concentration.

2.3.4. Selection of Wavelength. By appropriate dilutions of two standard drug solutions with ACN—double distilled water (70 : 30% V/V)—solutions containing 10 μg/mL of ATR and 10 μg/mL of FEL were scanned separately in the range of 200–400 nm to determine the wavelength of maximum absorption for both drugs (Figure 4).

2.3.5. Selection of Method and Wavelength. For quantitative estimation of ATR and FEL, simultaneous equation method

employing 245 nm and 268 nm as analytical wavelength was used. The two wavelengths were chosen from the overlain spectra of ATR and FEL. Overlain spectra of ATR and FEL are shown in Figure 5.

2.4. Procedure for Calculating Absorptivity of Both the Drugs at Selected Wavelengths. From standard drug solutions, six works in standard solutions with concentration of 20, 40, 60, 80, and 100 μg/mL for ATR and 2, 4, 6, 8, 10, and 12 μg/mL for FEL were prepared and scanned separately on the selected wavelengths for both the drugs. The absorptivity at selected wavelength was calculated.

2.5. Analysis of Tablet Formulation. Twenty tablets were powdered. Tablet formulation containing ATR 10 mg and FEL 10 mg was analyzed using this method. From the triturates of 3 tablets, an amount equivalent to 10 mg of ATR and 10 mg of FEL was weighed and dissolved in 10 mL of ACN—water (70 : 30% V/V)—and sonicated for 10 min. Then, the solution was filtered through Whatman filter paper number 41 and then final volume of the solution was made up to 100 mL with ACN—double distilled water (70 : 30% V/V)—to get a stock solution containing 100 μg/mL of ATR and 100 μg/mL FEL.

Appropriate aliquots of ATR and FEL within Beer's law limit were taken and analyzed by the proposed method using the procedure described earlier. The concentration of ATR and FEL present in the sample solution was calculated by using the simultaneous equation,

$$C_y = \frac{A_1 a_{x2} - A_2 a_{x1}}{a_{x2} a_{y1} - a_{x1} a_{y2}},$$

$$C_x = \frac{A_1 a_{y2} - A_2 a_{y1}}{a_{x2} a_{y1} - a_{x1} a_{y2}}, \tag{1}$$

where C_y is the concentration of ATR in gm/lit, C_x is the concentration of FEL in gm/lit, A_1 is the absorbance of sample solution at 268 nm, A_2 is the absorbance of sample solution at 245 nm, a_{x1} is the absorptivity of FEL at 268 nm, a_{y1} is the absorptivity of ATR at 268 nm, a_{x2} is the absorptivity of FEL at 245 nm, and a_{y2} is the absorptivity of ATR at 245 nm.

2.6. Method Validation

2.6.1. Linearity. In quantitative analysis, the calibration curve was constructed for both ATR and FEL after analysis of consecutively increased concentrations.

2.6.2. Recovery Studies. Accuracy of analysis was determined by performing recovery studies by spiking different concentrations of pure drug in the preanalyzed tablet samples within the analytical concentration range of the proposed method at three different sets at levels of 80, 100, and 120%. The added quantities of the individual drugs were estimated by above method. Intraday precision and interday precision have also been carried out.

TABLE 2: Absorptivity values for ATR and FEL.

Concentration (μg/mL)	Absorptivity of FEL		Concentration (μg/mL)	Absorptivity of ATR	
	268 nm	245 nm		268 nm	245 nm
2	0.00603	0.001399	20	0.08533	0.0254
4	0.00618	0.00148	40	0.08539	0.0266
6	0.00638	0.001462	60	0.08548	0.0282
8	0.00654	0.001495	80	0.08562	0.0299
10	0.00673	0.001516	100	0.08586	0.0331
12	0.00698	0.001543	—	—	—
Mean	**0.00759**	**0.001465**	Mean	**0.08553**	**0.0286**
S.D.	**0.44**	**1.03**	S.D.	**1.39**	**0.95**

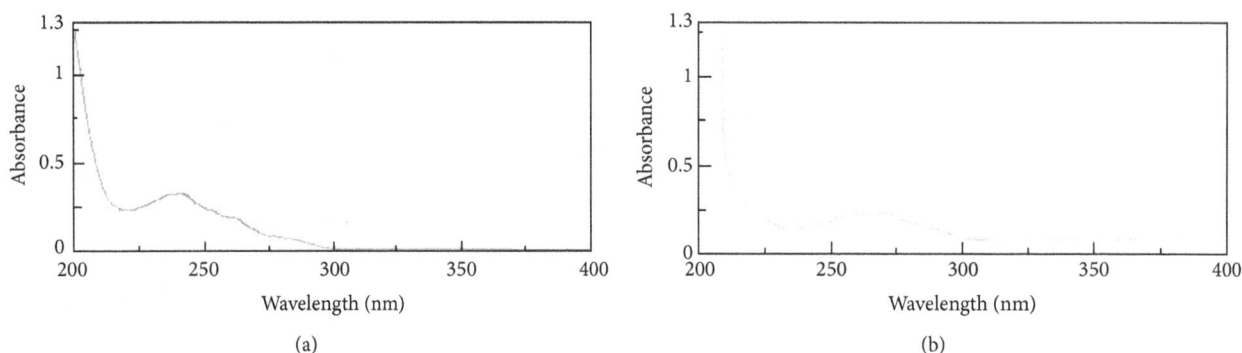

(a)

(b)

FIGURE 4: UV spectra of (a) ATR, (b) FEL in ACN: water (70 : 30% V/V).

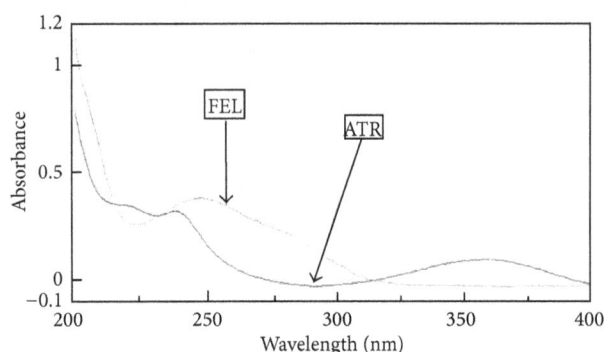

FIGURE 5: Overlain spectra of ATR and FEL.

2.6.3. Interday Precision. Analysis of drug was performed on two different days and the deviation in the results was observed. Results are shown in Table 4.

2.6.4. Intraday Precision. Analysis of drug was performed on the same day in morning and evening, and the deviation in the results was observed. Results are shown in Table 5.

2.6.5. Ruggedness. Ruggedness of the method was confirmed by the analysis of formulation that was done by the different analysts, using similar operational and environmental conditions.

3. Result and Discussion

Calibration curve of ATR and FEL was plotted by measuring the absorbance of prepared dilutions of the aforesaid different

TABLE 3: Results of analysis of laboratory samples.

Label claim (mg/tab)	% concentration estimated[*]
ATR	100.12 ± 1.03
FEL	99.98 ± 0.98

[*] Indicates $\pm$ SD ($n = 9$).

concentrations at their respective wavelength of maximum absorbance (Figures 6 and 7).

3.1. Linearity. Linear regression data showed a good linear relationship over a concentration range of 2 to 12 μg/mL for FEL and 20 to 100 μg/mL for ATR, whereas, Rajesh et al. demonstrated that linearity was within the range of 2–10 μg/mL for each atorvastatin calcium and felodipine [19]. Six point regression data at both wavelengths (268 nm 245 nm) were generated for FEL (Table 2).

3.2. Tablet Analysis. See Table 3.

3.3. Recovery Study. The added quantities of added drug were estimated by simultaneous equation (Table 4).

3.4. Interday Precision. Interday precision study was performed and method was found to be precise.

3.5. Intraday Precision. Intraday precision study was performed and method was found to be precise. Recoveries obtained for the two drugs do not differ significantly from 100%, which showed that there was no interference from

TABLE 4: Result of recovery studies and interday precision.

Day	Label claim (mg/tablet)	Amount added (%)	Total amount added (mg)	Concentration recovered[*] (mean ± SD)	% recovery estimated[*] (mean ± %SD)
Day 1	ATR 10	80	8	8.02 ± 0.34	100.22 ± 1.12
		100	10	10.90 ± 0.78	98.10 ± 0.98
		120	12	12.23 ± 0.89	103.91 ± 1.23
	FEL 10	80	8	8.17 ± 1.45	101.70 ± 1.18
		100	10	10.71 ± 1.78	101.68 ± 1.24
		120	12	12.99 ± 1.25	99.93 ± 0.98
Day 2	ATR 10	80	8	8.04 ± 1.11	101.11 ± 1.25
		100	10	10.08 ± 1.45	101.60 ± 1.17
		120	12	12.03 ± 0.34	100.64 ± 1.11
	FEL 10	80	8	8.17 ± 1.71	101.70 ± 0.65
		100	10	10.22 ± 1.12	97.76 ± 0.34
		120	12	12.54 ± 0.34	103.60 ± 0.32

[*]Indicates ± SD ($n = 3$).

TABLE 5: Result of intraday precision.

Day 1	Label claim (mg/tablet)	Amount added (%)	Total amount added (mg)	Concentration recovered[*] (mean ± SD)	% recovery estimated[*] (mean ± %SD)
Morning	ATR 10	80	8	8.04 ± 0.30	100.22 ± 1.12
		100	10	10.10 ± 0.34	97.10 ± 0.57
		120	12	12.13 ± 0.44	102.79 ± 1.52
	FEL 10	80	8	8.27 ± 1.47	101.80 ± 1.74
		100	10	10.56 ± 1.78	101.32 ± 1.04
		120	12	12.90 ± 1.28	99.99 ± 0.91
Evening	ATR 10	80	8	8.08 ± 1.18	101.64 ± 1.17
		100	10	10.04 ± 1.55	101.40 ± 1.87
		120	12	12.05 ± 0.28	101.55 ± 1.13
	FEL 10	80	8	8.11 ± 1.54	101.28 ± 0.45
		100	10	10.14 ± 1.17	99.68 ± 0.56
		120	12	12.99 ± 1.98	99.46 ± 0.98

[*]Indicates ± SD ($n = 3$).

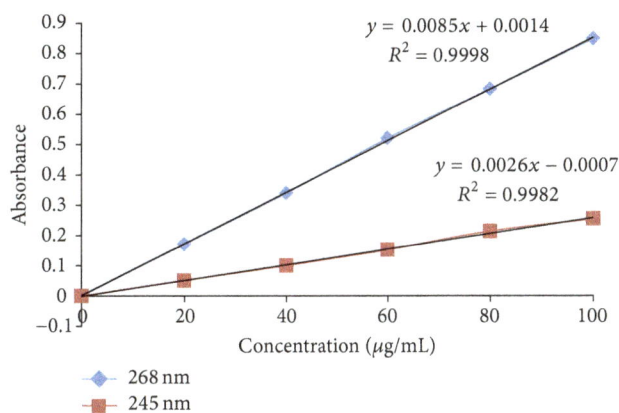

$y = 0.0085x + 0.0014$
$R^2 = 0.9998$

$y = 0.0026x - 0.0007$
$R^2 = 0.9982$

- 268 nm
- 245 nm

FIGURE 6: Calibration curve of ATR in ACN—water (70 : 30 v/v)—at 268 and 245 nm.

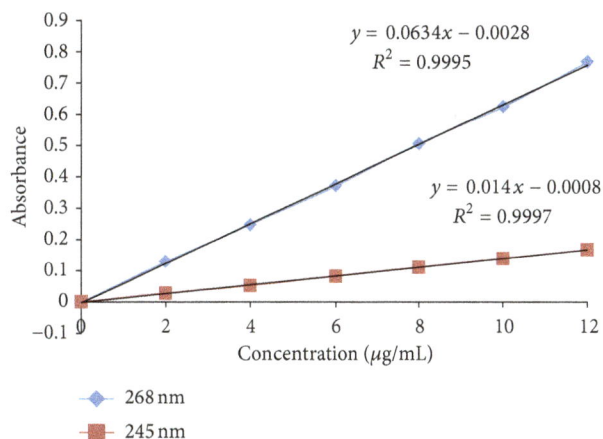

$y = 0.0634x - 0.0028$
$R^2 = 0.9995$

$y = 0.014x - 0.0008$
$R^2 = 0.9997$

- 268 nm
- 245 nm

FIGURE 7: Calibration curve of FEL in ACN—water (70 : 30 v/v)—at 268 and 245 nm.

common excipients used in the formulation indicating accuracy and reliability of the method (Table 5).

3.6. Ruggedness. Ruggedness of the method was tested using different chemical sources of acetonitrile and effects on

TABLE 6: Ruggedness of the method.

	Label claim (mg/tablet)	% recovery estimated* (mean ± %SD)	
		ATR	FEL
Source I	100/10	102.74 ± 0.35	101.18 ± 0.31
Source II	100/10	103.12 ± 0.35	103.18 ± 0.62

*Indicates ± SD ($n = 3$).

results were observed and shown in Table 6. From the result, it was found that method has ruggedness.

4. Conclusion

In present study, from the observation of the validation parameters, it can be concluded that the developed method is simple, accurate, reliable, and economical for the simultaneous quantitative estimation of atorvastatin calcium and felodipine from combined dosage form using UV spectrophotometric method.

Conflict of Interests

The authors declare that there is no conflict of interests regarding the publication of this paper.

References

[1] B. W. McCrindle, L. Ose, and A. D. Marais, "Efficacy and safety of atorvastatin in children and adolescents with familial hypercholesterolemia or severe hyperlipidemia: a multicenter, randomized, placebo-controlled trial," Journal of Pediatrics, vol. 143, no. 1, pp. 74–80, 2003.

[2] J. Villa and R. E. Pratley, "Ezetimibe/simvastatin or atorvastatin for the treatment of hypercholesterolemia in patients with the metabolic syndrome: the VYMET study," Current Diabetes Reports, vol. 10, no. 3, pp. 173–175, 2010.

[3] T. McCormack, P. Harvey, R. Gaunt, V. Allgar, R. Chipperfield, and P. Robinson, "Incremental cholesterol reduction with ezetimibe/simvastatin, atorvastatin and rosuvastatin in UK General Practice (IN-PRACTICE): randomised controlled trial of achievement of Joint British Societies (JBS-2) cholesterol targets," International Journal of Clinical Practice, vol. 64, no. 8, pp. 1052–1061, 2010.

[4] M. B. El-Hawary, M. T. khayall, and Z. Isaak, Hand Book of Pharmacology, The Scientific Book Center, S.O.P. Press, Cairo, Egypt, 1978.

[5] S. L. Thamake, S. D. Jadhav, and S. A. Pishawikar, "Development and validation of method for simultaneous estimation of atorvastatin calcium and ramipril from capsule dosage form by first order derivative spectroscopy," Asian Journal of Research in Chemistry, vol. 2, no. 1, pp. 52–53, 2009.

[6] R. Lakshmana, K. R. Rajeswari, and G. G. Sankar, "Spectrophotometric method for simultaneous estimation of atorvastatin and amlodipine in tablet dosage form," Research Journal of Pharmaceutical, Biological and Chemical Sciences, vol. 2, pp. 66–69, 2010.

[7] M. Saravanamuthukumar, M. Palanivelu, K. Anandarajagopal, and D. Sridharan, "Simultaneous estimation and validation of atorvastatin calcium and ubidecarenone (Coenzyme Q10) in combined tablet dosage form by RP-HPLC method," International Journal of Pharmacy and Pharmaceutical Sciences, vol. 2, no. 2, pp. 36–38, 2010.

[8] L. Joseph, M. George, and B. V. R. Rao, "Simultaneous estimation of atorvastatin and ramipril by RP-HPLC and spectroscopy," Pakistan Journal of Pharmaceutical Sciences, vol. 21, no. 3, pp. 282–284, 2008.

[9] L. Nováková, D. Šatínský, and P. Solich, "HPLC methods for the determination of simvastatin and atorvastatin," Trends in Analytical Chemistry, vol. 27, no. 4, pp. 352–367, 2008.

[10] B. G. Chaudhari, N. M. Patel, and P. B. Shah, "Stability indicating RP-HPLC method for simultaneous determination of atorvastatin and amlodipine from their combination drug products," Chemical and Pharmaceutical Bulletin, vol. 55, no. 2, pp. 241–246, 2007.

[11] A. Mohammadi, N. Rezanour, M. Ansari Dogaheh, F. Ghorbani Bidkorbeh, M. Hashem, and R. B. Walker, "A stability-indicating high performance liquid chromatographic (HPLC) assay for the simultaneous determination of atorvastatin and amlodipine in commercial tablets," Journal of Chromatography B: Analytical Technologies in the Biomedical and Life Sciences, vol. 846, no. 1-2, pp. 215–221, 2007.

[12] G. F. Patel, N. R. Vekariya, and R. B. Dholakiya, "Estimation of aspirin and atorvastatin calcium in combine dosage form using derivative spectrophotometric method," International Journal of Pharmaceutical Research, vol. 2, no. 1, pp. 62–66, 2010.

[13] N. R. Jadhav, R. S. Kambar, and S. J. Nadaf, "RP-HPLC method for simultaneous estimation of atorvastatin calcium and felodipine from tablet dosage form," Current Pharma Research, vol. 2, no. 4, pp. 637–642, 2012.

[14] A. J. López, V. Martínez, R. M. Alonso, and R. M. Jiménez, "High-performance liquid chromatography with amperometric detection applied to the screening of 1,4-dihydropyridines in human plasma," Journal of Chromatography, vol. 870, no. 1-2, pp. 105–114, 2010.

[15] A. B. Baranda, R. M. Jiménez, and R. M. Alonso, "Simultaneous determination of five 1,4-dihydropyridines in pharmaceutical formulations by high-performance liquid chromatography-amperometric detection," Journal of Chromatography A, vol. 1031, no. 1-2, pp. 275–280, 2004.

[16] J. Gottfries, J. Ahlbom, V. Harang et al., "Validation of an extended release tablet dissolution testing system using design and multivariate analysis," International Journal of Pharmaceutics, vol. 106, no. 2, pp. 141–148, 1994.

[17] B. Marciniec, E. Jaroszkiewicz, and M. Ogrodowczyk, "The effect of ionizing radiation on some derivatives of 1,4-dihydropyridine in the solid state," International Journal of Pharmaceutics, vol. 233, no. 1-2, pp. 207–215, 2002.

[18] K. Basavaiah, U. Chandrashekar, and H. C. Prameela, "Sensitive spectrophotometric determination of amlodipine and felodipine using iron(III) and ferricyanide," Farmaco II, vol. 58, no. 2, pp. 141–148, 2003.

[19] K. Rajesh, R. Rajalakshmi, S. Vijayaraj, and T. Sreelakshmi, "Simultaneous estimation of atorvastatin calcium and felodipine by UV-spectrophotometric method in formulation," Asian Journal of Research in Chemistry, vol. 4, no. 8, pp. 1202–1205, 2011.

2

Study of Swelling Properties and Thermal Behavior of Poly(N,N-Dimethylacrylamide-*co*-Maleic Acid) Based Hydrogels

Sadjia Bennour and Fatma Louzri

Laboratory of Polymer Materials, Faculty of Chemistry, University of Sciences and Technology Houari Boumediene, BP 32, El Alia, 16111 Algiers, Algeria

Correspondence should be addressed to Sadjia Bennour; sadjiabennour@yahoo.fr

Academic Editor: Alejandro Sosnik

Hydrogels copolymers N,N-dimethylacrylamide (DMA) and maleic acid (MA) were prepared by free-radical polymerization at 56°C in aqueous solution, using N,N-methylenebisacrylamide (NMBA) as cross-linking agent and potassium persulfate (KPS) as initiator. The effects of comonomer composition, cross-linker content, and variation of pH solutions on the swelling behavior of polymers were investigated. The obtained results showed an increase of the swelling of poly(N,N-dimethylacrylamide-*co*-maleic acid) (P(DMA-MAx)) as the content of maleic acid increases in the polymeric matrix, while they indicate a great reduction of the degree of swelling as the cross-linking agent ratio increases. It was also shown that the swelling of copolymer hydrogels increased with the increase of pH and the maximum extent was reached at pH 8.7 in all compositions. Fourier transform infrared spectroscopy (FTIR) revealed the existence of hydrogen bonding interactions between the carboxylic groups of MA and the carbonyl groups of DMA. Differential scanning calorimetry analysis (DSC) showed an increase of the glass-transition temperature (T_g) as concentrations of MA and NMBA increased. Thermogravimetric analysis (TGA) of copolymers was performed to investigate the degradation mechanism.

1. Introduction

Hydrogels are composed of hydrophilic homopolymer or copolymer network and can swell in the presence of water or physiological fluids. Chemical cross-links (covalent bonds) or physical junctions (e.g., secondary forces, crystallite formation, and chain entanglements) provide the hydrogels' unique swelling behavior and three-dimensional structure [1–3]. Volume changes in hydrogels occur in response to changing environmental conditions such as temperature [4–6], pH [7–9], solvent composition [10], and ionic strength [4, 11]. Hydrogels have been a topic of extensive research in the past decades and their properties as, for example, their high water content and the possible control over the swelling kinetics make them very attractive for biomedical applications [12–16]. The pH sensitive hydrogels containing pendant acidic or basic groups such as carboxylic acids, sulfonic acids, primary amines, or ammonium salts which change ionization in response to change in the pH have become the subject matter of major

interest for use as carriers in drug delivery research [17–19]. Poly(N,N-dimethylacrylamide) is a hydrophilic polymer, due to its remarkable properties, such as water solubility and biocompatibility, and it is very useful in the biomedical applications including polymer supports for protein synthesis, two-phase catalysts, and controlled drug delivery [20, 21]. Moreover, grafting of the N,N-dimethylacrylamide monomer onto substrates like poly(dimethyl-siloxane) produces a material with unique properties suitable for determining the diffusion characteristics of small molecule drugs, glucose, insulin, and albumin in vitro [22].

The present work was aimed to synthesize the cross-linked copolymers of N,N-dimethylacrylamide (DMA) with maleic acid (MA) and to investigate the swelling behavior of the copolymer composed of a nonionic monomer DMA and an ionizable monomer MA. In order to study the effects of comonomer composition and the cross-linker content on the swelling of polymers, the equilibrium swelling values of DMA homopolymer and DMA-MA copolymers in buffer

TABLE 1: Molar composition of the hydrogels.

DMA/MA mole ratio in feed	DMA/MA[a] mole ratio in copolymer	DMA (mmol)	MA (mmol)	KPS[b] (mol%)	NMBA[b] (mol%)	H_2O (mL)
100/0	100/0	40.4	0	0.036	0.32	4
91/9	92/8	36.4	3.45	0.036	0.32	4
82/18	85/15	32.3	6.89	0.036	0.32	4
73/27	71/29	28.3	10.35	0.036	0.32	4
64/36	67/33	24.3	13.79	0.036	0.32	4
54/46	62/38	20.2	17.24	0.036	0.32	4
91/9	92/8	36.4	3.45	0.036	0.16	4
91/9	92/8	36.4	3.45	0.036	0.48	4
91/9	92/8	36.4	3.45	0.036	0.64	4
91/9	92/8	36.4	3.45	0.036	1.28	4

[a]Determined by elemental analysis.
[b]mol% of the total mole number of monomers.

solutions with different pH values at room temperature were determined. The thermal stability of PDMA and P(DMA-MAx) was investigated by thermogravimetry.

2. Experimental

2.1. Materials. N,N-dimethylacrylamide (DMA) was supplied from Merck and purified with vacuum distillation prior to usage in order to remove the inhibitor. Maleic acid (MA, 99%) purchased from Fluka, N,N-methylenebisacrylamide (NMBA) from Fluka as a cross-linking agent, and potassium persulfate (KPS) from Merk as an initiator were used as received. Distilled water was used for the polymerization reactions and swelling studies.

The following pH-buffered solutions were used in order to examine the swelling behavior [23]: pH = 0.9, potassium chloride, $I = 0.14$ M; pH = 3.9, citric acid/phosphate, $I = 0.23$ M; pH = 6.9, citric acid/phosphate, $I = 0.49$ M; pH = 8.7, boric acid/NaOH, $I = 0.02$ M.

2.2. Synthesis of Hydrogels. Poly(N,N-dimethylacrylamide) (PDMA), poly(N,N-dimethylacrylamide-*co*-maleic acid) (P(DMA-MAx)) containing 8, 15, 29, 33, and 38 mol% of maleic acid, using NMBA (0.32 mol% of the total mole number of monomers) as cross-linking agent, and P(DMA-MA8) with varying compositions of NMBA (0.16, 0.48, 0.64, and 1.28 mol% of the total mole number of monomers) were synthesized by free-radical polymerization using KPS as initiator. The polymerization recipes are listed in Table 1.

Polymerization reactions were performed in glass tubes with the inner diameters of 1 cm and lengths of 10 cm. DMA and MA monomers with different molar ratios and cross-linker concentration were dissolved in water. After nitrogen bubbles for 20 min through the solution, the initiator KPS was added. Then, the polymerization was carried out at 56°C for 15 min. At the end of the reaction, the glass tubes were broken carefully without destroying the cylindrical hydrogels. The resulting hydrogels were sliced into small cylinders with lengths of 4 mm and then immersed in distilled water for

5 days, and the water was changed every 24 hours in order to remove the residual unreacted monomers. The resulting swollen gels were dried at room temperature for several days and then in a vacuum oven at 40°C until constant weight.

The compositions of the formed copolymer hydrogels were determined by elemental analysis for nitrogen (Microanalizador CHNOS-932 LECO).

2.3. Characterization

2.3.1. FTIR Measurements. Hydrogels samples were prepared by grinding the dry hydrogels with KBr and compressing the mixture to form disks and then dried in vacuum oven at 70°C for several days. FTIR spectra of these hydrogels were recorded on a model Perkin Elmer spectrophotometer at room temperature with a resolution of $2\,cm^{-1}$ and were averaged over 60 scans.

2.3.2. DSC Measurements. The glass transition temperatures (T_g) of PDMA and P(DMA-MAx) hydrogels were determined with a Perkin-Elmer DSC-7 differential scanning calorimeter at a heating rate of 10°C/min under a nitrogen atmosphere in the 50°C to 220°C temperature range.

2.3.3. Thermogravimetric Analysis. A thermogravimetric analysis of the hydrogels investigated was carried out on a TG SETARAM under nitrogen atmosphere from 25°C to 500°C at a heating rate of 10°C/min.

2.4. Swelling Measurements. Dried hydrogels were allowed to hydrate in buffer solutions with different pH values 0.9, 3.9, 6.9, and 8.7 at 16°C. After being fully hydrated, the samples were taken out and the excess water on their surface was gently removed by filter paper.

The weights of the hydrating samples were measured at timed intervals. The swelling ratio (SR) and equilibrium

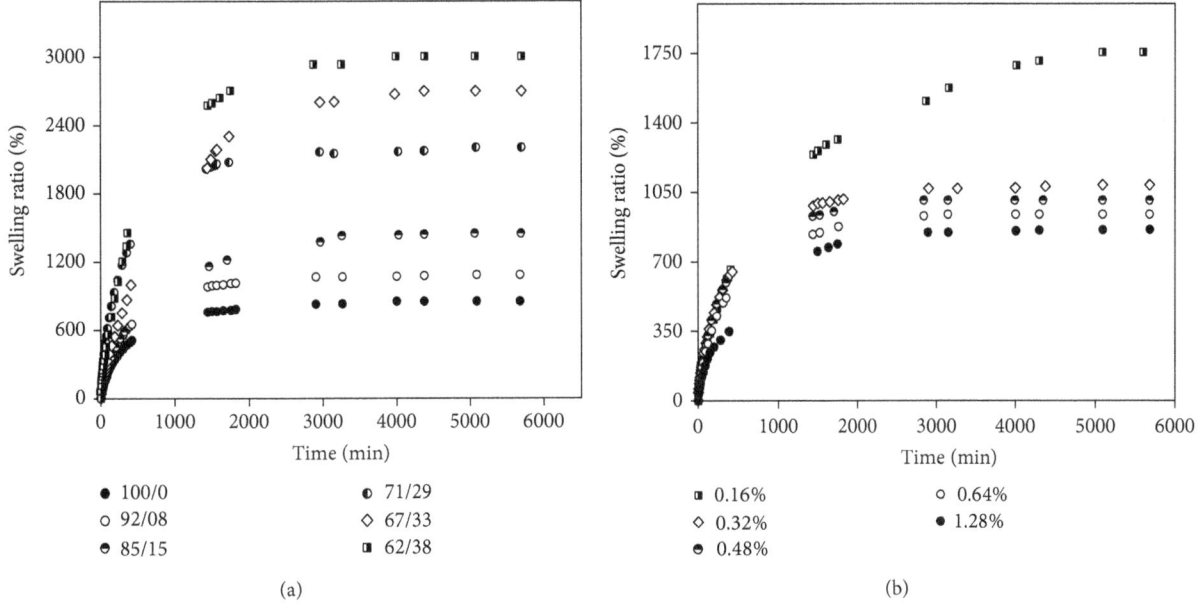

FIGURE 1: (a) Effect of maleic acid concentration on the dynamic swelling behavior of P(DMA-MAx) hydrogels in the media of pH 6.9 at 16°C (NMBA 0.32 mol% of monomers). (b) Effect of NMBA concentration on the dynamic swelling behavior of P(DMA-MA8) hydrogels in the media of pH 6.9 at 16°C.

swelling ratio (SR_e) are calculated by the following equations [14]:

$$SR\,(\%) = \left[\frac{W_s - W_d}{W_d} \right] \times 100,$$

$$SR_e\,(\%) = \left[\frac{W_e - W_d}{W_d} \right] \times 100,$$

(1)

where W_s is the weight of the swelled hydrogels at time t, W_d is the weight of the dried hydrogels, and W_e denotes the weight of the gels at equilibrium swelling.

3. Results and Discussion

3.1. Swelling Behavior

3.1.1. Effect of Maleic Acid Content. The effect of maleic acid composition on the swelling of the prepared P(DMA-MAx) hydrogels at different pHs (0.9, 3.9, 6.9, and 8.7) and at 16°C was investigated. For instance, the swelling curves of P(DMA-MAx) hydrogels with different compositions of MA at pH 6.9 are given in Figure 1(a). As seen in this figure, the swelling ratio increases with time, but, later, constant swelling values are observed. These swelling values can be considered as an equilibrium swelling ratio (SR_e).

It is well known that swelling is induced by the electrostatic repulsion of ionic charges present within the network. Thus, as the number of carboxylic groups increases on going from PDMA to P(DMA-MAx), the swelling increases too.

The values of equilibrium swelling ratio of DMA/MA hydrogels in different buffer solutions are reported in Table 2(a). As shown, the hydrophilic nature of the hydrogel

copolymer is enhanced by increasing the amount of maleic acid [24, 25].

3.1.2. Effect of Cross-linker Content. The effect of increasing the amount of the cross-linking agent NMBA on the swelling capacity of the P(DMA-MA8) in different buffer solutions (pH 0.9, 3.9, 6.9, and 8.7) was investigated.

It is well known that the cross-link density directly affects the mechanical deformation of hydrogels. As the ratio of the cross-linking agent NMBA varied from 0.16 to 1.28 mol% in P(DMA-MA8), the degree of swelling was greatly reduced (Figure 1(b)). It is expected that hydrogels with higher cross-link density would provide numerous water channels for the diffusion of water, but the water content decreases due to increased level of cross-linking [9, 26].

The values of equilibrium swelling ratio of P(DMA-MA8) hydrogels with different NMBA contents at different pHs are given in Table 2(b).

3.1.3. Effect of pH. Figure 2(a) represents pH dependence of the equilibrium swelling ratio for P(DMA-MAx) hydrogels at 16°C in buffer solution from pH 0.9 to 8.7. As shown in this figure, the equilibrium mass swelling percentage for pure PDMA is not affected by varying the pH of the swelling medium since PDMA is nonionic hydrogel and does not have any group that could be ionized in an aqueous solution. With the introduction of the MA groups into the main chain, the swelling ratios are enhanced as the pH values of buffer solutions are increased. The hydrogels exhibit lower swelling capability in an acid medium and higher degree in a basic medium. This is related to the fact that the carboxyl groups could accept or release protons in response to various pH

TABLE 2: (a) Equilibrium swelling ratios of P(DMA-MAx) hydrogels with different MA contents at different pH media. (b) Equilibrium swelling ratios of P(DMA-MA8) hydrogels with different NMBA contents at different pH media.

(a)

Hydrogels	Equilibrium swelling ratio (%)			
	pH = 0.9	pH = 3.9	pH = 6.9	pH = 8.7
PDMA	833	831	852	844
P(DMA-MA8)	870	901	1084	1526
P(DMA-MA15)	1200	1391	1448	3173
P(DMA-MA29)	1700	1836	2204	3799
P(DMA-MA33)	1900	2034	2699	4556
P(DMA-MA38)	2111	2514	3002	6961

(b)

Composition of NMBA (mol%)	Equilibrium swelling ratio (%)			
	pH = 0.9	pH = 3.9	pH = 6.9	pH = 8.7
0.16	1401	1535	1754	2637
0.32	870	901	1084	1526
0.48	685	781	1008	1436
0.64	600	697	938	1356
1.28	508	605	860	1288

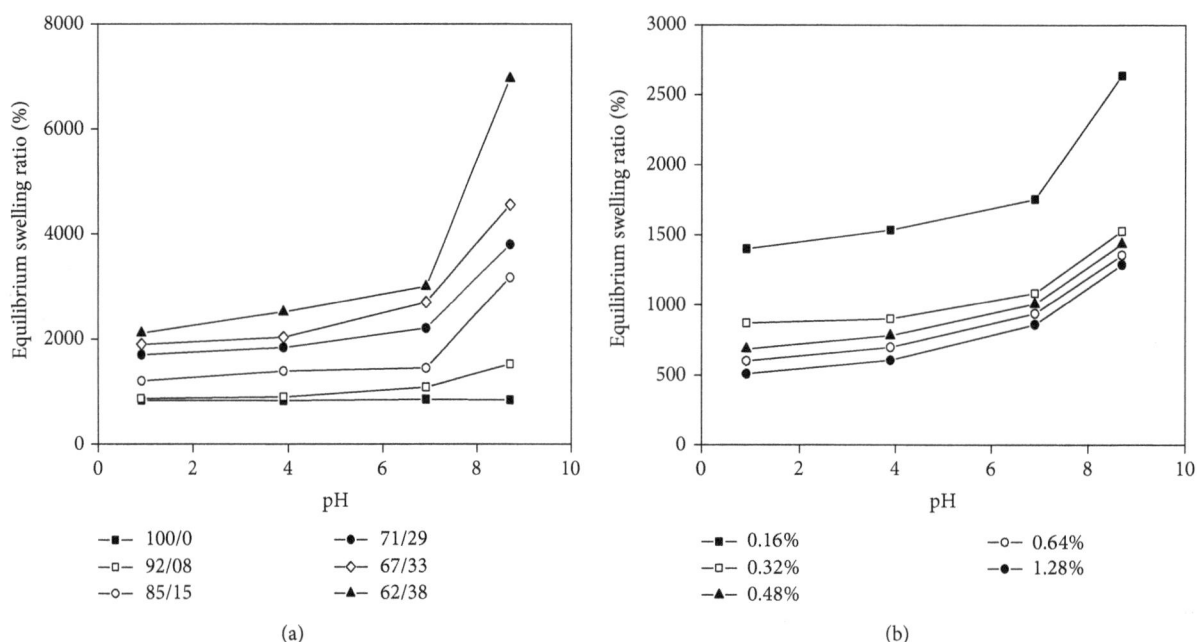

FIGURE 2: (a) Effect of pH on the equilibrium swelling ratio of P(DMA-MAx) hydrogels for different MA concentrations at 16°C (NMBA 0.32 mol% of monomers). (b) Effect of pH on the equilibrium swelling ratio of P(DMA-MA8) hydrogels for different NMBA compositions at 16°C.

aqueous media. At pH values lower than the pKa value of maleic acid, the first and second dissociation constants of MA are $pKa_1 = 1.85$ and $pKa_2 = 6.06$, respectively [27]; the carboxylic groups present within the network remain almost nonionized, thus imparting almost nonpolyelectrolyte type behavior to the hydrogel. This is ascribed to the fact that a lot of hydrogen bonds are formed due to the presence of carboxyl groups in the gel network. The hydrogen bond interaction would restrict the movement or relaxation of the gel network chains. As a result, a compact hydrogel network is formed leading to a lower swelling ratio. On the contrary, when the pH of the swelling medium was above the pKa of the gel, the ionization of the carboxylic acid groups of the gel occurred. Consequently, hydrogen bonds are broken and an electrostatic repulsion is generated among polymer networks. This repulsive force would push the network chain segments

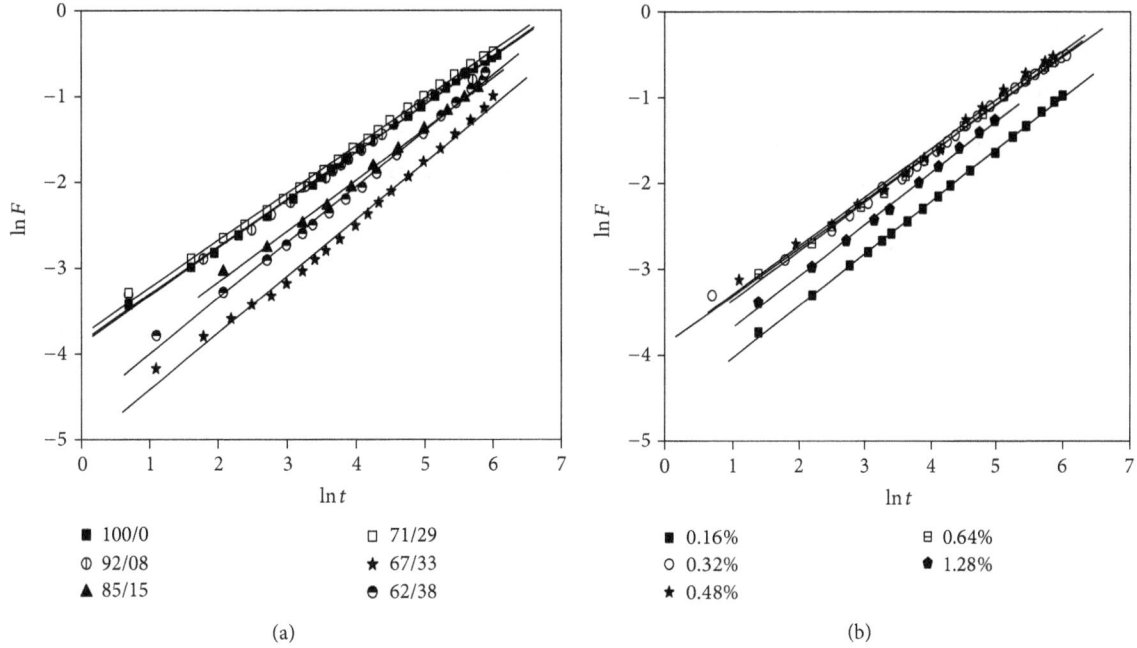

■ 100/0	□ 71/29
⊕ 92/08	★ 67/33
▲ 85/15	◕ 62/38

(a)

■ 0.16%	⊟ 0.64%
○ 0.32%	⬠ 1.28%
★ 0.48%	

(b)

FIGURE 3: (a) Dependence of $\ln F$ on $\ln t$ of P(DMA-MAx) hydrogels in the media of pH 6.9 with different compositions of MA at 16°C. (b) Dependence of $\ln F$ on $\ln t$ of P(DMA-MA8) in the media of pH 6.9 with different compositions of NMBA at 16°C.

apart and thus attract more water into the hydrogels, so a higher swelling ratio is observed [28–31].

The percentage equilibrium swelling of P(DMA-MA8) is decreased with increasing cross-linking agent concentration (Figure 2(b)). The increasing cross-link density restricts the swelling of the hydrogels.

3.1.4. Diffusion Process at Different pH. Analysis of the mechanisms of water diffusion in swelling polymeric systems has received considerable attention in recent years, because of the important applications in biomedical, pharmaceutical, environmental, and agricultural engineering fields.

To determine the nature of water diffusion into hydrogels, kinetic modeling was conducted based on Fickian diffusion law for the onset stage of swelling (the model is valid only for the first 60% of the swelling) [14, 32]:

$$F = \frac{M_t}{M_\infty} = kt^n, \quad (2)$$

where M_t is the total amount of water intake at time t, M_∞ is the total amount of water intake at an equilibrium state which is determined by a gravimetric method, k is a swelling coefficient which is a parameter correlated with the polymeric network structures, and n is an exponent characteristic of the swelling which represents solvent diffusion modes inside hydrogels. For Fickian kinetics in which the rate of penetrate diffusion is rate limiting, $n = 0.5$, whereas values of n between 0.5 and 1 indicate the contribution of non-Fickian processes such as polymer relaxation.

For example, Figures 3(a) and 3(b) can be used to elucidate dependence of $\ln F$ on $\ln(t)$ for various hydrogels samples with different MA and NMBA contents at pH = 6.9,

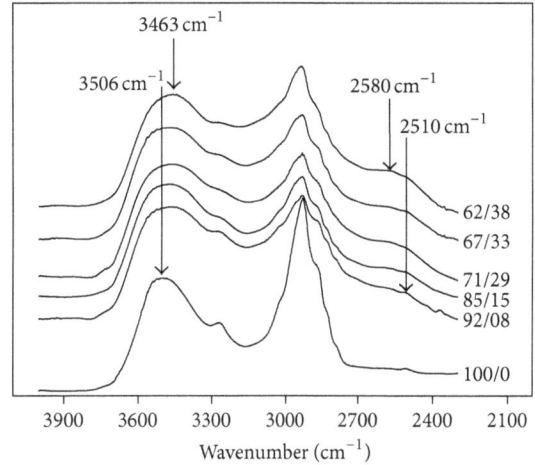

FIGURE 4: Scale-expanded infrared spectra of P(DMA-MAx) hydrogels in the hydroxyl region.

respectively. After having been fitted, the kinetic parameters k and n as well the correlation coefficient R^2 obtained are listed in Tables 3(a) and 3(b). The results indicate that the diffusion exponent for all samples ranges between 0.486 and 0.796, that is, non-Fickian or anomalous diffusion, and it occurs when the rate of permanent diffusion and the rate of polymer relaxation are comparable [32, 33].

3.2. FTIR Analysis. Figure 4 shows the scale-expanded infrared spectra of PDMA and P(DMA-MAx) recorded at room temperature in the 3900–2400 cm^{-1} region. The spectrum of pure PDMA shows a broad band at 3506 cm^{-1}, corresponded to hydrogen bonding between carbonyl groups

TABLE 3: (a) Swelling kinetic coefficients of P(DMA-MAx) hydrogels with different MA contents at different pH media. (b) Swelling kinetic coefficients of P(DMA-MA8) hydrogels with different NMBA contents at different pH media.

(a)

Medium	Hydrogels	R^2	n	$\ln k$
pH = 0.9	PDMA	0.9997	0.576	−3.428
	P(DMA-MA8)	0.9996	0.487	−3.142
	P(DMA-MA15)	0.9996	0.486	−3.405
	P(DMA-MA29)	0.9984	0.636	−4.912
	P(DMA-MA33)	0.9969	0.684	−4.760
	P(DMA-MA38)	0.9991	0.584	−4.174
pH = 3.9	PDMA	0.9998	0.594	−3.997
	P(DMA-MA8)	0.9993	0.582	−3.762
	P(DMA-MA15)	0.9989	0.586	−4.209
	P(DMA-MA29)	0.9978	0.685	−5.263
	P(DMA-MA33)	0.9986	0.768	−5.242
	P(DMA-MA38)	0.9992	0.661	−4.615
pH = 6.9	PDMA	0.9997	0.559	−3.883
	P(DMA-MA8)	0.9975	0.553	−3.858
	P(DMA-MA15)	0.9985	0.582	−4.293
	P(DMA-MA29)	0.9990	0.553	−3.782
	P(DMA-MA33)	0.9968	0.661	−5.075
	P(DMA-MA38)	0.9980	0.650	−4.648
pH = 8.7	PDMA	0.9991	0.553	−3.819
	P(DMA-MA8)	0.9991	0.564	−3.711
	P(DMA-MA15)	0.9988	0.668	−4.748
	P(DMA-MA29)	0.9991	0.738	−5.745
	P(DMA-MA33)	0.9991	0.734	−5.680
	P(DMA-MA38)	0.9981	0.796	−5.316

(b)

Medium	Composition of NMBA (mol%)	R^2	n	$\ln k$
pH = 0.9	0.16	0.9976	0.651	−4.617
	0.32	0.9996	0.487	−3.142
	0.48	0.9976	0.553	−3.665
	0.64	0.9964	0.564	−4.021
	1.28	0.9979	0.523	−3.555
pH = 3.9	0.16	0.9977	0.586	−4.675
	0.32	0.9993	0.582	−3.762
	0.48	0.9981	0.589	−4.081
	0.64	0.9985	0.575	−4.064
	1.28	0.9968	0.608	−4.410
pH = 6.9	0.16	0.9997	0.606	−4.633
	0.32	0.9975	0.553	−3.858
	0.48	0.9974	0.566	−3.858
	0.64	0.9989	0.571	−3.932
	1.28	0.9975	0.571	−4.191
pH = 8.7	0.16	0.9984	0.736	−4.974
	0.32	0.9991	0.564	−3.711
	0.48	0.9991	0.658	−4.333
	0.64	0.9971	0.631	−4.178
	1.28	0.9985	0.687	−4.429

and hydroxyl groups of water. When the MA groups are introduced into PDMA matrix, this later gradually shifts to lower wavenumbers 3463 cm^{-1}, indicating the presence of hydrogen bonding interactions between carboxylic group of MA and carbonyl of amide. Moreover, the shoulder observed around 2510 cm^{-1} (shifted to 2580 cm^{-1}) is characteristic of COOH dimers.

In the carbonyl stretching 1780–1560 cm^{-1} region, the infrared spectrum of PDMA recorded at room temperature shows a broad band at 1640 cm^{-1}, assigned to the free carbonyl groups of the amide. With the incorporation of the MA groups within PDMA chain, a new band appears at 1735 cm^{-1}, attributed to free carboxylic groups. In addition, the band at 1640 cm^{-1} shifts to lower wavenumbers 1635 cm^{-1}, corresponding to the DMA carbonyl group that is directly hydrogen bonded to the MA hydroxyl group (Figure 5(a)).

The infrared spectra of P(DMA-MA8) with different NMBA compositions recorded at room temperature in the carbonyl region are illustrated in Figure 5(b). We should notice a gradual shift of the characteristic band of the carboxyl-carbonyl amide interactions to higher wavenumbers with increasing NMBA content. This behavior reveals that hydrogen bonding associations are favored with the lower amount of NMBA.

3.3. Thermogravimetric Analysis. The thermogravimetric (TGA) and derivative thermogravimetric (DTGA) curves for PDMA and P(DMA-MAx) containing 8, 15, 29, 33, and 38 mol% of MA are represented in Figure 6(a). It is seen that all studied hydrogels degrade in two main stages.

Parameters, such as temperature of maximum degradation (determined considering the derivative curves) and percentage of mass loss in each stage of degradation for all studied systems, are summarized in Table 4(a). As observed, the mass loss for PDMA in stage 1 is apparently associated with adsorbed water. The second stage from 366 to 475°C corresponds to the effective degradation of the polymer.

For P(DMA-MAx), the percentage of mass loss at the range 5–16% in stage 1 can be assigned to the elimination of water adsorbed by the hydrophilic groups and probably to the elimination of hydroxyl groups due to the development of cyclic anhydride involving the liberation of water. Data from the literature indicated that with temperature increase (above 200°C), poly(styrene-*co*-maleic acid) with 42.8 mol% of maleic acid shows absorption bands at 1779 cm^{-1} and 1854 cm^{-1} which are typically associated with cyclic anhydride formation [34]. In the second stage, at higher temperatures decomposition of anhydride rings takes place and it is overlapped with degradation of the main chain.

The onset temperature of the second main degradation stage as well as temperature of maximum degradation for copolymers decreases with acid composition. The amount of residual weight at the end temperature of degradation increases with increasing acid content in P(DMA-MAx). The reason for incomplete degradation of copolymers is probably the thermal cross-linking induced by heating the sample during thermogravimetric analysis.

TABLE 4: (a) Thermogravimetric parameters for P(DMA-MAx) hydrogels with different compositions of MA. (b) Thermogravimetric parameters for P(DMA-MA8) hydrogels with different compositions of NMBA.

(a)

Hydrogels	Stage 1		Stage 2		
	ΔT^a (°C)	Δm^b (wt%)	ΔT^a (°C)	$T_{max}{}^c$ (°C)	Residual weight (wt%)
PDMA	55–170	4	366–475	447	0
P(DMA-MA8)	55–219	5.7	312–488	440	4.38
P(DMA-MA15)	53–220	9	281–485	428	6.44
P(DMA-MA29)	55–230	11	262–471	409	7.72
P(DMA-MA33)	56–224	13	254–489	406	9.42
P(DMA-MA38)	54–223	16	241–472	397	11.6

[a]Temperature range.
[b]Total weight loss percentage at the end of the step.
[c]Temperature maximum values of DTGA curves.

(b)

Composition of NMBA (mol%)	Stage 1		Stage 2		
	ΔT^a (°C)	Δm^b (wt%)	ΔT^a (°C)	$T_{max}{}^c$ (°C)	Residual weight (wt%)
0.16	52–220	5	279–467	434	4.13
0.32	53–219	5.7	281–479	441	4.36
0.48	58–220	4.58	277–478	442	3.71
0.64	58–222	5.11	283–486	449	3.59
1.28	59–211	4.38	285–496	485	3.47

[a]Temperature range.
[b]Total weight loss percentage at the end of the step.
[c]Temperature maximum values of DTGA curves.

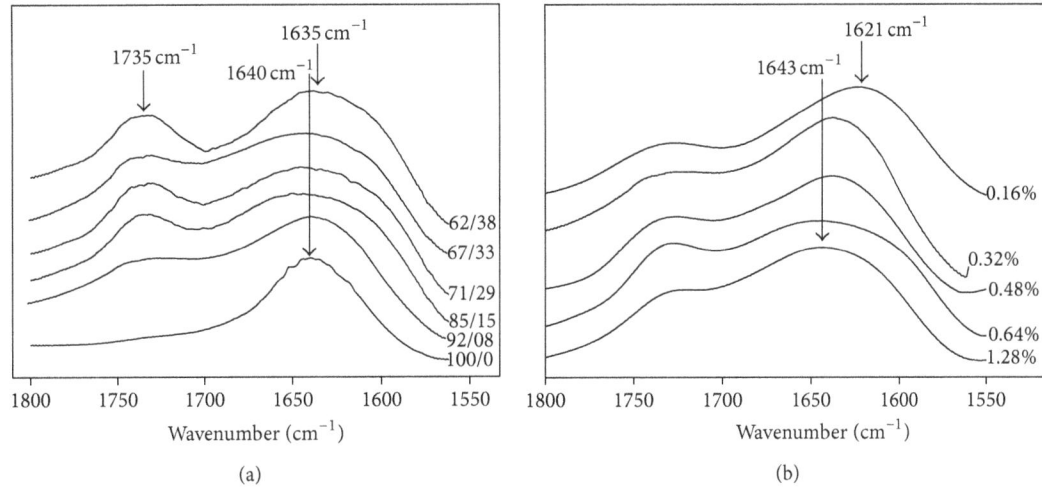

FIGURE 5: (a) Scale-expanded infrared spectra of P(DMA-MAx) hydrogels in the carbonyl region. (b) Scale-expanded infrared spectra of P(DMA-MA8) hydrogels with different compositions of NMBA in the carbonyl region.

The thermogravimetric (TGA) and derivative thermogravimetric (DTGA) curves for the copolymers P(DMA-MA8) with different amounts of NMBA are shown in Figure 6(b) while Table 4(b) summarizes their thermogravimetric parameters. The temperature of maximum degradation for these copolymers increases with NMBA content. This thermal stability is mainly due to an increase of the crosslink density.

3.4. DSC Analysis. Figure 7(a) shows thermograms of PDMA and P(DMA-MAx). The T_g values of PDMA and P(DMA-MAx) (x = 8, 15, 29, 33, and 38 mol%) are 126, 129, 136, 139, 143, and 158°C, respectively. It is clear that the introduction of maleic acidic species within PDMA backbone increases its T_g value. This fact is related to intermolecular hydrogen bonding between carboxyl and amide groups and intramolecular hydrogen bonding between carboxyl groups, which act as

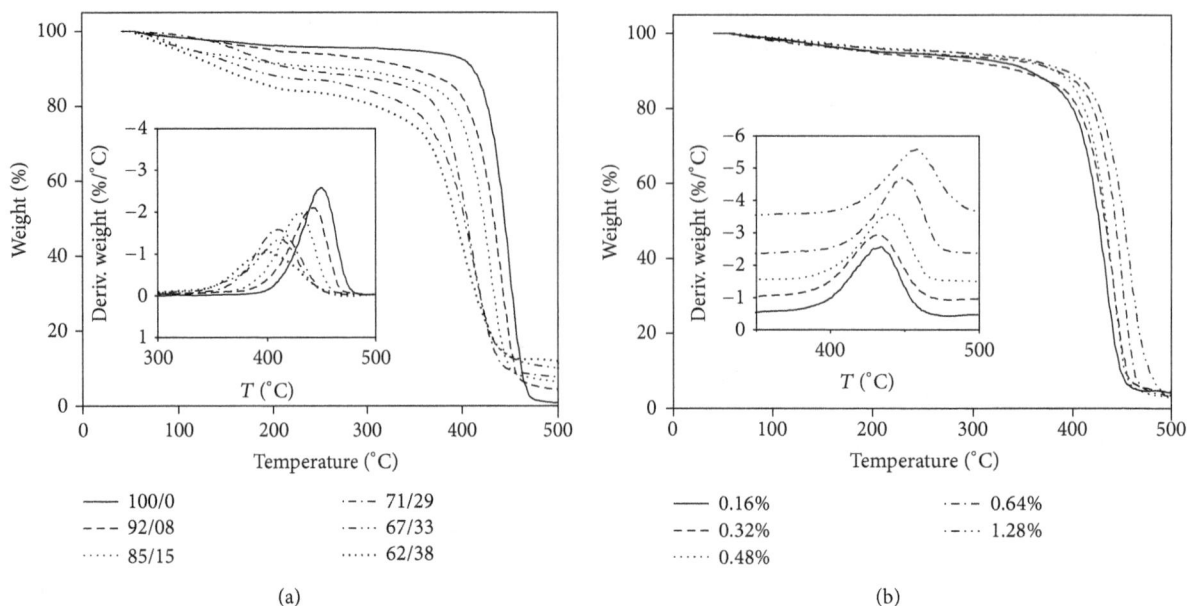

FIGURE 6: (a) TGA thermograms and DTGA curves of P(DMA-MAx) hydrogels. (b) TGA thermograms and DTGA curves of P(DMA-MA8) hydrogels with different compositions of NMBA.

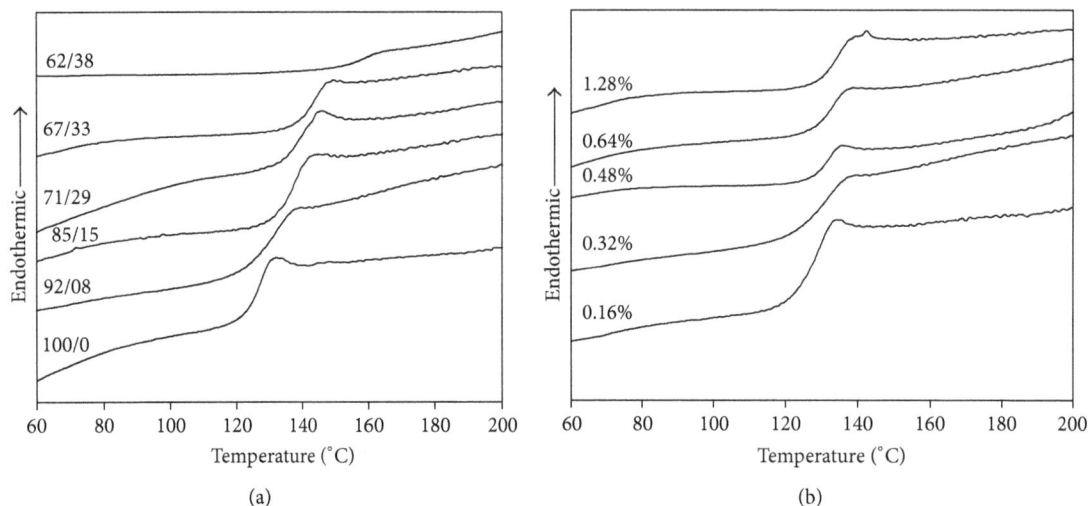

FIGURE 7: (a) DSC thermograms of P(DMA-MAx) hydrogels. (b) DSC thermograms of P(DAM-MA8) hydrogels with different compositions of NMBA.

additional cross-links, thus reducing subchain mobility and resulting in less flexible network.

The thermograms of P(DMA-MA8) with different amounts of NMBA are represented in Figure 7(b). The T_g values, in the range 127–134°C, are increased with increasing cross-linking agent content. The high cross-link density restricts the movement of polymer segment.

4. Conclusion

A series of pH-sensitive P(DMA-MAx) hydrogels were synthesized by free-radical polymerization in water using NMBA as cross-linker.

The equilibrium swelling ratio of hydrogels increases with the increase of MA content but reduces with the increase of NMBA concentration. The hydrogels have apparent pH-sensitive character. With increase in pH value from 0.9 to 8.7, the swelling ratio increased accordingly. In the diffusion transport mechanism study, the results indicate that the swelling exponent n for PDMA and DMA-MA copolymeric gels is in the range from 0.486 to 0.796. This implies that the swelling transport mechanism is a non-Fickian transport. The thermal stability of the copolymers is essentially associated with the DMA monomer, losing stability when the percentage of MA in the copolymer system increases. The described pH-sensitive hydrogel might have great potential application in drug delivery system.

Conflict of Interests

The authors declare that there is no conflict of interests regarding the publication of this paper.

References

[1] A. S. Hoffman, "Hydrogels for biomedical applications," *Advanced Drug Delivery Reviews*, vol. 54, no. 1, pp. 3–12, 2002.

[2] O. Wichterle and D. Lím, "Hydrophilic gels for biological use," *Nature*, vol. 185, no. 4706, pp. 117–118, 1960.

[3] D. M. García, J. L. Escobar, Y. Noa, N. Bada, E. Hernáez, and I. Katime, "Timolol maleate release from pH-sensible poly(2-hydroxyethyl methacrylate-co-methacrylic acid) hydrogels," *European Polymer Journal*, vol. 40, no. 8, pp. 1683–1690, 2004.

[4] M. Şen, Ö. Kantoğlu, and O. Güven, "The effect of external stimuli on the equilibrium swelling properties of poly(N-vinyl 2-pyrrolidone/itaconic acid) poly-electrolyte hydrogels," *Polymer*, vol. 40, no. 4, pp. 913–917, 1999.

[5] C. S. Brazel and N. A. Peppas, "Synthesis and characterization of thermo- and chemomechanically responsive poly(N-isopropylacrylamide-co-methacrylic acid) hydrogels," *Macromolecules*, vol. 28, no. 24, pp. 8016–8020, 1995.

[6] B. Yıldız, B. Işık, and M. Kış, "Synthesis of thermoresponsive N-isopropolylacrylamide-N-hydroxymethylacrylamide hydrogels by redox polymerization," *Polymer*, vol. 42, no. 6, pp. 2521–2529, 2001.

[7] B. Singh, N. Sharma, and N. Chauhan, "Synthesis, characterization and swelling studies of pH responsive psyllium and methacrylamide based hydrogels for the use in colon specific drug delivery," *Carbohydrate Polymers*, vol. 69, no. 4, pp. 631–643, 2007.

[8] S. P. Zhao, M. J. Cao, L. Y. Li, and W. L. Xu, "Synthesis and properties of biodegradable thermo- and pH-sensitive poly[(N-isopropylacrylamide)-co-(methacrylic acid)] hydrogels," *Polymer Degradation and Stability*, vol. 95, no. 5, pp. 719–724, 2010.

[9] K. Wang, S. Z. Fu, Y. C. Gu et al., "Synthesis and characterization of biodegradable pH-sensitive hydrogels based on poly(ε-caprolactone), methacrylic acid, and poly(ethylene glycol)," *Polymer Degradation and Stability*, vol. 94, no. 4, pp. 730–737, 2009.

[10] T. Çaykara and M. Doğmuş, "The effect of solvent composition on swelling and shrinking properties of poly(acrylamide-co-itaconic acid) hydrogels," *European Polymer Journal*, vol. 40, no. 11, pp. 2605–2609, 2004.

[11] Y. Luo, K. Zhang, Q. Wei, Z. Liu, and Y. Chen, "Poly(MAA-co-AN) hydrogels with improved mechanical properties for theophylline controlled delivery," *Acta Biomaterialia*, vol. 5, no. 1, pp. 316–327, 2009.

[12] S. K. Bajpai and S. Dubey, "In vitro dissolution studies for release of vitamin B_{12} from poly(N-vinyl-2-pyrrolidone-co-acrylic acid) hydrogels," *Reactive and Functional Polymers*, vol. 62, no. 1, pp. 93–104, 2005.

[13] B. Singh, N. Chauhan, and S. Kumar, "Radiation crosslinked psyllium and polyacrylic acid based hydrogels for use in colon specific drug delivery," *Carbohydrate Polymers*, vol. 73, no. 3, pp. 446–455, 2008.

[14] Y. Huang, H. Yu, and C. Xiao, "pH-sensitive cationic guar gum/poly (acrylic acid) polyelectrolyte hydrogels: swelling and in vitro drug release," *Carbohydrate Polymers*, vol. 69, no. 4, pp. 774–783, 2007.

[15] G. Ma, D. Yang, Q. Li et al., "Injectable hydrogels based on chitosan derivative/polyethylene glycol dimethacrylate/N,N-dimethylacrylamide as bone tissue engineering matrix," *Carbohydrate Polymers*, vol. 79, no. 3, pp. 620–627, 2010.

[16] J. Kopeček, "Hydrogel biomaterials: a smart future?" *Biomaterials*, vol. 28, no. 34, pp. 5185–5192, 2007.

[17] M. Sen and O. Güven, "Dynamic deswelling studies of poly(N-vinyl-2-pyrrolidone/itaconic acid) hydrogels swollen in water and terbinafine hydrochloride solutions," *European Polymer Journal*, vol. 38, no. 4, pp. 751–757, 2002.

[18] A. Vashist, Y. K. Gupta, and S. Ahmad, "Interpenetrating biopolymer network based hydrogels for an effective drug delivery system," *Carbohydrate Polymers*, vol. 87, no. 2, pp. 1433–1439, 2012.

[19] B. Singh, G. S. Chauhan, S. Kumar, and N. Chauhan, "Synthesis, characterization and swelling responses of pH sensitive psyllium and polyacrylamide based hydrogels for the use in drug delivery (I)," *Carbohydrate Polymers*, vol. 67, no. 2, pp. 190–200, 2007.

[20] S. Kondo, N. Nakashima, H. Hado, and K. Tsuoa, "Poly(N,N-dimethylamide-co-styrene)s as a highly efficient catalysts for two-phase reactions," *Journal of Polymer Science A: Polymer Chemistry*, vol. 28, no. 8, pp. 2229–2232, 1990.

[21] A. Valdebenito and M. V. Encinas, "Effect of solvent on the free radical polymerization of N,N-dimethylacrylamide," *Polymer International*, vol. 59, no. 9, pp. 1246–1251, 2010.

[22] W. L. F. Santos, M. F. Porto, E. C. Muniz, L. Olenka, M. L. Baessso, and A. C. Bento, "Poly(ethylene terephtalate) films modified with N, N'-dimethylacrylamide: incorporation of disperse dye," *Journal of Applied Polymer Science*, vol. 77, pp. 269–282, 2000.

[23] Alexeev, *Analyse Qualitative*, 4th edition, 1980.

[24] E. Karadağ and D. Saraydin, "Swelling of superabsorbent acrylamide/sodium acrylate hydrogels prepared using multifunctional crosslinkers," *Turkish Journal of Chemistry*, vol. 26, no. 6, pp. 863–875, 2002.

[25] E. Karadağ, Ö. B. Üzüm, and D. Saraydin, "Swelling equilibria and dye adsorption studies of chemically crosslinked superabsorbent acrylamide/maleic acid hydrogels," *European Polymer Journal*, vol. 38, no. 11, pp. 2133–2141, 2002.

[26] B. Yildiz, B. Işik, and M. Kiş, "Synthesis and characterization of thermoresponsive isopropolylacrylamide acrylamide hydrogels," *European Polymer Journal*, vol. 38, pp. 1343–1347, 2002.

[27] R. C. Weast, *Handbook of Chemistry and Physics*, The Chemical Rubber, Cleveland, Ohio, USA, 53rd edition, 1972.

[28] J. Zhang, L. Y. Chu, Y. K. Li, and Y. M. Lee, "Dual thermo- and pH-sensitive poly(N-isopropylacrylamide-co-acrylic acid) hydrogels with rapid response behaviors," *Polymer*, vol. 48, no. 6, pp. 1718–1728, 2007.

[29] M. Şen and A. Yakar, "Controlled release of antifungal drug terbinafine hydrochloride from poly(N-vinyl 2-pyrrolidone/itaconic acid) hydrogels," *International Journal of Pharmaceutics*, vol. 228, no. 1-2, pp. 33–41, 2001.

[30] S. Jin, M. Liu, F. Zhang, S. Chen, and A. Niu, "Synthesis and characterization of pH-sensitivity semi-IPN hydrogel based on hydrogen bond between poly(N-vinylpyrrolidone) and poly(acrylic acid)," *Polymer*, vol. 47, no. 5, pp. 1526–1532, 2006.

[31] B. Taşdelen, N. Kayaman-Apohan, O. Güven, and B. M. Baysal, "Preparation of poly(N-isopropylacrylamide/itaconic acid) copolymeric hydrogels and their drug release behavior," *International Journal of Pharmaceutics*, vol. 278, no. 2, pp. 343–351, 2004.

[32] P. L. Ritger and N. A. Peppas, "A simple equation for desciption of solute release I. Fickian and non-Fickian release from non-swellable devices in the form of slabs, spheres, cylinders or discs," *Journal of Controlled Release*, vol. 5, no. 1, pp. 23–36, 1987.

[33] P. L. Ritger and N. A. Peppas, "A simple equation for description of solute release I. Fickian and Non- Fickian release from swellable devices," *Journal of Controlled Release*, vol. 5, no. 1, pp. 37–42, 1987.

[34] M. Świtała-Zeliazkow, "Thermal degradation of copolymers of styrene with dicarboxylic acids—II: copolymers obtained by radical copolymerisation of styrene with maleic acid or fumaric acid," *Polymer Degradation and Stability*, vol. 91, no. 6, pp. 1233–1239, 2006.

3

Effect of C–O Bonding on the Stability and Energetics of High-Energy Nitrogen-Carbon Molecules $N_{10}C_2$ and $N_{16}C_2$

Douglas L. Strout

Department of Physical Sciences, Alabama State University, Montgomery, AL 36101, USA

Correspondence should be addressed to Douglas L. Strout; dstrout@alasu.edu

Academic Editor: Mohamed Sarakha

Molecules consisting of nitrogen have been the subject of much attention due to their potential as high-energy materials. Complex molecules consisting entirely of nitrogen can be subject to rapid decomposition, and therefore other atoms are incorporated into the structure to enhance stability. Previous studies have explored the incorporation of carbon atoms into otherwise all-nitrogen cages molecules. The current study involves two such cages, $N_{10}C_2$ and $N_{16}C_2$, whose structures are derived from N_{12} and N_{18}, respectively. The $N_{10}C_2$ and $N_{16}C_2$ cages in this study are modified by bonding groups O_3 and CO_3 to determine the effect on the relative energies between the isomers and on the thermodynamic energy release properties. Energetic trends for $N_{10}C_2$ and $N_{16}C_2$ are calculated and discussed.

1. Introduction

Molecules consisting entirely or predominantly of nitrogen have been the subject of much research because of their potential as high-energy materials. Decomposition reactions of the type $N_x \rightarrow (x/2)N_2$ can be exothermic by up to 50 kcal/mol per nitrogen atom (approximately 3.5 kilocalories per gram of material). Experimental synthetic successes in high-energy nitrogen materials include the N_5^+ and N_5^- ions [1–4] as well as various azido compounds [5–9] and even a network polymer of nitrogen [10]. Additionally, nitrogen-rich salts [11] and the N_7O^+ ion [12] have been achieved experimentally. The production of such a diverse group of nitrogen systems demonstrates the potential for such materials as novel high-energy molecules. Nitrogen-based energetic systems have also been the subject of much theoretical research. Theoretical studies of high-energy nitrogen include cyclic and acyclic compounds [13–20], as well as nitrogen cages [21–27]. Structures and thermodynamics of energetic nitrogen systems have been calculated for both small molecules and larger structures with up to seventy-two atoms.

Theoretical studies [28] of cage isomers of N_{24}, N_{30}, and N_{36} showed that the most stable isomers are narrow cylindrical structures consisting of bands of hexagons capped by triangle-pentagon endcaps in either D_{3d} or D_{3h} point group symmetry. A previous study [29] of molecules of $N_{22}C_2$ showed that the most stable isomer has a C_2 parallel to the long axis of the molecule, which allows the C_2 unit and its C=C double bond the most planar, ethylene-like environment. The least stable isomers have the C_2 unit in proximity to the triangular endcaps, where angle strain around the C=C double bond becomes a destabilizing factor. In the current study, the smaller analogues $N_{10}C_2$ and $N_{16}C_2$ are considered, with the bonding groups O_3 and CO_3 added to the C=C double bond. The addition of O_3 to C=C double bonds in cage molecules such as fullerenes is already well known [30], and the CO_3 bonding group could be achieved by reaction with the metastable carbon trioxide molecule [31] or other means involving carbon dioxide and oxygen. These bonding groups are chosen to determine the effects of C–O bonding on nitrogen-carbon molecules. The effects of these bonding groups on the relative isomer energies and on the heat of formation of the molecules are calculated and discussed.

2. Computational Methods

Geometries for all molecules in this study are optimized by the Hartree-Fock method, and Hartree-Fock vibrational frequencies are used to confirm each structure as a local minimum. Single energy points are calculated with

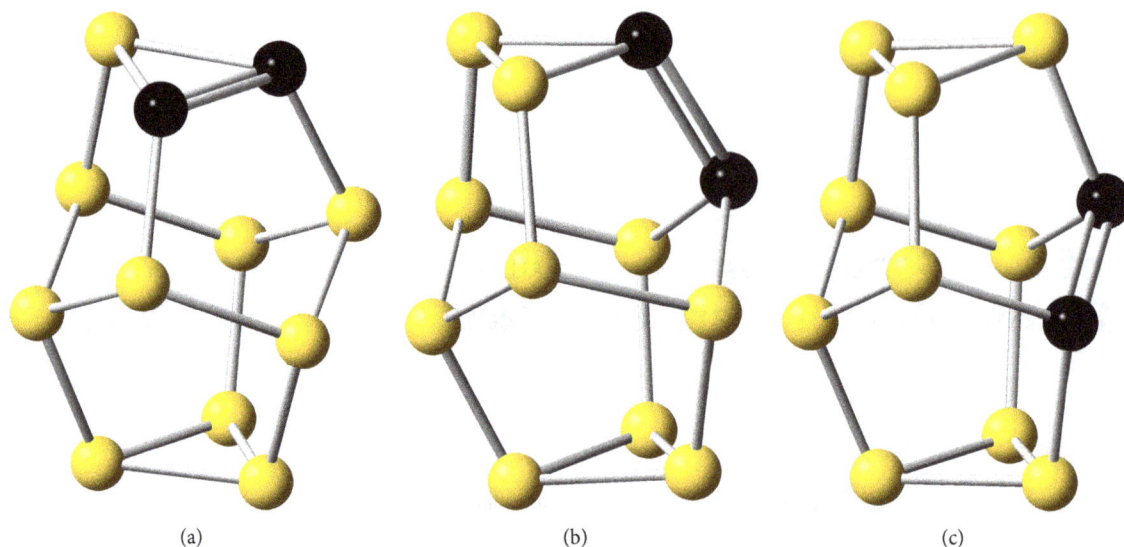

FIGURE 1: Isomers of $N_{10}C_2$: (a) isomer A, (b) isomer B, and (c) isomer C. Nitrogen atoms are shown in yellow, carbon atoms in black.

coupled-cluster theory [32, 33] (CCSD(T)). Hartree-Fock method is chosen for geometry optimization for two reasons: (1) previous calculations on nitrogen cages have shown that energy results are insensitive to the choice of optimized geometry and (2) other optimization methods, such as density functional theory, have been shown to be dissociative for certain nitrogen cages. The correlation-consistent cc-pVDZ atomic orbital basis set [34] of Dunning is used for all calculations in this study. Calculations have been carried out using the Gaussian 09 computational chemistry software [35].

3. Results and Discussion

3.1. $N_{10}C_2$. The $N_{10}C_2$ cage has three isomers, which are shown in Figure 1. These isomers are labeled A, B, and C and represent the three symmetry-independent substitutions of a C_2 unit into the structure of the N_{12} cage in D_{3d} symmetry. The O_3 and CO_3 adducts of each of these isomers are shown in Figure 2. Energies and vibrational frequencies have been calculated for each molecule, and the energies are shown in Table 1. For $N_{10}C_2$, isomer C has the lowest energy, mainly because of angle strain in the triangular endcaps since isomers A and B have at least one carbon atom in the triangle. However, the application of either O_3 or CO_3 bonding group causes a reversal, and isomer B becomes most stable. This is because the addition of O_3 or CO_3 converts the sp^2-hybridized carbon to sp^3, which relieves the ring strain to some extent. It is likely that the isomer A adducts also benefit from the adduct stabilization, but this cannot be confirmed since isomer A of $N_{10}C_2$ is not a local minimum on the potential energy surface.

What effect does the addition of O_3 or CO_3 have on the overall energetic properties of the molecules? Nitrogen-carbon cages tend to be highly energetic and have high heat of formation, but molecules with C–O bonds tend to have lower heat of formation (the heat of formation [36] of carbon dioxide, e.g., is −94.0 kcal/mol). Enthalpies of formation for

TABLE 1: Relative isomer energies (in kcal/mol) for $N_{10}C_2$ and its O_3 and CO_3 adducts. Energies calculated with CCSD(T)/cc-pVDZ method. Except where noted, all molecules are local minima confirmed by HF/cc-pVDZ vibrational frequencies.

	Isomer A	Isomer B	Isomer C
$N_{10}C_2$	Not minimum	+10.0	0.0
$N_{10}C_2O_3$	+9.3	−4.2	0.0
$N_{10}C_2CO_3$	+11.2	−5.9	0.0

TABLE 2: Enthalpies of formation (in kcal/mol and kcal/g) for $N_{10}C_2$ and its O_3 and CO_3 adducts. Energies calculated with CCSD(T)/cc-pVDZ method.

		ΔH_f (kcal/mol)	ΔH_f (kcal/g)
$N_{10}C_2$ (M = 164 g/mole)	Isomer A	Not minimum	Not minimum
	Isomer B	+528.3	+3.22
	Isomer C	+518.3	+3.16
$N_{10}C_2O_3$ (M = 212 g/mole)	Isomer A	+458.9	+2.16
	Isomer B	+445.4	+2.10
	Isomer C	+449.6	+2.12
$N_{10}C_2CO_3$ (M = 224 g/mole)	Isomer A	+338.2	+1.51
	Isomer B	+321.1	+1.43
	Isomer C	+327.0	+1.46

$N_{10}C_2$ and its O_3 and CO_3 adducts are shown in Table 2 and have been calculated using CCSD(T)/cc-pVDZ energies for the following chemical reactions:

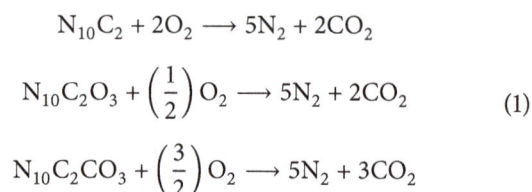

$$N_{10}C_2 + 2O_2 \longrightarrow 5N_2 + 2CO_2$$

$$N_{10}C_2O_3 + \left(\frac{1}{2}\right)O_2 \longrightarrow 5N_2 + 2CO_2 \qquad (1)$$

$$N_{10}C_2CO_3 + \left(\frac{3}{2}\right)O_2 \longrightarrow 5N_2 + 3CO_2$$

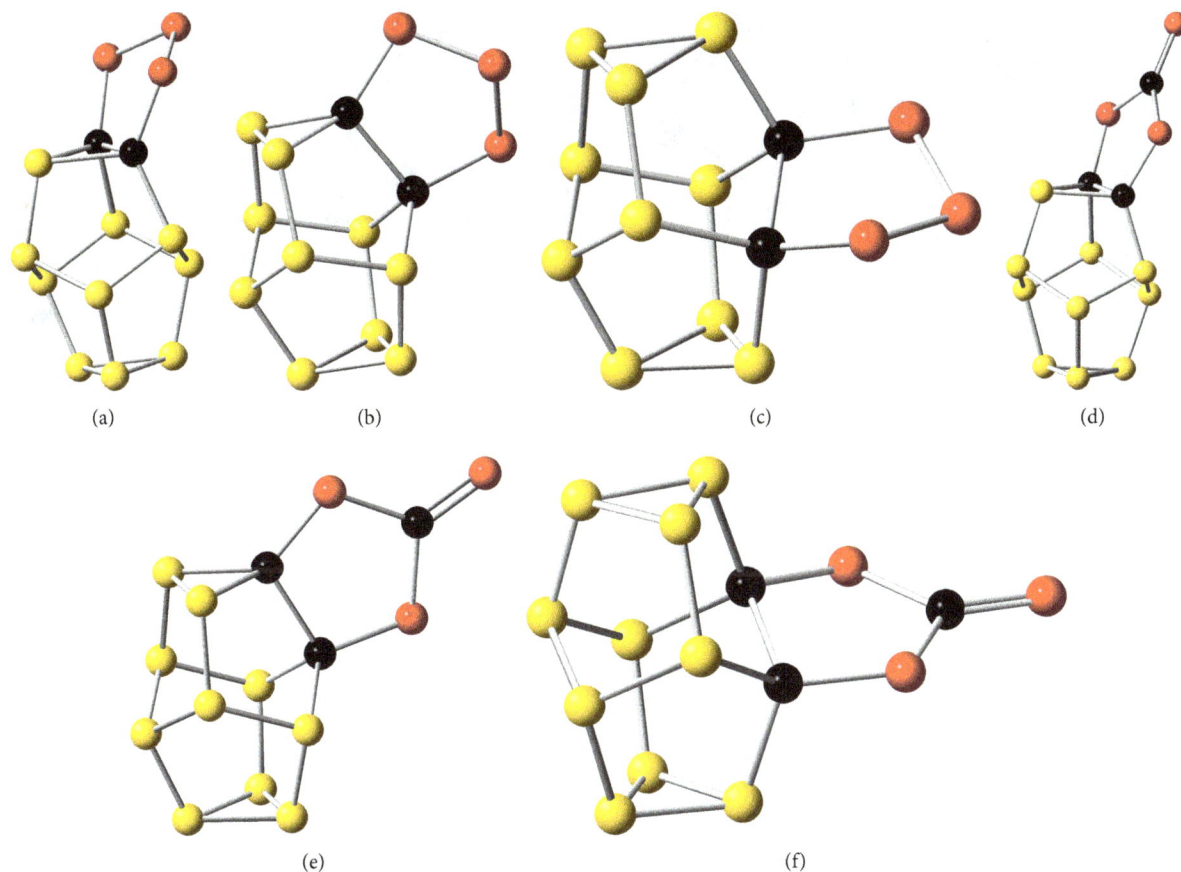

FIGURE 2: Adducts of $N_{10}C_2$: (a) isomer A + O_3, (b) isomer B + O_3, (c) isomer C + O_3, (d) isomer A + CO_3, (e) isomer B + CO_3, and (f) isomer C + CO_3. Nitrogen atoms are shown in yellow, carbon atoms in black, and oxygen atoms in red.

The results in Table 2 show that $N_{10}C_2$ is highly energetic, with a heat of formation above 3.0 kilocalories per gram of material. Such molecules would release a large amount of energy upon their decomposition. Introduction of the O_3 bonding group lowers the heat of formation to about 2.1 kilocalories per gram, and CO_3 lowers the energy even further, down to about 1.5 kilocalories per gram, which is still greater than the energy densities [37] of conventional explosives such as RDX and HMX, which are approximately 0.5–1.0 kcal/g. By the choice of adduct bonding group, the energetic properties of these molecules are essentially tunable, and the tradeoff between stability and energy release may be a significant consideration in the synthesis of these molecules.

3.2. $N_{16}C_2$.

The $N_{16}C_2$ cage has four isomers, which are shown in Figure 3. These isomers are labeled A, B, C, and D and represent the four symmetry-independent substitutions of a C_2 unit into the structure of the N_{18} cage in D_{3h} symmetry. The O_3 and CO_3 adducts of the four isomers are shown in Figure 4. Energies and vibrational frequencies have been calculated for each molecule, and the energies are shown in Table 3. For $N_{16}C_2$, isomer D is easily the most stable. Isomer D has a C_2 unit parallel to the C_3 axis of the molecule, and the arrangement of the four nitrogen atoms around the C_2 unit in isomer D provides an environment much

TABLE 3: Relative isomer energies (in kcal/mol) for $N_{16}C_2$ and its O_3 and CO_3 adducts. Energies calculated with CCSD(T)/cc-pVDZ method. Except where noted, all molecules are local minima confirmed by HF/cc-pVDZ vibrational frequencies.

	Isomer A	Isomer B	Isomer C	Isomer D
$N_{16}C_2$	Not minimum	+65.3	+28.8	0.0
$N_{16}C_2O_3$	+40.5	+21.8	+9.3	0.0
$N_{16}C_2CO_3$	+43.2	+20.6	+11.0	0.0

closer to the planar "ethylene-like" geometry preferred by sp^2-hybridized carbon atoms. Isomers A, B, and C all have a C_2 unit with significant angle strain or significant twisting of the four nitrogen atoms around the C_2. By contrast, $N_{10}C_2$ does not have this very stable isomer, which explains why $N_{10}C_2$ has isomer energies much closer together than does $N_{16}C_2$. $N_{16}C_2$ isomers show a rapid increase in energy as the C_2 is located closer to the triangular endcaps. In a manner similar to $N_{10}C_2$, the adduct energies in Table 3 show a substantial narrowing of the energy gaps between isomers, because of the relief of angle strain in isomers A and B. However, owing to the special stability of isomer D, the relief of ring strain does not cause a reversal of the ordering of the isomer energies.

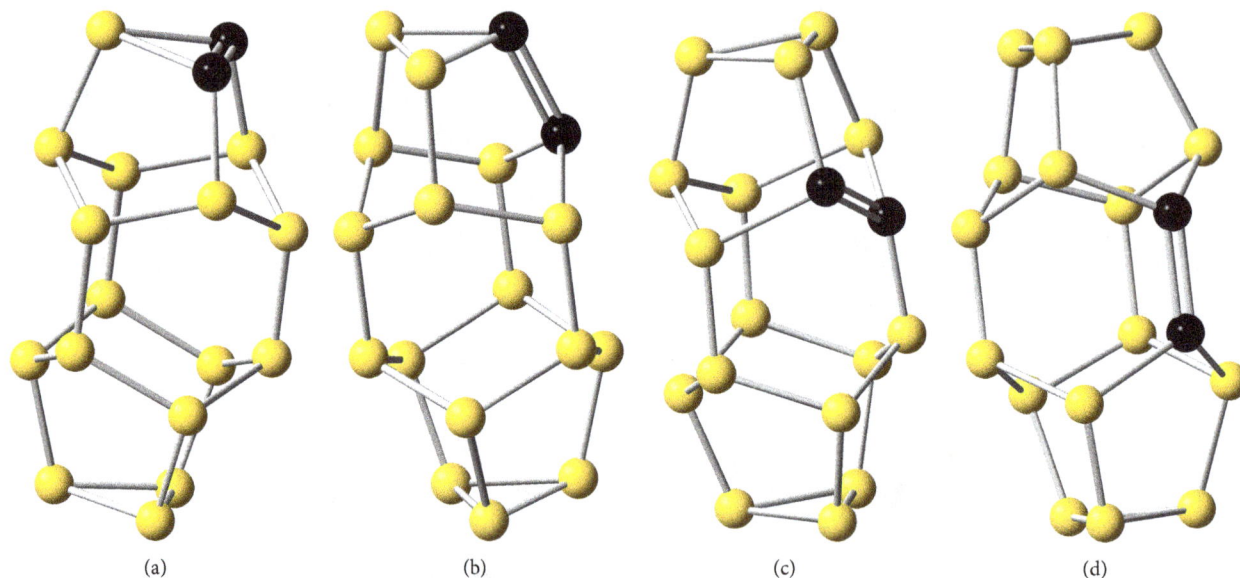

FIGURE 3: Isomers of $N_{16}C_2$: (a) isomer A, (b) isomer B, (c) isomer C, and (d) isomer D. Nitrogen atoms are shown in yellow, carbon atoms in black.

TABLE 4: Enthalpies of formation (in kcal/mol and kcal/g) for $N_{16}C_2$ and its O_3 and CO_3 adducts. Energies calculated with CCSD(T)/cc-pVDZ method.

		ΔH_f (kcal/mol)	ΔH_f (kcal/g)
	Isomer A	Not minimum	Not minimum
$N_{16}C_2$	Isomer B	+800.1	+3.23
(M = 248 g/mole)	Isomer C	+763.6	+3.08
	Isomer D	+734.8	+2.96
	Isomer A	+740.9	+2.50
$N_{16}C_2O_3$	Isomer B	+722.2	+2.44
(M = 296 g/mole)	Isomer C	+709.7	+2.40
	Isomer D	+700.4	+2.37
	Isomer A	+619.9	+2.01
$N_{16}C_2CO_3$	Isomer B	+597.3	+1.94
(M = 308 g/mole)	Isomer C	+587.7	+1.91
	Isomer D	+576.7	+1.87

Enthalpies of formation for $N_{16}C_2$ and its O_3 and CO_3 adducts are shown in Table 4 and have been calculated using CCSD(T)/cc-pVDZ energies for the following chemical reactions:

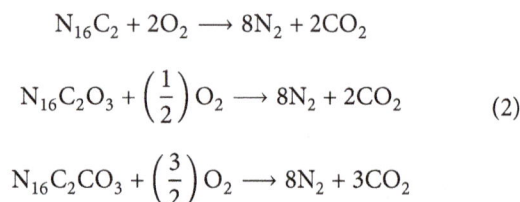

$$N_{16}C_2 + 2O_2 \longrightarrow 8N_2 + 2CO_2$$

$$N_{16}C_2O_3 + \left(\frac{1}{2}\right)O_2 \longrightarrow 8N_2 + 2CO_2 \qquad (2)$$

$$N_{16}C_2CO_3 + \left(\frac{3}{2}\right)O_2 \longrightarrow 8N_2 + 3CO_2$$

The first interesting result shown in Table 4 is that the $N_{16}C_2$ isomers have enthalpies of formation that are no higher than the corresponding isomers of $N_{10}C_2$ (about 3.0 kilocalories per gram), despite the fact that $N_{16}C_2$ has a higher proportionate nitrogen content. It appears that the increased size of the molecule, and correspondingly greater distance between the triangular endcaps, provides for a more relaxed, less strained structure that stabilizes the molecule. On the other hand, the higher nitrogen content of $N_{16}C_2$ does have an impact on the heat of formation of the $N_{16}C_2$ adducts. Whereas $N_{10}C_2$ and $N_{16}C_2$ have about the same heat of formation, $N_{16}C_2O_3$ and $N_{16}C_2CO_3$ have much higher heat of formation than their smaller counterparts. $N_{16}C_2O_3$ isomers have a heat of formation of 2.3–2.5 kilocalories per gram, and $N_{16}C_2CO_3$ has heat of formation of 1.9–2.0 kilocalories per gram. As with $N_{10}C_2$, $N_{16}C_2$ shows a variability of energetic properties depending on the nature of the bonding group added to the C=C double bond.

4. Conclusion

$N_{10}C_2$ and $N_{16}C_2$ are high-energy molecules whose properties can be varied by the addition of various bonding groups to the C=C double bond. Adducts of both $N_{10}C_2$ and $N_{16}C_2$ demonstrate changes in energetic properties that reflect both relief of ring strain with the structure and the stabilizing influence of the carbon-oxygen bond. The stability and energy-release properties of these molecules can be adjusted by the appropriate choice of adduct bonding group, which should provide a number of potential synthetic targets. Further studies involving multiple C_2 units in the structure and additional choices of adduct bonding group should reveal additional molecules with potential as high-energy materials.

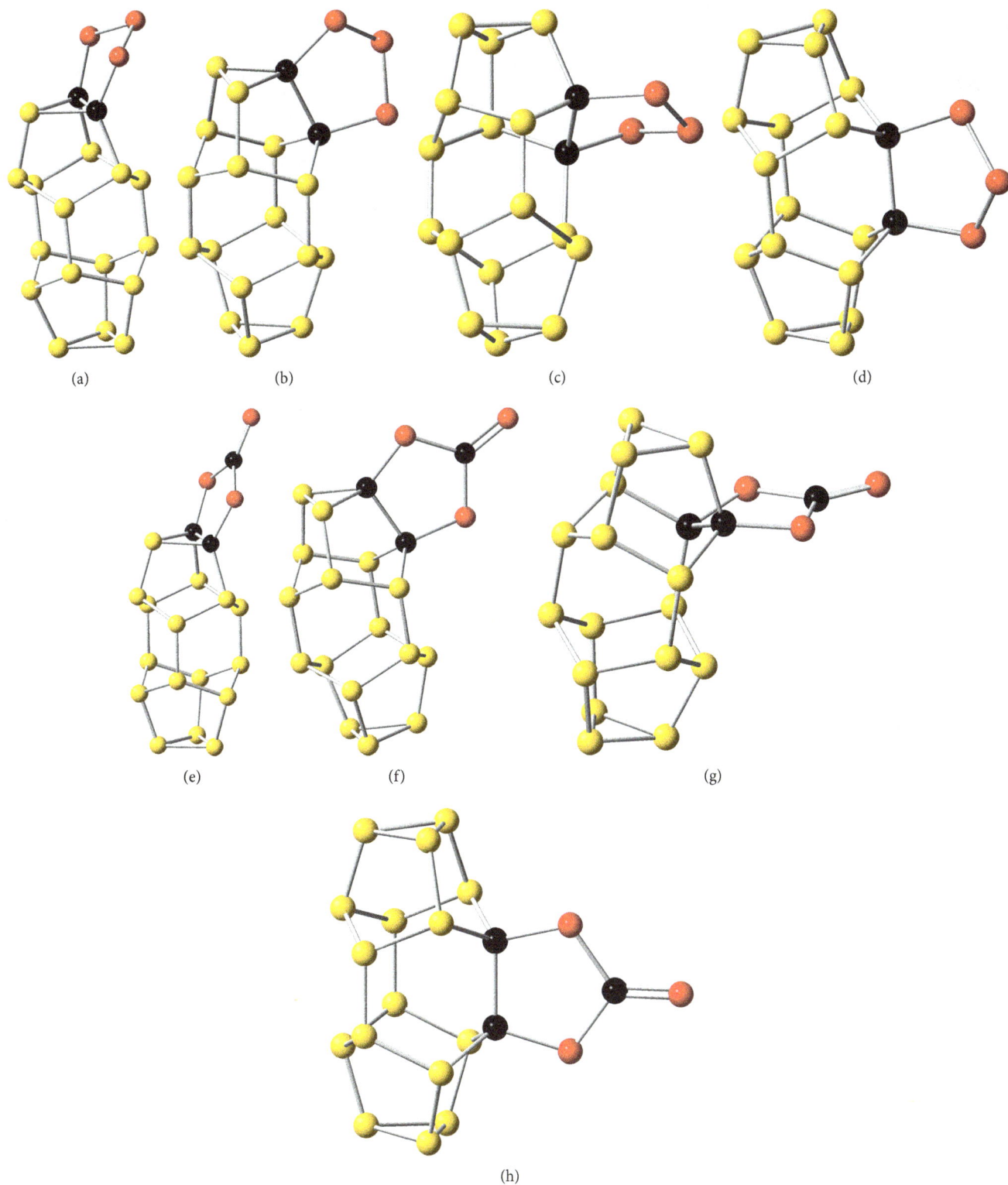

FIGURE 4: Adducts of $N_{16}C_2$: (a) isomer A + O_3, (b) isomer B + O_3, (c) isomer C + O_3, (d) isomer D + O_3, (e) isomer A + CO_3, (f) isomer B + CO_3, (g) isomer C + CO_3, and (h) isomer D + CO_3. Nitrogen atoms are shown in yellow, carbon atoms in black, and oxygen atoms in red.

Conflict of Interests

The author declares that there is no conflict of interests regarding the publication of this paper.

Acknowledgments

The Alabama Supercomputer Authority is gratefully acknowledged for a grant of computer time on the SGI Ultraviolet in Huntsville, AL. This work was supported by the National Science Foundation (NSF/HBCU-UP Grant 0505872). This work was also supported by the National Institutes of Health (NIH/NCMHD 1P20MD000547-01) and the Petroleum Research Fund, administered by the American Chemical Society (PRF 43798-B6). The taxpayers of the state of Alabama in particular and the United States in general are gratefully acknowledged.

References

[1] K. O. Christe, W. W. Wilson, J. A. Sheehy, and J. A. Boatz, "N_5^+: a novel homoleptic polynitrogen ion as a high energy density material," *Angewandte Chemie*, vol. 38, no. 13-14, pp. 2004–2009, 1999.

[2] A. Vij, J. G. Pavlovich, W. W. Wilson, V. Vij, and K. O. Christe, "Experimental detection of the pentaazacyclopentadienide (pentazolate) anion, cyclo-N_5^-," *Angewandte Chemie International Edition*, vol. 41, no. 16, pp. 3051–3054, 2002.

[3] R. N. Butler, J. C. Stephens, and L. A. Burke, "First generation of pentazole (HN_5, pentazolic acid), the final azole, and a zinc pentazolate salt in solution: a new N-dearylation of 1-(p-methoxyphenyl) pyrazoles, a 2-(p-methoxyphenyl) tetrazole and application of the methodology to 1-(p-methoxyphenyl) pentazole," *Chemical Communications*, no. 8, pp. 1016–1017, 2003.

[4] D. A. Dixon, D. Feller, K. O. Christe et al., "Enthalpies of formation of gas-phase N_3, N_3^-, N_5^+, and N5- from Ab Initio molecular orbital theory, stability predictions for $N_5^+ N_3^-$ and $N_5^+ N_5^-$, and experimental evidence for the instability of $N_5^+ N_3^-$," *Journal of the American Chemical Society*, vol. 126, no. 3, pp. 834–843, 2004.

[5] C. Knapp and J. Passmore, "On the way to "Solid Nitrogen" at normal temperature and pressure? Binary azides of heavier group 15 and 16 elements," *Angewandte Chemie International Edition*, vol. 43, pp. 4834–4836, 2004.

[6] R. Haiges, S. Schneider, T. Schroer, and K. O. Christe, "High-Energy-Density Materials: Synthesis and Characterization of $N_5^+[P(N_3)_6]^-$, $N_5^+[B(N_3)_4]^-$, $N_5^+[HF_2]^- \cdot n$ HF, $N_5^+[BF_4]^-$, $N_5^+[PF_6]^-$, and $N_5^+[SO_3F]^-$," *Angewandte Chemie International Edition*, vol. 43, pp. 4919–4924, 2004.

[7] M.-H. V. Huynh, M. A. Hiskey, E. L. Hartline, D. P. Montoya, and R. Gilardi, "Polyazido high-nitrogen compounds: hydrazo- and Azo-1,3,5-triazine," *Angewandte Chemie*, vol. 43, no. 37, pp. 4924–4928, 2004.

[8] T. M. Klapotke, A. Schulz, and J. McNamara, "Preparation, characterization, and ab initio computation of the first binary antimony azide, $Sb(N_3)_3$," *Journal of the Chemical Society, Dalton Transactions*, pp. 2985–2987, 1996.

[9] T. M. Klapotke, H. Noth, T. Schutt, and M. Warchhold, "Tetraphenylphosphonium hexaazidoarsenate(V): the first structurally characterized binary As^V–azide species," *Angewandte Chemie International Edition*, vol. 39, no. 12, pp. 2108–2109, 2000.

[10] M. I. Eremets, A. G. Gavriliuk, I. A. Trojan, D. A. Dzivenko, and R. Boehler, "Single-bonded cubic form of nitrogen," *Nature Materials*, vol. 3, no. 8, pp. 558–563, 2004.

[11] Y. Huang, Y. Zhang, and J. M. Shreeve, "Nitrogen-rich salts based on energetic nitroaminodiazido[1,3,5]triazine and Guanazine," *Chemistry*, vol. 17, no. 5, pp. 1538–1546, 2011.

[12] K. O. Christe, R. Haiges, W. W. Wilson, and J. A. Boatz, "Synthesis and properties of N_7O^+," *Inorganic Chemistry*, vol. 49, pp. 1245–1251, 2010.

[13] G. Chung, M. W. Schmidt, and M. S. Gordon, "An ab initio study of potential energy surfaces for Ng isomers," *Journal of Physical Chemistry A*, vol. 104, no. 23, pp. 5647–5650, 2000.

[14] D. L. Strout, "Acyclic N_{10} fails as a high energy density material," *The Journal of Physical Chemistry A*, vol. 106, no. 5, pp. 816–818, 2002.

[15] M. D. Thompson, T. M. Bledson, and D. L. Strout, "Dissociation barriers for odd-numbered acyclic nitrogen molecules N9 and N11," *Journal of Physical Chemistry A*, vol. 106, no. 29, pp. 6880–6882, 2002.

[16] Q. S. Li and Y. D. Liu, "Structures and stability of N_{11} cluster," *Chemical Physics Letters*, vol. 353, pp. 204–212, 2002.

[17] Q. S. Li, H. Qu, and H. S. Zhu, "Quantum chemistry calculations of nitrogen cages N_{10}," *Chinese Science Bulletin*, vol. 41, pp. 1184–1188, 1996.

[18] Q. S. Li and J. F. J. Zhao, "Theoretical study of potential energy surfaces for N_{12} clusters," *The Journal of Physical Chemistry A*, vol. 106, no. 21, pp. 5367–5372, 2002.

[19] H. Qu, Q. S. Li, and H. S. Zhu, "Quantum chemical calculations of nitrogen cages N_{12}," *Chinese Science Bulletin*, vol. 42, pp. 462–465, 1997.

[20] L. Gagliardi, S. Evangelisti, A. Bernhardsson, R. Lindh, and B. O. Roos, "Dissociation Reaction of N8 Azapentalene to 4N2: a Theoretical Study," *International Journal of Quantum Chemistry*, vol. 77, no. 1, pp. 311–315, 2000.

[21] L. Gagliardi, S. Evangelisti, P. O. Widmark, and B. O. Roos, "A theoretical study of the N8 cubane to N8 pentalene isomerization reaction," *Theoretical Chemistry Accounts*, vol. 97, no. 1–4, pp. 136–142, 1997.

[22] M. W. Schmidt, M. S. Gordon, and J. A. Boatz, "Cubic Fuels?" *International Journal of Quantum Chemistry*, vol. 76, no. 3, pp. 434–446, 2000.

[23] H. Zhou, N.-B. Wong, G. Zhou, and A. Tian, "Theoretical study on "multilayer" nitrogen cages," *The Journal of Physical Chemistry A*, vol. 110, no. 10, pp. 3845–3852, 2006.

[24] H. Zhou, N.-B. Wong, G. Zhou, and A. Tian, "What makes the cylinder-shaped N_{72} cage stable?" *The Journal of Physical Chemistry A*, vol. 110, no. 23, pp. 7441–7446, 2006.

[25] L. Y. Bruney, T. M. Bledson, and D. L. Strout, "What makes an N12 cage stable?" *Inorganic Chemistry*, vol. 42, no. 24, pp. 8117–8120, 2003.

[26] S. E. Sturdivant, F. A. Nelson, and D. L. Strout, "Trends in stability for N_{18} cages," *The Journal of Physical Chemistry A*, vol. 108, no. 34, pp. 7087–7090, 2004.

[27] D. L. Strout, "Cage isomers of N_{14} and N_{16}: nitrogen molecules that are not a multiple of six," *The Journal of Physical Chemistry A*, vol. 108, no. 49, pp. 10911–10916, 2004.

[28] D. L. Strout, "Isomer stability of N_{24}, N_{30}, and N_{36} cages: cylindrical versus spherical structure," *The Journal of Physical Chemistry A*, vol. 108, pp. 2555–2558, 2004.

[29] S. Jasper, A. Hammond, J. Thomas, L. Kidd, and D. L. Strout, "N22C2 versus N24: role of molecular curvature in determining isomer stability," *Journal of Physical Chemistry A*, vol. 115, no. 42, pp. 11915–11918, 2011.

[30] D. Heymann, S. M. Bachilo, and R. B. Weisman, "Ozonides, epoxides, and oxidoannulenes of C70," *Journal of the American Chemical Society*, vol. 124, no. 22, pp. 6317–6323, 2002.

[31] C. S. Jamieson, A. M. Mebel, and R. I. Kaiser, "Cover picture: identification of the D_{3h} isomer of carbon trioxide (CO_3) and its implications for atmospheric chemistry (ChemPhysChem 12/2006)," *ChemPhysChem*, vol. 7, no. 12, p. 2441, 2006.

[32] G. D. Purvis III and R. J. Bartlett, "A full coupled-cluster singles and doubles model: the inclusion of disconnected triples," *The Journal of Chemical Physics*, vol. 76, no. 4, pp. 1910–1918, 1982.

[33] G. E. Scuseria, C. L. Janssen, and H. F. Schaefer III, "An efficient reformulation of the closed-shell coupled cluster single and double excitation (CCSD) equations," *The Journal of Chemical Physics*, vol. 89, no. 12, pp. 7382–7387, 1988.

[34] T. H. Dunning Jr., "Gaussian basis sets for use in correlated molecular calculations. I. The atoms boron through neon and hydrogen," *The Journal of Chemical Physics*, vol. 90, no. 2, pp. 1007–1023, 1989.

[35] M. J. Frisch, G. W. Trucks, H. B. Schlegel et al., *Gaussian 09, Revision C.01*, Gaussian, Inc., Wallingford, Conn, USA, 2010.

[36] Data taken from Computational Chemistry Comparison and Benchmark Database, http://cccbdb.nist.gov.

[37] W. J. Evans, M. J. Lipp, C. S. Yoo et al., "Pressure-induced polymerization of carbon monoxide: disproportionation and synthesis of an energetic lactonic polymer," *Chemistry of Materials*, vol. 18, no. 10, pp. 2520–2531, 2006.

4

Detection and Estimation of *alpha*-Amyrin, *beta*-Sitosterol, Lupeol, and *n*-Triacontane in Two Medicinal Plants by High Performance Thin Layer Chromatography

Saikat S. Mallick and Vidya V. Dighe

Chemistry Department, Ramnarain Ruia College, Matunga East, Mumbai, Maharashtra 400 019, India

Correspondence should be addressed to Saikat S. Mallick; mallisaikat@gmail.com

Academic Editor: Liya Ge

A normal phase high performance thin layer chromatography (HPTLC) method has been developed and validated for simultaneous estimation of four components, *namely, alpha*-amyrin, *beta*-sitosterol, lupeol, and *n*-triacontane from two medicinally important plants, *Leptadenia reticulata* Wight & Arn. and *Pluchea lanceolata* (DC.) CB. Clarke. In Ayurveda, both plants have been reported to possess immunomodulatory activity. Chromatographic separation of the four components from the methanolic extracts of whole plant powders of *Leptadenia reticulata* Wight & Arn. and *Pluchea lanceolata* (DC.) CB. Clarke. was performed on TLC aluminium plates precoated with silica gel $60F_{254}$ using a suitable mobile phase. The densitometric scanning was done after derivatization at $\lambda = 580$ nm for α-amyrin, β-sitosterol, and lupeol, and at 366 nm for *n*-triacontane. The developed HPTLC method has been validated and used for simultaneous quantitation of the four components from the methanolic extracts of whole plant powders of *Leptadenia reticulata* Wight & Arn. and *Pluchea lanceolata* (DC.) CB. Clarke. The developed HPTLC method is simple, rapid, and precise and can be used for routine quality control.

1. Introduction

Herbal medicines have been used since ages to treat various ailments. Ayurveda is an Indian traditional system of medicine used since ancient times. It has a huge list of herbs used in various forms for treatment of different disease conditions. Owing to the medicinal properties attributed to herbal drugs, it is necessary to maintain their quality and purity, thereby justifying their acceptability in modern system of medicine. Standardisation of these herbal drugs is a challenge to the entire scientific fraternity. However, due to lack of suitable quality control and quality assurance standards for herbal drugs, it becomes difficult to ensure uniformity of their composition which in turn affects the efficacy of their final products. Analytical tools are important for qualitative, semiquantitative, and quantitative phytochemical analysis of herbal drugs and formulations. Chromatographic techniques such as high performance liquid chromatography (HPLC), high performance thin layer chromatography (HPTLC), and

gas chromatography (GC) are used to efficiently determine the quality of the herbs by developing fingerprints and estimation of biomarkers. Among the wide choice of chromatographic techniques, HPTLC is a simple, fast, and accurate technique for use, making it advantageous over others for quick assessment of a number of samples simultaneously [1]. In the present research work, HPTLC method has been developed as a quality control tool for whole plant powder of *L. reticulata* and *P. lanceolata*.

L. reticulata belonging to family Asclepiadaceae commonly known as "Jivanti" has been reported to possess antitumour and anticancer activity [2]. *P. lanceolata* belonging to family Asteraceae commonly known as "Rasna" is an important xerophytic medicinal herb. It is traditionally used for dyspepsia, bronchitis, and rheumatoid arthritis [3, 4]. In Ayurveda both plants have been reported to be used as a "Rasayana" for immunomodulatory activity [5]. Previous chemical studies of *L. reticulata* showed the presence of flavonoids, triterpenes, and steroids. It is also a rich source of biologically active cardiac and pregnane glycosides [2]. *P. lanceolata* contains high amounts of medicinally important secondary metabolites, namely, quercetin, *beta*-sitosterol, triterpenoids, and so forth [6, 7].

Among all phytochemicals, *beta*-sitosterol is a main phytosterol found in many plants. It has been reported to show anti-inflammatory, antineoplastic, antipyretic, and immunomodulating activity [8]. Also triterpenoids are among commonly present secondary metabolites in plants. *alpha*-Amyrin, a pentacyclic triterpenoid, has been reported to show anti-inflammatory properties. Lupeol has been reported to show anticarcinogenic and antitumour activity [9].

n-Triacontane has been reported to be present in number of plants including *L. reticulata*. *n*-Triacontane has also been reported to be tested for biological activity such as antibacterial, antidiabetic, and antitumor activity [10–12].

Some of the analytical methods for qualitative analysis of *alpha*-amyrin, *beta*-sitosterol, lupeol, and *n*-triacontane from other plant samples are discussed herewith. RP-HPTLC separation of twelve compounds including *alpha*-amyrin, lupeol, and *beta*-sitosterol from *Brassica oleracea* L., *Solanum lycopersicum* L., *Rosmarinus officinalis* L., *Salvia officinalis* L., and *Quercus robur* L. was carried out [9]. Another normal phase HPTLC technique was also reported for separation and determination of *alpha*-amyrin and lupeol from *Brassica oleracea* L. leaf extracts. Visual estimation after derivatization using anisaldehyde-sulphuric acid was carried out [13]. A method comprising capillary GC was used to accurately quantitate *alpha*-amyrin, *beta*-sitosterol, and lupeol from aerial part of *J. anselliana* as reported [14]. A gas chromatography-mass spectrometry (GC-MS) method was used for analysing compounds in *Salvia bicolour Desf.* extract. Among various compounds, lupeol and *beta*-sitosterol were determined [15]. A method was reported to study the chemical constituents of the essential oil of *Laggera pterodonta* (DC.) Sch. Bip. using GC-MS [16]. Percentage of *n*-triacontane in essential oil of *Laggera pterodonta* (DC.) Sch. Bip. was evaluated.

However, no method was applied for quantifying the presence of *alpha*-amyrin, *beta*-sitosterol, lupeol, and *n*-triacontane simultaneously from *L. reticulata* and *P. lanceolata*.

Hence, in present research work, a simple, rapid, precise, and accurate HPTLC method has been developed and validated using International Conference on Harmonization (ICH) guidelines for simultaneous determination and quantification of *alpha*-amyrin, *beta*-sitosterol, lupeol, and *n*-triacontane form dried whole plant powder of *L. reticulata*. and *P. lanceolata*.

2. Experimental Methods

2.1. Experimental Reagents.
The solvents, acetonitrile (purity 99.9%), petroleum ether (purity 99.8%), ethyl acetate (purity 99.0%), methanol (purity 98.9%), and chloroform (purity 99.9%) were obtained from E. Merck (India). The precoated TLC 60F$_{254}$ plates were obtained from E. Merck (India).

2.2. Reference Standards.
Reference standards *alpha*-amyrin (purity 99.3%), *beta*-sitosterol (purity 99.8%), lupeol (purity 99.7%), and *n*-triacontane (purity 98.9%) were procured from Sigma-Aldrich Chemie GmbH (Aldrich Division, Steinbeim, Germany).

2.3. Plant Materials.
Whole plant of *L. reticulata* was collected from Bhumel village, district of Nadiad, Gujarat. The plant material was authenticated from Agarkar Research Institute, Pune, India (Voucher no. WP-090). Whole plant of *P. lanceolata* was collected from Sonamukhi Nagar, Jodhpur, India. Herbarium of *P. lanceolata* was authenticated from Botanical Survey of India, Pune, India (Certificate no. BSI/WRC/Tech./2012/79).

The duplicate herbaria of both the plants were prepared and are preserved in Ramnarain Ruia College, Matunga, India, for future reference. Both plant materials were washed with water to remove soil particles, dried in shade, finely powdered and then sieved through BSS mesh size 85, and stored in an airtight container at room temperature (25 ± 2°C).

2.4. Preparation of Solutions

2.4.1. Preparation of Stock Solution of alpha-Amyrin, beta-Sitosterol, Lupeol, and n-Triacontane.
About 10.0 mg of *alpha*-amyrin was accurately weighed and transferred to 10.0 mL volumetric flask. 5.0 mL of methanol solution was added into the volumetric flask and sonicated in an ultrasonic bath (Model: TRANS-O-SONIC, Frequency: 50 Hz) for 5 minutes for complete dissolution of *alpha*-amyrin. The volume was then made up to the mark with methanol. A stock solution of *alpha*-amyrin with concentration of 1000 μg/mL was prepared. Similarly stock solutions of *beta*-sitosterol and lupeol were prepared. Stock solution of *n*-triacontane was prepared by taking accurately weighed 100 mg *n*-triacontane in a 10.0 mL volumetric flask and initially dissolving it in 5.0 mL chloroform as described above. The volume in the standard flask was then made up to the mark with methanol. A stock solution of *n*-triacontane with concentration of 10,000 μg/mL was prepared.

2.4.2. Preparation of Working Standard Solution of alpha-Amyrin (50.0 μg/mL), beta-Sitosterol (50.0 μg/mL), Lupeol (50.0 μg/mL), and n-Triacontane (500.0 μg/mL).
0.50 mL of above stock solution of each standard was then transferred to 10.0 mL volumetric flask and the contents of volumetric flask were diluted up to 10.0 mL by methanol to obtain mixture four standards of *alpha*-amyrin, *beta*-sitosterol, lupeol with concentration 50.0 μg/mL each, and standard of *n*-triacontane with concentration 500.0 μg/mL.

2.4.3. Preparation of Sample Solution.
About 1.0 g of dried powder of whole plant of *L. reticulata* was accurately weighed and transferred to a 100 mL stoppered conical flask. 25.0 mL of methanol : chloroform (1 : 1, v/v) was added to it and the flask was sonicated in an ultrasonic bath for 15 minutes. The flask was then shaken at 50 rpm, on a conical flask

shaker overnight at room temperature ($25 \pm 2°C$). Sample was filtered through Whatman filter paper no. 1. The filtrate was then finally filtered using $0.45 \, \mu m$ nylon filters (Millipore), collected in a beaker, and then evaporated on a hot water bath. The final volume was then made up to 10 mL with methanol : chloroform (1 : 1, v/v) in a 10 mL volumetric flask. The same procedure was followed for preparation of extract of whole plant of *P. lanceolata*.

2.4.4. Prederivatization Reagent. The standards and sample solution were applied in the form of a band on TLC plate. This plate was subjected to prederivatization by iodine. Iodine vapours were generated by heating iodine crystals in a closed stoppered flat bottom conical flask. The plate was then exposed to the iodine vapour in a dark enclosed chamber, for 10 minutes. After that, the plate was removed from the chamber and heated on Camag TLC plate heater at $100°C$ for 10 minutes, till the excessive iodine was removed. This prederivatized plate was finally used for development.

2.4.5. Preparation of Mobile Phase. The mobile phase used in the present research work for simultaneous quantification of *alpha*-amyrin, *beta*-sitosterol, and lupeol was prepared by mixing petroleum ether, ethyl acetate, and acetonitrile in the volume ratio of 8.2 : 1.8 : 0.1. During development of each plate, a fresh mobile phase was prepared.

2.4.6. Postderivatization Reagent. Anisaldehyde-sulphuric acid reagent was used as postderivatizing reagent. Anisaldehyde-sulphuric acid reagent was prepared by taking 10 mL of sulphuric acid, which was added to an ice cooled mixture of methanol (170.0 mL) and glacial acetic acid (20.0 mL). Further, 1.0 mL of anisaldehyde solution was added to the above mixture of methanol, glacial acetic acid, and sulphuric acid. The solvent mixture was thoroughly mixed by shaking and used as per requirement.

2.4.7. Chromatography. The Chromatography was performed on 20 cm × 10 cm TLC aluminum precoated silica gel $60F_{254}$ plate, with $200 \, \mu m$ layer thickness (E. Merck, Mumbai, India). Standard and sample solutions were applied to the plates as 8 mm bands, 6 mm apart from each other and 10 mm from bottom edge of the plate, under a continuous supply of nitrogen by means of a Camag Linomat V TLC sample applicator with a $100 \, \mu L$ syringe (Hamilton, Bonaduz, Switzerland). After the application, prederivatization was performed by exposing the plate to iodine vapour for 10 minutes. The prederivatized plate was developed vertically ascending in a twin-trough glass chamber (Camag, Switzerland) saturated with mobile phase comprising petroleum ether : ethyl acetate : acetonitrile (8.2 : 1.2 : 0.1 v/v/v). The optimized chamber saturation time for the mobile phase was 20 minutes at room temperature ($25 \pm 2°C$). The chromatographic run length was 90 mm from the bottom edge of the plate. After development, the plate was air dried for complete removal of mobile phase and derivatized by dipping the developed plate in anisaldehyde-sulphuric acid reagent for 2 seconds. The plate was then air-dried for complete

removal of anisaldehyde-sulphuric acid and heated at $110°C$ for 10 minutes. Densitometric scanning was then performed at $\lambda = 580 \, nm$ for *alpha*-amyrin, *beta*-sitosterol, and lupeol in reflectance/absorbance mode and $\lambda = 366 \, nm$ for *n*-triacontane using Camag TLC scanner 4 with winCATS software version 1.4.6. The slit dimension used was $6.0 \times 0.45 \, mm$ (micro) with scanning speed of 20 mm/sec, throughout the analysis.

2.5. Method Validation

2.5.1. Linear Working Range of alpha-Amyrin, beta-Sitosterol, Lupeol, and n-Triacontane. Determination of linear dynamic range concentration of *alpha*-amyrin, *beta*-sitosterol, lupeol, and *n*-triacontane was done by applying $2 \, \mu L$, $4 \, \mu L$, $6 \, \mu L$, $8 \, \mu L$, $10 \, \mu L$, $12 \, \mu L$, $14 \, \mu L$, $16 \, \mu L$, $18 \, \mu L$, $20 \, \mu L$, and $22 \, \mu L$ on TLC plate of working standard containing *alpha*-amyrin, *beta*-sitosterol, lupeol, and *n*-triacontane.

The peak areas obtained from densitograms for each applied concentration of *alpha*-amyrin, *beta*-sitosterol, lupeol, and *n*-triacontane were noted.

The calibration curves of all four standards were obtained by plotting graphs of mean peak areas of each standard versus corresponding concentration (Figure 1). The results, listed in Table 1, show that within the concentration range indicated, there was a good correlation between mean peak area and concentration of standards.

2.5.2. Limit of Detection (LOD) and Limit of Quantification (LOQ). The limit of detection (LOD) is defined as a peak, whose signal-to-noise (S/N) ratio is 3 : 1. The limit of quantification (LOQ) is defined as a peak, whose signal-to-noise (S/N) ratio is 10 : 1. The results are listed in Table 1.

2.5.3. System Suitability. System suitability was carried out to verify that resolution and reproducibility of the system were acceptable for the analysis. System suitability test was carried out by applying $6 \, \mu L$ standard solutions of *alpha*-amyrin, *beta*-sitosterol, lupeol, and *n*-triacontane on TLC plate in six replicates under specified chromatographic conditions. The chromatograms were recorded. The values of percent relative standard deviations of peak area and retention factor of standards were taken as an indicator of system suitability. Since the values of percent relative standard deviations of peak area were found to be less than 2 and peaks were well-resolved, the method was suitable for analysis.

2.5.4. Specificity. The specificity of the proposed HPTLC method was ascertained by comparing visible chromatograms of *alpha*-amyrin, *beta*-sitosterol, lupeol, and *n*-triacontane standards with those found in the sample. The chromatograms were compared by overlay. Good correlation was observed between chromatograms obtained from *alpha*-amyrin, *beta*-sitosterol, lupeol, and *n*-triacontane standards and samples at all R_f (*alpha*-amyrin—R_f 0.68, *beta*-sitosterol—R_f 0.48, lupeol—R_f 0.61, and *n*-triacontane—R_f 0.91) values, respectively.

TABLE 1: Method validation data for simultaneous quantification of *alpha*-amyrin, *beta*-sitosterol, lupeol, and *n*-triacontane from dried whole plant powder of *Leptadenia reticulata* (Retz.) Wight & Arn. and *Pluchea lanceolata* (DC.) CB. Clarke.

Parameters	Results			
	alpha-Amyrin	*beta*-Sitosterol	Lupeol	*n*-Triacontane
Linear range (µg/band)	0.1–1.0	0.1–1.0	0.1–1.0	1.0–10.0
Correlation coefficient (*r*)	0.999	0.999	0.999	0.999
LOD (µg/band)	0.03	0.03	0.03	0.30
LOQ (µg/band)	0.10	0.10	0.10	1.00
Stability of standard solution	Stable for minimum 48 hours	Stable for minimum 48 hours	Stable for minimum 48 hours	Stable for minimum 48 hours
System suitability (% R.S.D.)	Less than 2	Less than 2	Less than 2	Less than 2
Leptadenia reticulata (Retz.) Wight & Arn.				
Repeatability-% R.S.D. range (*n* = 3) (on the same day)	0.93 ± 0.01	0.99 ± 0.06	1.07 ± 0.06	1.23 ± 0.06
Intermediate precision % R.S.D range (*n* = 9) (percent R.S.D. for three successive days)	0.94 ± 0.04	0.99 ± 0.06	1.03 ± 0.10	0.93 ± 0.04
Assay (mg/2 g)	0.99 ± 0.01	1.18 ± 0.02	0.62 ± 0.06	0.65 ± 0.06
Percent recovery	99.01 ± 0.07	98.14 ± 0.04	98.60 ± 0.01	99.46 ± 0.05
Pluchea lanceolata (DC.) CB. Clarke.				
Repeatability-% R.S.D. range (*n* = 3) (on the same day)	1.00 ± 0.05	0.95 ± 0.02	1.00 ± 0.03	N.D.
Intermediate precision % R.S.D range (*n* = 9) (percent R.S.D. for three successive days)	0.96 ± 0.05	0.97 ± 0.06	0.98 ± 0.05	N.D.
Assay (mg/2 g)	0.72 ± 0.03	1.61 ± 0.06	0.17 ± 0.01	N.D.
Percent recovery	99.46 ± 0.01	99.25 ± 0.02	99.81 ± 0.05	N.D.

*Note: N.D.: not detected.

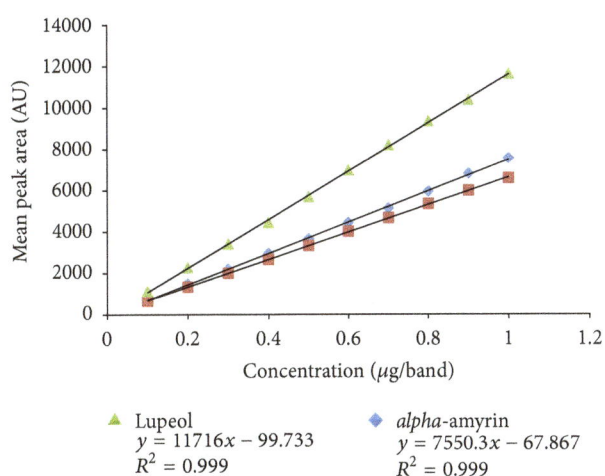

Lupeol
$y = 11716x − 99.733$
$R^2 = 0.999$

alpha-amyrin
$y = 7550.3x − 67.867$
$R^2 = 0.999$

beta-sitosterol
$y = 6622.2x + 17.067$
$R^2 = 0.999$

(a)

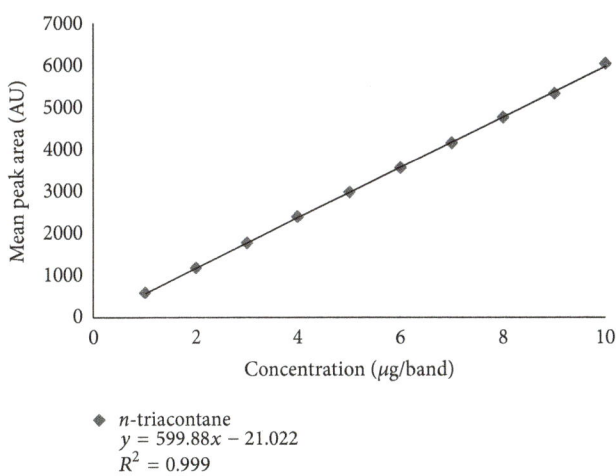

n-triacontane
$y = 599.88x − 21.022$
$R^2 = 0.999$

(b)

FIGURE 1: The figure represents a simultaneous plot of mean peak area v/s concentration of *alpha*-amyrin, *beta*-sitosterol, lupeol and *n*-triacontane standards, respectively.

2.5.5. Precision. The method was validated in terms of repeatability and intermediate precision.

The repeatability was evaluated in triplicates by applying extract of both plant materials on TLC plate on the same day, under the specified chromatographic conditions. The peak areas of *alpha*-amyrin, *beta*-sitosterol, lupeol, and *n*-triacontane were recorded and assessed.

The intermediate precision of the method was evaluated by analyzing the sample solution in triplicate on three different days, under the specified chromatographic conditions. The peak areas of *alpha*-amyrin, *beta*-sitosterol, lupeol, and *n*-triacontane were recorded and assessed.

The precision results were expressed as percentage relative standard deviations of peak areas of *alpha*-amyrin, *beta*-sitosterol, lupeol, and *n*-triacontane. The results, listed in Table 1, indicate that the proposed method is precise and reproducible.

2.5.6. Standard Stability. The stabilities of standard *alpha*-amyrin, *beta*-sitosterol, lupeol, and *n*-triacontane solution were determined by comparing the peak areas of the standard solution, and the stock solution was observed at different time intervals by spotting 6 μL of working standard solution on TLC plate for a period of minimum 48 hrs at room temperature. The results showed that the peak areas of *alpha*-amyrin, *beta*-sitosterol, lupeol, and *n*-triacontane almost remained unchanged (values of percent relative standard deviations were less than 2) over a period of 48 hrs, and no significant degradation was observed within the given period. Thus, the standard solutions of *alpha*-amyrin, *beta*-sitosterol, lupeol, and *n*-triacontane were stable for a minimum of 48 hrs.

2.5.7. Assay. The developed and validated HPTLC method was used for quantification of *alpha*-amyrin, *beta*-sitosterol, lupeol, and *n*-triacontane from the extract of dried whole plant powder of *L. reticulata* and *P. lanceolata*. 4.0 μL of extract of both plant materials was applied as bands on the same TLC plate. The plate was developed and scanned under the specified chromatographic conditions. The chromatograms were recorded. To check the repeatability of the method, assay experiment was repeated seven times and the values of mean standard deviation (S.D.) and percent relative standard deviation (%R.S.D.) were calculated. The results of assay experiment are shown in Table 1. The amounts of *alpha*-amyrin, *beta*-sitosterol, lupeol, and *n*-triacontane present in each sample solution were determined from the calibration curve, by using the peak areas of *alpha*-amyrin, *beta*-sitosterol, lupeol, and *n*-triacontane in the sample.

2.5.8. Accuracy. The accuracy of the method was established by performing recovery experiment by using standard addition method at three different levels. To accurately weighed, about 1.0 g of dried whole plant powder of *L. reticulata*, known amounts of standards *alpha*-amyrin, *beta*-sitosterol, lupeol, and *n*-triacontane were added and extracted. Each of the three different levels containing sample solution and standard was applied in seven replicates on the same plate. The plate was then developed and scanned under the specified chromatographic conditions, as described earlier. The *alpha*-amyrin, *beta*-sitosterol, lupeol, and *n*-triacontane contents were quantified by the proposed method and the percentage recovery was calculated. The same procedure was repeated for dried whole plant powder of *P. lanceolata*. except the addition of *n*-triacontane. The percent recovery values are shown in Table 1.

3. Results and Discussion

During HPTLC analysis, several different mobile phases were tried for separation of *alpha*-amyrin, *beta*-sitosterol, lupeol, and *n*-triacontane from other phytochemicals present in whole plant powder of *L. reticulata*. Good separation was achieved with the mobile phase comprising pet-ether: ethyl acetate: acetonitrile in the volume ratio of 8 : 2 : 0.1 along with prederivatization with iodine. Since, the phytochemicals, *alpha*-amyrin, *beta*-sitosterol, lupeol, and *n*-triacontane showed no UV and visible sensitivity on plate, the plate was postderivatized further with anisaldehyde-sulphuric acid reagent. The R_f values for *alpha*-amyrin, *beta*-sitosterol, lupeol, and *n*-triacontane were 0.60, 0.48, 0.69, and 0.91, respectively. Figure 2 shows typical HPTLC chromatograms of standard *alpha*-amyrin, standard *beta*-sitosterol, standard lupeol, and standard *n*-triacontane in extracts of dried whole plant powder of *L. reticulata* and dried whole plant powder of *P. lanceolata*.

Literature survey revealed that some of the related methods were reviewed. A qualitative normal phase HPTLC method was reported [13] for separating and determining *alpha*-amyrin and lupeol from *Brassica oleracea* L. leaf extracts. HPTLC silica gel 60F$_{254}$ plates were used as stationary phase with the mobile phase as *n*-hexane-ethyl acetate in a volume ratio of 5 : 1. Postderivatization was carried out using anisaldehyde-sulphuric acid reagent. Identification of *alpha*-amyrin and lupeol was carried out by visual comparison of the colour of *alpha*-amyrin and lupeol after derivatization. The reported method was unable to resolve the isomeric compounds *alpha*-amyrin and lupeol, because there was no significant difference in their R_f values. In a reported method [9] chromatographic separation of twelve compounds including *alpha*-amyrin, lupeol, and *beta*-sitosterol from *Brassica oleracea* L., *Solanum lycopersicum* L., *Rosmarinus officinalis* L., *Salvia officinalis* L., and *Quercus robur* L. was studied. The study described a combination of two RP-HPTLC methods for a qualitative determination of twelve phytochemicals (*alpha*-amyrin, *beta*-amyrin, delta-amyrin, lupeol, lupenone, lupeol acetate, cycloartenol, cycloartenol acetate, ursolic acid, oleanolic acid, stigmasterol, and *beta*-sitosterol) and evaluation of their presence in different plant extracts. In the study, RP-HPTLC was used to analyse the phytochemicals. Experiment was performed on RP-HPTLC plates, using the combination of two mobile phases to isolate compounds; these were further identified using RP-HPLC method. The reported methods were only used for qualitative screening and identification of these compounds. Also capillary GC was used for quantitation of *alpha*-amyrin, *beta*-sitosterol, and lupeol from aerial part of *J. anselliana*. Solid phase extraction

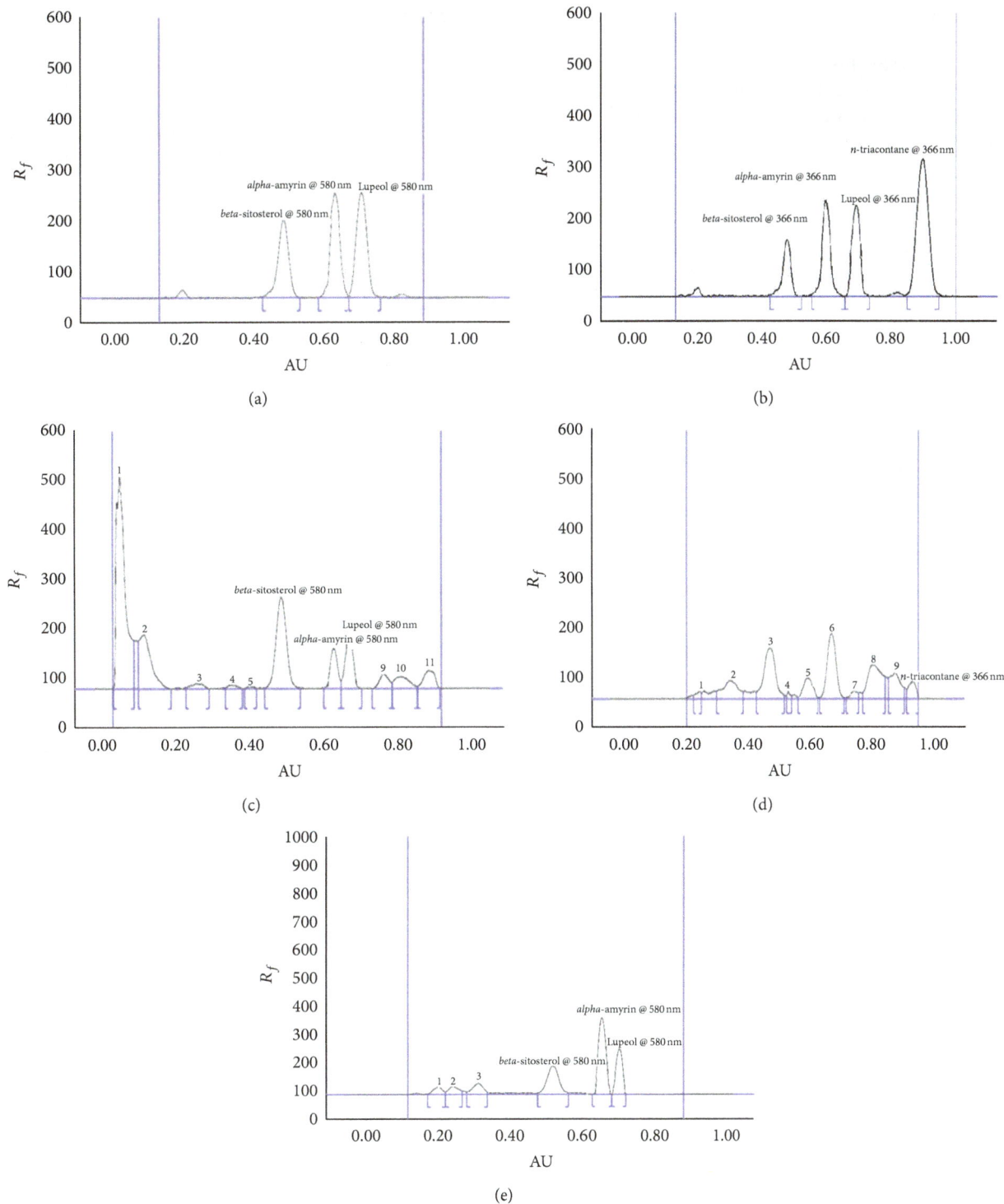

FIGURE 2: (a) and (b) represent HPTLC chromatogram of standard obtained at λ = 580 nm and λ = 366 nm. (c) and (d) represent HPTLC chromatogram of sample *L. reticulata* obtained at λ = 580 nm and λ = 366 nm. (e) represents HPTLC chromatogram of sample *P. lanceoata* obtained at λ = 580 nm.

was used to remove the matrix [15]. A GC/MS method was reported for analysing compounds in *Salvia bicolour Desf.* extract. *beta*-Sitosterol and lupeol were detected. The retention time observed was 41.04 mins for *beta*-sitosterol and 41.5 mins for lupeol [14]. Both methods were time consuming. Another method has been reported [16] to study the chemical constituents of the essential oil of Laggera pterodonta (DC.) Sch. Bip. using GC-MS. *n*-Triacontane was detected using Helium as carrier gas and injection done at elevated temperature of 250°C. The flow rate of 1.61 mL/min was used. Retention time observed for *n*-triacontane was 30.217. *n*-Triacontane was found to be 43.18% approximately. No HPTLC method has been reported for quantification of *n*-triacontane from whole plant powder of *L. reticulata* and *P. lanceolata*.

Therefore, in the present research work, in order to standardize the plants with these markers, a precise and accurate HPTLC method for simultaneous estimation of *alpha*-amyrin, *beta*-sitosterol, lupeol, and *n*-triacontane from the extract of whole plant powder of *L. reticulata* and *P. lanceolata* was developed. The present developed method is advantageous compared to the above reported methods as it uses a simple prederivatization technique of iodination which resolved the isomeric compounds, *alpha*-amyrin and lupeol. Also, alkanes are generally high molecular weight compounds and difficult to be analysed on TLC quantitatively. *n*-Triacontane which is an alkane has also been quantified. The mobile phase comprising petroleum ether: ethyl acetate: acetonitrile in the volume ratio of 8.2 : 1.8 : 0.1 helped in resolving the phytoconstituents without the interference from sample matrix, with the development time of less than 10 minutes which helped in reduction of analysis time. Finally the developed plates were derivatized by anisaldehyde-sulphuric acid reagent. The developed method was validated following ICH guidelines criteria and is economical, simple, and rapid which can be easily performed at any laboratory conditions with specified parameters of HPTLC.

4. Conclusion

The developed HPTLC technique is simple, precise, specific, and accurate, which can be used for the routine quality control analysis and simultaneous quantitative determination of *alpha*-amyrin, *beta*-sitosterol, lupeol, and *n*-triacontane from the whole plant powder of *L. reticulata* and *P. lanceolata*. The method can be applied to effectively quantitate the presence of *alpha*-amyrin, *beta*-sitosterol, lupeol, and *n*-triacontane in other samples as well.

Conflict of Interests

The authors declare that there is no conflict of interests regarding the publication of this paper.

Acknowledgments

The authors wish to thank the Chemistry Department, Ruia College, India, for providing lab facility. The authors are also grateful to Anchrom HPTLC Technologists and special thanks go to their application specialist team, Mumbai. Also they are thankful to Dr. T. Prajapati, President of Cultivator Natural products, Dr. P. S. Nagar, MS University, and Dr. M. Parabia, (Dept. of Bioscience), Veer Narmad South Gujarat University.

References

[1] E. Reich and A. Schibli, *High-Performance Thin-Layer Chromatography for the Analysis of Medicinal Plants*, Thieme Medical, 2011.

[2] S. Srivastav, D. Deepak, and A. Khare, "Three novel pregnane glycosides from Leptadenia reticulata Wight and Arn," *Tetrahedron*, vol. 50, no. 3, pp. 789–798, 1994.

[3] V. N. Dwivedi, *Bhav Prakash Nighantu*, Hindi Translational Motilal Banarsidas Banaras, 1949.

[4] Anonymous, *The Wealth of India Raw Materials Publication and Information Directorate*, CSIR, New Delh, India, 1969.

[5] H. Wagner, *Immunomodulatory Agents from Plants*, pp. 291-292, Birkhäuser, Basel, Switzerland, 1st edition, 1998.

[6] B. S. Kaith, "Neolupeol and anti-inflammatory activity of Pluchea lanceolata," *International Journal of Pharmacognosy*, vol. 34, pp. 73–75, 1995.

[7] M. Ali, N. Ali Siddiqui, and R. Ramachandram, "Phytochemical investigation of aerial parts of *Pluchea lanceolata* C.B. Clarke," *Indian Journal of Chemistry B*, vol. 40, no. 8, pp. 698–706, 2001.

[8] L. Fraile, E. Crisci, L. Córdoba, M. A. Navarro, J. Osada, and M. Montoya, "Immunomodulatory properties of beta-sitosterol in pig immune responses," *International Immunopharmacology*, vol. 13, no. 3, pp. 316–321, 2012.

[9] M. Martelanc, I. Vovk, and B. Simonovska, "Separation and identification of some common isomeric plant triterpenoids by thin-layer chromatography and high-performance liquid chromatography," *Journal of Chromatography A*, vol. 1216, no. 38, pp. 6662–6670, 2009.

[10] "n-Triacontane," http://pubchem.ncbi.nlm.nih.gov/summary/summary.cgi?cid=12535#x299.

[11] C. P. Khare, *Indian Herbal Remedies: Rational Western Therapy, Ayurvedic , and Other Traditional Usage and Botany*, Springer, 2004.

[12] D. Mammen, M. Daniel, and R. T. Sane, "Seasonal and geographical variations in chemical constituents of *Leptadenia reticulata*," *International Journal of Pharmaceutical Sciences Review and Research*, vol. 4, no. 2, pp. 111–116, 2010.

[13] M. Martelanc, I. Vovk, and B. Simonovska, "Determination of three major triterpenoids in epicuticular wax of cabbage (*Brassica oleracea* L.) by high-performance liquid chromatography with UV and mass spectrometric detection," *Journal of Chromatography A*, vol. 1164, no. 1-2, pp. 145–152, 2007.

[14] T. A. Ibrahim, "Chemical composition and biological activity of extracts from *Salvia bicolor* desf. growing in Egypt," *Molecules*, vol. 17, no. 10, pp. 11315–11334, 2012.

[15] D. S. S. Kpoviéssi, F. Gbaguidi, J. Gbénou et al., "Validation of a method for the determination of sterols and triterpenes in the aerial part of *Justicia anselliana* (Nees) T. Anders by capillary gas chromatography," *Journal of Pharmaceutical and Biomedical Analysis*, vol. 48, no. 4, pp. 1127–1135, 2008.

[16] E. H. Omoregie, K. F. Oluyemisi, O. S. Koma, and O. J. Ibumeh, "Chemical constituents of the essential oil of laggera pterodonta (DC.) sch. bip. from north-central Nigeria," *Journal of Applied Pharmaceutical Science*, vol. 2, no. 8, pp. 198–202, 2012.

5

Synthesis and Antimicrobial Studies of Pyrimidine Pyrazole Heterocycles

Rakesh Kumar,[1] **Jyoti Arora,**[1,2] **Sonam Ruhil,**[3] **Neetu Phougat,**[3]
Anil K. Chhillar,[3] **and Ashok K. Prasad**[2]

[1] *Department of Chemistry, Bio-organic Laboratory, Kirori Mal College, University of Delhi, Delhi 110 007, India*
[2] *Department of Chemistry, Bio-organic Laboratory, University of Delhi, Delhi 110 007, India*
[3] *Centre for Biotechnology, Maharshi Dayanand University, Rohtak 124 001, India*

Correspondence should be addressed to Rakesh Kumar; rakeshkp@email.com

Academic Editor: Adriana I. Segall

Prompted from the diversity of the wider use and being an integral part of genetic material, an effort was made to synthesize pyrimidine pyrazole derivatives of pharmaceutical interest by oxidative cyclization of chalcones with satisfactory yield and purity. A novel series of 1,3-dimethyl-6-hydroxy-2,4-dioxo-5-(1'-phenyl-3'-aryl-1H-pyrazol-5'-yl)-1,2,3,4-tetrahydropyrimidines (**5a–d**) and 1,3-diaryl-6-hydroxy-4-oxo-2-thioxo-5-(1'-phenyl-3'-aryl-1H-pyrazol-5'-yl)-1,2,3,4-tetrahydropyrimidines (**5e–l**) has been synthesized. The structures of these compounds were established on the basis of FT-IR, ^{1}H NMR, ^{13}C NMR, and mass spectral analysis. All the synthesized compounds were screened for their antimicrobial activity against bacteria and fungi. Among all the compounds, **5g** was found to be the most active as its MIC was 31.25 μg/mL against *S. aureus* and *B. cereus*. The compounds **5h**, **5c**, and **5e** also possess antibacterial activity with MIC values as 62.50, 125.00, and 500.00 μg/mL, respectively. The compounds **5c** and **5j** were found to have antifungal activity against *Aspergillus* spp. As antifungal drugs lag behind the antibacterial drugs, therefore we tried *in vitro* combination of these two compounds with standard antifungal drugs (polyene and azole) against *Aspergillus* spp. The combination of ketoconazole with **5c** and **5j** showed synergy at 1:8 (6.25:50.00 μg/mL) and 1:4 (25:100 μg/mL) against *A. fumigatus* (ITCC 4517) and *A. fumigatus* (VPCI 190/96), respectively.

1. Introduction

Nitrogen heterocycles are of special interest as they constitute an important class of natural and nonnatural products, many of which exhibit useful biological activities. Pyrimidine, being an integral part of DNA and RNA, imparts diverse pharmacological properties, such as bactericide, fungicide, vermicide, insecticide, and anticancer and antiviral agents [1]. Certain pyrimidine derivatives are also known to display antimalarial, antifilarial, and antileishmanial activities [2].

The pyrazole derivatives are well known to have antimicrobial [3], antifungal [4], antitubercular [5], anticancer [6], analgesic [7], anti-inflammatory [8], antipyretic [9], anticonvulsant [10], antidepressant [11], muscle relaxing [12], antiulcer [13], antiarrhythmic [14], and antidiabetic [15] activities. With growing application of their synthesis and bioactivity, chemists and biologists in recent years have

directed considerable attention to the study of pyrazole derivatives. In view of the above mentioned importance of pyrimidines and pyrazoles, we tried to accommodate these moieties in a single molecular framework to synthesize the linked heterocycles for enhancing biological activity.

2. Results and Discussion

2.1. Chemistry. (*E*)-1-(1',3'-Dimethyl-6'-hydroxy-2',4'-dioxo-1',2',3',4'-tetrahydropyrimidin-5'-yl)-3-aryl-prop-2-ene-1-ones (**4a–d**) and (*E*)-1-(1',3'-diaryl-6'-hydroxy-4'-oxo-2'-thiooxo-1',2',3',4'-tetrahydropyrimidin-5'-yl)-3-aryl-prop-2-ene-1-ones (**4e–l**) were synthesized by the Claisen condensation of 5-acetyl barbituric/thiobarbituric acid (**2a–c**) with aromatic aldehydes **3a–d** in methanol in the presence of NaOH as a base at 60°C [16]. Further, cyclocondensation of propenones **4a–l** with phenylhydrazine in acidic condition in dioxane as solvent yielded 1,3-dimethyl-6-hydroxy-2,4-dioxo-5-(1′-phenyl-3′-aryl-1H-pyrazol-5′-yl)-1,2,3,4-tetrahydropyrimidines (5a–d) and 1,3-diaryl-6-hydroxy-4-oxo-2-thioxo-5-(1′-phenyl-3′-aryl-1H-pyrazol-5′-yl)-1,2,3,4-tetrahydropyrimidines (5e–l) in 46–81% yields [17]. Structure and yield of compounds, that is, 4a–l and 5a–l, are listed in Table 1. The 5-acetyl-1,3-diarylthiobarbituric acids (2b–c) in turn were synthesized by the following known method from 1,3-diarylthiobarbituric acids (1b–c) and acetic anhydride [18, 19] (Scheme 1).

All the compounds synthesized were characterized by IR, ^{1}H NMR, ^{13}C NMR, and mass spectroscopy. Spectroscopic data was in complete agreement with the structures assigned for these compounds. IR spectrum of cyclized derivatives of barbituric acid (5a–d) showed band in the region of 1700–1740 cm^{-1} for carbonyl group (at C2). The other carbonyl group at C4 showed band in the region of 1640–1699 cm^{-1}, whereas cyclized derivatives of thiobarbituric acid (5e–l) showed band in the region of 1050–1100 cm^{-1}, which indicates the presence of thiocarbonyl group (at C2) and other carbonyl groups at C4 showed band in the region of 1625–1680 cm^{-1}. Frequency band of OH group appears at 3200–3450 cm^{-1} in the compounds 5a–l. In ^{1}H NMR spectra, chemical shift values of all the compounds were in accordance with the expected values. Aromatic protons of compounds 4a–l resonated in the region of δ 6.86–7.82. Two doublets of α-H (attached to C2) at δ 8.06 (J = 16.11 Hz) and β-H (attached to C3) at 8.41 (J = 16.11 Hz), respectively, of 4f demonstrate the formation of α, β-unsaturated carbonyl moiety and J = 16.11 Hz indicates that the ethylene moiety in the enone linkage is in trans confirmation in the chalcone. Disappearance of these doublets in 5f indicates the absence of chalcone moiety. All other phenyl protons in the compounds 5a–l appeared in the aromatic region at δ 6.82–7.58. In ^{13}C NMR of 4f, all the characteristic peaks were in good agreement with the proposed structure. Carbonyl carbon at C-1 and C-4′ appeared at δ 184.6 and 168.2, respectively. The characteristic peak of C=S appeared at δ 178.9. The C-2 and C-3 carbons appeared at δ 114.5 and 139.7, respectively. The OCH$_3$ carbon appeared at δ 55.4. The aromatic carbons attached to OCH$_3$ (i.e., C-4″) appeared at δ 162.9. The other aromatic carbons of 4f resonated in the region of δ 127.3–131.7 and the aromatic carbon attached to nitrogen appeared at 148.6. In ^{13}C NMR of 5f, disappearance of peak at δ 184.6 indicates the cyclization of chalcone. Carbonyl carbon at C-4 appeared at δ 164.0. The characteristic peak of C=S appeared at δ 180.0. C-6 carbon appeared at δ 160.8. The pyrazole carbon at C-4′ appeared at δ 88.3 [20]. The OCH$_3$ carbon appeared at δ 55.3. The aromatic carbon attached to OCH$_3$, that is, C-4″, appeared at δ 161.4. The other aromatic carbons resonated in the region of δ 127.4–129.5 and the aromatic carbon attached to nitrogen appeared at δ 147.13. Details of ^{1}H NMR and ^{13}C NMR spectra of 5a–l are given in experimental section.

2.2. Biology

2.2.1. Antifungal Activity.
The antifungal activity against Aspergillus spp. was evaluated by different methods [21–23],

that is, disc diffusion assay (DDA), microbroth dilution assay (MDA), and percent spore germination inhibition (PSGI). The Minimum Inhibitory Concentration (MIC) values of Amphotericin B (Amp B) and Nystatin (NYS) against all the three Aspergillus species were found to be 0.75 μg/disc and 1.00 μg/disc, respectively, by DDA and 1.95 μg/mL and 3.90 μg/mL, respectively, by MDA and PSGI. The MIC of 5c compound was 46.75 μg/disc against all the tested isolates of Aspergillus spp. by DDA, whereas 5j compound possesses a slight higher MIC against A. flavus and A. niger, that is, 187.5 μg/disc, but against A. fumigatus it possesses the same MIC, that is, 46.75 μg/disc. The MIC of 5c by MDA and PSGI was found to be 250.0 μg/mL against A. fumigatus and 500.0 μg/mL against A. flavus and A. niger. The MIC of 5j was found to be 500.0 μg/mL against A. fumigatus and A. flavus and 1000 μg/mL against A. niger, respectively, by MDA and PSGI (Table 2).

Results revealed that the synthesized compounds 5c and 5j exhibited mild antifungal activity which is lower than the standard drugs. Some other substituted pyrimidines and pyrazoles have earlier been reported as potent antifungal agents against a number of pathogenic fungi alone and in combination [24, 25].

As these compounds showed promising activity, so we further tried these compounds in combination with standard antifungal drugs to evaluate their synergistic behaviour, if any. Therefore, the compounds 5c and 5j were tried for in vitro combination with polyenes and azoles.

(1) In Vitro Combination Study of Pyrimidine Pyrazole Analogues (5c and 5j) with Antifungal Drugs.
Among the human pathogenic species of Aspergillus, A. fumigatus is the primary causative agent of human infection followed by A. flavus and A. niger [23]. Therefore, A. fumigatus [ITCC 4517 (IARI, Indian Agricultural Research Institute, Delhi), ITCC 1634, clinical isolate VPCI 190/96 (VPCI, Vallabhbhai Patel Chest Institute, Delhi)] was selected for in vitro combination study of pyrimidine pyrazole analogues with antifungal drugs. The data of in vitro combination was analysed by Fraction Inhibitory Concentration Index (FICI) model [24] and summarized in Tables 3–6.

(1.1) In Vitro FIC Index of 5c with Polyene (Amp B, NYS) and Azole (KTZ, FLZ).
In combination of Amp B and NYS with 5c, the FICI values were found to be in the range of 0.8 to 1.03; indifference (IND) was declared against A. fumigatus strains (Table 3). The MIC end point value of 5c reduced from 314.98 to 7.87/12.50 μg/mL. But the MIC of Amp B and NYS almost remains the same, that is, 1.96 μg/mL and 3.12 μg/mL.

The combination of 5c with KTZ and FLZ reduced the MIC end point value of KTZ and FLZ from 39.37 to 7.87 μg/mL and from 314.98 to 62.90 μg/mL, respectively, against A. fumigatus strains. The MIC of 5c reduced from 314.98 to 62.90 μg/mL in combination with KTZ and with FLZ it gets reduced to 251.98. Depending upon FICI model indifference (IND) and synergy (SYN) was observed. The combination of KTZ with 5c showed SYN against only one strain of A. fumigatus, that is, ITCC 4517, at 1:8 (6.25:50.00 μg/mL, FICI = 0.40) (Table 4). The FICI (GM)

TABLE 1: Structure, formula, and yields of compounds **5a–l**.

Entry	R	X	Ar	Chalcone (**4a–l**)	Cyclised product (**5a–l**)	Yield (%)
2a, 3a, 4a, 5a	–CH₃	O	p-CH₃			66
2a, 3b, 4b, 5b	–CH₃	O	p-OCH₃			66
2a, 3c, 4c, 5c	–CH₃	O	p-Br			80
2a, 3d, 4d, 5d	–CH₃	O	p-Cl			66

TABLE 1: Continued.

Entry	R	X	Ar	Chalcone (4a–l)	Cyclised product (5a–l)	Yield (%)
2b, 3a, 4e, 5e	C_6H_6	S	p-CH$_3$			64
2b, 3b, 4f, 5f	C_6H_6	S	p-OCH$_3$			80
2b, 3c, 4g, 5g	C_6H_6	S	p-Br			46
2b, 3d, 4h, 5h	C_6H_6	S	p-Cl			78

TABLE 1: Continued.

Entry	R	X	Ar	Chalcone (4a–l)	Cyclised product (5a–l)	Yield (%)
2c, 3a, 4i, 5i	o-OCH₃C₆H₅	S	p-CH₃			68
2c, 3b, 4j, 5j	o-OCH₃C₆H₅	S	p-OCH₃			68
2c, 3c, 4k, 5k	o-OCH₃C₆H₅	S	p-Br			80
2c, 3d, 4l, 5l	o-OCH₃C₆H₅	S	p-Cl			81

TABLE 2: Antifungal activity of pyrimidine pyrazole analogues.

Compound	DDA (μg/disc)			MIC MDA (μg/mL)			PSGI (μg/mL)		
	A. flavus	A. niger	A. fumigatus	A. flavus	A. niger	A. fumigatus	A. flavus	A. niger	A. fumigatus
5c	46.75	46.75	46.75	500	500	250	500	500	250
5e	—	—	—	—	—	—	—	—	—
5f	—	—	—	—	—	—	—	—	—
5g	—	—	—	—	—	—	—	—	—
5h	—	—	—	—	—	—	—	—	—
5i	—	—	—	—	—	—	—	—	—
5j	187.5	187.5	46.75	500	1000	500	500	1000	500
5k	—	—	—	—	—	—	—	—	—
5l	—	—	—	—	—	—	—	—	—

—: no significant inhibition.

TABLE 3: *In vitro* combination of compound **5c** with polyene (AmpB, NYS) against *A. fumigatus*.

Strain	MIC (μg/mL)					FIC index		Interpretation	
	5c	AmpB	NYS	AmpB + 5c	NYS + 5c	AmpB + 5c	NYS + 5c	AmpB + 5c	NYS + 5c
ITCC 4517	250	1.95	3.9	1.56/6.25	3.12/12.50	0.83 (0.8 + 0.03)	0.85 (0.8 + 0.05)	IND	IND
ITCC 1634	500	1.95	3.9	3.12/12.50	3.12/12.50	1.62 (1.6 + 0.02)	0.83 (0.8 + 0.03)	IND	IND
VPCI 190/96	250	1.95	3.9	1.56/6.25	3.12/12.50	0.83 (0.8 + 0.03)	0.85 (0.8 + 0.05)	IND	IND
Geometric mean	314.98	1.95	3.9	1.96/7.87	3.12/12.50	1.03 (1.00 + 0.3)	0.83 (0.8 + 0.03)	IND	IND

IND: indifference.

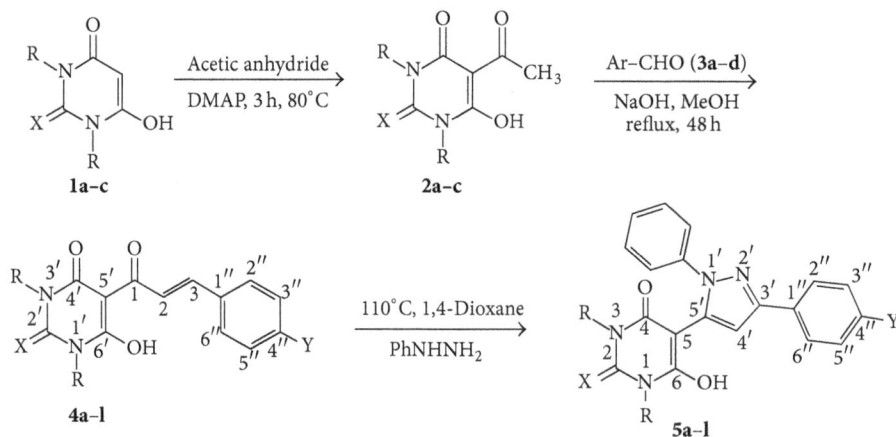

Compounds 1–5	R	X
1a, 2a, 4a–d, 5a–d	–CH$_3$	O
1b, 2b, 4e–h, 5e–h	–C$_6$H$_5$	S
1c, 2c, 4i–l, 5i–l	o-OCH$_3$C$_6$H$_4$	S

Compounds 3–5	Y
3a, 4a, 4e, 4i, 5a, 5e, 5i	CH$_3$
3b, 4b, 4f, 4j, 5b, 5f, 5j	OCH$_3$
3c, 4c, 4g, 4k, 5c, 5g, 5k	Br
3d, 4d, 4h, 4l, 5d, 5h, 5l	Cl

SCHEME 1: Synthesis of pyrimidine pyrazole derivatives.

TABLE 4: *In vitro* combination of compound **5c** with azole (KTZ, FLZ) against *A. fumigatus*.

| Strain | MIC (μg/mL) | | | | | FIC index | | Interpretation | |
	5c	KTZ	FLZ	KTZ + **5c**	FLZ + **5c**	KTZ + **5c**	FLZ + **5c**	KTZ + **5c**	FLZ + **5c**
ITCC 4517	250	31.25	500	6.25/50.0	50.0/200.0	0.40 (0.2 + 0.2)	0.9 (0.1 + 0.8)	SYN	IND
ITCC 1634	500	62.50	250	25.0/100.0	100.0/400.0	0.60 (0.4 + 0.2)	1.2 (0.4 + 0.8)	IND	IND
VPCI 190/96	250	31.25	250	12.50/50.0	50.0/200.0	0.60 (0.4 + 0.2)	1.0 (0.2 + 0.8)	IND	IND
Geometric mean	314.98	39.37	314.98	7.87/62.9	62.9/251.98	0.52 (0.32 + 0.20)	1.0 (0.2 + 0.8)	IND	IND

IND: indifference; SYN: synergy.

TABLE 5: *In vitro* combination of compound **5j** with polyene (AmpB, NYS) against *A. fumigatus*.

| Strain | MIC (μg/mL) | | | | | FIC index | | Interpretation | |
	5j	AmpB	NYS	AmpB + **5j**	NYS + **5j**	AmpB + **5j**	NYS + **5j**	AmpB + **5j**	NYS + **5j**
ITCC 4517	250.00	1.95	3.90	1.56/6.25	3.12/12.50	0.83 (0.8 + 0.03)	0.85 (0.8 + 0.05)	IND	IND
ITCC 1634	500.00	1.95	3.90	3.12/12.50	6.25/25.00	1.62 (1.6 + 0.02)	1.65 (1.6 + 0.05)	IND	IND
VPCI 190/96	500.00	1.95	3.90	1.56/3.12	3.12/12.50	0.81 (0.8 + 0.01)	0.83 (0.8 + 0.03)	IND	IND
Geometric mean	396.85	1.95	3.90	1.96/6.24	3.93/15.70	1.03 (1.00 + 0.3)	1.05 (1.00 + 0.5)	IND	IND

IND: indifference.

values for the rest of combination were 0.52 and 1.02; IND occurred.

Since the MIC value of KTZ is significantly reduced in combination with **5c**, this compound may be a potential candidate for further research and may be developed as a potential candidate to be used in combination therapy against fungal infections.

*(1.2) In Vitro FIC Index of **5j** with Polyene (Amp B, NYS) and Azole (KTZ, FLZ).* The MIC (GM) end point value of Amp B and NYS in combination with **5j** remains almost the same, that is, 1.96 and 3.93 μg/mL, respectively. But the MIC (GM) end point value of **5j** in combination with Amp B and NYS reduced from 396.84 to 6.24 and 15.70 μg/mL, respectively. The FICI (GM) values were found to be 1.03 and 1.05 with Amp B and NYS combination with **5j**; showed IND against the tested strain (Table 5).

The combination of azole (KTZ and FLZ) with **5j** reduced the MIC (GM) end point value of KTZ from 39.33 to 25.00 and from 314.98 to 62.99 μg/mL of FLZ. The MIC (GM) end point value of **5j** reduced from 396.85 to 100 μg/mL with KTZ and 251.98 μg/mL with FLZ. But this reduction is not as much significant as the combination of KTZ with **5j**, which showed synergy against only one *A. fumigatus* VPCI 190/96, that is, 1 : 4 (25 : 100 μg/mL). The FICI (GM) values for the other combinations were 0.70 and 0.83; indifference was declared (Table 6).

2.2.2. Antibacterial Activity. Among all the analogues the most active compound was **5g** whose MIC was 31.25 μg/mL against *S. aureus* and *B. cereus* and the second and third most active compounds were **5h** and **5c** which showed MIC at 62.50 μg/mL against *B. cereus* and *S. aureus* and 125 μg/mL against *S. aureus*, respectively. The other two compounds **5e** and **5j** showed activity at 500 μg/mL against *S. aureus* and *E. coli*, respectively. Erythromycin was used as a standard drug (Table 7).

It has already been reported that the pyrimidine pyrazole analogues have strong antibacterial activity against a number of pathogenic bacteria [26]. Therefore, we have tried to evaluate their *in vitro* antibacterial potential against gram positive as well as gram negative bacteria.

The compound **5g** showed potent antibacterial activity against gram positive bacteria *S. aureus* and *B. cereus*. These results suggest that there may be a useful practical application from the chemistry of pyrimidine pyrazole analogues.

3. Experimental

3.1. General. All reagents were of commercial grade and were used as received. Solvents were dried and purified using standard techniques. ^{1}H-NMR (400 MHz) and ^{13}C-NMR (100.5 MHz) were recorded on JNM ECX-400P (Jeol, USA) spectrometer using TMS as an internal standard. Chemical shifts are reported in parts per million (ppm). Mass spectra were recorded on API-2000 mass spectrometer. IR absorption spectra were recorded in the 400–4000 cm^{-1} range on a Perkin-Elmer FT-IR spectrometer model 2000 using KBr pallets. Melting points were determined using Buchi M-560 and are uncorrected. These reactions were monitored by thin layer chromatography (TLC), on aluminium plates coated with silica gel 60 F$_{254}$ (Merck). UV radiation and iodine were used as the visualizing agents. Column chromatography was performed on silica gel (100–200 mesh).

3.2. General Procedure for the Synthesis of Chalcone Analogues (4a–l). A solution of **2a–c** (1 mmol) and corresponding aryl aldehydes **3a–d** (1 mmol) in 20 mL of methanol was treated with sodium hydroxide as base at 60°C. The reaction mixture was refluxed for 50 h. After completion of reaction, it was concentrated and extracted with chloroform (3 × 20 mL). The combined organic extract was dried over anhydrous sodium sulphate and concentrated under reduced pressure. The crude product was purified by column chromatography.

TABLE 6: *In vitro* combination of compound **5j** with azole (KTZ and FLZ) against *A. fumigatus*.

Strain	MIC (μg/mL)					FIC index		Interpretation	
	5j	KTZ	FLZ	KTZ + 5j	FLZ + 5j	KTZ + 5j	FLZ + 5j	KTZ + 5j	FLZ + 5j
ITCC 4517	250.00	31.25	500.00	25.00/100.00	50.00/200.00	1.20 (0.8 + 0.4)	0.90 (0.1 + 0.8)	IND	IND
ITCC 1634	500.00	62.50	250.00	25.00/100.00	100.00/400.00	0.60 (0.4 + 0.2)	1.20 (0.4 + 0.8)	IND	IND
VPCI 190/96	500.00	31.25	250.00	25.0/100.00	50.00/200.00	0.50 (0.4 + 0.1)	0.60 (0.2 + 0.4)	SYN	IND
Geometric mean	396.85	39.37	314.90	25.00/100.00	62.99/251.98	0.7 (0.5 + 0.2)	0.83 (0.2 + 0.63)	IND	IND

IND: indifference.

TABLE 7: Antibacterial activity of pyrimidine pyrazole analogues.

Compound	Bacterial pathogens (MIC μg/mL)						
	S. aureus	*B. cereus*	*E. coli*	*S. typhi*	*M. luteus*	*B. pumilus*	*B. subtilis*
5c	125	—	—	—	—	—	—
5e	500	—	—	—	—	—	—
5f	—	—	—	—	—	—	—
5g	31.25	31.25	—	—	—	—	—
5h	62.50	62.50	—	—	—	—	—
5i	—	—	—	—	—	—	—
5j	—	—	—	—	—	—	—
5k	—	—	—	—	—	—	—
5l	—	—	—	—	—	—	—
ERY	15.62	15.62	7.81	31.25	15.62	15.62	15.62

ERY: erythromycin (standard drug).
—: no significant inhibition.

3.2.1. (E)-1-(1',3'-Dimethyl-6'-hydroxy-2',4'-dioxo-1',2',3',4'-tetrahydropyrimidin-5'-yl)-3-(p-tolyl)-prop-2-ene-1-one, (4a). The product was obtained as mentioned in general procedure from **2a** and **3a** as yellow solid in 72% yield; M.p. 187.0°C; **IR** ν_{max} (cm^{-1}) = 1662, 1718 (C=O), 2925 (C–H), 3432 (OH); ^{1}**H NMR** (400 MHz, CDCl$_3$): δ (ppm) 2.34 (3H, s, –CH$_3$), 3.20 (6H, s, N–CH$_3$), 7.29 (2H, d, J = 7.32 Hz, ArH), 7.62 (2H, d, J = 8.04 Hz, ArH), 7.95 (1H, d, J = 16.11 Hz, α–H) 8.45 (1H, d, J = 16.11 Hz, β–H), 16.94 (1H, s, –OH); ^{13}**C NMR** (100 MHz, CDCl$_3$): δ (ppm) 21.05, 27.67, 27.84, 119.10, 126.37, 128.43, 129.00, 129.98, 131.85, 139.38, 141.81, 145.65, 149.84, 154.69, 163.22, 165.83, 182.02.

3.2.2. (E)-1-(1',3'-Diphenyl-6'-hydroxy-4'-oxo-2'-thiooxo-1',2',3',4'-tetrahydropyrimidin-5'-yl)-3-(p-tolyl)-prop-2-ene-1-one, (4e). The product was obtained as mentioned in general procedure from **2b** and **3a** as yellow solid in 67% yield; M.p. 284.6°C; **IR** ν_{max} (cm^{-1}) = 1039 (C=S), 1690 (C=O), 2924 (C–H), 3433 (OH); ^{1}**H NMR** (400 MHz, CDCl$_3$): δ (ppm) 2.38 (3H, s, –CH$_3$), 7.18 (2H, d, J = 8.05 Hz, ArH), 7.28–7.31 (2H, m, ArH), 7.45–7.58 (10H, m, ArH), 8.09 (1H, d, J = 15.38 Hz, α-H), 8.51 (1H, d, J = 16.84 Hz, β-H), 16.79 (1H, s, –OH); ^{13}**C NMR** (100 MHz, CDCl$_3$): δ (ppm) 21.68 (–CH$_3$), 119.15, 128.55, 128.63, 128.77, 129.11, 129.55, 129.61, 129.67, 129.81, 131.82, 139.72, 142.96, 148.62, 159.94, 168.63, 178.74 (C=S), 185.18.

3.2.3. (E)-1-(1',3'-Bis(2''-methoxyphenyl)-6'-hydroxy-4'-oxo-2'-thiooxo-1',2',3'4'-tetrahydro pyrimidin-5'-yl)-3-(p-tolyl)-prop-2-ene-1-one, (4i). The product was obtained as mentioned in general procedure from **2c** and **3a** as yellow solid in 68% yield; M.p. 220.5°C; IR ν_{max} (cm^{-1}) = 1025 (C=S), 1663 (C=O), 2926 (C–H), 3434 (OH); ^{1}H NMR (400 MHz, CDCl$_3$): δ (ppm) 2.35 (3H, s, –CH$_3$), 3.84 (6H, s, –OCH$_3$), 7.03–7.10 (4H, m, ArH), 7.16 (2H, d, J = 7.32 Hz, ArH), 7.21–7.26 (2H, m, ArH), 7.44 (2H, d, J = 8.05 Hz, ArH), 7.55 (2H, d, J = 8.05 Hz, ArH), 8.03 (1H, d, J = 16.11 Hz, α–H), 8.51 (1H, d, J = 16.11 Hz, β-H), 16.84 (1H, s, –OH); ^{13}C NMR (100 MHz, CDCl$_3$): δ (ppm) 21.63 (–CH$_3$), 56.14 (–OCH$_3$), 119.57, 121.06, 121.19, 128.57, 129.53, 129.72, 129.82, 130.22, 130.59, 131.96, 142.46, 147.83, 159.53, 168.40, 178.54 (C=S), 184.94.

3.3. General Procedure for the Synthesis of Pyrimidine Pyrazole Heterocycles (5a–l). To the mixture of corresponding chalcone **4a-l** (1 mmol) and phenylhydrazine (1.5 mmol) in 20 mL of 1,4-dioxane, 2 drops of acetic acid were added. The reaction mixture was refluxed at 110°C overnight. After completion of reaction as monitored by TLC, reaction mixture was concentrated and extracted with chloroform (3 × 20 mL). The combined organic extract was dried over anhydrous sodium sulphate and concentrated under reduced pressure. The crude product was purified by column chromatography (40% Ethyl acetate: pet ether).

3.3.1. 1,3-Dimethyl-6-hydroxy-2,4-dioxo-5-(1'-phenyl-3'-(p-tolyl)-1H-pyrazol-5'-yl)-1,2,3,4 tetrahydropyrimidine, (5a). The product was obtained as mentioned in general procedure from **4a** as white solid; M.p. 150–152°C; IR ν_{max} (cm^{-1}) 1646, 1702, (2 × C=O), 2924 (C–H), 3210 (–OH); ^{1}H NMR (400 MHz, CDCl$_3$): δ (ppm) 2.34 (3H, s, –CH$_3$), 3.28 (3H, s, N–CH$_3$), 3.36 (3H, s, N–CH$_3$), 6.93 (2H, d, J = 8.08 Hz, ArH), 7.06 (1H, t, ArH), 7.18 (2H, d, J = 8.08 Hz, ArH), 7.25–7.30 (4H, m, Pyrazole H, ArH), 12.85 (1H, s, –OH); ^{13}C NMR (100 MHz, CDCl$_3$): δ (ppm) 21.04 (–CH$_3$), 27.51 (N–CH$_3$), 107.42, 115.89, 126.00, 129.46, 129.57, 129.79, 139.19, 147.81, 153.50, 159.21, 165.28; ESI-MS m/z: [M + H]$^+$ = 388.1.

3.3.2. 1,3-Dimethyl-6-hydroxy-2,4-dioxo-5-(1'-phenyl-3'-(p-methoxyphenyl)-1H-pyrazol-5'-yl)-1,2,3,4-tetrahydropyrimidine, (5b). The product was obtained as mentioned in general procedure from **4b** aslight brown solid, M.p. 170–171°C; IR ν_{max} (cm^{-1}) 1647, 1716 (2 × C=O), 2924 (C–H), 3245 (–OH); ^{1}H NMR (400 MHz, CDCl$_3$): δ (ppm) 3.35 (3H, s, N–CH$_3$), 3.39 (3H, s, N–CH$_3$), 3.80 (3H, s, –OCH$_3$), 6.89–6.94 (4H, m, ArH), 7.07 (1H, t, ArH), 7.25–7.33 (5H, m, pyrazole H, ArH), 12.83 (1H, s, –OH); ^{13}C NMR (100 MHz, CDCl$_3$): δ (ppm) 27.65 (N–CH$_3$), 29.67 (N–CH$_3$), 55.33 (–OCH$_3$), 86.08, 114.45, 116.27, 124.13, 127.47, 129.54, 132.39, 148.08, 151.88, 159.61, 161.54, 162.08, 165.49; ESI-MS m/z: [M + H]$^+$ = 404.1.

3.3.3. 1,3-Dimethyl-6-hydroxy-2,4-dioxo-5-(1'-phenyl-3'-(p-bromo)-1H-pyrazol-5'-yl)-2,4-dioxo-1,2,3,4-tetrahydropyrimidine, (5c). The product was obtained as mentioned in general procedure from **4c** as white solid, M.p. 197–198°C; IR ν_{max} (cm^{-1}) 1699, 1734 (2 × C=O), 2925 (C–H), 3417 (–OH); ^{1}H NMR (400 MHz, CDCl$_3$): δ (ppm) 3.28 (3H, s, N–CH$_3$), 3.35 (3H, s, N–CH$_3$), 6.92 (2H, d, J = 8.05 Hz, ArH), 7.09 (1H, t, ArH), 7.27–7.31 (5H, m, Pyrazole H, ArH), 7.52 (2H, d, J = 8.05 Hz, ArH), 12.82 (1H, s, –OH); ^{13}C NMR (100 MHz, CDCl$_3$): δ (ppm) 27.84 (N–CH$_3$), 86.14, 116.18, 122.26, 124.36, 127.87, 129.66, 132.29, 139.77, 147.76, 151.45, 161.45, 161.95, 164.99; ESI-MS m/z: [M + H]$^+$ = 453.3.

3.3.4. 1,3-Dimethyl-6-hydroxy-2,4-dioxo-5-(1'-phenyl-3'-(p-chloro)-1H-pyrazol-5'-yl)-1,2,3,4-tetrahydropyrimidine, (5d). The product was obtained as mentioned in general procedure from **4d** as light brown solid, M.p. 121–123°C; IR ν_{max} (cm^{-1}) 1642, 1702 (2 × C=O), 2923 (C–H), 3319 (–OH); ^{1}H NMR (400 MHz, CDCl$_3$): δ (ppm) 3.28 (3H, s, N–CH$_3$), 3.36 (3H, s, N–CH$_3$), 6.92 (2H, d, J = 8.05 Hz, ArH), 7.07–7.11 (2H, m, ArH), 7.24–7.48 (6H, m, ArH, pyrazole H), 12.83 (1H, s, –OH); ^{13}C NMR (100 MHz, CDCl$_3$): δ (ppm) 27.71 (N–CH$_3$), 87.69, 116.27, 119.88, 121.09, 124.36, 125.04, 127.56, 129.00, 129.33, 129.49, 130.04, 134.16, 139.09, 147.62, 151.71, 161.40; ESI-MS m/z: [M + H]$^+$ = 408.1.

3.3.5. 1,3-Diphenyl-6-hydroxy-4-oxo-2-thiooxo-5-(1'-phenyl-3'-(p-tolyl)-1H-pyrazol-5'-yl)-1,2,3,4-tetrahydropyrimidine, (5e). The product was obtained as mentioned in general procedure from **4e** as light brown solid, M.p. 131–133°C; IR ν_{max} (cm^{-1}) 1071 (C=S), 1676 (C=O), 2924 (C–H), 3245 (–OH); ^{1}H NMR (400 MHz, CDCl$_3$): δ (ppm) 2.35 (3H, s, –CH$_3$), 6.90 (2H, d, J = 8.08 Hz, ArH), 7.06 (1H, t, ArH), 7.17–7.30 (9H, m, ArH), 7.32–7.38 (3H, m, ArH), 7.40–7.53 (5H, m, Pyrazole H, ArH), 12.82 (1H, s, –OH); ^{13}C NMR (100 MHz, CDCl$_3$): δ (ppm) 21.24 (N–CH$_3$), 88.32, 124.27, 125.55, 128.32, 128.70, 128.86, 129.38, 129.55, 129.82, 137.48, 138.25, 139.53, 140.35, 146.93, 160.75, 161.33, 163.97, 180.24 (C=S); ESI-MS m/z: [M + H]$^+$ = 528.1.

3.3.6. 1,3-Diphenyl-6-hydroxy-4-oxo-2-thiooxo-5-(1'-phenyl-3'-(p-methoxyphenyl)-1H-pyrazol-5'-yl)-1,2,3,4-tetrahydropyrimidine, (5f). The product was obtained as mentioned in general procedure from **4f** as green solid, M.p. 120–121°C; IR ν_{max} (cm^{-1}) 1030 (C=S), 1676 (C=O), 2924 (C–H), 3214 (–OH); ^{1}H NMR (400 MHz, CDCl$_3$): δ (ppm) 3.82 (3H, s, –OCH$_3$), 6.90–6.93 (4H, m, ArH), 7.08 (1H, t, ArH), 7.21–7.40 (10H, m, ArH), 7.43–7.58 (5H, m, Pyrazole H, ArH), 12.83 (1H, s, –OH); ^{13}C NMR (100 MHz, CDCl$_3$): δ (ppm) 55.32 (–OCH$_3$), 88.35, 114.49, 116.41, 127.45, 128.68, 129.37, 129.44, 129.47, 129.53, 131.2, 139.76, 140.32, 147.13, 159.80, 160.86, 161.49, 164.09, 180.03 (C=S); ESI-MS m/z: [M + H]$^+$ = 545.4.

3.3.7. 1,3-Diphenyl-6-hydroxy-4-oxo-2-thiooxo-5-(1'-phenyl-3'-(p-bromo)-1H-pyrazol-5'-yl)-1,2,3,4-tetrahydropyrimidine, (5g). The product was obtained as mentioned in general procedure from **4g** as light green solid, M.p. 209–210°C; IR ν_{max} (cm^{-1}) 1071 (C=S), 1675 (C=O), 2925 (C–H), 3182 (–OH); ^{1}H NMR (400 MHz, CDCl$_3$): δ (ppm) 6.89 (2H, d, J = 8.05 Hz, ArH), 7.08 (1H, t, ArH), 7.26–7.28 (4H, m, ArH), 7.33 (2H, d, J = 7.32 Hz, ArH), 7.40 (2H, d, J = 8.05 Hz, ArH), 7.44–7.57 (9H, m, Pyrazole H, ArH), 12.80 (1H, s, –OH); ^{13}C NMR (100 MHz, CDCl$_3$): δ (ppm) 88.36, 106.31, 116.31, 122.11, 124.93, 125.37, 127.83, 128.65, 129.39, 129.46, 129.66, 132.33, 136.11, 139.09, 139.89, 146.74, 160.55, 161.35, 164.18, 179.92 (C=S); ESI-MS m/z: [M + H]$^+$ = 592.0.

3.3.8. 1,3-Diphenyl-6-hydroxy-4-oxo-2-thiooxo-5-(1'-phenyl-3'-(p-chloro)-1H-pyrazol-5'-yl)-1,2,3,4-tetrahydropyrimidine, (5h). The product was obtained as mentioned in general procedure from **4h** as dark green solid, M.p. 207–208°C; IR ν_{max} (cm^{-1}) 1089 (C=S), 1675 (C=O), 2924 (C–H), 3198 (–OH); ^{1}H NMR (400 MHz, CDCl$_3$): δ (ppm) 6.89 (2H, d, J = 8.05 Hz, ArH), 7.08 (1H, t, ArH), 7.18–7.29 (5H, m, ArH), 7.32–7.40 (6H, m, ArH), 7.44–7.57 (6H, m, Pyrazole H, ArH), 12.80 (1H, s, –OH); ^{13}C NMR (100 MHz, CDCl$_3$): δ (ppm) 88.50, 105.95, 116.30, 127.51, 128.64, 129.39, 129.47, 129.65, 134.23, 138.66, 139.48, 139.89, 146.59, 160.96, 161.35, 164.21, 179.94 (C=S); ESI-MS m/z: [M + H]$^+$ = 548.1.

3.3.9. 1,3-Bis(2''-methoxyphenyl)-6-hydroxy4-oxo-2-thioxo-5-(1'-phenyl-3'-(p-tolyl)-1H-pyrazol-5'-yl)-1,2,3,4-tetrahydropyrimidine, (5i). The product was obtained as mentioned in general procedure from **4i** as light brown solid, M.p. 107–109°C; IR ν_{max} (cm^{-1}) 1075 (C=S), 1677 (C=O), 2925 (C–H), 3302 (–OH); ^{1}H NMR (400 MHz, CDCl$_3$): δ (ppm) 2.36 (3H, s, –CH$_3$), 3.85 (6H, s,–OCH$_3$), 6.91 (2H, d, J = 7.32 Hz, ArH), 7.00–7.12 (6H, m, ArH), 7.18–7.30 (5H, m, ArH),

7.30–7.43 (5H, m, Pyrazole H, ArH), 12.89 (1H, s, –OH); ^{13}C NMR (100 MHz, CDCl$_3$): δ (ppm) 29.61 (–CH$_3$), 56.01 (–OCH$_3$), 85.53, 105.37, 112.20, 112.72, 115.94, 121.02, 125.57, 128.51, 129.56, 129.67, 129.79, 129.99, 140.16, 148.36, 154.66, 178.37 (C=S); ESI-MS m/z: [M + H]$^+$ = 588.1.

3.3.10. 1,3-Bis(2″-methoxyphenyl)-6-hydroxy-4-oxo-2-thio-oxo-5-(1′-phenyl-3′-(p-methoxyphenyl)-1H-pyrazol-5′-yl)-1,2,3,4-tetrahydropyrimidine, (5j). The product was obtained as mentioned in general procedure from **4j** as light green solid, M.p. 127–129°C; IR ν_{max} (cm^{-1}) 1074 (C=S), 1627 (C=O), 2926 (C–H), 3422 (–OH); ^{1}H NMR (400 MHz, CDCl$_3$): δ (ppm) 3.83 (9H, s, –OCH$_3$), 6.83–6.92 (7H, m, ArH), 7.01–7.11 (7H, m, ArH), 7.20–7.32 (4H, m, pyrazole H, ArH), 12.80 (1H, s, –OH); ^{13}C NMR (100 MHz, CDCl$_3$): δ (ppm) 55.32 (–OCH$_3$), 88.15, 112.78, 113.53, 114.45, 116.33, 121.01, 121.53, 121.81, 124.32, 127.53, 129.16, 129.46, 129.84, 130.18, 146.75, 154.93, 159.55, 167.76, 185.32 (C=S); ESI-MS m/z: [M + H]$^+$ = 604.1.

3.3.11. 1,3-Bis(2″-methoxyphenyl)-6-hydroxy-4-oxo-2-thio-oxo-5-(1′-phenyl-3′-(p-bromo)-1H-pyrazol-5′-yl)-1,2,3,4-tetrahydropyrimidine, (5k). The product was obtained as mentioned in general procedure from **4k** as light yellow solid, M.p. 112-113°C; IR ν_{max} (cm^{-1}) 1044 (C=S), 1674 (C=O), 2925 (C–H), 3287 (–OH); ^{1}H NMR (400 MHz, CDCl$_3$): δ (ppm) 3.78 (6H, s, –OCH$_3$), 6.82–6.84 (2H, m, ArH), 6.93–7.04 (5H, m, ArH), 7.12–7.25 (6H, m, ArH), 7.32–7.39 (3H, m, ArH), 7.45–7.47 (2H, m, pyrazole H, ArH), 12.77 (1H, s, –OH); ^{13}C NMR (100 MHz, CDCl$_3$): δ (ppm) 56.36 (–OCH$_3$), 88.36, 116.26, 121.03, 121.13, 127.90, 128.00, 129.60, 129.70, 129.89, 130.25, 132.30, 139.30, 139.36, 146.77, 154.51, 159.70, 160.94, 163.93, 179.53 (C=S); ESI-MS m/z: [M + H]$^+$ = 652.0.

3.3.12. 1,3-Bis(2″-methoxyphenyl)-4-oxo-2-thiooxo-6-hydroxy-5-(1′-phenyl-3′-(p-chloro)-1H-pyrazol-5′-yl)-1,2,3,4-tetrahydropyrimidine, (5l). The product was obtained as mentioned in general procedure from **4l** as light yellow solid, M.p. 232-233°C; IR ν_{max} (cm^{-1}) 1043 (C=S), 1654 (C=O), 2927 (C–H), 3437 (–OH); ^{1}H NMR (400 MHz, CDCl$_3$): δ (ppm) 3.85 (6H, s, –OCH$_3$), 6.89–6.90 (2H, m, ArH), 6.98–7.09 (5H, m, ArH), 7.23–7.28 (5H, m, ArH), 7.33–7.45 (6H, m, pyrazole H, ArH), 12.83 (1H, s, –OH); ^{13}C NMR (100 MHz, CDCl$_3$): δ (ppm) 56.19 (–OCH$_3$), 88.18, 112.23, 116.06, 120.91, 124.52, 127.60, 129.34, 129.59, 129.88, 134.26, 138.88, 154.89, 160.60, 163.97, 185.19 (C=S); ESI-MS m/z: [M + H]$^+$ = 609.3.

3.4. Antifungal Susceptibility Test. The pathogenic isolates of *Aspergillus fumigatus* (ITCC 4517 (IARI, Indian Agricultural Research Institute Delhi), ITCC 1634 (IARI, Delhi), clinical isolate 190/96 (VPCI, Vallabhbhai Patel Chest Institute Delhi)), *Aspergillus flavus* (clinical isolate 223/96 (VPCI, Delhi)), and *Aspergillus niger* (clinical isolate 56/96 (VPCI, Delhi)) were employed in the current study. These pathogenic species of *Aspergillus*, namely, *A. fumigatus, A. flavus,* and

A. niger, were cultured in laboratory on Sabouraud dextrose (SD) agar plates. The plates were inoculated with stock cultures of *A. fumigatus, A. flavus,* and *A. niger* and incubated in a BOD incubator at 37°C. The spores were harvested from 96 h cultures and suspended homogeneously in phosphate buffer saline (PBS). The spores in the suspension were counted and their number was adjusted to 10^8 spores/mL before performing the experiments. The antifungal activity of compounds was analysed by MDA, DDA, and PSGI. Each assay was repeated at least three times on different days. Amp B was used as a standard drug in antifungal susceptibility test.

3.4.1. Disc Diffusion Assay (DDA). The disc diffusion assay was performed in radiation sterilized petri plates (10.0 cm diameter, Tarsons). The SD agar plates were prepared and plated with a standardized suspension of 1×10^8 spore/mL of *Aspergillus* spp. Then, plates were allowed to dry and discs {(5.0 mm in diameter) of Whatman filter paper number 1} were placed on the surface of the agar. The different concentrations of compounds in the range of 750–1.0046 µg were impregnated on the discs. An additional disc for solvent (DMSO) was also placed on agar plate. The plates were incubated at 37°C and examined at 24 h, 48 h for zone of inhibition, if any, around the discs. The concentration, which developed the zone of inhibition of at least 6.0 mm diameter, was taken as end point (Minimum Inhibitory Concentration, MIC).

3.4.2. Percent Spore Germination Inhibition Assay (PSGI). Different concentrations of the test compounds in 90.0 µL of culture medium were prepared in 96-well flat-bottomed microculture plates (Tarson) by double dilution method. Each well was then inoculated with 10.0 µL of spore suspension (100 ± 5 spores). The plates were incubated at 37°C for 16 h and then examined for spore germination under inverted microscope (Nikon, diphot). The number of germinated and nongerminated spores was counted. The lowest concentration of the compound, which resulted in >90% inhibition of germination of spores in the wells, was considered as MIC$_{90}$.

3.4.3. Microbroth Dilution Assay (MDA). The test was performed in 96-well culture plates (Tarson). Various concentrations of compounds in the range of 1250–4.3 µg/mL were prepared in 90.0 µL of culture medium by double dilution method. Each well was inoculated with 10 µL of spore suspension (1×10^8 spore/mL) and incubated for 48 h at 37°C. After 48 h, the plates were assessed visually. The optically clear well was taken as end point, MIC.

3.5. Antifungal Drugs and Pyrimidine Pyrazole Analogues Checkerboard Testing. *In vitro* combination of pyrimidine pyrazole analogues was studied with antifungal drug AmpB (Himedia) and NYS (Himedia). The starting range of final concentration was taken as approximate one fold higher than individual MIC to compute all *in vitro* interactions (Antagonistic; Synergy, SYN; and Indifference, IND). The final concentrations of antifungal agents which ranged from 3.125 to 0.02 μg/mL for Amp B, 6.25 to 0.09 for NYS, and 400 to 3.125 μg/mL for **5c**, **5j** were taken. Aliquots of 45 μL of each drug at a concentration four times the targeted final were dispensed in the wells in order to obtain a two-dimensional checkerboard (8 × 8 combination) [27]. Each well then was inoculated with 10 μL of spore suspension (1 × 10^8 spore/mL). The plates were incubated at 37°C for 48 h. The plates were then assessed visually. The optically clear well was taken as end point, MIC.

3.6. Drug Interaction Modelling. The drug interaction was determined by the most popular FICI model. The FICI represents the sum of the FICs (Fraction Inhibitory Concentration) of each drug tested. The FIC of a drug was defined as MIC of a drug in combination divided by MIC of the same drug alone (MIC of drug in combination/MIC of drug alone), FICI = 1 (revealed indifference), FICI ≤ 0.5 (revealed synergy), and FICI > 4 (revealed antagonism) [28].

3.7. Antibacterial Susceptibility Test. The antibacterial activity of compound was analysed by microbroth dilution Resazurin based assay [29]. Each assay was repeated at least three times on different days. The different pathogenic species of bacteria, *Staphylococcus aureus* (MTCC number 3160), *Bacillus cereus* (MTCC number 10085), *Escherichia coli* (MTCC number 433), *Salmonella typhi* (MTCC number 733), *Micrococcus luteus* (MTCC number 8132), *Bacillus pumilis* (MTCC number 2299), and *Bacillus subtilis* (MTCC number 8142), were cultured in Luria broth. Using aseptic techniques, a single colony was transferred into a 100 mL Luria broth and placed in incubator at 35°C. After 12–18 h of incubation, the culture was centrifuged at 4000 rpm for 5 minutes. The supernatant was discarded and pellet was resuspended in 20 mL PBS and centrifuged again at 4000 rpm for 5 min. This step was repeated until the supernatant was clear. The pellet was then suspended in 20 mL PBS. The optical density of the bacteria was recorded at 500 nm and serial dilutions were carried out with appropriate aseptic techniques until the optical density was in the range of 0.5–1.0, representing 5 × 10^6 CFU/mL.

3.7.1. Resazurin Based Microtitre Dilution Assay. Resazurin based MDA was performed in 96-well plates under aseptic conditions. The concentrations of compounds in the range of 2000–7.8 μg/mL were prepared in 100 μL of culture medium by serial dilution method. 10 μL of Resazurin indicator solution (5X) was added in each well. Finally, 10 μL of bacterial suspension was added (5 × 10^6 CFU/mL) to each well to achieve a concentration of 5 × 10^5 CFU/mL. Each plate had a set of controls: a column with erythromycin as positive control. The plates were prepared in triplicate and incubated at 37°C for 24 h. The colour change was then assessed visually. The lowest concentration at which colour change occurred was taken as the MIC value.

4. Conclusion

In search of novel antimicrobial molecules, we came across that pyrimidine pyrazole heterocycles can be of interest as **5g** showed significant antibacterial activity. The compounds **5e** and **5h** also showed moderate antibacterial activity. **5j** showed moderate antifungal activity. Out of all heterocycles, **5c** possesses both antifungal and antibacterial activity. Our studies showed that these novel heterocycles can supplement the existing antifungal therapy. Monotherapy can be replaced by combination therapy. Therefore **5c**, **5g**, and **5j** might be of great interest for the development of novel antimicrobial molecule.

Conflict of Interests

The authors declare that there is no conflict of interests regarding the publication of this paper.

Acknowledgments

The authors would like to thank Council of Scientific and Industrial Research (CSIR), New Delhi, and Defence Research and Development Organisation (DRDO) for the financial support.

References

[1] A. L. Stuart, N. K. Ayisi, G. Tourigny, and V. S. Gupta, "Antiviral activity, antimetabolic activity and cytotoxicity of 3*t*-substituted deoxypyrimidine nucleosides," *Journal of Pharmaceutical Sciences*, vol. 74, no. 3, pp. 246–249, 1985.

[2] A. Agarwal, N. Goyal, P. M. S. Chauhan, and S. Gupta, "Dihydropyrido[2,3-d]pyrimidines as a new class of antileishmanial agents," *Bioorganic and Medicinal Chemistry*, vol. 13, no. 24, pp. 6678–6684, 2005.

[3] R. E. Mitchell, D. R. Greenwood, and V. Sarojini, "An antibacterial pyrazole derivative from *Burkholderia glumae*, a bacterial pathogen of rice," *Phytochemistry*, vol. 69, no. 15, pp. 2704–2707, 2008.

[4] R. Basawaraj, B. Yadav, and S. S. Sangapure, "Synthesis of some 1H-pyrazolines bearing benzofuran as biologically active agents," *Indian Journal of Heterocyclic Chemistry*, vol. 11, no. 1, pp. 31–34, 2001.

[5] K. T. Ashish and M. Anil, "Synthesis and antifungal activity of 4-substituted-3,7-dimethylpyrazolo [3,4-e] [1,2,4] triazine," *Indian Journal of Chemistry B*, vol. 45, p. 489, 2006.

[6] B. P. Chetan and V. V. Mulwar, "Synthesis and evaluation of certain pyrazolines and related compounds for their anti tubercular, anti bacterial and anti fungal activities," *Indian Journal of Chemistry B*, vol. 44, article 232, 2000.

[7] K. S. Nimavat and K. H. Popat, "Synthesis, anticancer, antitubercular and antimicrobial activities of 1-substituted, 3-aryl-5-(3′-bromophenyl) pyrazoline," *Indian Journal of Heterocyclic Chemistry*, vol. 16, p. 333, 2007.

[8] R. H. Udupi, A. R. Bhat, and K. Krishna, "Synthesis and investigation of some new pyrazoline derivatives for their antimicrobial, anti inflammatory and analgesic activities," *Indian Journal of Heterocyclic Chemistry*, vol. 8, p. 143, 1998.

[9] F. R. Souza, V. T. Souza, V. Ratzlaff et al., "Hypothermic and antipyretic effects of 3-methyl- and 3-phenyl-5-hydroxy-5-trichloromethyl-4,5-dihydro-1H-pyrazole-1-carboxyamides in mice," *European Journal of Pharmacology*, vol. 451, no. 2, pp. 141–147, 2002.

[10] K. Ashok, Archana, and S. Sharma, "Synthesis of potential quinazolinyl pyrazolines as anticonvulsant agents," *Indian Journal of Heterocyclic Chemistry*, vol. 9, p. 197, 2001.

[11] M. Abdel-Aziz, G. E. A. Abuo-Rahma, and A. A. Hassan, "Synthesis of novel pyrazole derivatives and evaluation of their antidepressant and anticonvulsant activities," *European Journal of Medicinal Chemistry*, vol. 44, no. 9, pp. 3480–3487, 2009.

[12] L. A. Elvin, E. C. John, C. G. Leon, J. L. John, and H. E. Reiff, "Synthesis and muscle relaxant property of 3-amino-4-aryl pyrazoles," *Journal of Medicinal Chemistry*, vol. 7, no. 3, pp. 259–268, 1964.

[13] G. Doria, C. Passarotti, R. Sala et al., "Synthesis and antiulcer activity of (E)-5-[2-(3-pyridyl) ethenyl]-1 H,7 H-pyrazolo [1,5-a] pyrimidine-7-ones," *Farmaco*, vol. 41, p. 417, 1986.

[14] W. H. Robert, "The antiarrhythmic and antiinflammatory activity of a series of tricyclic pyrazoles," *Journal of Heterocyclic Chemistry*, vol. 13, no. 3, pp. 545–553, 2009.

[15] R. Soliman, H. Mokhtar, and H. F. Mohamed, "Synthesis and antidiabetic activity of some sulfonylurea derivatives of 3,5-disubstituted pyrazoles," *Journal of Pharmaceutical Sciences*, vol. 72, no. 9, pp. 999–1004, 1983.

[16] R. Kumar, J. Arora, A. K. Prasad, N. Islam, and A. K. Verma, "Synthesis and antimicrobial activity of pyrimidine chalcones," *Medicinal Chemistry Research*, vol. 22, no. 11, pp. 5624–5631, 2013.

[17] A. Solankee, S. Lad, S. Solankee, and G. Patel, "Chalcones, pyrazolines and aminopyrimidines as antibacterial agents," *Indian Journal of Chemistry B*, vol. 48, article 1442, 2009.

[18] B. S. Jursic and D. M. Neumann, "Preparation of 5-formyl- and 5-acetylbarbituric acids, including the corresponding Schiff bases and phenylhydrazones," *Tetrahedron Letters*, vol. 42, no. 48, pp. 8435–8439, 2001.

[19] F. S. Crossley, E. Miller, W. H. Hartung, and M. L. Moore, "Thiobarbiturates. III. Some N-substituted derivatives," *Journal of Organic Chemistry*, vol. 5, no. 3, pp. 238–243, 1940.

[20] P. Cabildo, R. M. Claramunt, and J. Elguero, " ^{13}C NMR chemical shifts of N-unsubstituted and N-methyl-pyrazole derivatives," *Organic Magnetic Resonance*, vol. 22, no. 9, pp. 603–607, 1984.

[21] V. Yadav, J. Gupta, R. Mandhan et al., "Investigations on anti-Aspergillus properties of bacterial products," *Letters in Applied Microbiology*, vol. 41, no. 4, pp. 309–314, 2005.

[22] S. Ruhil, M. Balhara, S. Dhankhar, M. Kumar, V. Kumar, and A. K. Chhillar, "Advancement in infection control of opportunistic pathogen (Aspergillus spp.): adjunctive agents," *Current Pharmaceutical Biotechnology*, vol. 14, no. 2, pp. 226–232, 2013.

[23] T. R. T. Dagenais and N. P. Keller, "Pathogenesis of *Aspergillus fumigatus* in invasive aspergillosis," *Clinical Microbiology Reviews*, vol. 22, no. 3, pp. 447–465, 2009.

[24] J. Smith and D. Andes, "Therapeutic drug monitoring of antifungals: pharmacokinetic and pharmacodynamic considerations," *Therapeutic Drug Monitoring*, vol. 30, no. 2, pp. 167–172, 2008.

[25] S. Bondock, W. Khalifa, and A. A. Fadda, "Synthesis and antimicrobial activity of some new 4-hetarylpyrazole and furo[2,3-c]pyrazole derivatives," *European Journal of Medicinal Chemistry*, vol. 46, no. 6, pp. 2555–2561, 2011.

[26] K. S. Jain, T. S. Chitre, P. B. Miniyar et al., "Biological and medicinal significance of pyrimidines," *Current Science*, vol. 90, no. 6, pp. 793–803, 2006.

[27] E. M. O'Shaughnessy, J. Meletiadis, T. Stergiopoulou, J. P. Demchok, and T. J. Walsh, "Antifungal interactions within the triple combination of amphotericin B, caspofungin and voriconazole against Aspergillus species," *Journal of Antimicrobial Chemotherapy*, vol. 58, no. 6, pp. 1168–1176, 2006.

[28] S. Ruhil, M. Balhara, S. Dhankhar, V. Kumar, and A. K. Chhillar, "Invasive aspergillosis: adjunctive combination therapy," *Mini-Reviews in Medicinal Chemistry*, vol. 12, no. 12, pp. 1261–1272, 2012.

[29] S. Dhankhar, M. Kumar, S. Ruhil, M. Balhara, and A. K. Chhillar, "Analysis toward innovative herbal antibacterial & antifungal drugs," *Recent Patents on Anti-Infective Drug Discovery*, vol. 7, no. 3, pp. 242–248, 2012.

6

Adsorption Studies of Lead by *Enteromorpha* Algae and Its Silicates Bonded Material

Hassan H. Hammud,[1,2] **Ali El-Shaar,**[2] **Essam Khamis,**[3] **and El-Sayed Mansour**[4]

[1] *Chemistry Department, Faculty of Science, King Faisal University, Al-Ahsa 31982, Saudi Arabia*
[2] *Chemistry Department, Faculty of Science, Beirut Arab University, Beirut, Lebanon*
[3] *City of Scientific Research & Technological Applications, Borg Al Arab, Alexandria, Egypt*
[4] *Faculty of Science, Alexandria University, Alexandria, Egypt*

Correspondence should be addressed to Hassan H. Hammud; hhammud@yahoo.com

Academic Editor: Ioannis Konstantinou

Lead adsorption by green *Enteromorpha* algae was studied. Adsorption capacity was 83.8 mg/g at pH 3.0 with algae (E) and 1433.5 mg/g for silicates modified algae (EM). FTIR and thermal analysis of algae materials were studied. Thomas and Yoon-Nelson column model were best for adsorbent (E) and algae after reflux (ER) and Yan model for (EM) with capacity 76.2, 71.1, and 982.5 mg/g, respectively. (ER) and (EM) show less swelling and better flow rate control than (E). Nonlinear methods are more appropriate technique. Error function calculations proved valuable for predicting the best adsorption isotherms, kinetics, and column models.

1. Introduction

The contamination of wastewater and soil with heavy metal ions is a complex problem, since these metals are toxic in both their elemental and chemically combined forms. Natural water is contaminated with several heavy metals due to their widespread use in industry and agriculture arising mostly from mining wastes and industrial discharges. From an environmental protection point of view, heavy metal ions should be removed at the source in order to avoid pollution of natural waters and subsequent metal accumulation in the food chain. In fact, removal of this contamination has received much attention in recent years [1, 2].

Lead being one of the "big three" toxic heavy metals, it is of profound concern as a toxic waste and contaminant of surface waters as it becomes concentrated throughout the food chain to humans [3]. Lead damages different body organs (central and peripheral nervous systems and kidney); also, lead has a teratogenic effect, causing stillbirth in women and affecting the fetus [4].

Conventional methods for removal are chemical precipitation, chemical oxidation, chemical reduction, ion exchange, filtration, electrochemical treatment, and evaporation [5, 6]. These methods often are very expensive.

Alternative method for heavy metal removal was developed in the last past decade and known as biosorption. Marine algae, an abundant renewable natural biomass, have been used as dead nonliving materials for removal of heavy metals [7, 8]. In addition, algae were found accumulating heavy metal in their habitat and are thus used as heavy metal pollution monitors in fresh and salty water such as river, sea, and ocean. They have been also used in on-site bioremediation of polluted natural water [9, 10].

Furthermore, the search for a low-cost and easily available adsorbent has led to the investigation of materials of agricultural and biological origin (bacteria, fungi, yeast, and algae can remove heavy metals from a solution in considerable quantities), along with industrial byproducts, as potential metal sorbents. These low-cost and nonconventional adsorbents include agricultural wastes like natural compost, Irish peanut, peanut shell, coconut shell, bone, and biomass such as *Aspergillus terreus* and *Mucus remanianus*, polymerized onion skin with formaldehyde and EDTA, modified cellulosic material, and natural materials such as hair and cattails

(*Typha* plant), waste wool, peanut skin, modified barks, barely straw, low rank coal, soils, human hair, peat moss, fertilizer waste slurry, waste tire rubber, and tea leaves [11–22].

Biosorption process for metal removal has a performance comparable to a commercial competitor, ion exchange treatment. While commercial ion exchange resins are rather costly, the price tag of biosorbents can be an order of magnitude cheaper (1/10 of the ion exchange resin cost) [23, 24].

Some of the key features of biosorption compared to conventional processes include competitive performance, heavy metal selectivity, cost-effectiveness, regenerative, process equipment known, no sludge generation, and metal recovery possible [25].

In the literature, there are divergent mechanisms explaining the metal uptake by marine algae. A semispeculative model of the structure of the cell walls of the algae has been proposed recently. This could suggest that there are two common moieties to which the uptake ability of taxonomically different algal biomass was attributed: sulfated ester polysaccharides (fucoidans, carrageenans, and galatians and xylans) and polyuronides (galacturonic, glucuronic, guluronic, and mannuronic acids). The sulfate and carboxylic acid groups could be considered as responsible for the bulk of metal uptake sorption [26]. In general, the algae are very soft, with the tendency to disintegrate which prevents the follow-up experiments even in laboratory column [27]. Also, the small size of algae particles with low density, mechanical strength, and rigidity affects the performance of algae. Immobilization of algae within a matrix (e.g., silicates) through chemical pretreatment can overcome these problems and allow industrial application [28–30].

The present work is focused on the ability of green algae *Enteromorpha* collected from Lebanese coast to adsorb Pb^{2+} from aqueous medium at 25°C. Comparative studies are made between nonmodified *Enteromorpha* algae (E), algae residue after reflux (ER), and algae modified with silicates (EM).

Optimum conditions for adsorption (mass, pH, concentration, and residence time) are considered in batch experiments. Also column application for algae materials was studied for maximum capacity per one gram.

Infrared spectra and thermogravimetric analysis of *Enteromorpha* algae were detected before and after adsorption.

2. Theoretical

2.1. Isotherm and Column Models.
The equations and parameters of the isotherms adsorption models (Langmuir, Freundlich, Redlich-Peterson, Temkin, Elovich, and Dubinin-Radushkevich), column kinetic models (Thomas, Yoon-Nelson, and Yan model) are described in detail in our previous work [31–34] and references therein, Tables 1 and 2.

In general all calculations are based on the following two terms: q_e (mg/g), the equilibrium metal uptake (mass of adsorbed pollutant (mg) per mass of adsorbent (g)) and C_e (mg/L) or (ppm), the equilibrium ion concentration of pollutant remaining in solution.

TABLE 1: Nonlinear equations of different adsorption isotherms.

Isotherm model	Equation
Langmuir	$Q_e = \dfrac{Q^o K_L C_e}{1 + K_L C_e}$
Freundlich	$Q_e = K_F (C_e)^{1/n}$
Redlich-Peterson	$Q_e = \dfrac{A_{RP} C_e}{1 + K_{RP} C_e^{\beta}}$
Temkin	$Q_e = B_T \ln (K_T) + B_T \ln (C_e)$
Elovich	$\dfrac{Q_e}{Q_E} = K_E C_e e^{-(Q_e/Q_E)}$
Dubinin-Radushkevich	$Q_e = Q_m \exp^{\beta[RT \ln(1+(1/C_e))]^2}$

2.2. Error Analysis.
The computational part of the obtained experimental measurements is introduced. To obtain the curve which best fits the data in least-squares sense, the following minimization problem is solved by calculating error parameters R^2, χ^2, and SSE. Error calculations were used for the fitting quality of the isotherm and column models to the experimental data, Table 3 [31–34], where $x\text{data}_i$ and $y\text{data}_i$ are the experimental measurements and $f(x, x\text{data}_i)$ is a function of the linear or nonlinear curve proposed to fit the given data. The $x\text{data}_i$ is the equilibrium concentration C_e and $y\text{data}_i$ is dependent on q_e.

The squared correlation coefficient R^2 is defined in Table 3, R being the correlation between the experimental data and the used model, where $X = y\text{data}_i = y_{e,\exp}$ and $Y = f(x, x\text{data}_i) = y_{e,\text{cal}}$.

The sum of squares of errors SSE and chi-square test χ^2 depend on the total deviation of the calculated values from the fit to the experimental values in Table 3.

If data from the model are similar to the experimental data, χ^2 and SSE will be a small number. On the other hand, the closer R^2 value to 1.00 the better is the fitting and the more suitable is the model. Thus, the best model chosen in each case is the one giving highest R^2 and lowest SSE and χ^2.

3. Experimental

3.1. Chemicals.
All reagents are from Fluka: lead nitrate $Pb(NO_3)_2$, sodium silicates for chemical modification of algae, sodium acetate, and acetic acid for preparation of acetate buffer.

3.2. Algae Collection.
The raw algae *Enteromorpha intestinalis* was harvested from the Lebanese coast, washed with tap and deionized water in order to remove extra salts, sun dried, and grounded to particle size (0.5 mesh). Finally, the fine powder is oven dried at 60°C for 24 hours to give algae material (E) for further use in metal uptake study.

3.3. Preparation of Algae Residue (ER), Modified Algae (EM), and Swedish Wood Sawdust (SWS)

3.3.1. Preparation of Algae Residue Material (ER).
Enteromorpha algae (J) powder 10 gm was refluxed in deionised

TABLE 2: Nonlinear and linear equations of different kinetic column models.

Kinetic models	Nonlinear equations	Linear equations
Thomas model	$\dfrac{C_e}{C_o} = \dfrac{1}{1 + \exp\left((K_{Th}/Q)\left(q_T m - C_o V_{eff}\right)\right)}$	$\ln\left[\left(\dfrac{C_o}{C_e}\right) - 1\right] = \left(\dfrac{K_{Th} q_T M}{Q}\right) - (K_{Th} C_o t)$
Yoon-Nelson model	$\dfrac{C_e}{C_o} = \dfrac{1}{1 + e^{K_{YN}(\tau - t)}}$	$\ln\left(\dfrac{C_e}{C_o - C_e}\right) = K_{YN} t - \tau K_{YN}$
Yan et al. model [42]	$\dfrac{C_e}{C_o} = 1 - \dfrac{1}{1 + \left(Q^2 t / K_y q_y m\right)^{K_y C_o / Q}}$	$\ln\left[\left(\dfrac{C_e}{C_o - C_e}\right)\right] = \left(\dfrac{K_y C_o}{Q}\right) \ln\left(\dfrac{Q^2}{K_y q_y m}\right) + \left(\dfrac{K_y C_o}{Q}\right) \ln t$

TABLE 3: Error function and squared correlation coefficient for best-fitting model estimation.

$$R^2 = \frac{\left(\sum (X - \bar{X})(Y - \bar{Y})\right)^2}{\sum (X - \bar{X})^2 (Y - \bar{Y})^2} \qquad \text{Chi-square} = \chi^2 = \sum \frac{\left(y_{e,exp} - y_{e,cal}\right)^2}{y_{e,cal}} \qquad \text{SSE} = \sum \left(y_{e,exp} - y_{e,cal}\right)^2$$

water 100 mL for 15 min giving green suspension; the residue material was separated from the green solution by filtration and then dried to give *Enteromorpha* algae after reflux (ER) in 80% yield.

3.3.2. Preparation of Swedish Wood Sawdust (SWS).
Swedish wood sawdust was ground to 0.5 mesh particle size washed with deionized water and dried similar to algae.

3.3.3. Preparation of Sodium Silicate Algae Material (EM).
Enteromorpha powder (E) was washed with HCl (0.1 M) and water, respectively. The material was then separated by centrifuging. Sulfuric acid (120 mL, 5%) was mixed with enough sodium silicate (Na_2SiO_3) (6%) to raise the pH of the solution to 2.0. At pH 2.0, the washed biomass was added to the solution and stirred for 10 minutes. Additional amount of sodium silicate (6%) was then slowly added to raise the pH of the solution to 7.0. The resulting polymer gel was washed with water and oven dried at 60°C for one week. The dried polymer was ground by mortar and pestle and sieved to be ready for further application [34].

3.4. Equipment and Instruments.
The potentiometric measurements were carried out using Denver Instrument Model 225 pH-ion selective electrode meter fitted with a combined glass electrode (reading to ±0.01 pH unit). The reaction flask was kept constant at 25°C (±0.1°C) by using a thermostat Model Heto HMT 200. The shaker (Wiggen Hauser OS-150, Germany) and centrifuge (Sigma 203) were used for agitation experiment. Lead content was analyzed using Buck Scientific Atomic absorption spectrometer. Infrared data were collected on a Shimadzu 8300 FTIR spectrophotometer using KBr pellet method. Thermogravimetric-differential scanning calorimetry (TG-DSC) curve was recorded on SETARAM LABSYS thermal analyzer in the flow of N_2 within the 25–900°C temperature range, with a heating rate of 3°C/min [31–33].

3.5. Effect of Mass of Algae.
The effect of mass of algae (E) on metal uptake was studied in batch system. 25 mL solution of 300 ppm lead (at pH 4.00) in 50 mL Erlenmeyer flask was shaken with different masses of algae for 2 min at 200 r.p.m

and left to stand for 24 hrs in water bath at 25°C and then analyzed for the remaining lead [35].

3.6. pH Effect.
The mass of algae (E) used in this experiment is the optimum mass 0.3 g done in batch system using 50 mL Erlenmeyer flasks with a reaction volume of 25 mL lead solution (300 mg/L) at 25°C. The mixture was shaken for 2 min (at 200 r.p.m) and left to stand for 24 hrs, at different pH using hydrochloric acid, sodium acetate buffer, and sodium hydroxide to cover the pH range 3.0–7.5 (3, 3.6, 4, 5, 6.08, 7.08, and 7.50). In this experiment the pH value does not exceed 7.50 due to the precipitation of lead as lead hydroxide [36].

3.7. Effect of Lead Concentration.
This experiment was done as above at only pH 3.0 (optimum value) using sodium acetate buffer but with different lead concentrations (10, 25, 50, 100, 150, 200, 250, 300, 400, 500, and 1000 ppm).

3.8. Effect of Residence Time.
This experiment was done at 25°C with 25 mL lead (200 ppm) and 0.3 g algae (E) in 50 mL Erlenmeyer flasks at pH 3.0 (using sodium acetate buffer) and shaken at 200 r.p.m for 2 minutes with variable standing time (2, 5, 10, 15, 20, 30, 45, 60, 90, and 120 minutes) [36].

3.9. Swelling Characteristics.
Swelling characteristics (distention index, swelling ratio, and volume of absorbed solvent) were obtained from the weights and volumes of dry and swollen particles for algae alone (E), algae residue after reflux (ER), algae modified with silicates (EM), and Swedish wood sawdust (SWS). Dry particles were swollen in cylinders with deionized water and degassed under lower pressure. The volume was measured after periodically till two hours [27].

The distention index (DI) was calculated from the ratio V_s/W_D, where V_s is the volume of the particles after swelling and W_D is the weight of the dry particles. The swelling ratio is W_s/W_D, where W_s is the weight of swollen particles. The volume of absorbed solvent (V_{AS}) was calculated from the ratio $(W_s - W_D)/W_D$.

3.10. Procedure for Column Utilization (Breakthrough Curve).
The column used has a diameter of 2 cm and a length of

TABLE 4: Thermal analysis of algae materials: *Enteromorpha* algae (E), *Enteromorpha* algae with adsorbed Lead (E-Pb), and *Enteromorpha* algae modified with sodium silicates (EM).

Sample	Peak (°C) (temperature range)	% Experimental mass loss	Enthalpy (J/g)
(E)	84.10	12.77	238.50
	174.82	3.34	8.22
	215.12	33.88	−17.31
	231.41	—	23.98
	442.21	16.22	541.42
	726.16	20.91	254.63
(E-Pb)	73.61	17.83	234.55
	169.58	—	19.72
	221.14	39.14	−30.03
	428.15	15.57	−7.14
	656.85	21.91	161.65
(EM)	(25–180)	4.0	—
	(180.0–750.0)	13.0	41.59
	290.56	—	—
	756.85	2.0	13.43

44 cm. It was packed uniformly with 1.0 g of green *Enteromorpha* algae powder (E) or *Enteromorpha* algae powder after reflux (ER) or modified *Enteromorpha* algae with sodium silicates (EM). The column dead volume was 30 mL. The lead solution (200 ppm) at pH 3.0 was drained through the column at a constant rate 2.5 mL/min in the case of (E) and (ER). The lead solution (1000 ppm) was eluted at a rate of 2.0 mL/min in the case of column packed with (EM). Collected aliquots of similar volume were collected from the column and analyzed for lead [37].

4. Result and Discussion

4.1. Infrared Spectroscopy. FTIR spectra of *Enteromorpha* algae (E) show strong stretch at 3395 cm^{-1} due to –NH of amino group, strong stretch at 1646.5 cm^{-1} and a weaker ones at 1436 cm^{-1} due to carboxylate group, and strong bending vibration at 1105 and 1158 cm^{-1} due to C–O of ether and alcoholic group, respectively [38]. FTIR spectra of *Enteromorpha* algae with adsorbed lead also show many peaks similar to free algae spectra. However the peak due to –NH stretch has been shifted to lower wave number 3291 cm^{-1} indicating involvement of binding of lead to –NH group. The bands due to carboxylate group show a little shift to higher wave number (1651 cm^{-1}) compared to free algae peak, indicating the involvement of carboxylate and ester groups in the binding with lead.

FTIR of (EM) is similar to that of (E) but with increasing peak strength at 1100 cm^{-1} and at 3437 cm^{-1} due to Si–O–Si and SiO–H stretching vibrations, while FTIR of (ER) has no noticeable difference from that of (E).

4.2. Thermal Analysis. Thermogravimetry analysis (TGA) and differential scanning calorimetry (DSC) curves were recorded for algae materials, which were heated to 800°C under N$_2$, where all relevant weight loss was complete. The results of thermal analysis (% mass loss at each temperature and associated enthalpy) of free *Enteromorpha* algae (E) and algae with adsorbed lead (E-Pb) and modified algae with sodium silicates (EM) are listed in Table 4. The two TGA and dTG curves reveal difference between loss in weight with temperatures for both materials (E) and (E-Pb), with greatest difference occurring about 220°C where free algae show a loss of 33.88% at 215.12°C (exothermic) and 231.41°C (endothermic) while algae with biosorbed lead show a loss of 39.14% at 221.14°C (exothermic). The 5.26% difference in mass loss between the two could be attributed to the adsorbed amount of lead. This peak is shifted to higher temperature (290.56°C) in the case of (EM) with endothermic peak, indicating chemical binding of algae to sodium silicates. The amount of bonded algae is about 19.0%.

The sharp decrease at 726.16°C (20.91%) for (E), 656.85°C (21.91%) for (E-Pb), and 756.85°C for (EM), respectively, is associated with high endothermic heat energy attributed to the conversion of CaCO$_3$(s) to CaO(s) and CO$_2$(g). The weight loss occurring before this step (50 to 600°C) is due to organic materials volatilization and decomposition.

4.3. Batch Studies

4.3.1. Effect of Mass of Algae. The effect of mass of algae on metal uptake was studied in batch system for initial lead concentration (300 ppm) and contact time (24 h) in water bath at 25°C. There was a steep increase in % extraction of lead as the mass of algae increased from 0.05 g to 0.30 g (from 53% to 82%). For biomass weight greater than 0.30 g there were no significant increases in the metal uptake. Thus the optimum mass of algae (E) and (ER) was 0.30 g corresponding to a maximum % metal uptake for a volume reaction of 25 mL

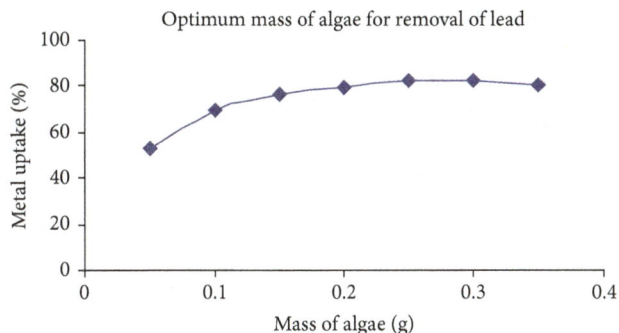

FIGURE 1: Effect of mass of *Enteromorpha* algae on metal uptake for a 25 mL lead solution (300 ppm).

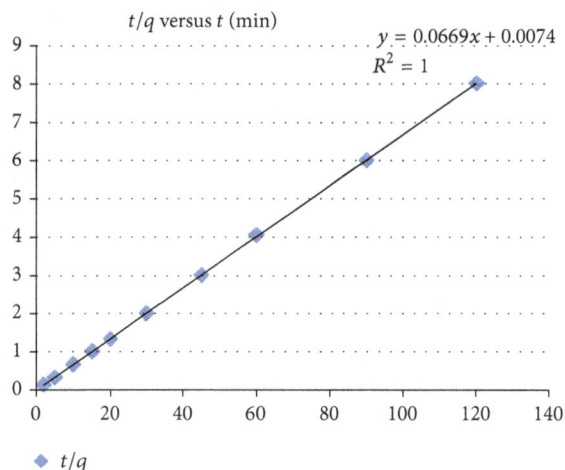

FIGURE 2: Pseudosecond order kinetics for adsorption of lead onto *Enteromorpha* algae (E) at 25°C.

lead (300 ppm), Figure 1, while the optimum mass and reaction volume are 0.1 g and 100 mL for (EM).

4.3.2. pH Effect. Experiments concerning the effect of pH on the sorption were carried out within a pH range that avoids the metal precipitation as hydroxide. The standard lead solution (300 ppm) was prepared in acetate buffer to cover a pH range 3.0–7.5. The data of metal uptake by 0.30 g algae in a 25 mL standard lead solution was studied for different pH values. It was found that lead shows maximum binding to the biomass at an optimum pH 3.0 with a metal uptake of 86.62% and capacity of 21.55 mg/g. There is no change in metal uptake for pH values between 3.6 and 7.5.

4.3.3. Effect of Lead Concentration. The experimental effect of lead concentration on metal uptake was done in batch system concentrations of 0.3 gm at optimum pH 3.0 (using sodium acetate buffer) for a residence time of 24 hrs and different lead concentrations in the range (10–1000 ppm). There was a sharp increase in adsorption as lead concentration increases from 10 to 150 ppm and reaches a maximum at 200 ppm lead with a % metal uptake of 95.50% and a metal uptake capacity q of 15.29 mg/g algae. For lead solution greater than 200 ppm, the algae become saturated and a decrease in % uptake was noticed. Thus the optimum concentration was found to be 200 ppm.

4.3.4. Effect of Residence Time. Previous work showed that sorption of heavy metal ions by lived algae followed a two-step mechanism where the metal ion was physico/chemically uptaken onto the surface of the algae before being taken up biologically into the cell. The first step, known as a passive transport, took place quite rapidly, that is, within 20–30 min, whilst the second biological step or active transport could take much longer time to complete. In this case, since the algae were dried and biological functions were no longer active, the sorption could only take place on the surface of the cell. Therefore the sorption equilibrium took place quickly within 20 min and no further sorption was observed thereafter [39, 40].

The effect of residence time on metal uptake was studied with 0.30 g algae in 25 mL lead solution 200 (ppm) at

optimum pH 3.0 using sodium acetate buffer and variable residence time (2–120 min), in which each flask was periodically shaken at 200 r.p.m for 2 min. This experiment shows that algae have bound most of the metal after 20 min and that equilibrium was reached after maximum metal uptake 88.94% (capacity 14.49 mg/g). This result is in parallel with those previously obtained with different algae and fungi for heavy metal uptake [37].

The linear equation for pseudosecond order model [41] is as follows:

$$\frac{t}{q_t} = \frac{1}{k_2 \left(q_e\right)^2} + \frac{t}{q_e}, \tag{1}$$

where q_e and q_t in mg/g are the amount of lead adsorbed at equilibrium and at time t, respectively. k_2 (g·mg^{-1}·min^{-1}) is the rate constant of pseudosecond order adsorption. The plot of t/q_t versus t (min) gave a linear relationship from which q_e was determined from the slope and k_2 from the intercept of the plot, Figure 2. The values of k_2, q_e, and R^2 are 0.605, 14.95, and 1.0, respectively. $R^2 = 1.0$ for the plot clearly indicates that adsorption kinetics is best described as pseudosecond order.

4.3.5. Swelling Behavior. The volume of swollen particles increased during the first 20 minutes and then it remained constant during 2 hours. The values of DI, Q, and V_{AS} of the biomass particles are presented in Table 5. The values for algae modified with silicates (EM) and algae after reflux (ER) indicate an improvement of stability and mechanical properties of the biomass and better flow rate [27].

4.3.6. Adsorption Isotherm Calculations. Adsorption is usually modeled by isotherms which relate the relative concentrations of solute adsorbed to the solid (Q_e) and in solution (C_e). The equilibrium data were analyzed using most commonly used nonlinear isotherms equations: Langmuir, Freundlich, Redlich-Peterson, Temkin, Elovich, and Dubinin-Radushkevich models, Table 1, Figure 3 [31–33].

TABLE 5: Swelling characteristics of different algae biomass.

Type of biomass	Particle size (mesh)	W_D (g)	W_s (g)	Q	V_s cm^3	DI cm^3/g	V_{AS}
Enteromorpha algae (E)	0.5	1.442	10.41	7.219	10	6.934	6.219
Enteromorpha algae after reflux (ER)	0.5	1.442	6.876	4.775	7	4.166	3.775
Enteromorpha algae modified with silicates (EM)	0.5	4.320	11.54	3.687	10.5	2.430	1.671
Swedish wood sawdust (SWS)	0.5	1.442	8.92	6.185	10	6.935	5.185

TABLE 6: Results of nonlinear approach of Langmuir, Freundlich, Temkin, Elovich, Redlich-Peterson, and Dubinin-Radushkevich isotherm constants for the adsorption of lead at 25°C onto (a) raw *Enteromorpha* (E); (b) *Enteromorpha* modified with sodium silicates (EM).

(a)

Sorbent (E)								
Langmuir			Freundlich			Temkin		
Q^0 calc (mg/g)	K_L (L/mg)	R^2	K_F (L/g)	n	R^2	B_T (J/mol)	K_T (L/mg)	R^2
83.82	0.0105	0.879	0.3411	0.8576	0.9593	14.342	0.143	0.8562
χ^2	SSE		χ^2	SSE		χ^2	SSE	
20.66	165.29		6.970	55.76		24.651	197.21	
Elovich			Redlich-Peterson			Dubinin-Radushkevich		
Q_E (mg/g)	K_E (L/mg)	R^2	A_{RP} (L/mg)$^\beta$	β_{RP}	K_{RP} (L/g)	Q_m (mg/g)	β (mol^2/J^2)	E (J/mol)
42.8513	−0.00786	0.52311	1.3644	0.5525	0.1754	37.92	−8 × 10^{-5}	79.06
χ^2	SSE		R^2	χ^2	SSE	R^2	χ^2	SSE
0.02838	0.2270		0.8160	36.16	253.14	0.9532	8.0182	64.15

(b)

Sorbent (EM)								
Langmuir			Freundlich			Temkin		
Q^0 calc (mg/g)	K_L (L/mg)	R^2	K_F (L/mg)	n	R^2	B_T (J/mol)	K_T (L/mg)	R^2
1433.46	0.00157	0.9872	6.7298	1.3555	0.9925	186.746	0.0373	0.9260
χ^2	SSE		χ^2	SSE		χ^2	SSE	
415.336	3322.7		241.09	1928.74		2392	19136	
Elovich			Redlich-Peterson			Dubinin-Radushkevich		
Q_E (mg/g)	K_E (L/mg)	R^2	A_{RP} (L/mg)$^\beta$	β_{RP}	K_{RP} (L/g)	Q_m (mg/L)	β (mol^2/kj^2)	E (J/mol)
754.41	0.00366	0.7515	10.4550	0.313	0.9915	557.44	−0.00244	14.32
χ^2	SSE		R^2	χ^2	SSE	χ^2	SSE	R^2
0.0607	0.4853		0.9924	282.1	1974.5	5440.24	43522	0.8317

The data is well fitted with *Langmuir model* indicating the presence of chemical monolayer adsorption onto a surface with a finite number of similar active sites. The calculated maximum sorbate uptake Q^0 (mg/g) is 83.82 for nonmodified algae (E) and 1433.46 mg/g for algae modified with sodium silicates (EM). The K_L and R_L values are also calculated where $R_L = 1/(1 + K_L C_0)$ and K_L is a coefficient related to the affinity between the sorbent and sorbate. C_0 is the maximum initial concentration. The R_L value implies that adsorption to be unfavorable for ($R_L > 1$), linear for ($R_L = 1$), and favorable for ($O < R_L < 1$), or irreversible ($R_L = 0$). Adsorption of lead is favorable since the obtained value of R_L for the materials (E) is (0.0872) and for (EM) is (0.389), Table 6. The ability of *Freundlich model* was examined. It is the best suited model in our study indicating that sorption occurs onto a heterogeneous surface. It was also reported that Freundlich model was applicable in adsorption studies of lead with *Enteromorpha* algae [37]. It is of importance to realize that the values of $1/n$ greater than unity in the case of (E) algae indicate formation of multilayer of metal on the surface of biomass, Table 6.

Redlich-Peterson model is also suitable model, where A_{RP}, K_{RP}, and β are parameters to be estimated, Table 1. The obtained β values lie in the middle between 0 and 1 indicating that adsorption includes both Freundlich and Langmuir features, Table 6. *Temkin* model explains sorbent/sorbate interactions in relation to heat of adsorption, where B_T is a factor related to the heat of adsorption and K_T is Temkin equilibrium constant (L/mg). The B_T for (EM) 186.75 J/mol is much greater than 14.34 J/mol for (E), Tables 1 and 6.

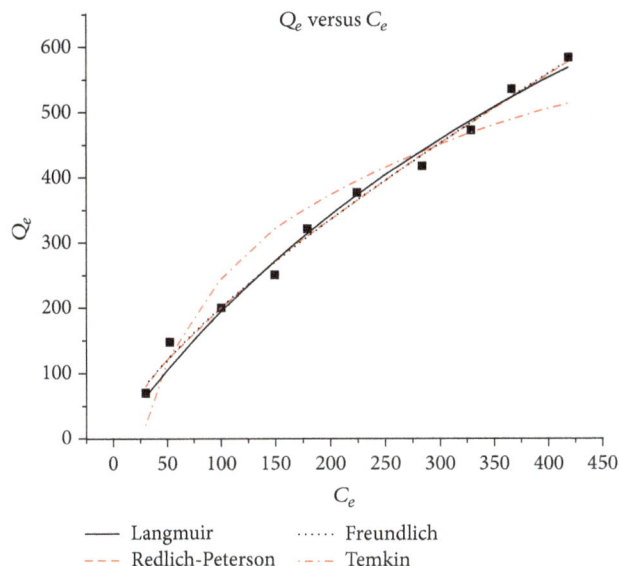

FIGURE 3: The curves of the four nonlinear isotherm models with the experimental data at $t = 25°C$, for adsorption of lead onto *Enteromorpha* modified with sodium silicates (EM).

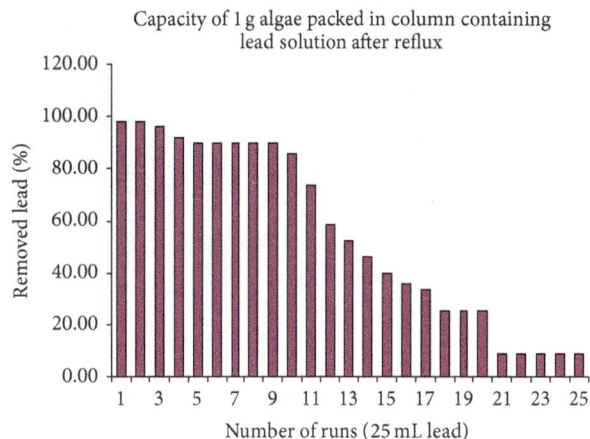

FIGURE 4: Saturation of the column after passage of 25 aliquots of 25 mL lead (200 ppm) for a column filled with 1 g *Enteromorpha* algae residue after reflux (ER).

Elovich model deals with multilayer adsorption, based on a kinetic principle that adsorption site increase exponentially with adsorption, where K_E is Elovich equilibrium constant and Q_E is Elovich maximum adsorption capacity. The low R^2 value indicates than Elovich model is not applicable to lead adsorption onto (E) and (EM) in Tables 1 and 6.

Dubinin-Radushkevich isotherm, where Q_m is the maximum amount of ions sorbed onto algae (mg/g), β (mol^2/kJ^2) is a constant related to the sorption energy $E = 1/\sqrt{-2\beta}$ in kJ/mol, Table 1. The obtained values of E 14.32 kJ/mol for modified algae (EM) and 79.06 kJ/mol for nonmodified algae (E) are greater than 8 kJ/mol, indicating that the adsorption process is chemical in nature, Table 6.

4.3.7. Comparative with Literatures.

However, the results obtained for metal uptake (q) of lead cations using dead algae *E. intestinalis* (E) in agitation experiments at optimum conditions show a medium uptake of *E. intestinalis* for these cations in comparison to the other published data, Table 7 [43–47, 49–52]. However, the highest adsorption capacity 1433 mg/g was obtained for *Enteromorpha* algae modified with sodium silicates (EM) indicating the presence of strong electrostatic force of attraction between lead ions and binding-sites on the surface of the silicates-algae modified material. This indicates that silicates support greatly improves the adsorption capacity compared to nonmodified algae.

4.4. Column Studies.

The batch experiments have demonstrated the ability of green algae *Enteromorpha* to bind lead ions from solution. A practical approach to decontaminate metal-polluted water with the biomass would be to pass the contaminated water through a column contain the biomass. This experiment was performed using different biomass in which the biomass was packed in the column by rewetting it in deionized water.

4.4.1. Column Capacity.

The breakthrough curve of lead was determined using column packed uniformly with 1.0 g of *Enteromorpha* algae materials (E), (ER), or (EM). Aliquots were collected from the column and analyzed for lead till reaching saturation, Figure 4. The column capacity q_e was calculated from the area A under the breakthrough curve (Figure 5) according to the equations presented in Table 8.

(i) The experiment using column filled with (E) showed that 67.4 mg/g of lead was removed after passage of 525 mL solutions of lead (200 ppm) and that the percent removal is 64.2%. This result is in parallel with those recently obtained with different microalgae *Chlamydomonas reinhardtii* for heavy metal uptake [45].

(ii) In the case of column filled with (ER), saturation occurs after passage of 625 mL of lead (200 ppm). The column capacity and percent removal are 66.3 mg/g and 53.04%, respectively.

(iii) Column studies using sodium silicates modified algae (EM) indicate higher adsorption capacity and % removal (230.20 mg/g and 60.6%, resp.). Saturation occurred after passage of 380 mL of lead (1000 ppm).

4.4.2. Column Models.

Three kinetic coulmn models (Thomas, Yoon-nelson, and Yan and Clark) (Table 2) were used to analyze column performance using the data for the adsorption of lead (1000 ppm) with 1 gm (EM) materials.

Thomas model assumes negligible axial dispersion in the column adsorption since the rate driving force obeys the second order reversible reaction kinetics. K_{Th} (mL/mg·min) is Thomas rate constant, q_T (mg/g) is the equilibrium adsorption capacity, m (g) is the amount of adsorbent in the column, Q (mL/min) is the volumetric flow rate, and C_o (mg/L) is the initial concentration of the adsorbate in the feed solution [48].

Yoon-Nelson developed a model based on the assumption that the rate of decrease in the probability of adsorption for

TABLE 7: Lead uptake: optimum results obtained from batch experiments, comparison with literature.

Name of algae—color	pH	Contact time (min)	Metal uptake (q) mg/g	Reference
E. intestinalis (E)—green	3.0	5–20	83.82	Present work
E. intestinalis modified with sodium silicates (EM)—green	7.0	60	1433.46	Present work
Caulerpa lentillifera—green	5.0	10–20	28.71	[43]
Cystoseira baccata—brown	4.5	—	186	[44]
Chlamydomonas reinhardtii—green	5.0	60	96.30	[45]
Sargassum sp.—brown	5.0	60	183.37	[46]
Ulva lactuca—green	4-5	5–30	126.5	[47]
Enteromorpha spp. (living)—green	—	0–5	20.72	[28]
Anabaena sphaerica—blue green	3.0	90	121.95	[52]
Mixture of Ulva lactuca, Jania Rubens, and Sargassum (green, red, and brown)	4.0	120	281.80	[36]
Ecklonia radiata—brown	5.0	10	282	[49]
Codiumtaylori—green	3.5	—	130.00	[50]
Sargassum natans—brown	3.5	—	220.00	[51]

TABLE 8: Column adsorption capacity and parameters calculation of a column loaded with (a) 1 gm of (E), (b) 1 gm of (ER), eluted with MG 200 ppm at a flow rate Q of 2.5 mL/min, and (c) with 1 gm (EM) eluted with MG 1000 ppm at Q = 2.0 mL/min.

Parameter	$A =$	$q_{total} =$	$m_{total} =$	% total removal =	$q_e =$	$C_e =$
Equation	$\int_{t=o}^{t=t_{total}} C_{ads} dt$	$\dfrac{QA}{1000}$	$\dfrac{C_o Q t_{total}}{1000}$	$\dfrac{q_{total}}{m_{total}} \times 100$	$\dfrac{q_{tot}}{m}$	$C_o - \dfrac{A}{t_{total}}$
Unit	mg·min/L	mg	mg	%	mg/g	mg/L
(a) (E)	26948	67.37	105	64.16	67.37	71.68
(b) (ER)	26520	66.30	125	53.04	66.30	93.92
(c) (EM)	115095	230.20	380	60.58	230.19	394.24

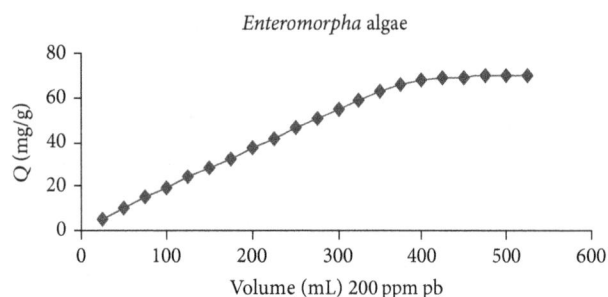

FIGURE 5: Breakthrough curve: capacity of 1 gram of *Enteromorpha* algae (E) at pH 3 and flow rate 2.5 mL/min with 200 ppm lead solution (particle size 0.5 mesh).

each adsorbate molecule is proportional to the probability of adsorbate adsorption and the probability of adsorbate breakthrough on the adsorbent. K_{YN} (min^{-1}) is the rate constant and τ_{YN} (min) is the time required for 50% adsorbate breakthrough [53].

Yan et al. model is described in Table 2, where K_Y (L/min·mg) is the kinetic rate constant for Yan et al. model and q_Y (mg/g) is the maximum adsorption capacity of adsorbent estimated by Yan model [42].

Thomas, Yoon-Nelson, and Yan and Clark models were applied using linear and nonlinear regression statistical techniques according to the equations listed in Table 2. The values

of the maximum capacity q, the rate constant K, and error functions for each model are presented in Table 9. The nonlinear methods having higher R^2 values would be more appropriate techniques in predicting the adsorption column models, for nonlinear methods.

Comparing the values of χ^2, SSE, and R^2, we find that Thomas and Yoon-Nelson models describe better the adsorption behavior than Yan model for column packed with (E) and (ER). However, Yan model was found the best for column packed with (EM). The predicted capacity value is $q_T = q_{YN} = 76.2$ mg/g for column (E) and =71.1 mg/g in the case of column (ER), while the capacity q_Y is equal to 982.5 for column (EM). Again algae anchoring onto silicates improved the performance capacity of column more than 10 times compared to nonmodified algae materials. Application of Yoon-Nelson model indicated that τ the half-life of adsorbate breakthrough is equal to 152.35 min and 142.10 min for column (E) and (ER), respectively using nonlinear approach, Table 9.

5. Conclusion

(i) The amount of lead uptake increased steeply by increasing the weight of *Enteromorpha intestinalis* algae and reached equilibrium state after 20 minutes in batch experiment with g at 25°C. The algae show also high adsorption of lead at optimum conditions

TABLE 9: Statistical evaluation of parameters of a column loaded with 1 gm (EM) and eluted with 1000 ppm lead using linear and nonlinear fit of breakthrough data to Thomas, Yoon-nelson, and Yan and Clark kinetic models.

Adsorbent		Linear approach				Nonlinear approach				
Thomas model		K_{Th} mL·(mg·min)$^{-1}$	q_T (mg·g^{-1})	R^2	SD	K_{Th} mL·(mg·min)$^{-1}$	q_T (mg·g^{-1})	R^2	SSE	χ^2
	(E)	1.72×10^{-4}	69.25	0.8305	0.8988	3.1×10^{-4}	76.18	0.9720	0.07145	3.8×10^{-3}
	(ER)	1.32×10^{-4}	72.14	0.9677	0.3529	1.4×10^{-4}	71.08	0.9827	0.0496	0.0022
	(EM)	7.15×10^{-6}	375.9	0.6994	0.2715	6.0×10^{-6}	396.78	0.7422	0.0397	0.0023
Yoon-Nelson model		K_{YN} (min^{-1})	τ_{YN} (min) q_{YN} (mg·g^{-1})	R^2	SD	K_{YN} (min^{-1})	τ_{YN} (min) q_{YN} (mg·g^{-1})	R^2	SSE	χ^2
	(E)	0.0344	138.50 / 69.26	0.8305	0.8988	0.0615	152.35 / 76.18	0.9720	0.07145	0.0038
	(ER)	0.0264	144.18 / 72.10	0.9651	0.3529	0.0280	142.10 / 71.05	0.9819	0.04839	0.0022
	(EM)	0.0072	187.96 / 375.92	0.6994	0.2715	0.0061	198.40 / 396.80	0.7422	0.03969	0.0023
Yan's model		K_Y L·(mg·min)$^{-1}$	q_Y (mg·g^{-1})	R^2	χ^2	K_Y L·(mg·min)$^{-1}$	q_Y (mg·g^{-1})	R^2	SSE	χ^2
	(E)	4.0×10^{-5}	21.83	0.6412	1.709	1.2×10^{-4}	7.58	0.9658	0.0874	0.0046
	(ER)	3.3×10^{-5}	23.73	0.8525	0.0526	5.0×10^{-5}	17.16	0.9818	0.0484	0.0022
	(EM)	1.5×10^{-6}	817.34	0.9365	0.0156	1.0×10^{-6}	982.46	0.9387	0.0094	5.5×10^{-4}

adsorbent (0.3 g), pH (3.0), and lead concentration (200 ppm) and followed Freundlich model adsorption isotherm. The Langmuir adsorption capacity of lead was 83.8 mg/g using nonmodified algae (E) and 1433.5 mg/g using algae modified with silicates (EM). Comparative study of thermal analysis and FTIR of algae materials clearly proves the presence of adsorbed lead in the algae materials.

(ii) Breakthrough area method indicated that 64.2% and 53.0% of lead can be removed from 200 ppm eluting solution using column filled with (E) and algae after reflux (ER), respectively, while, for column filled with sodium silicates modified algae (EM), the % removal was 60.6% of lead from 1000 ppm. Thomas and Yoon-Nelson models describe better the adsorption behavior than Yan model for column packed with (E) and (ER). The predicted capacity value was 76.2 mg/g for column (E) and 71.1 mg/g in the case of column (ER). However, Yan model was the best for column packed with (EM), with a capacity 982.5 mg/g, indicating that sodium silicates modification of algae improved the performance capacity of column. Moreover, (ER) and (EM) show less swelling and better flow rate control than (E).

(iii) The calculation of various error functions proved to be valuable for predicting the best adsorption isotherms, kinetics, and column models. The nonlinear methods would be more appropriate techniques in predicting the best adsorption models.

Conflict of Interests

The authors declare that there is no conflict of interests regarding the publication of this paper.

References

[1] J. Dojlido and G. A. Best, *Chemistry of Water and Water Pollution*, Ellis Horwood, New York, NY, USA, 1993.

[2] A. Stafiej and K. Pyrzynska, "Adsorption of heavy metal ions with carbon nanotubes," *Separation and Purification Technology*, vol. 58, no. 1, pp. 49–52, 2008.

[3] R. H. Crist, K. Oberholser, J. McGarrity, D. R. Crist, J. K. Johnson, and J. M. Brittsan, "Interaction of metals and protons with algae. 3. Marine algae, with emphasis on lead and aluminum," *Environmental Science and Technology*, vol. 26, no. 3, pp. 496–502, 1992.

[4] B. Volesky and I. Prasetyo, "Cadmium removal in a biosorption column," *Biotechnology and Bioengineering*, vol. 43, no. 11, pp. 1010–1015, 1994.

[5] J. R. Boulding, *EPA Environmental Engineering Sourcebook*, Ann Arbor Press, 1996.

[6] R. H. Christ, J. R. Martin, and D. R. Christ, "Ionic mechanisms for heavy metal removal as sulphides and hydroxides," in *Mineral Bioprocessing*, R. W. Smith and M. Misera, Eds., pp. 275–287, The Mineral, Metals & Materials Society, Warrendable, Pa, USA, 1991.

[7] R. Ofer, A. Yerachmiel, and Y. Shmuel, "Marine macroalgae as biosorbents for cadmium and nickel in water," *Water Environment Research*, vol. 75, no. 3, pp. 246–253, 2003.

[8] N. Kuyucak and B. Volesky, "Biosorbents for recovery of metals from industrial solutions," *Biotechnology Letters*, vol. 10, no. 2, pp. 137–142, 1988.

[9] M. T. K. Tsui and W.-X. Wang, "Temperature influences on the accumulation and elimination of mercury in a freshwater cladoceran, Daphnia magna," *Aquatic Toxicology*, vol. 70, no. 3, pp. 245–256, 2004.

[10] G. Lozano, A. Hardisson, A. J. Gutiérez, and M. A. Lafuente, "Lead and cadmium levels in coastal benthic algae (seaweeds) of Tenerife, Canary Islands," *Environment International*, vol. 28, no. 7, pp. 627–631, 2003.

[11] X. Sort and J. M. Alcañiz, "Effects of sewage sludge amendment on soil aggregation," *Land Degradation & Development*, vol. 10, no. 1, pp. 3–12, 1999.

[12] M. Hanbali, H. Holail, and H. Hammud, "Remediation of lead by pretreated red algae: adsorption isotherm, kinetic, column modeling and simulation studies," *Green Chemistry Letters and Reviews*, vol. 7, no. 4, pp. 342–358, 2014.

[13] I. Sastre, M. A. Vicente, and M. C. Lobo, "Influence of the application of sewage sludges on soil microbial activity," *Bioresource Technology*, vol. 57, no. 1, pp. 19–23, 1996.

[14] ISSS Working Group RB, *World Reference Base for Soil Resources: Atlas*, ISRIC-FAO-ISSS Acco, Leuven, Belgium, 1998.

[15] USEPA, "Method 3051a: Microwave assisted acid dissolution of sediments, Sludges, soils, and oils," 2nd ed U.S. Gov. Print. Office, Washington, DC, USA, 1997.

[16] M. Schnitzer and S. U. Khan, *Soil Organic Matter*, Elsevier, Amsterdam, The Netherlands, 1978.

[17] G. Gascó, M. J. Martínez-Iñigo, and M. C. Lobo, "Soil organic matter transformation after a sewage sludgeapplication," *Electronic Journal of Environmental, Agricultural and Food Chemistry*, vol. 3, no. 4, pp. 716–722, 2004.

[18] T. C. Tan, C. K. Chia, and C. K. Teo, "Uptake of metal ions by chemically treated human hair," *Water Research*, vol. 19, no. 2, pp. 157–162, 1985.

[19] B. Coupal and J. M. Lalancette, "The treatment of waste waters with peat moss," *Water Research*, vol. 10, no. 12, pp. 1071–1076, 1976.

[20] S. K. Srivastava, R. Tyagi, and N. Pant, "Adsorption of heavy metal ions on carbonaceous material developed from the waste slurry generated in local fertilizer plants," *Water Research*, vol. 23, no. 9, pp. 1161–1165, 1989.

[21] A. G. Rowley, F. M. Husband, and A. B. Cunningham, "Mechanisms of metal adsorption from aqueous solutions by waste tyre rubber," *Water Research*, vol. 18, no. 8, pp. 981–984, 1984.

[22] D. K. Singh, D. P. Tiwari, and D. W. Saksena, "Removal of Pb from aqueous Solution by chemically treated used tea leaves," *Water Research*, vol. 3, pp. 169–177, 1993.

[23] P. Donghee, *Chromium removal by biosorption using Ecklonia seaweed Biomass [M.S. thesis]*, POSTECH, Gyeongsangbuk-do, Republic of Korea, 2002.

[24] Y. S. Yun, D. Park, J. M. Park, and B. Volesky, "Biosorption of trivalent chromium on the brown seaweed biomass," *Environmental Science and Technology*, vol. 35, no. 21, pp. 4353–4358, 2001.

[25] N. Kuyucak and B. Volesky, in *Biosorption of Heavy Metals*, B. Volesky, Ed., pp. 173–198, CRC Press, Boca Raton, Fla, USA, 1990.

[26] H. A. Waldern, *Stofen Sub-Clinic Lead Poisoning*, Academic Press, New York, NY, USA, 1974.

[27] Z. R. Holan, B. Volesky, and I. Prasetyo, "Biosorption of cadmium by biomass of marine algae," *Biotechnology and Bioengineering*, vol. 41, no. 8, pp. 819–825, 1993.

[28] I. Tüzün, G. Bayramoğlu, E. Yalçın, G. Başaran, G. Çelik, and M. Y. Arıca, "Equilibrium and kinetic studies on biosorption of Hg(II), Cd(II) and Pb(II) ions onto microalgae Chlamydomonas reinhardtii," *Journal of Environmental Management*, vol. 77, no. 2, pp. 85–92, 2005.

[29] M. Spinti, H. N. Zhuang, and E. M. Trujillo, "Evaluation of immobilized biomass beads for removing heavy metals from wastewaters," *Water Environment Research*, vol. 67, no. 6, pp. 943–952, 1995.

[30] R. S. Laxman and S. More, "Reduction of hexavalent chromium by *Streptomyces griseus*," *Minerals Engineering*, vol. 15, no. 11, pp. 831–837, 2002.

[31] H. H. Hammud, L. Fayoumi, H. Holail, and M. E. El-Sayed, "Biosorption studies of methylene blue by mediterranean algae carolina and its chemically modified forms. Linear and non-linear models' prediction based on statistical error calculation," *International Journal of Chemistry*, vol. 3, no. 4, pp. 147–163, 2011.

[32] H. H. Hammud, M. Chahine, B. El Hamaoui, and Y. Hanifehpour, "Lead uptake by new silica-carbon nanoparticles," *European Journal of Chemistry*, vol. 4, no. 4, pp. 432–440, 2013.

[33] I. Abbas, H. H. Hammud, and H. Shamsaldeen, "Calix[4]pyrrole macrocycle: extraction of fluoride anions from aqueous media," *European Journal of Chemistry*, vol. 3, no. 2, pp. 156–162, 2012.

[34] H. H. Hammud, M. E. Mansour, S. Shaalan, E. Khamis, and A. El-Shaar, "Adsorption of mercuric ion by marine algae enteromorpha," *International Journal of Applied Chemistry*, vol. 2, no. 2, pp. 87–102, 2006.

[35] J. L. Zhou, R. J. Kiff, and J. Chem, "The uptake of copper from aqueous solution by immobilized fungal biomass," *Journal of Chemical Technology and Biotechnology*, vol. 52, no. 3, pp. 317–330, 1991.

[36] A. A. Hamdy, "Removal of Pb^{2+} by biomass of marine algae," *Current Microbiology*, vol. 41, no. 4, pp. 239–245, 2000.

[37] A. A. Hamdy, "Biosorption of heavy metals by marine algae," *Current Microbiology*, vol. 41, no. 4, pp. 232–238, 2000.

[38] P. Ahuja, R. Gupta, and R. K. Saxena, "Sorption and desorption of cobalt by *Oscillatoria anguistissima*," *Current Microbiology*, vol. 39, no. 1, pp. 49–52, 1999.

[39] H. Kojima and K. Y. Lee, *Photosynthetic Microorganisms in Environmental Biotechnology*, Springer, Hong Kong, 2001.

[40] B. Volesky, *Biosorption of Heavy Metals*, CRC Press, Boca Raton, Fla, USA, 1990.

[41] Y. S. Ho and G. McKay, "Pseudo-second order model for sorption processes," *Process Biochemistry*, vol. 34, no. 5, pp. 451–465, 1999.

[42] G. Y. Yan, T. Viraraghavan, and M. Chen, "A new model for heavy metal removal in a biosorption column," *Adsorption Science & Technology*, vol. 19, no. 1, pp. 25–43, 2001.

[43] P. Pavasant, R. Apiratikul, V. Sungkhum, P. Suthiparinyanont, S. Wattanachira, and T. F. Marhaba, "Biosorption of Cu^{2+}, Cd^{2+}, Pb^{2+}, and Zn^{2+} using dried marine green macroalga *Caulerpa lentillifera*," *Bioresource Technology*, vol. 97, no. 18, pp. 2321–2329, 2006.

[44] P. Lodeiro, J. L. Barriada, R. Herrero, and M. E. Sastre de Vicente, "The marine macroalga *Cystoseira baccata* as biosorbent for cadmium(II) and lead(II) removal: kinetic and equilibrium studies," *Environmental Pollution*, vol. 142, no. 2, pp. 264–273, 2006.

[45] R. Herrero, P. Lodeiro, C. Rey-Castro, T. Vilariño, and M. E. Sastre de Vicente, "Removal of inorganic mercury from aqueous solutions by biomass of the marine macroalga *Cystoseira baccata*," *Water Research*, vol. 39, no. 14, pp. 3199–3210, 2005.

[46] P. X. Sheng, Y.-P. Ting, J. P. Chen, and L. Hong, "Sorption of lead, copper, cadmium, zinc, and nickel by marine algal biomass: characterization of biosorptive capacity and investigation of mechanisms," *Journal of Colloid and Interface Science*, vol. 275, no. 1, pp. 131–141, 2004.

[47] R. Jalali, H. Ghafourian, Y. Asef, S. J. Davarpanah, and S. Sepehr, "Removal and recovery of lead using nonliving biomass of

marine algae," *Journal of Hazardous Materials*, vol. 92, no. 3, pp. 253–262, 2002.

[48] J. R. Rao and T. Viraraghavan, "Biosorption of phenol from an aqueous solution by *Aspergillus niger* biomass," *Bioresource Technology*, vol. 85, no. 2, pp. 165–171, 2002.

[49] J. T. Matheickal and Q. Yu, "Biosorption of lead from aqueous solutions by marine algae *Ecklonia radiata*," *Water Science and Technology*, vol. 34, no. 9, pp. 1–7, 1996.

[50] B. Volesky and Z. R. Holan, "Biosorption of heavy metals," *Biotechnology Progress*, vol. 11, no. 3, pp. 235–250, 1995.

[51] Z. R. Holan and B. Volesky, "Biosorption of lead and nickel by biomass of marine algae," *Biotechnology and Bioengineering*, vol. 43, no. 11, pp. 1001–1009, 1994.

[52] A. M. Abdel-Aty, N. S. Ammar, H. H. Abdel Ghafar, and R. K. Ali, "Biosorption of cadmium and lead from aqueous solution by fresh water alga *Anabaena sphaerica* biomass," *Journal of Advanced Research*, vol. 4, no. 4, pp. 367–374, 2013.

[53] Y. H. Yoon and J. H. Nelson, "Application of gas adsorption kinetics I. A theoretical model for respirator cartridge service life," *The American Industrial Hygiene Association Journal*, vol. 45, no. 8, pp. 509–516, 1984.

7

Determination of Tannins of Three Common *Acacia* Species of Sudan

Isam Eldin Hussein Elgailani and Christina Yacoub Ishak

Department of Chemistry, Faculty of Science, University of Khartoum, P.O Box 321, 11115 Khartoum, Sudan

Correspondence should be addressed to Isam Eldin Hussein Elgailani; gailani23@hotmail.com

Academic Editor: Armando Zarrelli

The objective of this study is to analyze and compare tannins of three common *Acacia* species of Sudan, since vegetable tannins are important in leather industry. *Acacia nilotica* and *Acacia seyal* samples were collected from Sunt Forest in Khartoum State, while *Acacia senegal* samples were collected from the Debabat Forest in South Kordofan State. Bark samples from bulk collections of the three *Acacia* species were extracted with boiled deionized water. The amount of tannins present in these bulk samples was determined by Folin-Denis method for total phenolic materials, followed by precipitation with hide-powder. The difference between the amount of phenolic materials present before and after addition of hide-powder represents the amount of tannins present. The percentage of tannins in the leaves, bark, and mature and immature fruits of collections of individuals of *Acacia* species was estimated; mature and immature fruits of *Acacia nilotica* contain tannins (22.15% and 22.10%, resp.). The leaves of *Acacia nilotica* and *Acacia seyal* contain tannins (11.80% and 6.30%, resp.). The barks of *Acacia seyal*, *Acacia nilotica*, and *Acacia senegal* contain tannins (12.15%, 10.47%, and 3.49%, resp.).

1. Introduction

Tannins are amorphous, astringent substances occurring widely in the bark, wood, leaves, and resinous exudations of plants [1, 2]. They are water-soluble phenolic compounds which occur widely in vascular plants [3]. The term was introduced by Seguim in 1796 to describe the substances present in a number of vegetable extracts which possessed the property of converting animal skins into leather [4]. Most authors prefer to speak of "tannin extracts" rather than "tannin." The tannins are colourless and noncrystalline substances which form colloidal solutions in water; these solutions have an astringent taste [5, 6]. The astringency of tannins, that is, their efficiency as precipitants of proteins in the mouth causing the sensation of astringency [7–9], is determined by their reaction with salivary proteins in the oral cavity [10]. Astringency and tanning properties are associated with the higher molecular weight proanthocyanidins (condensed tannins) [11]. Tannins are polymeric phenolic compounds with numerous hydroxyl groups and quite diverse in chemical structure [12, 13]. Hydrolysis of some of the tannins yields

the simple, seven-carbon gallic acid and others give ellagic acid or other phenolic acids [14, 15]. Tannins are generally divided into hydrolyzable and condensed tannins. Molecular weight as high as 20,000 has been reported for condensed tannins. The molecular weights of hydrolyzable tannins range from 500 to 5,000 [13, 16–18]. Beside the variation from plant to plant, and from one part of a plant to another, the concentration of tannins in any one organ varies with time [19]. The use of vegetable tannins to tan hides and produce leather predates written history. Hides are usually tanned by either a mineral [20] or vegetable process, depending on the type of animal and the extended use of the leather. In Sudan, approximately 11,400,000 kg of cattle hides and 3,750,000 kg of sheep hides are tanned each year (by both processes). Vegetable tannins consumption in Sudan varies between 350 and 400 tons per year, and a large proportion is locally produced as the *Acacia* species is more distributed in Sudan (data were obtained from the National Centre for Leather Technology and Khartoum Tannery). One of the best sources of tannins is *Acacia* species which belong to family of Leguminosae in plant kingdom. There are about 800 species of the genus *Acacia*. They are abundant in savannas and arid regions [21]. The commercial wattle grown in Kenya (*Acacia mearnsii*) is a well-known tannin-rich species and tannin-based adhesive [22, 23]. Tannins are complexed with the proteins of the hide and become an integral part of the final product. The ability of tannins to complex with proteins is largely responsible for the production of leather from hide [24]. In this work, we initiated this study in order to identify sources of tannins that grow in Sudan and to determine the amount of tannins present and the distribution of these compounds in different parts of the plants involved. Three common species, *Acacia nilotica*, *Acacia seyal*, and *Acacia senegal*, were selected for study.

2. Material and Methods

2.1. Study Area. *Acacia nilotica* and *Acacia seyal* samples were collected from the Sunt Industrial and Tourism Centre (Sunt Forest), about 1 kilometer south of the White Nile Bridge near the junction of White Nile and Blue Nile on the eastern bank of the White Nile River at Khartoum State, while the *Acacia senegal* samples were collected from Debabat Forest in South Kordofan State at West of Sudan.

2.2. Sampling Methodology. Samples of leaves, bark, and mature and immature fruits from individual collections of *Acacia nilotica* (leaves, bark, and mature and immature fruits), *Acacia seyal* (leaves and bark), and *Acacia senegal* (bark only) were used to determine the tannins. Bark was removed from wood before drying. Plant materials were taken from several trees in each instance.

2.3. Chemicals and Reagents. All chemicals and reagents used in this study were of analytical grade.

2.4. Extraction of Bark Samples. Air-dried bark samples (from bulk collections) were ground in a Wiley mill (2 mm screen). A portion (40 g) was extracted with boiled deionized water (200 mL). The samples were filtered (Whatman 1 paper, 18.5 cm disc) and the residual material was rinsed with additional water (2 × 50 mL). Extracts were transferred to a tarred, round-bottomed flask and concentrated under vacuum by rotary evaporator to form a thick extract. The sample extracts were then dried in a vacuum oven at 60°C until a solid material was obtained.

2.5. Determination of Total Phenolic Compounds. Total phenolic compounds were measured in plant samples by the Folin-Denis method [25]. Folin-Denis reagent was prepared by mixing $Na_2WO_4 \cdot 2H_2O$ (20.66 g), dodeca-molybdophosphoric acid (4.13 g), phosphoric acid (85%, 10 mL), and water (150 mL) and permitting the mixture to reflux for two hours. The resulting solution was then diluted to 500 mL. Sodium carbonate solution was prepared by dissolving Na_2CO_3 (106 g) in 1000 mL of water. Solution of tannic acid (6% water content) was then prepared by dissolving the tannin (250 mg) in double distilled water (500 mL). A small amount (2-3 drops) of sodium azide solution (0.1%) was

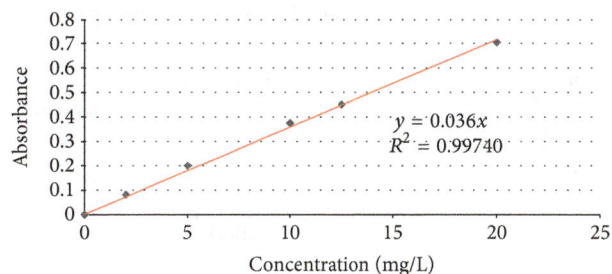

FIGURE 1: Absorbance of solutions of tannic acid as a function of concentrations.

added to prevent contamination by fungi and bacteria. Before use, aliquot of this solution was diluted 1 : 100 with double distilled water.

Folin-Denis reagent (2 mL) was added to an aliquot (2 mL) of the 1 : 100 dilution described above. The solution was shaken vigorously and allowed to stand for three minutes. Sodium carbonate solution (2 mL) was added and the sample again was shaken and allowed to stand for two hours. At that time, the sample was centrifuged at low speed until particulate materials have been removed. The absorbance was measured at 725 nm by UV/VIS Spectrophotometer (Perkin-Elmer 551). A blank was also analyzed in each instance. By a series of dilutions of the tannin solutions prepared (1 : 250, 1 : 50, 1 : 40, 1 : 25, and 1 : 10), a curve was plotted for tannic acid (Figure 1).

The phenolic content of each sample was measured by the Folin-Denis method as described above. Duplicate determinations were carried out for each sample (Tables 1 and 2).

2.6. Determination of Tannins by the Hide-Powder Method. The determination of tannins consists of 4 steps: (i) measurement of total phenolic material in plant samples extracts by Folin-Denis method, (ii) preparation of hydrated, chromed, hide-powder, (iii) absorption of tannins onto hide-powder, and (iv) determination of phenolic materials in the solution remaining after step (iii).

Hydrated hide-powder used in these analyses was prepared from air-dried hide-powder (brought from the National Center for Leather Technology, Khartoum). Sufficient air-dried hide-powder to yield 3.0 g oven-dried hide-powder was used for each analysis performed. The amount of hide-powder necessary to perform the desired number of analyses was allowed to stand with 10 times its weight of distilled water (30 min., 25°C) and was stirred 3 or 4 times during this period. Chromium potassium sulphate (chrome alum, 3% aqueous solution, 1 g/mL hide-powder) was added and the mixture was stirred each 15 minutes for two hours, allowed to stand overnight, and then filtered through a piece of unbleached, white cotton cloth and was squeezed or pressed until the hydrated powder contained about 75% water (when new cloth was used, it was washed to remove sizing and other extraneous materials). The percentage of water in the hydrated hide-powder was determined by weight (4 times the weight of the total hide-powder plus the weight

TABLE 1: Absorbance of solution of *Acacia* species parts before precipitation of tannins with hydrated-hide powder (at 725 nm).

Species	Plant part	Absorbance		
		I	II	Average value
Acacia nilotica	Leaves	0.251	0.255	0.253
Acacia seyal	Leaves	0.128	0.126	0.127
Acacia nilotica	Bark	0.205	0.202	0.203
Acacia seyal	Bark	0.232	0.232	0.232
Acacia senegal	Bark	0.081	0.081	0.081
Acacia nilotica	Mature fruits	0.461	0.463	0.462
Acacia nilotica	Immature fruits	0.439	0.444	0.442

TABLE 2: Absorbance of solutions of *Acacia* species parts after precipitation of tannins with hydrated-hide powder (at 725 nm).

Species	Plant part	Absorbance		
		I	II	Average value
Acacia nilotica	Leaves	0.053	0.053	0.053
Acacia seyal	Leaves	0.308	0.309	0.308
Acacia nilotica	Bark	0.239	0.241	0.240
Acacia seyal	Bark	0.273	0.276	0.274
Acacia senegal	Bark	0.340	0.338	0.339
Acacia nilotica	Mature fruits	0.060	0.065	0.063
Acacia nilotica	Immature fruits	0.056	0.056	0.055

of cloth). The mass of hide-powder was then broken up and redigested with water 4 times (15 minutes each) in amount of water 15 times the weight of hide-powder used. After the final wash, the hide-powder was squeezed to 72.5% water content (determined by weight). Hide-powder prepared in this manner should be refrigerated and used the same day as prepared.

2.6.1. Determination of Water in Hydrated Hide-Powder Samples. An aliquot (10 g) of hydrated hide-powder was removed and placed in an oven (98°C) and dried for 17 hours. The difference in weight was used to calculate the percentage water in the sample.

2.6.2. Precipitation of Tannins with Hydrated Hide-Powder. Freshly prepared, hydrated hide-powder, equivalent to 3.0 g oven-dried hide-powder, was weighed and added to an Erlenmeyer flask (150 mL). Solutions for the determination of tannins by the hide-powder method were prepared by dissolving commercial tannin samples as above in doubled distilled water (500 mg/L). Solutions of plant extracts were prepared in a similar manner. Solution should be stored refrigerated and small amount (2-3 drops) of 0.1% sodium azide solution was added to prevent fungal or bacterial contamination. Solutions were at room temperate when used.

Aliquots of tannin solutions (50 mL) were removed and added to the flasks that contained preweighed hide-powder samples (10.9 g for the conditions outlined above). The flasks were then shaken for ten minutes and the hide-powder was removed by filtration. The mixture was filtered into a flask with plastic Buchner funnel (7.0 cm, Whatman 1 paper) under vacuum. The flask and the sample were washed with double distilled water (10 mL). Cloudy solutions were refiltered. After filtration, the filtrate (about 60 mL) was quantitatively transferred to a volumetric flask and adjusted to 100 mL. Blanks were run with distilled water and with hydrated hide-powder. Aliquots (2 mL) were then removed from each sample and residual phenolic materials determined by the Folin-Denis method (Table 2).

2.7. Determination of Tannins in Leaves, Bark, and Mature and Immature Fruits. Samples of leaves, bark, and mature and immature fruits material of *Acacia nilotica*, *Acacia seyal*, and *Acacia senegal* were ground and extracted. 10.0 g of the quantity ground was extracted with double distilled water (100 mL) in an Erlenmeyer flask (150 mL) by mechanical stirring and heating. The mixture was heated to boiling for 10 minutes and filtered (Whatman 1, 18.5 cm). The flask and filtered material were then rinsed with additional water (15 mL) and the volume was adjusted to 100 mL. Aliquots of this initial extract were then removed and diluted to an appropriate concentration for Folin-Denis analysis, which was carried out in the manner previously described.

For the determination of tannins by the hide-powder method an aliquot of the initial extract was diluted as above (to give 0.05–0.2 absorbance units after precipitation with hide-powder). An aliquot of this solution (50 mL) was then utilized as previously described.

3. Results and Discussions

Folin-Denis reagent is a mixture of phosphotungstic phosphomolybdic acids, and when this labile complex acid is reduced by phenols, a blue tungstic oxide is obtained [26],

TABLE 3: Tannins percentage of *Acacia* species parts.

Species	Plant part	% Phenolics before precipitation	% Phenolics after precipitation	% Tannins (hide-powder)
Acacia nilotica	Leaves	14.00	2.20	11.80
Acacia seyal	Leaves	7.00	0.69	6.31
Acacia nilotica	Bark	11.00	0.53	10.47
Acacia seyal	Bark	12.75	0.60	12.15
Acacia senegal	Bark	4.25	0.76	3.49
Acacia nilotica	Mature fruits	24.75	2.60	22.15
Acacia nilotica	Immature fruits	24.50	2.40	22.10

*Total phenolic content relative to tannic acid.

which was measured spectrophotometrically at 725 nm [18, 24]. The amount of tannins is determined by the preparation of tannins solutions and absorption of the tannins on hydrated, chromed hide-powder. The difference in the phenolic materials, as indicated by the Folin-Denis analysis before and after treatment with hydrated, chromed hide-powder, is utilized to measure tannins. Phenolic compounds of different structures give response to Folin-Denis reagent. All results in this study have been expressed in terms of tannic acid equivalents. Inspection of curve for tannic acid suggests that this technique may underestimate the quantity of tannins present.

As indicated by the Folin-Denis method, the major portion of phenolics in these *Acacia species* consists of tannins (Table 3). Small differences in the amount of residual phenolic materials after precipitation of tannins with hide-powder have a relatively small effect on calculation of the percentage of tannins present in the original plant material.

Comparison of the amount of tannins present in the leaves, bark, and mature and immature fruits of the species was made with aqueous extracts of samples collected from single individual (Table 3). Total phenolic materials were richest in mature and immature fruits, leaves, and bark of *Acacia nilotica* and *Acacia seyal*. They were lower in the bark of *Acacia senegal*. The same trends are observed for tannins. Additional studies will be necessary to estimate variation within and among population of each species. The results showed that mature and immature fruits of *Acacia nilotica* had the highest percentage of tannins (22.15% and 22.10%, resp.), while the leaves and the bark of the same species had 11.8% and 10.47%, respectively. The leaves and the bark of *Acacia seyal* had intermediate values (6.32% and 12.15%, resp.). The bark of *Acacia senegal* had much lower percentage of tannins (3.49%) (Table 3).

4. Conclusions

We can conclude that, among the three *Acacia species* studied, *Acacia nilotica* is the richest in tannins content, and within the *Acacia nilotica* parts, mature and immature fruits were the highest in tannins content, while the barks of the three *Acacia species* were the least. Folin-Denis method for total phenolic materials, followed by precipitation of tannins by hide-powder, is a suitable procedure for evaluation of tannins content.

Recommendations that could be drawn from this study are that additional studies will be necessary to estimate variation within and among population of each species. Suggestions are made for further studies on the possible mode of linkage and the conformation of the molecules of tannins, since condensed tannins occur in plants in various stages of polymerization, and this is very important for the synthesis of tannins. Since tannins are antifungal, antibacterial, and antiviral agents, further studies may also be required to apply them in medicines in wide range.

Conflict of Interests

The authors declare that there is no conflict of interests regarding the publication of this paper.

Acknowledgments

Special thanks are to the Department of Chemistry, University of Khartoum, where the current evaluation and investigation have been carried out, for laboratory facilities and valuable assistance in the use of various types of equipment. Thanks are extended to the National Centre for Leather Technology, Khartoum Tannery and White Nile Tannery, for continuous help during the course of the study.

References

[1] J. Read, *A Text-Book of Organic Chemistry*, G. Bell and Sons, London, UK, 1946.

[2] J. Zivkovic, I. Mujic, Z. Zekovic, G. Nikolic, S. Vidovic, and A. Mujic, "Extraction and analysis of condensed tannins in *Castanea sativa* Mill," *Journal of Central European Agriculture*, vol. 10, no. 3, pp. 283–288, 2009.

[3] J. S. Martin and M. M. Michael, "Tannin assays in ecological studies Precipitation of ribulose-1,5-bisphosphate carboxylase/oxygenase by tannic acid, quebracho, and oak foliage extracts," *Journal of Chemical Ecology*, vol. 9, no. 2, pp. 285–294, 1983.

[4] T. Swain, *Plant Biochemistry*, Academic Press, London, UK, 1965.

[5] P. Bernfield, *Biogenesis of Natural Compounds*, Pergamon Press, New York, NY, USA, 1963.

[6] I. L. Finar, *Organic Chemistry, Volume 2*, Longman, London, UK, 1964.

[7] E. C. Bate-Smith, "Haemanalysis of tannins: the concept of relative astringency," *Phytochemistry*, vol. 12, no. 4, pp. 907–912, 1973.

[8] E. Haslam, "Polyphenol protein interactions," *Biochemical Journal*, vol. 139, no. 1, pp. 285–288, 1974.

[9] E. C. Bate-Smith, "Astringent tannins of the leaves of *Geranium* species," *Phytochemistry*, vol. 20, no. 2, pp. 211–216, 1981.

[10] T. N. Asquith, J. Uhlig, H. Mehansho, L. Putman, D. M. Carlson, and L. Butler, "Binding of condensed tannins to salivary proline-rich glycoproteins: the role of carbohydrate," *Journal of Agricultural and Food Chemistry*, vol. 35, no. 3, pp. 331–334, 1987.

[11] Z. Czochanska, L. Y. Foo, R. H. Newman, and L. J. Porter, "Polymeric proanthocyanidins. Stereochemistry, structural units, and molecular weight," *Journal of the Chemical Society, Perkin Transactions*, vol. 1, pp. 2278–2286, 1980.

[12] V. A. Greulach, *Plant Function and Structure*, Collier Macmillan International Edition, New York, NY, USA, 1973.

[13] A. E. Hagerman, "Tannin—protein interactions," *ACS Symposium Series*, vol. 506, pp. 236–247, 1992.

[14] E. Sondheimer and J. B. Simeone, *Chemical Ecology*, Academic Press, New York, NY, USA, 1970.

[15] T. Okuda and H. Ito, "Tannins of constant structure in medicinal and food plants-hydrolyzable tannins and polyphenols related to tannins," *Molecules*, vol. 16, no. 3, pp. 2191–2217, 2011.

[16] K. Nakanishi, T. Goto, S. Ito, S. Natori, and S. Nozoe, *Natural Products Chemistry*, Academic Press, New York, NY, USA, 1975.

[17] T. C. Somers, "Wine tannins—isolation of condensed flavonoid pigments by gel-filtration," *Nature*, vol. 209, no. 5021, pp. 368–370, 1966.

[18] J. S. Martin and M. M. Martin, "Tannin assays in ecological studies: lack of correlation between phenolics, proanthocyanidins and protein-precipitating constituents in mature foliage of six oak species," *Oecologia*, vol. 54, no. 2, pp. 205–211, 1982.

[19] M. Thomas, S. L. Ranson, and J. A. Richardson, *Plant Physiology*, Longman, London, UK, 1973.

[20] M. Sathiyamoorthy, V. Selvi, D. Mekonnen, and S. Habtamu, "Preparation of eco-friendly leather by process modifications to make pollution free tanneries," *Journal of Engineering Computers & Applied Sciences*, vol. 2, no. 5, pp. 17–22, 2013.

[21] K. O. Rachie and National Academy of Sciences (NAS), *Tropical Legumes: Resources for the Future*, National Academy of Science, Washington, DC, USA, 1979.

[22] J. Z. A. Mugedo and P. G. Waterman, "Sources of tannin: alternatives to wattle (*Acacia mearnsii*) among indigenous Kenyan species," *Economic Botany*, vol. 46, no. 1, pp. 55–63, 1992.

[23] T. Sellers and G. D. Miller, "Laboratory manufacture of high moisture southern pine strandboard bonded with three tannin adhesive types," *Forest Products Journal*, vol. 54, no. 12, pp. 296–301, 2004.

[24] D. S. Seigler, S. Seilheimer, J. Keesy, and H. F. Huang, "Tannins from four common *Acacia* species of Texas and Northeastern Mexico," *Economic Botany*, vol. 40, no. 2, pp. 220–232, 1986.

[25] T. Swain and J. L. Goldstein, "The quantitative analysis of phenolic compounds," in *Methods in Polyphenol Chemistry*, J. B. Pridham, Ed., Permagon, New York, NY, USA, 1964.

[26] F. D. Snell and C. T. Snell, *Colorimetric Methods of Analysis*, vol. 2, D. Van Nostrand Company, New York, NY, USA, 1937.

8

Increasing the Stability of Metal-Organic Frameworks

Mathieu Bosch,[1] **Muwei Zhang,**[1] **and Hong-Cai Zhou**[1,2]

[1] *Department of Chemistry, Texas A&M University, College Station, TX 77842, USA*
[2] *Department of Materials Science and Engineering, Texas A&M University, College Station, TX 77842, USA*

Correspondence should be addressed to Hong-Cai Zhou; zhou@mail.chem.tamu.edu

Academic Editor: Qiaohui Fan

Metal-organic frameworks (MOFs) are a new category of advanced porous materials undergoing study by many researchers for their vast variety of both novel structures and potentially useful properties arising from them. Their high porosities, tunable structures, and convenient process of introducing both customizable functional groups and unsaturated metal centers have afforded excellent gas sorption and separation ability, catalytic activity, luminescent properties, and more. However, the robustness and reactivity of a given framework are largely dependent on its metal-ligand interactions, where the metal-containing clusters are often vulnerable to ligand substitution by water or other nucleophiles, meaning that the frameworks may collapse upon exposure even to moist air. Other frameworks may collapse upon thermal or vacuum treatment or simply over time. This instability limits the practical uses of many MOFs. In order to further enhance the stability of the framework, many different approaches, such as the utilization of high-valence metal ions or nitrogen-donor ligands, were recently investigated. This review details the efforts of both our research group and others to synthesize MOFs possessing drastically increased chemical and thermal stability, in addition to exemplary performance for catalysis, gas sorption, and separation.

1. Introduction

Metal-organic frameworks are composed of metal-containing secondary building units (SBUs) connected by rigid or semirigid polytopic organic linkers. Depending on the geometry and connectivity of the SBUs, this can often create a structure with inherent porosity, with the void volume in the framework initially filled by solvent molecules [1]. These frameworks are often robust enough to survive a desolvation process termed activation by heating and/or vacuum, resulting in materials with extremely high surface areas. Some representative MOFs include MOF-5, MIL-101, HKUST-1, PCN-14, and UiO-66 [2–6]. MOF development has experienced a rapid expansion after the discovery of MOF-2 in 1998 [7]. Even though MOF-2 only possesses a modest porosity, it is one of the earliest MOFs that were demonstrated to have permanent porosity, as it was stable enough to survive solvent exchange with chloroform followed by vacuum desolvation or activation.

Soon after, MOF-5 was published, which is composed of $Zn_4(\mu_4-O)$ SBUs (Figure 1) and benzene-1,4-dicarboxylate (BDC) linkers in a **pcu** topology. Subsequent sorption measurements showed an exceptional surface area, breaking all porosity records at the time. However, while MOF-5 is thermally stable to approximately 300°C, it is not water stable and thus cannot survive long after exposure to humid air [8]. The thermal and chemical stability of MOFs are both of great interest to researchers for multiple reasons. First, a MOF must be stable enough to undergo characterization. For example, a MOF that decomposes quickly is difficult to characterize by X-ray diffraction (XRD), and thus its crystal structure is difficult to determine. A MOF that collapses upon solvent removal cannot be said to be practically "porous," because its experimental surface area and gas sorption cannot be determined. Furthermore, many applications of MOFs, such as gas storage and separation, hinge on their chemical stability, as materials that are not air-stable are often much less cost-effective than materials that may have inferior absolute performance but that are more robust. Some applications may also rely on thermal stability, such as catalysis [9].

The vulnerability of MOFs typically lies in the lability of ligand-metal bonds. According to ligand field theory, because

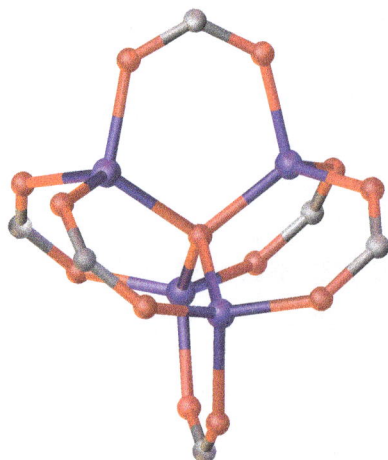

FIGURE 1: The $Zn_4(\mu_4\text{-}O)$ unit of MOF-5 shown coordinated by 6 carboxylates [30].

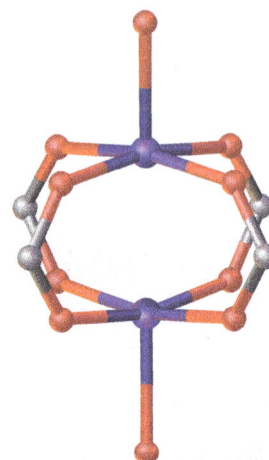

FIGURE 2: The copper paddlewheel unit of the PCN-6X series, found in many MOFs. Typically, it is coordinated equatorially by 4 ligand carboxylates, while the axial ligands are solvent molecules that may be removed by activation. The zinc paddlewheel is almost identical, but unlike the copper paddlewheel, attempted activation usually causes framework collapse.

Zn^{2+} is a transition metal ion with d^{10} electron configuration, it experiences no ligand field stabilization energy overall. Thus, its ligand environment will be controlled primarily by steric factors. This favors a tetrahedral environment but is not heavily destabilized when this environment is disturbed during ligand exchange. Facile ligand exchange allows the formation of a crystalline framework with high surface area and favors the formation of larger single crystals, allowing for more precise structural characterization through single-crystal XRD [10]. However, it also lowers the chemical stability of the resulting MOF, as easily exchanged carboxylates will be displaced by water or other nucleophiles, and if too many of the linking carboxylates are displaced, the framework will collapse. Even if the bond strength is high, if the energy barrier for ligand substitution is low, the MOF will not be chemically stable.

In general, MOFs with SBUs consisting of tetrahedral Zn^{2+} are not chemically stable. However, MOFs containing Zn^{2+} in a different coordination environment, such as MOF-69, have been shown to be more stable than those with purely tetrahedral Zn^{2+} ions. MOF-69 contains both tetrahedral and octahedral Zn centers which share oxygen to form infinite columns and exhibit chemical stability to exchange with a variety of solvents [11].

In order to increase the chemical stability of MOFs, early effort involved the employment of SBUs containing Cu^{2+} ions with d^9 configuration. Compared to a d^{10} transition metal like Zn, metal ions with d^9 configuration possess ligand field stabilization energy, regardless of what coordination environment those ions adopt. This increased ligand field stabilization energy should make the ligand substitution processes less favorable. HKUST-1, composed of copper paddlewheel SBUs (Figure 2) capped by axial water ligands and 1,3,5-benzene-tricarboxylate (BTC), displays an increased chemical stability and shelf-life over MOF-5 [12, 13]. When this material is activated, the water ligand on its axial position will be removed, leaving a relatively stable square planar coordination geometry on cupric SBUs in the activated

sample. However, this material was reported to be unstable to direct contact with water, indicating only a limited increase in stability [14].

Our group has reported the PCN-6X series of (3,24) connected isoreticular MOFs using copper paddlewheel SBUs and extended trigonal planar ligands with isophthalate groups as linkers on each arm, which exhibited both surface areas of up to $5109 \, m^2 \, g^{-1}$ and moderately good stability [15]. The framework structure itself can also impact the stability, as in PCN-61 where the mesoporous structure is stabilized by areas of less open connectivity, that is, the microwindows into the mesopores [16]. The control of framework topology and how it might impact framework stability were comprehensively reviewed elsewhere and will not be further covered here [17]. In general, the chemical stabilities of Cu-paddlewheel based MOFs were improved compared to $Zn_4(\mu_4\text{-}O)$ based MOFs but are still insufficient for certain applications. For example, HKUST-1 was reported to undergo framework collapse under steaming conditions at temperatures above 343 K [12].

2. Framework Templating, Metal-Ion Metathesis, and High-Valent MOFs

Our group has explored a technique called framework templating, in which single crystals of an MOF using a more labile metal (such as zinc) are synthesized, which are then metal exchanged with a less labile metal (such as copper) that is still stable in the coordination environment produced in the Zn MOF. The lability of the metal-ligand bond is decreased and the stability is enhanced by replacing, for example, the zinc in a paddlewheel SBU with copper. Many frameworks based on Zn paddlewheel SBUs also possess an isostructural MOF with Cu paddlewheels [18]. Consequently, a zinc based MOF,

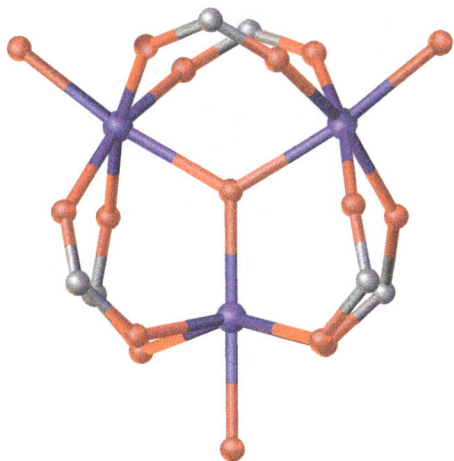

FIGURE 3: The $Cr_3(\mu_3\text{-}O)$ SBU of MIL-101 shown bound by 6 carboxylates and 3 solvent oxygen atoms.

PCN-921, by metal exchange with copper, was transformed to an isostructural MOF named PCN-922 via a single-crystal-to-single-crystal transformation. Zinc-based PCN-921 collapsed upon activation, while copper-based PCN-922 exhibited a BET surface area of 2006 $m^2\,g^{-1}$ after activation, showing a permanent porosity [19]. Interestingly, it also changed its color from green to blue, which is consistent with copper's change in coordination environment from a distorted square-pyramidal geometry with an axially coordinated solvent molecule to a square planar one stabilized by the Jahn-Teller effect due to its d^9 configuration [20]. On the other hand, Zn^{2+}, upon loss of the axial solvent ligand, presumably twists into a tetrahedral geometry which destroys the framework [21]. Metal exchange like this is also known as metal-ion metathesis and has been used to synthesize many MOFs which were not able to be synthesized directly [22].

To further increase the chemical stability of MOFs, researchers can go further along the path of using high-valence metal ions, such as Cr^{3+}, Fe^{3+}, and Zr^{4+} [23]. With all the coordination environments being equal, an increased charge will decrease lability simply by increasing the electrostatic interaction between the metal ions and the ligands. This trend can also be rationalized by the hard/soft acid-base principle, where soft acids like low-oxidation state metals form less stable coordination bonds with harder bases like the oxygen donors on carboxylate ligands. It is not just the charge of the metal ion that increases stability but also the charge density. Small, hard ions with high charge density, such as Cr^{3+} or Zr^{4+}, are able to bond more strongly to carboxylates than larger, soft ions like Zn^{2+} could [24]. This was exploited in the synthesis of MIL-101, which has a complex structure consisting of large clusters of four smaller $Cr_3(\mu_3\text{-}O)$ SBUs (Figure 3) joined by BDC linkers. This produces a framework with both a high BET surface area (4100 $m^2\,g^{-1}$) and very high chemical stability, being stable for several months in air and also being stable to various solvents and conditions [4].

Among metal units of a particular charge and coordination number, different metals can have ligand-metal exchange constants that are slower, which would result in a more chemically stable MOF. For example, $Ni(bdc)(ted)_{0.5}$ (BDC = 1,4-Benzenedicarboxylate, TED triethylene diamine) was shown to undergo slower ligand substitution with water vapor than $Cu(bdc)(ted)_{0.5}$, $Zn(bdc)(ted)_{0.5}$, or $Co(bdc)(ted)_{0.5}$, even though these MOFs were isostructural [13]. This is analogous to the differences found in the water exchange rate constants among the metal ions, though carboxylate-water exchanges will have different values than the known water-water exchange rate constants. Al^{3+} has a lower water exchange constant than Fe^{3+}, and Cr^{3+} is lower still, and so MOFs based on Cr^{3+} and Al^{3+} should be more water stable, all other things being equal, than isostructural MOFs based on Fe^{3+} or Ti^{3+}.

Increased chemical stability is also reported in UiO-66 and its isoreticular derivatives [3]. UiO-66 possesses 12-connected Zr_6O_8 SBUs in which the Zr^{4+} ions have stronger interactions with carboxylate ligands than copper or zinc, and thus these SBUs are less vulnerable to ligand substitution [25]. Additionally, the larger, more highly connected cluster means the framework suffers proportionally less disconnection if substitution was to occur. In general, higher nuclearity in the metal-containing unit of MOFs can increase stability in this manner. This is also shown by the increased stability of infinite chain SBUs. For example, as discussed earlier, MOF-69 is more chemically stable than Zn^{2+} MOFs containing tetrahedral Zn^{2+} paddlewheels and other discrete SBUs [11]. MIL-140, with its infinite Zr-oxide chains, as or more water stable than MOFs with a discrete Zr_6 SBU such as the UiO-66 series [26, 27].

However, the increased chemical stability reported for Cr^{3+}, Zr^{4+}, and other MOFs comes at a cost: the decreased lability that causes the higher chemical stability also makes it more difficult for the researchers to obtain high-quality single crystals for single crystal XRD purposes, typically producing only microcrystalline powders. Even Cu^{2+} MOFs are more difficult to crystallize than labile but unstable Zn^{2+} ones [19]. Lability and a relative equilibrium of ligand substitution are essential to the formation of single crystals. If the ligands bind very strongly to the metals with slow exchange and the equilibrium shifted toward precipitation, any nucleation will resort in the formation of either microcrystalline powder, or of amorphous products of typically low porosity and little use. Both laboratory research and industrial application of MOFs rely on accurate characterization of the products, and single-crystal XRD requires relatively large single crystals. Scientifically, development of MOFs requires study and understanding of the relationships between the structure of the MOF and its particular properties, which can only happen when the structure is well understood. Undoubtedly, many MOF and other products that may have had great application potential have been shelved or thrown away because of an inability to adequately characterize them, so synthesis of single-crystalline products over microcrystalline powders is very often highly preferred in a research setting.

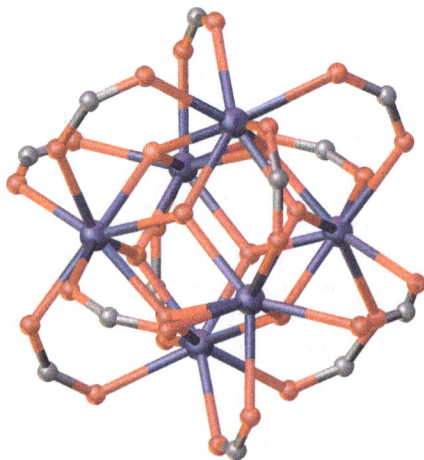

FIGURE 4: The Zr_6O_8 "brick" of UiO-66 shown coordinated by 12 carboxylates.

3. Modulated Synthesis to Increase Crystal Size and Crystallinity of High-Valence MOFs

It is desirable to find suitable reaction conditions that give the MOF ligands high lability during synthesis, while producing a framework with extremely strong bonds and low ligand lability after synthesis is complete. In order to produce large single crystals with high chemical stability despite the fact that the metals producing this stability are correlated with decreased lability and lower crystal size and crystallinity, modulated synthesis, originally developed by Tsuruoka et al. [10] and very successfully used by Behrens [28], should be employed. In this approach, nonbridging ligands are used to influence crystal growth. The addition of modulating reagents, such as monocarboxylic acids, can allow the formation of large single crystals of a MOF that otherwise may have only been synthesized as a powder [24, 29].

MOFs containing high-valence Zr^{4+} cations such as UiO-66 (Figure 4), PIZOF-1, and PCN-222 have been demonstrated to be stable towards air and water [3, 29, 32]. Additionally, zirconium is attractive as a MOF component due to its high abundance and low cost [24]: both are important when designing novel catalysts or sorbents. Note that much said about zirconium-based MOFs can also be applied to isostructural hafnium based MOFs due to the two elements' high chemical similarity, but Hf-based MOFs are less studied due to that element's higher expense. As discussed above, there were previously few examples of Zr MOFs being isolated in a single-crystalline state, due to the fact that using less labile metal-ligand bonds simultaneously increases chemical stability and hinders the crystallization process. Through the use of carefully tuned modulated synthesis, our group has recently synthesized and characterized single crystals of several Zr-MOFs for both gas sorption and catalysis study [32–34]. This is done by adding a varying amount of acetic, trifluoroacetic, or other monocarboxylic acid that can both change the pH of the solution and bind competitively to the metal ions during crystal formation [10, 28]. Equilibria in chemical reactions are governed not only by the energetic

differences between the different atoms and molecules, but also by the concentrations of the products and reactants. By simply introducing an excess amount of modulating ligand, we can drive the equilibria of crystal formation away from overly quick structure formation and towards the consecutive binding, release, and rebinding of ligands that allow ordered and large single crystal formation. However, since the mechanisms of crystal formation and solution equilibria under these conditions are difficult to predict, the exact amount and type of competing reagent must be tuned over many trials to produce large single crystals, as opposed to microcrystalline powders or no product at all. Catalysis is a possible application of MOFs that is especially demanding of chemical and thermal stability [35]. The MOF catalyst must not only possess appropriate sites, such as Lewis acidic or basic sites that are well exposed in accessible pores, but it must also be stable to the solvents, reagents, and temperatures that the reaction demands [36]. Based on synthesis conditions, the same catalytic cluster can exhibit different connectivity without compromising its stability. For instance, besides 12-connected Zr_6O_8 in UiO-66, 8-connected (in PCN-222 and PCN-521), and 6-connected (in PCN-225) Zr_6O_8 has also been found. Furthermore, by varying reaction conditions, MOFs based on cubic Zr_8O_6 have also been isolated in our laboratory.

In PCN-222, our group synthesized single crystals of an extremely stable Zr-based MOF with porphyrin-containing ligands that can themselves bind different metal ions, enabling a variety of catalytic activities, which was enabled both by the actual catalytic sites and by the high pore size and water and temperature stability of this MOF [32]. PCN-222 is stable not only to air and boiling water, but also to immersion in concentrated HCl for 24 hours. Similar Zr-porphyrin MOFs PCN-224 and PCN-225 exhibited different catalytic activity [33] or pH-dependent fluorescence [34]. Additionally, by using similar Zr SBUs along with tetrahedral ligands, stability was maintained alongside a higher surface area (BET 3411 $m^2 g^{-1}$) in PCN-521 (Figure 5), which mimicked the topology of fluorite by exploiting the cubic nature of the 8-connected Zr_6O_8 clusters in combination with the 4-connected tetrahedral linkers [24]. All of these Zr-MOFs exhibited high chemical stability due to the strong Zr–O bonds, and all were synthesized as single crystals by carefully varying the type and amount of modulating reagent. However, while many Zr-MOFs have demonstrated stability to neutral or acidic aqueous solutions, their stability towards base was lower, with only NO_2-functionalized UiO-66 retaining its crystallinity in pH 14 aqueous NaOH among tagged UiO-66 derivatives and PCN-225 being stable from pH 1–11 [34, 37].

Behrens et al. developed a series of porous interpenetrated zirconium-organic frameworks (PIZOFs,) using Zr^{4+} and $HO_2C[PE-P(R^1,R^2)-EP]CO_2H$ linkers. These MOFs demonstrated high porosity, tunability, and high stability against moisture and temperature. The very long organic linkers in these MOFs retain chemical stability, unlike UiO-67 and other UiO-66 derivatives with extended ligands. This is an example of modulated synthesis being used to prepare single crystals of a high-valence metal containing MOF,

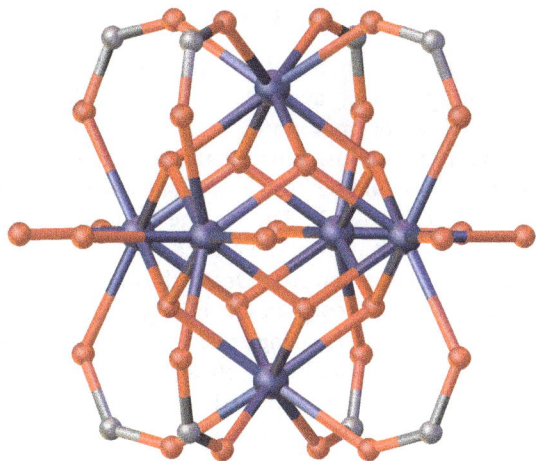

FIGURE 5: The Zr_6O_8 unit of PCN-521, which is similar to that of UiO-66 except that it is 8-connected instead of 12 connected, having a "cubic" symmetry that is compatible with the tetrahedral ligands used. The equatorial Zr atoms are coordinated by hydroxyl groups on the periphery of the unit [24].

with single-crystal X-ray diffraction studies produced for PIZOF-1, -2, and -8 after they were synthesized with benzoic acid as modulating reagent [29]. Other examples of water-stable Zr^{4+}-based MOFs after the use of modulating reagents have also been published [38].

4. N-Donor Ligands

So far, carboxylates are the most commonly used ligands in MOF synthesis. Utilization of high-valence metals as hard acids appears to be the most straightforward approach for the construction of stable MOFs by taking advantage of the existing ligand database. In addition to this approach, the interactions between softer ligands (such as imidazolates, triazolates, tetrazolates, and other nitrogen containing heterocycle-incorporated ligands) with softer metal ions (such as Zn^{2+} and Co^{2+}) can also be exploited in stable MOF synthesis. An early example of this by the long group was the synthesis of a framework using Mn^{2+} and 1,3,5-benzenetristetrazolate [39]. Detailed stability measurements were not conducted, but a crystal remained single through activation at 150°C allowing single-crystal XRD structural determination of the desolvated framework, implying high stability. Chen et al. published Zn^{2+} MOFs with imidazolate ligands and zeolite topology that possessed high thermal stability, Eddaoudi et al. developed zeolite-like MOFs using indium and bis(bidentate) imidazoledicarboxylic acid ligands, and the Yaghi group also introduced zeolitic imidazolate frameworks (ZIFs), using soft Zn^{2+} and Co^{2+} and imidizolate linkers to construct a wide variety of highly stable frameworks that mimic zeolite topology, due to the metal ions adopting a tetrahedral environment while the imidazolates link them at angles similar to the oxides in zeolite minerals [40–42].

The Rosi group synthesized a mesoporous framework bio-MOF-100, which is, to our knowledge, the highest BET surface area material known using N-donor ligands with 4,300 m^2 g^{-1} [43]. However, this material uses soft nitrogen containing adenine only to build Zn^{2+} SBUs with the SBUs linked by biphenyl dicarboxylate (BPDC) ligands. Thus, its overall stability would actually be expected to be much lower than that of Zn^{2+} MOFs linked exclusively by soft N-donor ligands, and this was confirmed by the fact that "gentle" supercritical CO_2 activation methods were required to achieve high porosity, with "harsh" conventional activation methods producing low N_2 adsorption consistent with collapse of the framework. This is in contrast with Bio-MOF-1 developed by the same group, which survived activation and demonstrated porosity at 125°C [44]. This is likely partially due simply to its relatively lower porosity compared to bio-MOF-100.

Similarly, our group reported another adenine-incorporated MOF, PCN-530, which consists of 2 distinct Zn-adenine SBUs forming 1D zinc-adenine chains linked by 4,4',4''-s-triazine-2,4,6-triyl-tribenzoate (TATB) ligands [45]. Even though the porosity of this framework is not impressive due to the utilization of low symmetry ligands, this framework has demonstrated a significantly improved stability over the traditional Zn MOFs.

Using 1,3,5-tris(1H-pyrazol-4-yl)benzene (H_3BTP) and Ni, Co, Zn, and Co salts, Long et al. synthesized several different frameworks that are very stable both thermally and chemically compared to most carboxylate-based MOFs, due to the less labile linkages between the pyrazolate-based ligand and metal,. $Ni_3(BTP)_2$ in particular was stable to both 430°C in air and immersion in boiling acid (pH = 2) or base (pH = 14), while retaining crystallinity and possessing a BET surface area of 1650 m^2 g^{-1}. This framework is both extremely stable and can expose Ni metal sites, making it promising for catalysis, even for reactions in harsh conditions [46]. The same group later reported $Fe_2(BDP)_3$, which has a BET surface area of 1230 m^2 g^{-1} and is based on a benzene dipyrazolate ligand. This material was stable to 280°C in air and boiling in aqueous acid (pH = 2) and base (pH = 10) for two weeks and was proven to be useful in separation of hexane isomers. Though the BET surface areas of these frameworks are not record breaking, their extreme stability while retaining high porosity makes them extremely promising materials that showcase the capabilities of N-based MOFs [47].

5. Ligands with Superhydrophobicity

Our group also attempted a method of synthesizing moisture-stable MOFs that did not involve altering the metal-containing units at all: synthesis of superhydrophobic MOFs by functionalizing the ligands with pendant hydrophobic groups. By partially filling the pores with $-O^n$Hex groups attached to [1,1':4',1'']terphenyl-3,3',5,5'-tetracarboxylic acid (TPTC) ligands, the water stability was drastically increased as compared to the nonfunctionalized copper paddlewheel containing NOTT-10X structure or even compared to ligands functionalized with shorter hydrophobic groups [31]. Predictably, some porosity was sacrificed, but a moderate BET surface area of 1083 m^2 g^{-1} was retained. Most interesting was the superhydrophobic behavior of Cu_2TPTC-O^nHex,

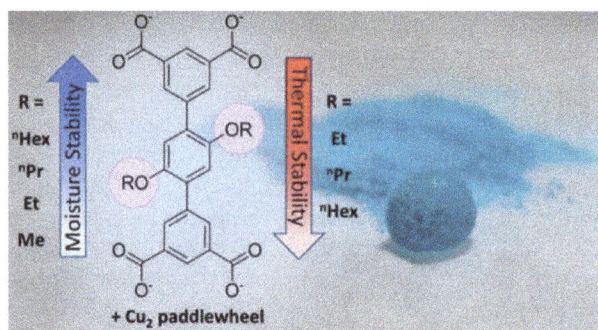

FIGURE 6: Superhydrophobicity shown in $Cu_2TPTC-O^nHex$. (Reprinted with permission from [31]. Copyright 2013 American Chemical Society).

which completely resisted any absorption of water under standard conditions. However, this behavior was inversely proportional to the thermal stability of the material, with a loss of crystallinity found in the $-O^nhex$ variant at 292°C as compared to 315°C for the $-O^nEt$ variant (Figure 6).

The Omary group developed the fluorous MOFs FMOF-1 and FMOF-2, which showed no water adsorption found near 100% relative humidity and no solvent adsorption after being immersed in distilled water for extended periods, despite its large channels that show high and selective adsorption of C_6-C_8 hydrocarbons [48]. FMOF-1 is built from 3,5-bis(trifluoromethyl)-1,2,4-triazolate and Ag^+, and FMOF-2 is produced by annealing FMOF-1 followed by resynthesis from a toluene and acetonitrile solution, so these can also be counted as examples of stable N-based MOFs. Serre et al. studied a series of MIL-88 derivatives with many different functional groups attached to the terephthalate and 4,4'-biphenyl dicarboxylate ligands of the Fe^{3+} MOFs, including various fluorinated ligands. Permanent porosity was retained in the $2CF_3$ versions of MIL-88B, with a BET surface area of 330 $m^2 g^{-1}$, while very little surface area was found in the 4F derivative, likely because the pore size in that version is too small to accommodate N_2 molecules. This study mainly covered how the functional groups changed the flexibility and swelling of the framework, but it also showed that most functional groups (except for the BDC-2OH linker) did not cause a large change in the thermal stability of the framework [49].

Another example of superhydrophobic N-based MOFs stable to moisture was the Ni- bis-pyrazolate MOFs developed by Padial et al. [50]. These were designed to take advantage of the more stable bonds between azolates and softer "borderline" metal ions [51]. Their hydrophobicity was tuned by changing the ligand length to change pore size, and by using trifluoromethyl and methyl-functionalized ligands. $[Ni_8(L5-CF_3)_6]$, the most hydrophobic MOF presented, effectively captured a flow of diethylsulfide, a hydrophobic and volatile organic compound, under 80% relative humidity, unlike similar nonhydrophobic MOFs. High water stability was further demonstrated by water adsorption-desorption isotherms.

6. Conclusion and Perspective

MOF research, though based on decades of research on coordination polymers and traditional porous materials (such as zeolites or mesoporous silica), blossomed after the discovery of porous, functionalizable Zn^{2+} and Cu^{2+} frameworks that could be grown as single crystals and thus easily characterized [2, 5]. Due to the limited stability of the early frameworks, researchers tried various methods to create more robust frameworks. One of the most popular and successful methods, incorporation of high-valence metals, had the downside of decreasing crystal size and crystallinity [3]. This has caused difficulties in structural characterization that have been partially surmounted through the use of novel techniques [52] for deeper elucidation of MOF structures [53]. A modulated synthesis strategy for the growth of larger single crystals of stable high-valence MOFs can also be used to allow structural characterization through single crystal X-ray diffraction [10, 28].

As discussed in a recent review of our work and other works on the rational design of MOFs [36], it should also be noted that modulated synthesis techniques have also been reported to increase the porosity and catalytic activity of UiO-66, due to the introduction of defects that leave coordinated modulator in place of some proportion of the linking ligands [54, 55]. However, it is possible that this defect creation may reduce the stability of the resulting MOF as well. Our group has used these techniques to synthesize a variety of Zr-MOFs of lower cost and high stability and porosity, as well as exploring the use of soft N-donor and hydrophobic ligands to increase the stability of MOFs using bivalent metals. Many promising recently reported MOFs for gas storage and catalysis have used high-valence metals that have low ligand exchange rates, and so promising avenues of near future MOF research likely involve the use of modulating reagents, metal exchange, and other new techniques to synthesize highly porous and stable MOFs from other readily available high-valence metals, such as iron, chromium, or aluminum.

So far, most stability measurements on MOFs have been concentrated on their thermal and water stabilities, but further avenues of possible research are to determine their stabilities towards contaminants such as salts, body fluids, or corrosive molecules such as H_2S or NH_3. Some studies of this sort have been done, especially on MOFs designed for drug delivery, and have been reviewed elsewhere [56]. As future MOFs are developed for more widespread application, their stability towards more varied chemicals and contaminants should be determined, and application-built MOFs should be rationally designed to be stable towards conditions present in their targeted environment.

Conflict of Interests

The authors declare that there is no conflict of interests regarding the publication of this paper.

Acknowledgments

This work was supported by the Center for Gas Separations Relevant to Clean Energy Technologies, an Energy Frontier

Research Center funded by the U.S. Department of Energy (DOE), Office of Science, Office of Basic Energy Sciences under Award no. DE-SC0001015, part of the Hydrogen and Fuel Cell Program under Award no. DE-FC36-07G017033, part of the Methane Opportunities for Vehicular Energy (MOVE) Program, an ARPA-e project under Award no. DE-AR0000249. Mathieu Bosch also acknowledges the Texas A&M graduate merit fellowship.

References

[1] H.-C. Zhou, J. R. Long, and O. M. Yaghi, "Introduction to metal-organic frameworks," *Chemical Reviews*, vol. 112, no. 2, pp. 673–674, 2012.

[2] H. Li, M. Eddaoudi, M. O'Keeffe, and O. M. Yaghi, "Design and synthesis of an exceptionally stable and highly porous metal-organic framework," *Nature*, vol. 402, no. 6759, pp. 276–279, 1999.

[3] J. H. Cavka, S. Jakobsen, U. Olsbye et al., "A new zirconium inorganic building brick forming metal organic frameworks with exceptional stability," *Journal of the American Chemical Society*, vol. 130, no. 42, pp. 13850–13851, 2008.

[4] G. Ferey, C. Mellot-Draznieks, C. Serre et al., "A chromium terephthalate-based solid with unusually large pore volumes and surface area," *Science*, vol. 309, no. 5743, pp. 2040–2042, 2005.

[5] S. S.-Y. Chui, S. M.-F. Lo, J. P. H. Charmant, A. G. Orpen, and I. D. Williams, "A chemically functionalizable nanoporous material [Cu$_3$(TMA)$_2$(H$_2$O)$_3$]$_n$," *Science*, vol. 283, no. 5405, pp. 1148–1150, 1999.

[6] S. Ma, J. M. Simmons, D. Sun, D. Yuan, and H. Zhou, "Porous metal-organic frameworks based on an anthracene derivative: syntheses, structure analysis, and hydrogen sorption studies," *Inorganic Chemistry*, vol. 48, no. 12, pp. 5263–5268, 2009.

[7] H. Li, M. Eddaoudi, T. L. Groy, and O. M. Yaghi, "Establishing microporosity in open metal-organic frameworks: gas sorption isotherms for Zn(BDC) (BDC = 1,4-benzenedicarboxylate)," *Journal of the American Chemical Society*, vol. 120, no. 33, pp. 8571–8572, 1998.

[8] H. Li, W. Shi, K. Zhao, H. Li, Y. Bing, and P. Cheng, "Enhanced hydrostability in Ni-doped MOF-5," *Inorganic Chemistry*, vol. 51, no. 17, pp. 9200–9207, 2012.

[9] J. T. Hupp and K. R. Poeppelmeler, "Chemistry: better living through nanopore chemistry," *Science*, vol. 309, no. 5743, pp. 2008–2009, 2005.

[10] T. Tsuruoka, S. Furukawa, Y. Takashima, K. Yoshida, S. Isoda, and S. Kitagawa, "Nanoporous nanorods fabricated by coordination modulation and oriented attachment growth," *Angewandte Chemie*, vol. 48, no. 26, pp. 4739–4743, 2009.

[11] N. L. Rosi, M. Eddaoudi, J. Kim, M. O'Keeffe, and O. M. Yaghi, "Infinite secondary building units and forbidden catenation in metal-organic frameworks," *Angewandte Chemie International Edition*, vol. 41, no. 2, pp. 284–287, 2002.

[12] D. Mustafa, E. Breynaert, S. R. Bajpe, J. A. Martens, and C. E. A. Kirschhock, "Stability improvement of Cu3(BTC)2 metal-organic frameworks under steaming conditions by encapsulation of a Keggin polyoxometalate," *Chemical Communications*, vol. 47, no. 28, pp. 8037–8039, 2011.

[13] K. Tan, N. Nijem, P. Canepa et al., "Stability and hydrolyzation of metal organic frameworks with paddle-wheel SBUs upon hydration," *Chemistry of Materials*, vol. 24, no. 16, pp. 3153–3167, 2012.

[14] P. Küsgens, M. Rose, I. Senkovska et al., "Characterization of metal-organic frameworks by water adsorption," *Microporous and Mesoporous Materials*, vol. 120, no. 3, pp. 325–330, 2009.

[15] D. Q. Yuan, D. Zhao, D. Sun, and H. Zhou, "An isoreticular series of metal-organic frameworks with dendritic hexacarboxylate ligands and exceptionally high gas-uptake capacity," *Angewandte Chemie*, vol. 49, no. 31, pp. 5357–5361, 2010.

[16] D. Zhao, D. Yuan, D. Sun, and H.-C. Zhou, "Stabilization of metal-organic frameworks with high surface areas by the incorporation of mesocavities with microwindows," *Journal of the American Chemical Society*, vol. 131, no. 26, pp. 9186–9188, 2009.

[17] M. Li, D. Li, M. O'Keeffe, and O. M. Yaghi, "Topological analysis of metal–organic frameworks with polytopic linkers and/or multiple building units and the minimal transitivity principle," *Chemical Reviews*, vol. 114, no. 2, pp. 1343–1370, 2013.

[18] M. Zhang, Y.-P. Chen, and H.-C. Zhou, "Structural design of porous coordination networks from tetrahedral building units," *CrystEngComm*, vol. 15, no. 45, pp. 9544–9552, 2013.

[19] Z. Wei, W. Lu, H. Jiang, and H. Zhou, "A route to metal-organic frameworks through framework templating," *Inorganic Chemistry*, vol. 52, no. 3, pp. 1164–1166, 2013.

[20] M. V. Veidis, G. H. Schreiber, T. E. Gough, and G. J. Palenik, "Jahn-Teller distortions in octahedral copper(II) complexes," *Journal of the American Chemical Society*, vol. 91, no. 7, pp. 1859–1860, 1969.

[21] M. K. Bhunia, J. T. Hughes, J. C. Fettinger, and A. Navrotsky, "Thermochemistry of paddle wheel MOFs: Cu-HKUST-1 and Zn-HKUST-1," *Langmuir*, vol. 29, no. 25, pp. 8140–8145, 2013.

[22] Y. Kim, S. Das, S. Bhattacharya et al., "Metal-ion metathesis in metal-organic frameworks: a synthetic route to new metal-organic frameworks," *Chemistry*, vol. 18, no. 52, pp. 16642–16648, 2012.

[23] H. Chevreau, T. Devic, F. Salles, G. Maurin, N. Stock, and C. Serre, "Mixed-linker hybrid superpolyhedra for the production of a series of large-pore iron(III) carboxylate metal-organic frameworks," *Angewandte Chemie*, vol. 52, no. 19, pp. 5056–5060, 2013.

[24] M. Zhang, Y.-P. Chen, M. Bosch et al., "Symmetry-guided synthesis of highly porous metal-organic frameworks with fluorite topology," *Angewandte Chemie International Edition*, vol. 53, no. 3, pp. 815–818, 2014.

[25] L. Valenzano, B. Civalleri, S. Chavan et al., "Disclosing the complex structure of UiO-66 metal organic framework: a synergic combination of experiment and theory," *Chemistry of Materials*, vol. 23, no. 7, pp. 1700–1718, 2011.

[26] J. B. Decoste, G. W. Peterson, H. Jasuja, T. G. Glover, Y. Huang, and K. S. Walton, "Stability and degradation mechanisms of metal-organic frameworks containing the Zr$_6$O$_4$(OH)$_4$ secondary building unit," *Journal of Materials Chemistry A*, vol. 1, no. 18, pp. 5642–5650, 2013.

[27] V. Guillerm, F. Ragon, M. Dan-Hardi et al., "A series of isoreticular, highly stable, porous zirconium oxide based metal-organic frameworks," *Angewandte Chemie*, vol. 51, no. 37, pp. 9267–9271, 2012.

[28] A. Schaate, P. Roy, A. Godt et al., "Modulated synthesis of Zr-based metal-organic frameworks: from nano to single crystals," *Chemistry A*, vol. 17, no. 24, pp. 6643–6651, 2011.

[29] A. Schaate, P. Roy, T. Preuße, S. J. Lohmeier, A. Godt, and P. Behrens, "Porous interpenetrated zirconium-organic frameworks (PIZOFs): a chemically versatile family of metal-organic frameworks," *Chemistry*, vol. 17, no. 34, pp. 9320–9325, 2011.

[30] O. V. Dolomanov, L. J. Bourhis, R. J. Gildea, J. A. K. Howard, and H. Puschmann, "OLEX2: a complete structure solution, refinement and analysis program," *Journal of Applied Crystallography*, vol. 42, no. 2, pp. 339–341, 2009.

[31] T. A. Makal, X. Wang, and H. C. Zhou, "Tuning the moisture and thermal stability of metal-organic frameworks through incorporation of pendant hydrophobic groups," *Crystal Growth & Design*, vol. 13, no. 11, pp. 4760–4768, 2013.

[32] D. W. Feng, Z. Y. Gu, J. R. Li, H. L. Jiang, Z. W. Wei, and H. C. Zhou, "Zirconium-metalloporphyrin PCN-222: mesoporous metal-organic frameworks with ultrahigh stability as biomimetic catalysts," *Angewandte Chemie*, vol. 51, no. 41, pp. 10307–10310, 2012.

[33] D. W. Feng, W. C. Chung, Z. W. Wei et al., "Construction of ultrastable porphyrin zr metal-organic frameworks through linker elimination," *Journal of the American Chemical Society*, vol. 135, no. 45, pp. 17105–17110, 2013.

[34] H. L. Jiang, D. W. Feng, K. C. Wang et al., "An exceptionally stable, porphyrinic Zr metal-organic framework exhibiting pH-dependent fluorescence," *Journal of the American Chemical Society*, vol. 135, no. 37, pp. 13934–13938, 2013.

[35] M. Yoon, R. Srirambalaji, and K. Kim, "Homochiral metal-organic frameworks for asymmetric heterogeneous catalysis," *Chemical Reviews*, vol. 112, no. 2, pp. 1196–1231, 2012.

[36] M. Zhang, M. Bosch, T. Gentle, and H.-C. Zhou, "Rational design of metal–organic frameworks with anticipated porosities and functionalities," *CrystEngComm*, vol. 16, pp. 4069–4083, 2014.

[37] M. Kandiah, M. H. Nilsen, S. Usseglio et al., "Synthesis and stability of tagged UiO-66 Zr-MOFs," *Chemistry of Materials*, vol. 22, no. 24, pp. 6632–6640, 2010.

[38] V. Bon, V. Senkovskyy, I. Senkovska, and S. Kaskel, "Zr(IV) and Hf(IV) based metal-organic frameworks with reo-topology," *Chemical Communications*, vol. 48, no. 67, pp. 8407–8409, 2012.

[39] M. Dinča, A. Dailly, Y. Liu, C. M. Brown, D. A. Neumann, and J. R. Long, "Hydrogen storage in a microporous metal-organic framework with exposed Mn2+ coordination sites," *Journal of the American Chemical Society*, vol. 128, no. 51, pp. 16876–16883, 2006.

[40] R. Banerjee, A. Phan, B. Wang et al., "High-throughput synthesis of zeolitic imidazolate frameworks and application to CO_2 capture," *Science*, vol. 319, no. 5865, pp. 939–943, 2008.

[41] X.-C. Huang, Y.-Y. Lin, J.-P. Zhang, and X.-M. Chen, "Ligand-directed strategy for zeolite-type metal-organic frameworks: zinc(II) imidazolates with unusual zeolitic topologies," *Angewandte Chemie*, vol. 45, no. 10, pp. 1557–1559, 2006.

[42] Y. Liu, V. C. Kravtsov, R. Larsen, and M. Eddaoudi, "Molecular building blocks approach to the assembly of zeolite-like metal-organic frameworks (ZMOFs) with extra-large cavities," *Chemical Communications*, no. 14, pp. 1488–1490, 2006.

[43] J. An, O. K. Farha, J. T. Hupp, E. Pohl, J. I. Yeh, and N. L. Rosi, "Metal-adeninate vertices for the construction of an exceptionally porous metal-organic framework," *Nature Communications*, vol. 3, article 604, 2012.

[44] J. An, S. J. Geib, and N. L. Rosi, "Cation-triggered drug release from a porous zinc-adeninate metal-organic framework," *Journal of the American Chemical Society*, vol. 131, no. 24, pp. 8376–8377, 2009.

[45] M. Zhang, W. Lu, J.-R. Li et al., "Design and synthesis of nucleobase-incorporated metal-organic materials," *Inorganic Chemistry Frontiers*, vol. 1, no. 2, pp. 159–162, 2014.

[46] V. Colombo, S. Galli, H. J. Choi et al., "High thermal and chemical stability in pyrazolate-bridged metal-organic frameworks with exposed metal sites," *Chemical Science*, vol. 2, no. 7, pp. 1311–1319, 2011.

[47] Z. R. Herm, B. M. Wiers, J. A. Mason et al., "Separation of hexane isomers in a metal-organic framework with triangular channels," *Science*, vol. 340, no. 6135, pp. 960–964, 2013.

[48] C. Yang, U. Kaipa, Q. Z. Mather et al., "Fluorous metal-organic frameworks with superior adsorption and hydrophobic properties toward oil spill cleanup and hydrocarbon storage," *Journal of the American Chemical Society*, vol. 133, no. 45, pp. 18094–18097, 2011.

[49] P. Horcajada, F. Salles, S. Wuttke et al., "How linker's modification controls swelling properties of highly flexible iron(III) dicarboxylates MIL-88," *Journal of the American Chemical Society*, vol. 133, no. 44, pp. 17839–17847, 2011.

[50] N. M. Padial, E. Quartapelle Procopio, C. Montoro et al., "Highly hydrophobic isoreticular porous metal-organic frameworks for the capture of harmful volatile organic compounds," *Angewandte Chemie*, vol. 52, no. 32, pp. 8290–8294, 2013.

[51] J. P. Zhang, Y. B. Zhang, J. B. Lin, and X. M. Chen, "Metal azolate frameworks: from crystal engineering to functional materials," *Chemical Reviews*, vol. 112, no. 2, pp. 1001–1033, 2012.

[52] Y. B. Zhang, J. Su, H. Furukawa et al., "Single-crystal structure of a covalent organic framework," *Journal of the American Chemical Society*, vol. 135, no. 44, pp. 16336–16339, 2013.

[53] X. Q. Kong, H. X. Deng, F. Y. Yan et al., "Mapping of functional groups in metal-organic frameworks," *Science*, vol. 341, no. 6148, pp. 882–885, 2013.

[54] H. Wu, Y. S. Chua, V. Krungleviciute et al., "Unusual and highly tunable missing-linker defects in zirconium metal-organic framework UiO-66 and their important effects on gas adsorption," *Journal of the American Chemical Society*, vol. 135, no. 28, pp. 10525–10532, 2013.

[55] F. Vermoortele, B. Bueken, G. Le Bars et al., "Synthesis modulation as a tool to increase the catalytic activity of metal-organic frameworks: The unique case of UiO-66(Zr)," *Journal of the American Chemical Society*, vol. 135, no. 31, pp. 11465–11468, 2013.

[56] P. Horcajada, R. Gref, T. Baati et al., "Metal-organic frameworks in biomedicine," *Chemical Reviews*, vol. 112, no. 2, pp. 1232–1268, 2012.

9

Cobalt(II) Chloride Hexahydrate as an Efficient and Inexpensive Catalyst for the Preparation of Biscoumarin Derivatives

Mohammad Reza Nazarifar

Young Researchers and Elites Club, Shiraz Branch, Islamic Azad University, Shiraz, Iran

Correspondence should be addressed to Mohammad Reza Nazarifar; mohammadreza.nazarifar@gmail.com

Academic Editor: Georgia Melagraki

Cobalt(II) chloride hexahydrate (CoCl$_2$·6H$_2$O) has been found to be an efficient catalyst for the one-pot synthesis of biscoumarin derivatives through a combination of aromatic aldehydes and 4-hydroxycoumarin in aqueous media at 70°C. Several types of aromatic aldehyde, containing electron-withdrawing groups as well as electron-donating groups, were used in the reaction and in all cases the desired products were synthesized successfully. The present approach offers remarkable advantages such as short reaction times, excellent yields, straightforward procedure, easy purification, environment friendliness, and low catalyst loading.

1. Introduction

Coumarin derivatives, especially biscoumarins, are important compounds in organic synthesis due to their wide spectrum of pharmacological properties such as antifungal, anti-HIV, anticancer, anticoagulant, antithrombotic, antimicrobial, and antioxidant [1–5]. These compounds are also utilized as urease inhibitors [6].

A number of methods have been reported for the synthesis of these compounds in the presence of various catalysts like molecular iodine [7], sodium dodecyl sulfate (SDS) [8], tetrabutylammonium bromide (TBAB) [9], ([MIM(CH$_2$)$_4$SO$_3$H][HSO$_4$]) [10], tetrabutylammonium hexatungstate ([TBA]$_2$[W$_6$O$_{19}$]) [11], sulfated titania (TiO$_2$/SO$_4^{2-}$) [12], ruthenium(III) chloride hydrate (RuCl$_3$·nH$_2$O) [13], n-dodecylbenzene sulfonic acid (DBSA) [14], and silica chloride nanoparticles (nano SiO$_2$Cl) [15]. However, these methods suffer from one or more disadvantages such as low yields of products, long reaction times, use of expensive catalyst, toxic solvents, or harsh reaction conditions. Therefore, introducing a clean procedure by the use of green and environmentally friendly catalyst with high catalytic activity, moderate temperature, and short reaction time accompanied with excellent yield for the production of biscoumarin derivatives is needed.

We hoped to develop a more general protocol for the efficient synthesis of biscoumarin derivatives via CoCl$_2$·6H$_2$O, which have recently attracted much attention as catalyst to organic synthesis due to their low toxicity and easy availability [16–18].

2. Results and Discussion

We herein present efficient and eco-friendly procedure for the synthesis of biscoumarin derivatives (3 a–m) by three-component condensation of 4-hydroxycoumarin (1) and aromatic aldehyde (2) catalyzed by CoCl$_2$·6H$_2$O in water-ethanol solvent system 70°C (Scheme 1).

For this study, a reaction between 4-hydroxycoumarin (2 mmol) and 3-nitrobenzaldehyde (1 mmol) was examined as the model reaction. Initial studies showed that better results could be obtained in the presence of (10 mol%) CoCl$_2$·6H$_2$O in aqueous ethanol (1:1, H$_2$O : EtOH) at 70°C.

To optimize the mol% of catalyst, the above reaction was performed with different mol% of CoCl$_2$·6H$_2$O such as 5, 10, 15, 20, and 25 mol%. The results are summarised in Table 1 which shows that the reaction catalysed by about 10 mol% CoCl$_2$·6H$_2$O results in the highest yield (Table 1, entry 2). In the presence of less than this amount, the yield decreased

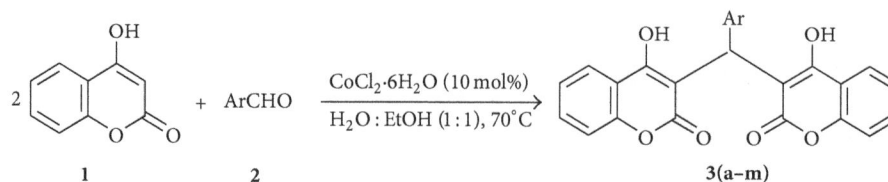

SCHEME 1: Synthesis of biscoumarins.

TABLE 1: Optimization of the catalysed model reaction for synthesis of 3,3'-(3-nitrobenzylidene)-bis-(4-hydroxycoumarin).

Entry	Catalyst	Amount of catalyst (mol%)	Time (min)	Yield (%)[a]
1	$CoCl_2 \cdot 6H_2O$	5	8	89
2	$CoCl_2 \cdot 6H_2O$	10	2	98
3	$CoCl_2 \cdot 6H_2O$	15	2	98
4	$CoCl_2 \cdot 6H_2O$	20	2	98
5	$CoCl_2 \cdot 6H_2O$	25	2	96

All reactions were carried out in aqueous ethanol at 70°C.
[a]Isolated yields.

(Table 1, entry 1). When the amount of $CoCl_2 \cdot 6H_2O$ was increased over 10 mol%, neither the yield nor the reaction time was improved (Table 1, entry 3).

To study the effect of temperature on this synthesis, we also performed four experiments in aqueous ethanol at room temperature, 50, and 70 (Celsius degrees) and under reflux condition (Table 2). It was observed that the yield of the product is maximum at 70°C (Table 2, entry 3).

During the optimization of the reaction condition, various solvents were also screened to test their efficiency and the results are summarized in Table 3. The highest reaction activity was achieved in the system using aqueous ethanol (1:1, H_2O:EtOH) as a solvent in comparison to other solvents under similar reaction conditions (Table 3, entry 5).

With these encouraging results in hand, the generality of this reaction was examined using various aromatic aldehydes containing electron-donating as well as electron-withdrawing groups. In all cases, the reactions gave the corresponding products in good yields and short reaction times without formation of any byproducts (Table 4). Substituents on the aromatic ring had no obvious effect on yield or reaction time under the above optimal conditions.

In order to assess the efficiency of this methodology, the obtained result from the reaction of 3-nitrobenzaldehyde with 4-hydroxycoumarin by this method has been compared with those of the previously reported methods. As demonstrated in Table 5, the use of $CoCl_2 \cdot 6H_2O$ leads to an improved protocol in terms of compatibility with environment, reaction time, yield of the product, and amount of the catalyst when compared with other catalysts.

3. Experimental

3.1. Materials and Methods.
All reagents were purchased from Fluka, Merck, and Aldrich with high-grade quality and used without any purification. The reactions were monitored by TLC. Visualisation of the developed chromatogram was performed by UV light (254 nm). All yields refer to isolated products after purification. Products were characterized by comparison with authentic samples and by spectroscopy data (IR, ^{1}H NMR spectra). IR spectra were recorded from KBr disk on the FT-IR Bruker Tensor 27. ^{1}H NMR spectra were recorded on a Bruker Avance 400 MHz spectrometer using TMS as an internal standard (DMSO-d_6 solution). Melting points were measured by using the capillary tube method with IA 9000 series thermal analyser.

3.2. General Procedure for the Synthesis of Biscoumarin Derivatives.
A mixture of the 4-hydroxycoumarin (2 mmol), aromatic aldehyde (1 mmol), and $CoCl_2 \cdot 6H_2O$ (10 mol%) was stirred in 5 mL aqueous ethanol (1:1, H_2O:EtOH) 70°C for the appropriate time. Completion of the reaction was indicated by TLC. After the completion, the reaction mixture was filtered off and washed with n-hexane (2×5 mL) to obtain pure products. As the catalyst is completely soluble in distilled water, it was easily separated from the reaction mixture. All of the products are known compounds and were characterized by IR and ^{1}H NMR spectroscopic data and their melting points are compared with reported values.

3.3. Selected Spectral Data

3.3.1. 3,3'-(4-Chlorobenzylidene)-bis-(4-hydroxycoumarin) (Table 4, Entry 7):
IR(KBr). 3420 (OH), 2923 (C–H stretching), 1668 (–C=O stretching of –COOR group), 1606 (–C=C stretching), 1563, 1490 (C=C– stretching of aromatic ring), 765 (C–H out of plane bending) cm^{-1}; ^{1}H NMR (400 MHz, DMSO-d_6): δ 6.63 (s, 1H, CH), 7.16–7.90 (m, 12H, ArH), 7.90–9 (m, 2H, OH).

3.3.2. 3,3'-(3-Nitrobenzylidene)-bis-(4-hydroxycoumarin) (Table 4, Entry 9):
IR(KBr). 3424 (OH), 2925 (C–H stretching), 1655 (–C=O stretching of –COOR group), 1616 (–C=C stretching), 1564, 1494 (C=C– stretching of aromatic ring),

TABLE 2: Optimisation of temperature for synthesis of 3,3'-(3-nitrobenzylidene)-bis-(4-hydroxycoumarin) using $CoCl_2 \cdot 6H_2O$ (10 mol%) as catalyst in aqueous ethanol.

Entry	Catalyst	Temperature °C	Time (min)	Yields (%)
1	$CoCl_2 \cdot 6H_2O$	Room temperature	4	85
2	$CoCl_2 \cdot 6H_2O$	Reflux	2	98
3	$CoCl_2 \cdot 6H_2O$	70	2	98
4	$CoCl_2 \cdot 6H_2O$	50	4	95

Isolated yield of the pure compound.

TABLE 3: Effect of solvents in reaction of 3-nitrobenzaldehyde and 4-hydroxycoumarin catalyzed by $CoCl_2 \cdot 6H_2O$.

Entry	Catalyst	Solvent	Time (min)	Yields (%)[a]
1	$CoCl_2 \cdot 6H_2O$	EtOH	4	95
2	$CoCl_2 \cdot 6H_2O$	H_2O	3	97
3	$CoCl_2 \cdot 6H_2O$	H_2O : EtOH (1 : 2)	4	96
4	$CoCl_2 \cdot 6H_2O$	H_2O : EtOH (2 : 1)	3	98
5	$CoCl_2 \cdot 6H_2O$	H_2O : EtOH (1 : 1)	2	98
6	$CoCl_2 \cdot 6H_2O$	Solvent free	2 h	Trace

All reactions were catalyzed by $CoCl_2 \cdot 6H_2O$ at 70°C.
[a]Isolated yields.

TABLE 4: $CoCl_2 \cdot 6H_2O$ catalyzed synthesis of biscoumarin derivatives[a].

Entry	Ar	Product	Time (min)	Yield[b] (%)	mp (°C) Reported [ref]	mp (°C) This work
1	C_6H_5	3a	2	92	229–231 [10]	230–232
2	$4\text{-}FC_6H_4$	3b	2	97	212–214 [15]	211-212
3	$4\text{-}BrC_6H_4$	3c	5	96	266–268 [15]	264–266
4	$4\text{-}CNC_6H_4$	3d	5	96	240–242 [13]	240–242
5	$2\text{-}ClC_6H_4$	3e	3	93	201–203 [10]	200–202
6	$3\text{-}ClC_6H_4$	3f	3	94	221–223 [10]	221–223
7	$4\text{-}ClC_6H_4$	3g	1	97	261–263 [10]	256–258
8	$2\text{-}O_2NC_6H_4$	3h	3	94	198–200 [10]	195–197
9	$3\text{-}O_2NC_6H_4$	3i	1	98	235 [19]	234–236
10	$4\text{-}O_2NC_6H_4$	3j	1	98	232–234 [20]	238–239
11	$4\text{-}CH_3C_6H_4$	3k	5	92	271–273 [11]	266–268
12	$4\text{-}CH_3OC_6H_4$	3l	4	94	251–253 [11]	246–248
13	$3\text{-}CH_3OC_6H_4$	3m	5	92	238 [21]	238–240
14	CH_3CHO	—	120	—		

[a]Reaction conditions: 4-hydroxycoumarin (2 mmol), aromatic aldehyde (1 mmol), $CoCl_2 \cdot 6H_2O$ (10 mol%), and aqueous ethanol (5 mL), at 70°C.
[b]Yields refer to isolated products.

TABLE 5: Comparison of the efficiency of $CoCl_2 \cdot 6H_2O$ with other reported catalysts in the synthesis of 3,3'-(3-nitrobenzylidene)-bis-(4-hydroxycoumarin).

Entry	Catalyst	Conditions	Time	Yield (%) [ref]
1	$[P_4VPy\text{-}BuSO_3H]Cl\text{-}X(AlCl_3)$, 0.07 mmol	Toluene, 90°C	0.5 (h)	96 [22]
2	SBPPSP, 0.06 g	EtOH/H_2O, reflux	15 (min)	94 [20]
3	SDS, 20 mol%	H_2O, 60°C	2.15 (h)	95 [8]
4	Nano-SiO_2Cl, 75 mg	CH_2Cl_2, 40°C	3.5 (h)	90 [15]
5	$[TBA]_2[W_6O_{19}]$, 0.08 mmol	EtOH, reflux	7 (min)	85 [11]
6	$NaHSO_4 \cdot SiO_2$ or Indion 190 resin, 150 mg	Toluene, 100°C	30 (min)	90 [19]
7	$CoCl_2 \cdot 6H_2O$, 10 mol%	EtOH/H_2O, 70°C	2 (min)	98 [present work]

762 (C–H out of plane bending) cm^{-1}; ^{1}H NMR (400 MHz, DMSO-d$_6$): δ 6.39 (s, 1H, CH), 7.28–8.04 (m, 12H, ArH), 8.04–9.52 (m, 2H, OH).

3.3.3. 3,3′-(4-Methoxybenzylidene)-bis-(4-hydroxycoumarin) (Table 4, Entry 12).

IR(KBr): 3443 (OH), 2926 (C–H stretching), 1668 (–C=O stretching of –COOR group), 1606 (–C=C stretching), 1563, 1510 (C=C– stretching of aromatic ring), 767 (C–H out of plane bending) cm^{-1}; ^{1}H NMR (400 MHz, DMSO-d$_6$): δ 3.71 (s, 3H, CH$_3$O), 6.31 (s, 1H, CH), 6.80–7.93 (m, 12H, ArH), 8.16–8.78 (m, 2H, OH).

4. Conclusion

In conclusion, we have developed a green, practical, and facile approach for the preparation of biscoumarin derivatives through the three-component reaction of 4-hydroxycoumarin and aromatic aldehydes using a catalytic amount of CoCl$_2$·6H$_2$O as an efficient and inexpensive catalyst. The distinguished advantages of this procedure are (i) simple experimental procedure, (ii) mild reaction conditions, (iii) high to excellent yields of products, (iv) short reaction times, (v) and utilization of an inexpensive and readily available catalyst.

Conflict of Interests

The author declares that there is no conflict of interests regarding the publication of this paper.

Acknowledgment

The author gratefully acknowledges the financial support from the Research Council of Islamic Azad University, Shiraz Branch.

References

[1] I. Kostova, I. Manolov, and G. Momekov, "Cytotoxic activity of new neodymium (III) complexes of bis-coumarins," *European Journal of Medicinal Chemistry*, vol. 39, no. 9, pp. 765–775, 2004.

[2] I. Manolov, C. Maichle-Moessmer, and N. Danchev, "Synthesis, structure, toxicological and pharmacological investigations of 4-hydroxycoumarin derivatives," *European Journal of Medicinal Chemistry*, vol. 41, no. 7, pp. 882–890, 2006.

[3] N. Hamdi, M. C. Puerta, and P. Valerga, "Synthesis, structure, antimicrobial and antioxidant investigations of dicoumarol and related compounds," *European Journal of Medicinal Chemistry*, vol. 43, no. 11, pp. 2541–2548, 2008.

[4] C. C. Chiang, J. F. Mouscadet, H. J. Tsai, C. T. Liu, and L. Y. Hsu, "Synthesis and HIV-1 integrase inhibition of novel bis- or tetra-coumarin analogues," *Chemical and Pharmaceutical Bulletin*, vol. 55, no. 12, pp. 1740–1743, 2007.

[5] D. Završnik, S. Muratović, S. Špirtović, D. Softić, and M. Medić-Šarić, "The synthesis and antimicrobial activity of some 4-hydroxycoumarin derivatives," *Bosnian Journal of Basic Medical Sciences*, vol. 8, no. 3, pp. 277–281, 2008.

[6] K. M. Khan, S. Iqbal, M. A. Lodhi, G. M. Maharvi, M. I. Choudhary, and S. Perveen, "Biscoumarin: new class of urease inhibitors; economical synthesis and activity," *Bioorganic and Medicinal Chemistry*, vol. 12, no. 8, pp. 1963–1968, 2004.

[7] M. Kidwai, V. Bansal, P. Mothsra et al., "Molecular iodine: a versatile catalyst for the synthesis of bis(4-hydroxycoumarin) methanes in water," *Journal of Molecular Catalysis A: Chemical*, vol. 268, no. 1-2, pp. 76–81, 2007.

[8] H. Mehrabi and H. Abusaidi, "Synthesis of biscoumarin and 3,4-dihydropyrano[c]chromene derivatives catalysed by sodium dodecyl sulfate (SDS) in neat water," *Journal of the Iranian Chemical Society*, vol. 7, no. 4, pp. 890–894, 2010.

[9] J. M. Khurana and S. Kumar, "Tetrabutylammonium bromide (TBAB): a neutral and efficient catalyst for the synthesis of biscoumarin and 3,4-dihydropyrano[c]chromene derivatives in water and solvent-free conditions," *Tetrahedron Letters*, vol. 50, no. 28, pp. 4125–4127, 2009.

[10] N. Tavakoli-Hoseini, M. M. Heravi, F. F. Bamoharram, A. Davoodnia, and M. Ghassemzadeh, "An unexpected tetracyclic product isolated during the synthesis of biscoumarins catalyzed by [MIM(CH$_2$)$_4$SO$_3$H][HSO$_4$]: characterization and X-ray crystal structure of 7-(2-hydroxy-4-oxo-4H-chromen-3-yl)-6H,7H-chromeno[4,3-b]chromen-6-one," *Journal of Molecular Liquids*, vol. 163, no. 3, pp. 122–127, 2011.

[11] A. Davoodnia, "A highly efficient and fast method for the synthesis of biscoumarins using tetrabutylammonium hexatungstate [TBA]$_2$[W$_6$O$_{19}$] as green and reusable heterogeneous catalyst," *Bulletin of the Korean Chemical Society*, vol. 32, no. 12, pp. 4286–4290, 2011.

[12] B. Karmakar, A. Nayak, and J. Banerji, "Sulfated titania catalyzed water mediated efficient synthesis of dicoumarols—a green approach," *Tetrahedron Letters*, vol. 53, no. 33, pp. 4343–4346, 2012.

[13] K. Tabatabaeian, H. Heidari, A. Khorshidi, M. Mamaghani, and N. O. Mahmoodi, "Synthesis of biscoumarin derivatives by the reaction of aldehydes and 4-hydroxycoumarin using ruthenium(III) chloride hydrate as a versatile homogeneous catalyst," *Journal of the Serbian Chemical Society*, vol. 77, no. 4, pp. 407–413, 2012.

[14] B. Pawar, V. Shinde, and A. Chaskar, "n-Dodecylbenzene sulfonic acid (DBSA) as a novel Brønsted acid catalyst for the synthesis of bis(indolyl)methanes and bis(4-hydroxycoumarin-3-yl)methanes in water," *Green and Sustainable Chemistry*, vol. 3, no. 2, pp. 56–60, 2013.

[15] R. Karimian, F. Piri, A. A. Safari, and S. J. Davarpanah, "One-pot and chemoselective synthesis of bis(4-hydroxycoumarin) derivatives catalyzed by nano silica chloride," *Journal of Nanostructure in Chemistry*, vol. 3, pp. 52–57, 2013.

[16] S. Velusamy, J. S. K. Kumar, and T. Punniyamurthy, "Cobalt(II) catalyzed tosylation of alcohols with p-toluenesulfonic acid," *Tetrahedron Letters*, vol. 45, no. 1, pp. 203–205, 2004.

[17] S. Velusamy, S. Borpuzari, and T. Punniyamurthy, "Cobalt(II)-catalyzed direct acetylation of alcohols with acetic acid," *Tetrahedron*, vol. 61, no. 8, pp. 2011–2015, 2005.

[18] A. T. Khan, T. Parvin, and L. H. Choudhury, "A simple and convenient one-pot synthesis of benzimidazole derivatives using cobalt(II) chloride hexahydrate as catalyst," *Synthetic Communications*, vol. 39, no. 13, pp. 2339–2346, 2009.

[19] V. Padalkar, K. Phatangare, S. Takale, R. Pisal, and A. Chaskar, "Silica supported sodium hydrogen sulfate and Indion 190 resin: an efficient and heterogeneous catalysts for facile synthesis of bis-(4-hydroxycoumarin-3-yl) methanes," *Journal of Saudi Chemical Society*, 2012.

[20] K. Niknam and A. Jamali, "Silica-bonded N-propylpiperazine sodium n-propionate as recyclable basic catalyst for synthesis of 3,4-dihydropyrano[c]chromene derivatives and biscoumarins," *Chinese Journal of Catalysis*, vol. 33, no. 11, pp. 1840–1849, 2012.

[21] P. Singh, P. Kumar, A. Katyal et al., "Phosphotungstic acid: an efficient catalyst for the aqueous phase synthesis of bis-(4-hydroxycoumarin-3-yl)methanes," *Catalysis Letters*, vol. 134, no. 3-4, pp. 303–308, 2010.

[22] K. P. Boroujeni and P. Ghasemi, "Synthesis and application of a novel strong and stable supported ionic liquid catalyst with both Lewis and Brønsted acid sites," *Catalysis Communications*, vol. 37, pp. 50–54, 2013.

Effect of γ-Irradiation and Calcination Temperature of Nanosized ZnO/TiO$_2$ System on Its Structural and Electrical Properties

Abdelrahman A. Badawy,[1] **Shaymaa E. El-Shafey,**[2]
Suzan Abd El All,[3] **and Gamil A. El-Shobaky**[2]

[1] *Physical Chemistry Department, Center of Excellence for Advanced Science, Renewable Energy Group, National Research Center, Dokki, Cairo 12622, Egypt*
[2] *Physical Chemistry Department, National Research Center, Dokki, Cairo 12622, Egypt*
[3] *Radiation Physics Department, National Center for Radiation Research and Technology (NCRRT), Nasr City, Cairo 11762, Egypt*

Correspondence should be addressed to Abdelrahman A. Badawy; aabadawy107@yahoo.com

Academic Editor: Fabien Grasset

ZnO/TiO$_2$ powders were synthesized by sol-gel method using ammonium hydroxide. The effects of calcination temperature (500–1000°C) and gamma rays (with doses from 25 to 150 kGy) on the phases present and their electrical properties were investigated. The results revealed that heating the system investigated at 500°C led to the formation of ZnTiO$_3$-rohom and TiO$_2$-rutile. The degree of crystallinity of the phases produced increased by increasing the calcination temperature. When heating at 1000°C, ZnTiO$_3$-rohom turned to ZnTiO$_3$-cubic but the rutile phase remained stable. γ-Irradiation decreased considerably the crystallite size of the rutile phase from 146 to 63 nm and that of ZnTiO$_3$-cubic decreased from 101 to 39 nm. This treatment led also to the creation of holes in the matrix of irradiated solids which increased the mobility of charge carriers (electrons) leading to a significant increase in the electrical conductivity reaching to 10^2 to 10^3-fold.

1. Introduction

Fundamental studies concerning the phase diagram and characterization of ZnO-TiO$_2$ system have been published since 1960s. This system still attracts the attention of researchers because of its importance in practical applications [1–5]. TiO$_2$ ceramics have been investigated for diverse applications in the optical and semiconductors industries because of their interesting semiconducting and dielectric properties. Semiconducting titania had especially been employed in producing different electronic devices, including oxygen sensors, varistors, and current collecting electrodes in Na-S batteries [6–8]. TiO$_2$-ZnO-based compounds have been employed as promising catalysts in some chemical industries. Many attempts have been made to improve the photoelectrochemical (PEC) conversion efficiency of TiO$_2$ ceramics by treatment with certain compounds including

Al$_2$O$_3$ [9–14] and BeO [15]. The base of the phase diagram for this system was established by Dulin and Rase [1], who have reported that there are three compounds existing in the ZnO-TiO$_2$ system, including Zn$_2$TiO$_4$ (cubic), ZnTiO$_3$ perovskite metatitanate (hexagonal), and Zn$_2$Ti$_3$O$_8$ (cubic) [16]. ZnTiO$_3$ has perovskite structure and could be considered as a useful candidate for microwave resonator materials [17] and gas sensor [18] and as a catalyst for oxidation of ethanol NO, CO, and so forth . . ., [2] and colour pigments [11]. Yamaguchi et al. [3] claimed that Zn$_2$TiO$_4$ can be easily prepared by conventional solid-state reaction between 2 moles of ZnO and 1 mole of TiO$_2$. However, preparation of pure ZnTiO$_3$ from a mixture of equimolar proportion of zinc and titanium oxides has not been successful because the produced compound undergoes partial decomposition by heating at 900°C into a mixture of ZnO and rutile.

It is well known that the properties of materials depend on their synthesis processes and calcination conditions. For $ZnTiO_3$, its physicochemical properties are influenced by their preparation conditions. In general, there are some methods to prepare $ZnTiO_3$ powder, including conventional solid-state reaction [1], sol-gel method [19], ..., and so forth. The solid-state reaction method has some drawbacks including high temperature, large particle size, and limited degree of chemical homogeneity. The chemical solution methods can provide products of fine and homogeneous particles with high specific surface area. The processing of complex oxide ceramics using sol-gel techniques has been extensively studied [20–23]. The Pechini method is a conventional approach to prepare powder consisting of oxides of transition metals. This method requires first dissolution of hydrous oxides or alkoxides of element in polyhydroxy alcohol such as ethylene glycol, with a chelating agent, such as citric acid. Subsequently thermal treatment is performed at relatively low temperatures leading to the formation of highly homogenous powder [24].

In addition, much attention has been paid to their electrical properties leading to numerous applications as solid oxides fuel cells (SOFCs) [25]. Some progress of microwave devices in the area of mobile telephones and satellite communications brought the need for development of microwave dielectrics with low dielectric loss, high dielectric constant, and low-temperatures coefficient of resonant frequencies. It has been demonstrated that zinc titanates are good dielectric materials for microwave devices [26–29]. So, they are nowadays widely applied as dielectric resonators and filters [25].

Damage of solids caused by ionizing radiations occurs mostly below the surface of the material and often creates features very small in size (nm to μm in scale). The characteristics of the damage can be either microstructure (dislocation loops, voids, and structural disordering) or microchemical (segregation) or both (precipitation of secondary phase) [30]. In certain materials a permanent change in their electrical conductivity may be induced by radiation damage of the crystal.

In this paper, the authors have attempted to prepare $ZnTiO_3/TiO_2$ composite ceramic materials by sol-gel method. The role of calcination temperature and exposure to different doses of γ-rays of the obtained solids on their structural and electrical properties were investigated. The techniques employed were DTA, XRD, SEM, and electrical conductivity.

2. Experimental Details

2.1. Materials. TiO_2/ZnO mixed solids were prepared using the following method: certain amounts of $ZnClO_4$ and $TiClO_4$ with stoichiometry (Zn/Ti = 1:1) were dissolved in distilled water stirred for 3 hours. NH_4OH solution was added dropwise to mixed solutions as a precipitating agent. The precipitation process was carried out at room temperature and a pH of about 8. The precipitate was dried at 70°C for two days and the large agglomerates were pulverized in an agate mortar. The ceramic composite was obtained after heating in air for 6 hours at temperatures between 500

and 1000°C. Systematic studies concerning the role of calcination temperature and doses of gamma rays on structural and electric properties of the treated solids were investigated.

The irradiation process was performed in air, at room temperature, where a cooling system was used in the irradiation chamber to avoid heating of the samples during irradiation. The irradiated solids were kept in sealed tubes for 3 weeks before carrying out any measurements. The doses of γ-rays were 25, 50, 100, and 150 kGy.

2.2. Techniques. X-ray powder diffractograms of nonirradiated and variously irradiated samples calcined at temperatures within 500 and 1000°C were obtained using a Bruker diffractometer (Bruker D8 advance target). The patterns were run with Cu Kα1 with secondly monochromator (λ = 1.5405 Å) at 40 kV and 40 mA. The crystallite size of crystalline phases present in different solids investigated was calculated from line broadening profile analysis of the main diffraction lines of crystalline phases present using the Scherer equation [18]:

$$d = \frac{K\lambda}{\beta_{1/2}}\cos\theta, \tag{1}$$

where d is the mean crystallite diameter, λ is the wave length of X-ray beam, K is the Scherer constant (0.89), $\beta_{1/2}$ is full width at half maximum {FWHM} of the main diffraction peaks in radians, and θ is the diffraction angle. The scanning rate was 8° and 0.8° in 2θ/min in phase identification and line broadening profile analysis, respectively.

Differential thermal analysis (DTA) measurements were carried out by Shimadzu instrument (DTA-50) calibrated through the melting points of indium and tin. The thermogravimetric analysis (TGA) was recorded using the (TGA-50) system in the presence of air, within a temperature range from room temperature up to 600°C at a heating rate of 30°C min^{-1}. The nature and morphological feature of composite ceramic were studied by scanning electron microscope SEM (JOEL, JSM).

The electrical conductivity measurements were performed using a locally designed electrical circuit, where the samples were pressed under hydraulic pressure of 9 ton cm^{-2} in the form of discs of a cross-sectional area of 0.6129 cm^2. A Tennelec TC 952 power supply was used to apply DC voltage across the sample discs, and the resulting current was measured using a digital electrometer (Model 6517A, manufactured by Keithely, USA).

3. Results and Discussion

3.1. Thermal Behavior of Composite Powders. It is well know that the sol-gel technique has the unique advantage of providing extremely small and uniform particle size for the precursor powders. The prepared powders were analyzed using DTA-TGA for burnout behaviors.

Figure 1 shows TG-DTA curves of the prepared noncalcined mixed solids, being dried at 70°C and having equimolar ratio of ZnO and TiO_2 existing as an amorphous powder. The DTA curve shows an endothermic reaction at ~200°C, which

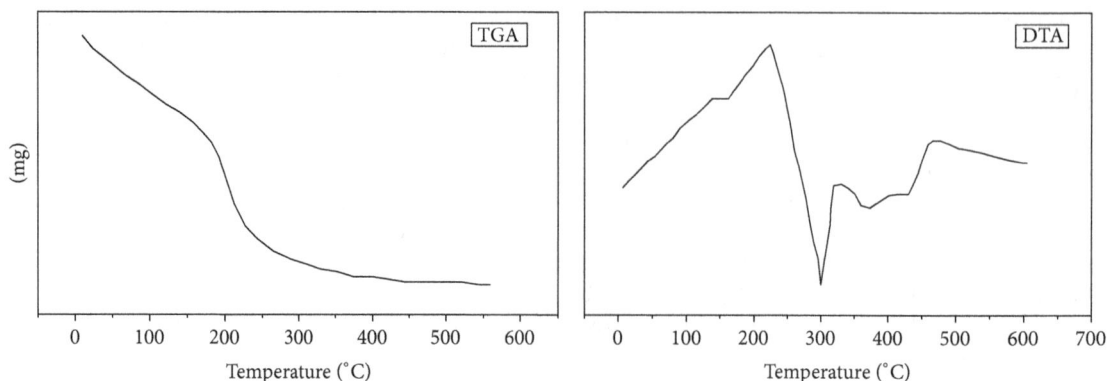

FIGURE 1: TGA-DTA curves of ZnO/TiO_2 systems.

is attributed to the loss of the remaining absorbed water as shown in Figure 1. This finding agrees with the weight loss given by TGA data as shown in Figure 1, an endothermic peak at above 200°C could be attributed to the formation of TiO_2/ZnO particles. A broad endothermic peak due to the crystallization of TiO_2/ZnO is observed at about 470°C, resulting from the dehydroxylation of titanium and zinc hydroxides into TiO_2 and ZnO phases, respectively.

3.2. XRD Investigation of ZnO/TiO_2 System Calcined at Different Temperatures.

X-ray diffractograms of ZnO/TiO_2 system calcined at 500, 600, 800, and 1000°C were determined. Figure 2 shows the recorded diffractograms. Examination of Figure 2 shows the following: (i) the mixed solids calcined at 500, 600, and 800°C consisted of TiO_2-rutile as a major phase together with $ZnTiO_3$-rohom. The absence of any diffraction line of ZnO phase in different solids investigated calcined at 500–1000°C might suggest its possible presence in dissolved form in TiO_2 lattice forming solid solution. The dissolution of ZnO in TiO_2 lattice stabilized the rutile structure playing almost the role of Al_2O_3 dissolved in TiO_2[9]. This conclusion is reached by phase transformation of TiO_2 from rutile to anatase by heating a temperature below 1000°C. (ii) The increase in the calcination temperature within 500–1000°C brought about an effective increase in the degree of crystallinity of the phases present. (iii) Increasing the calcination temperature of the system investigated up to 1000°C led to a phase transition process of $ZnTiO_3$-rohom into cubic $ZnTiO_3$ structure. (iv) The average crystallite size of TiO_2-rutile in solids calcined at 500–1000°C was calculated from the Scherrer equation [18]. The computed values of crystallite size of TiO_2-rutile in mixed solids calcined at 500, 600, 800, and 1000°C measured 63, 74, 82, and 146 nm, respectively. (vi) The crystallite size of the $ZnTiO_3$-rohom in mixed solids calcined as 500, 600, and 800°C measured 83, 78, and 74 nm, respectively while the crystallite size of $ZnTiO_3$-cubic present in mixed solid calcined at 1000°C reached 101 nm. These results show clearly an effective sintering of the rutile phase due to increasing its calcination temperature within 500–1000°C.

3.3. Effect of Gamma Radiation on Structural Characteristics of ZnO/TiO_2 System.

ZnO/TiO_2 samples calcined at 1000°C and exposed to γ-rays with absorbed doses from 25 to 150 kGy were analyzed by XRD technique. The recorded diffractograms are illustrated in Figure 3. It is clear from this figure that γ-rays did not induce any change in the present crystalline phases. The crystallite size of detected phases (TiO_2-rutile and $Zn\,TiO_3$) was measured from XRD data applying the Scherrer equation [19–29]. The computed values of the crystallite size of rutile phase were found to be equal to 146, 63, 100, and 102 nm for nonirradiated sample and those exposed to 25, 50, 100, and 150, respectively. These values show clearly that γ-irradiation induced progressive decreasing of the crystallite size of rutile phase. The lower value of this latter was observed at 50 kGy. The decrease in the crystallite size reached about 57% in case of $ZnTiO_3$-cubic formed by heating ZnO/TiO_2 at 1000°C. Its crystallite size measured 101 nm and decreased considerably after irradiation at 25, 50, 100, and 150 kGy to reach 63, 51, 49, and 39 nm, respectively. Similar results have been reported in case of Co_3O_4/Al_2O_3 system calcined at 650°C [30] by exposure to γ-rays with high doses (50 MGy). After irradiation, the crystallite size of Co_3O_4 decreased effectively via splitting its crystallites to an extent proportional to the absorbed dose. It fell to a minimum value and the increased with the dose upon 50 MGy. Such a very high dose might increase the degree of aggregation of small-sized particles.

3.4. SEM Micrographs of the Powder.

Figure 4 shows SEM pictures realized for a nonirradiated sample of ZnO/TiO_2 and samples irradiated at doses of 25, 50, 100, and 150 kGy. These pictures revealed that, before irradiation and at relatively low doses, there were few small black holes dispersed in the whole mass of the samples. It appears from Figure 4 that the concentration of these holes increased by increasing the dose. These created holes or cavities in the irradiated solids might reflect an order-disorder transition from the cation and not to a phase shift or due to charged oxygen vacancies, which are systematically generated by irradiation in the oxides. These created holes or cavities are expected to increase

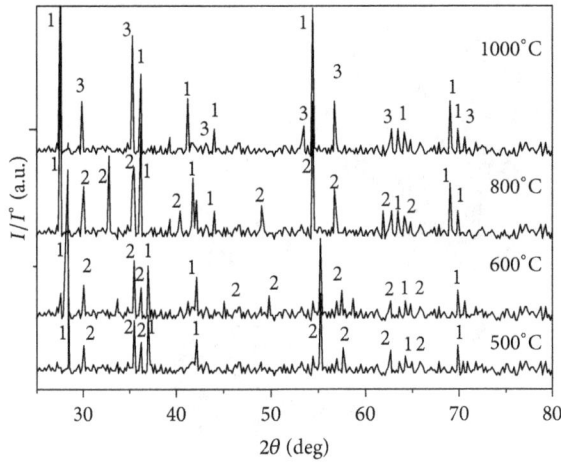

FIGURE 2: XRD diffractograms of ZnO_2/TiO_2 powders calcined at temperatures of 500–1000°C. $1 \rightarrow TiO_2$-rutile, $2 \rightarrow ZnTiO_3(R)$, $3 \rightarrow ZnTiO_3$ (c).

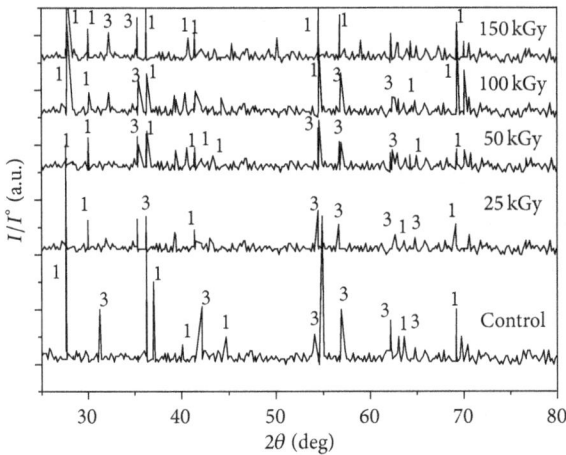

FIGURE 3: XRD diffractograms of effect of γ ray on the ZnO/TiO$_2$ solids calcined at 1000°C. $1 \rightarrow TiO_2$ (rutile), $3 \rightarrow ZnTiO_3$ (c).

the electrical conductivity of the irradiated solids via increasing the mobility of charge carriers.

3.5. Effect of Calcination Temperature of ZnO/TiO$_2$ System on Its Electrical Properties. The electrical conductivity (σ) of ZnO/TiO$_2$ system calcined at temperature within 500–1000°C was investigated. It was measured for various solids at temperatures between 25 and 150°C. Figure 5(a) shows log σ as a function of $1/T$ for variously calcined solids. It appears that these solids behave as a semiconductor. In fact, one can observe that the conductivity exhibits Arrhenius behavior, which can be expressed as

$$\sigma = \sigma_0 \exp\left(-\frac{\Delta E}{kT}\right), \qquad (2)$$

where σ is the electrical conductivity at a given absolute temperature T, σ_0 is the preexponential factor of the Arrhenius equation, ΔE is the activation energy of electric conductivity

of the sample under test, and k is the Boltzmann constant. From this relation the activation energy can be calculated and the results obtained are illustrated in Figure 5(b).

As both TiO$_2$ and ZnO are known to show n-type semiconducting character, the variation of conductivity can be influenced by the change in concentration and mobility of charge carriers (electrons) induced by the possible defect reactions. Furthermore, it has been reported that because TiO$_2$ contains interstitial channels in the c direction, certain transition metal cations diffuse through these channels into lattices. The diffused cations are located preferentially in interstitial and/or in host cation positions via substituting some of the host lattice cations [31]. Thus, the decrease of conductivity with increasing the sintering temperature, despite the increase in crystallite size, can be explained by substitution of Zn^{+2} (0.74 Å) by Ti^{+4} (0.68 Å) with subsequent decrease in the concentration of electrons.

It is seen from Figure 5(a) that σ, measured at different temperatures, decreases progressively as a function of calcination temperature of the system investigated. A higher calcination temperature increased the degree of crystallinity of the phases present as shown in XRD Section 3.2. The observed increase in the degree of crystallinity of the phases present may act as an energy barrier opposing the flow of the electric current leading to a decrease in σ value with subsequent increase in the activation energy. The decrease in σ value due to increasing the calcination temperature might also result from occupation of cationic vacancies by which decreases the diffusion of oxygen anions [10].

In certain materials a permanent change in electrical conductivity may take place as a result of a possible induced damage due to exposure to ionizing radiations.

3.6. Effect of γ Irradiation on Electrical Properties of ZnO/TiO$_2$ System Calcined at 1000°C. The electrical conductivity of ZnO/TiO$_2$ sample calcined at 1000°C and exposed to γ-rays with different absorbed doses was measured at temperatures between 25 and 150°C. The results obtained are graphically illustrated in Figure 6(a). It is clearly shown from this figure that σ increases progressively as a function of the absorbed dose of γ-rays. This increase in σ value attained 10^2–10^3-fold. The change in the activation energy of electric conductivity (ΔE_σ) as a function of the absorbed dose of γ-rays was calculated and the results obtained are given in Figure 6(b). The measurable increase in σ value and the subsequent decrease in (ΔE_σ) might be attributed to possible lattice defects via displacing of atoms from their equilibrium position together with creation of an increased number of holes throughout the whole mass of the irradiated solids (c.f. Figure 4). The produced holes facilitate diffusion of charge carriers across the matrices of the irradiated solids resulting in an effective increase in the electrical conductivity with subsequent decrease in (ΔE_σ).

4. Conclusions

The following are the main conclusions that may be drawn from the results obtained.

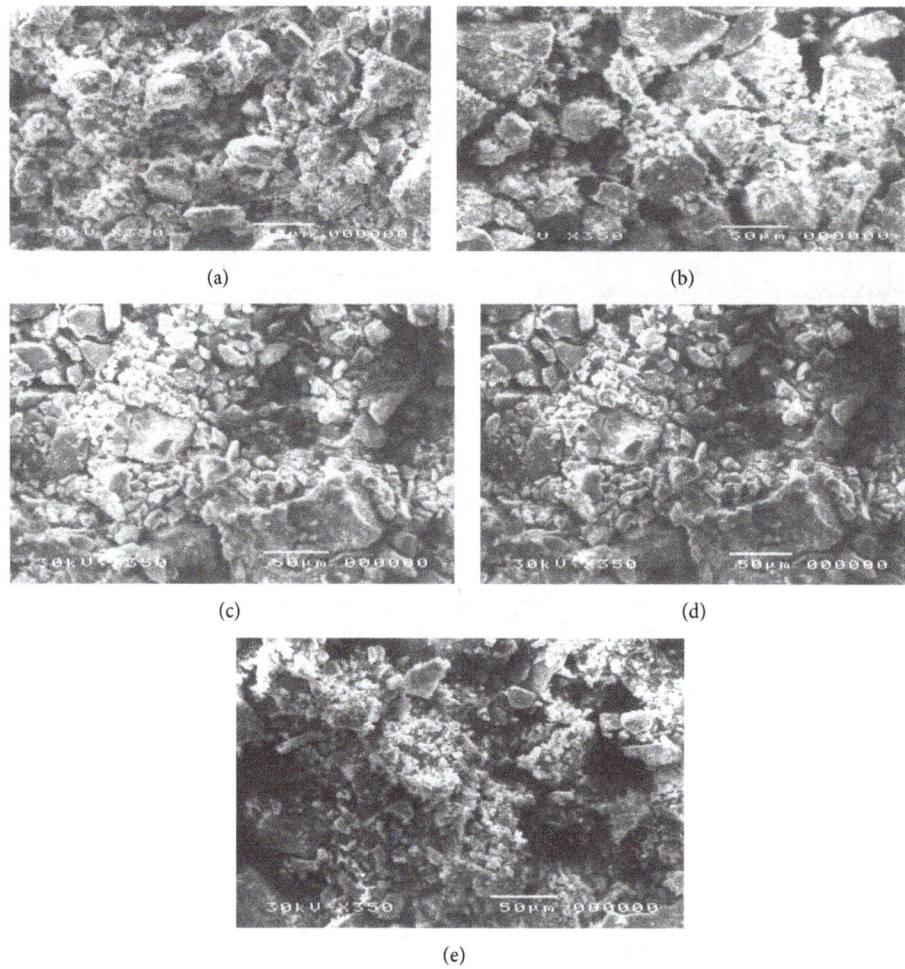

FIGURE 4: Scanning electron micrographs of irradiated samples (a) control, (b) 25 kGy, (c) 50 kGy, (d) 100 kGy, and (e) 150 kgy. All images were visible with aid of magnifying glass.

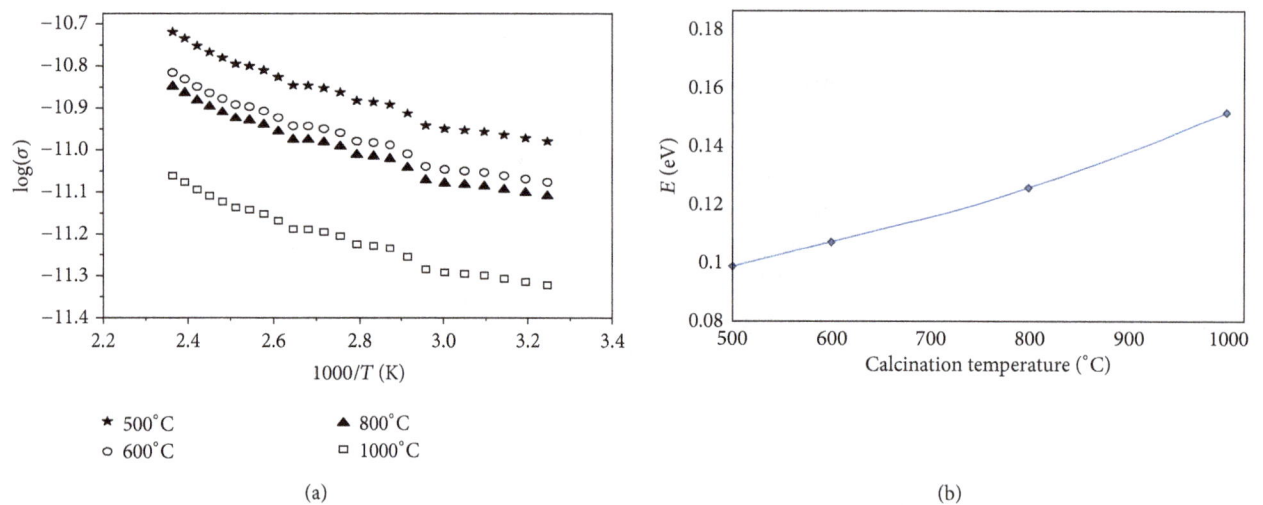

FIGURE 5: (a) Effect of sintering temperature on the electrical conductivity of ZnO/TiO$_2$ system calcined at different temperatures. (b) Effect of sintering temperature on the activation energy of electrical conductivity of ZnO/TiO$_2$ system calcined at different temperatures.

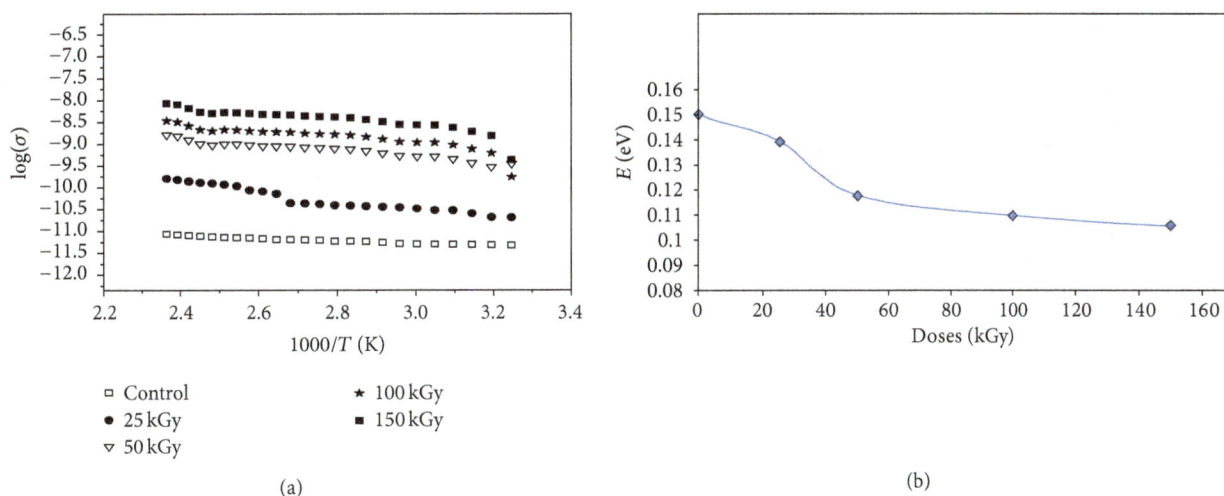

FIGURE 6: (a) Relationship between log σ and $1/T$ for ZnO/TiO$_2$ calcined at 1000°C and exposed to different doses of γ rays. (b) Effect of γ-rays doses on the activation energy of the electrical conductivity of ZnO/TiO$_2$ system calcined at 1000°C.

(1) Heating ZnO/TiO$_2$ system, prepared by sol-gel at 500–900°C, turned to ZnTiO$_3$-rohom and TiO$_2$-rutile.

(2) The produced phases existed as nano solids.

(3) The degree of crystallinity and crystallite size of ZnTiO$_3$-rohom and TiO$_2$-rutile increased by increasing the calcination temperature within 500–800°C.

(4) Increasing the calcination temperature of the system investigated up to 1000°C converted ZnTiO$_3$ from rohom structure to cubic phase together with a thermally stable TiO$_2$-rutile. The outstanding thermal stability of the rutile phase up to 1000°C has been attributed to a possible dissolution of some ZnO in the lattice of TiO$_2$-rutile forming solid solution.

(5) Gamma irradiation of the system investigated calcined at 1000°C decreased considerably the crystallite size of ZnTiO$_3$-cubic from 101 to 39 nm and that of rutile phase from 146 to 63 nm.

(6) Gamma irradiation increased considerably the electrical conductivity of irradiated solids with subsequent decrease in the activation energy of electrical conductivity.

Conflict of Interests

The authors declare that there is no conflict of interests regarding the publication of this paper.

References

[1] F. H. Dulin and D. E. Rase, "Phase equilibria in the system ZnO-TiO$_2$," *Journal of the American Ceramic Society*, vol. 43, pp. 125–131, 1960.

[2] S. F. Bartram and R. A. Slepetys, "Compound formation and crystal structure in the system ZnO-TiO$_2$," *Journal of the American Ceramic Society*, vol. 44, pp. 493–498, 1961.

[3] O. Yamaguchi, M. Morimi, H. Kawabata, and K. Shimizu, "Formation and Transformation of ZnTiO$_3$," *Journal of the American Ceramic Society*, vol. 70, no. 5, pp. 97–98, 1987.

[4] A. Baumgarte and R. Blachnik, "Isothermal sections in the systems ZnOAO$_2$Nb$_2$O$_5$ (A Ti, Zr, Sn) at 1473 K," *Journal of Alloys and Compounds*, vol. 210, no. 1-2, pp. 75–81, 1994.

[5] U. Steinike and B. Wallis, "Formation and structure of Ti-Zn-oxides," *Crystal Research and Technology*, vol. 32, no. 1, pp. 187–193, 1997.

[6] K. H. Yoon, J. Cho, and D. H. Kang, "Physical and photo-electrochemical properties of the TiO$_2$-ZnO system," *Materials Research Bulletin*, vol. 34, no. 9, pp. 1451–1461, 1999.

[7] R. Laishram, O. P. Thakur, D. K. Bhattacharya, and Harsh, "Dielectric and piezoelectric properties of la doped lead zinc niobate-lead zirconium titanate ceramics prepared from mechano-chemically activated powders," *Materials Science and Engineering B*, vol. 172, no. 2, pp. 172–176, 2010.

[8] W. Tai, "Photoelectrochemical properties of ruthenium dye-sensitized nanocrystalline SnO$_2$:TiO$_2$ solar cells," *Solar Energy Materials and Solar Cells*, vol. 76, no. 1, pp. 65–73, 2003.

[9] S. A. El All and G. A. El-Shobaky, "Structural and electrical properties of γ-irradiated TiO$_2$/Al$_2$O$_3$ composite prepared by sol-gel method," *Journal of Alloys and Compounds*, vol. 479, no. 1-2, pp. 91–96, 2009.

[10] R. Kant, K. Singh, and O. P. Pandey, "Structural and ionic conductive properties of Bi$_4$V$_{2-x}$Ti$_x$O$_{11-\delta}$ (0 ≤ x ≤ 0.4) compound," *Materials Science and Engineering: B*, vol. 158, pp. 63–68, 2009.

[11] L. Khemakhem, A. Maalej, A. Kabadou, A. Ben Salah, A. Simon, and M. Maglione, "Dielectric ferroelectric and piezoelectric properties of BaTi$_{0.975}$(Zn$_{1/3}$Nb$_{2/3}$)$_{0.025}$O$_3$ ceramic," *Journal of Alloys and Compounds*, vol. 452, no. 2, pp. 441–445, 2008.

[12] Y. Ku, Y.-H. Huang, and Y.-C. Chou, "Preparation and characterization of ZnO/TiO$_2$ for the photocatalytic reduction of Cr(VI) in aqueous solution," *Journal of Molecular Catalysis A: Chemical*, vol. 342-343, pp. 18–22, 2011.

[13] E. Yun, J. W. Jung, and B. C. Lee, "Effects of O$_2$ fraction on the properties of Al-doped ZnO thin films treated by high-energy electron beam irradiation," *Journal of Alloys and Compounds*, vol. 496, no. 1-2, pp. 543–547, 2010.

[14] K. G. Chandrappa and T. V. Venkatesha, "Generation of Co_3O_4 microparticles by solution combustion method and its Zn-Co_3O_4 composite thin films for corrosion protection," *Journal of Alloys and Compounds*, vol. 542, pp. 68–77, 2012.

[15] Y.-S. Chang, Y.-H. Chang, I.-G. Chen, G.-J. Chen, and Y.-L. Chai, "Synthesis and characterization of zinc titanate nanocrystal powders by sol-gel technique," *Journal of Crystal Growth*, vol. 243, no. 2, pp. 319–326, 2002.

[16] H. T. Kim, S. Nahm, J. D. Byun, and Y. Kim, "Low-fired (Zn,Mg)TiO_3 microwave dielectrics," *Journal of the American Ceramic Society*, vol. 82, no. 12, pp. 3476–3480, 1999.

[17] H. Obayashi, Y. Sakurai, and T. Gejo, "Perovskite-type oxides as ethanol sensors," *Journal of Solid State Chemistry*, vol. 17, no. 3, pp. 299–303, 1976.

[18] B. D. Cullity, *Publishing Cos*, pp. 102–105, Addison-Wesley, Reading, Mass, USA, 2nd edition, 1978.

[19] X. H. Zeng, Y. Y. Liu, X. Y. Wang, W. C. Yin, L. Wang, and H. Guo, "Preparation of nanocrystalline $PbTiO_3$ by accelerated sol-gel process," *Materials Chemistry and Physics*, vol. 77, no. 1, pp. 209–214, 2003.

[20] X. W. Wang, Z. Y. Zhang, and S. X. Zhou, "Preparation of nanocrystalline $SrTiO_3$ powder in sol-gel process," *Materials Science and Engineering B*, vol. 86, no. 1, pp. 29–33, 2001.

[21] Z. Surowiak, M. F. Kupriyanov, and D. Czekaj, "Properties of nanocrystalline ferroelectric PZT ceramics," *Journal of the European Ceramic Society*, vol. 21, no. 10-11, pp. 1377–1381, 2001.

[22] H. Fan and H. Kim, "Microstructure and electrical properties of sol-gel derived $Pb(Mg_{1/3}Nb_{2/3})_{0.7}Ti_{0.3}O_3$ thin films with single perovskite phase," *Japanese Journal of Applied Physics*, vol. 41, pp. 6768–6772, 2002.

[23] Y. S. Chang, I. G. Chen, and G. J. Chen, "Synthesis and characterization of zinc titanate doped with magnesium," *Solid State Communications*, vol. 128, no. 5, pp. 203–208, 2003.

[24] N. Obradović, N. Labus, T. Srećković, D. Minić, and M. M. Ristić, "Synthesis and characterization of zinc titanate nanocrystal powders obtained by mechanical activation," *Science of Sintering*, vol. 37, no. 2, pp. 123–129, 2005.

[25] H. T. Kim, Y. H. Kim, and J. D. Byun, "Phase transformation and thermal stability in zinc magnesium titanates," *Journal of the Korean Physical Society*, vol. 32, no. 1, pp. S159–S161, 1998.

[26] H. T. Kim, Y. H. Kim, and J. D. Byun, "Microwave dielectric properties of magnesium modified zinc titanates," *Journal of the Korean Physical Society*, vol. 32, no. 1, pp. 346–348, 1998.

[27] A. Golovchansky, H. T. Kim, and Y. Kim, "Zinc titanates dielectric ceramics prepared by sol-gel process," *Journal of the Korean Physical Society*, vol. 32, no. 3, pp. S1167–S1169, 1998.

[28] H. T. Kim, J. D. Byun, and Y. Kim, "Microstructure and microwave dielectric properties of modified zinc titanates (I)," *Materials Research Bulletin*, vol. 33, no. 6, pp. 963–973, 1998.

[29] L. M. Wang, "Applications of advanced electron microscopy techniques to the studies of radiation effects in ceramic materials," *Nuclear Instruments and Methods in Physics Research B: Beam Interactions with Materials and Atoms*, vol. 141, no. 1-4, pp. 312–325, 1998.

[30] O. M. Hemeda and M. El-Saadawy, "Effect of gamma irradiation on the structural properties and diffusion coefficient in Co-Zn ferrite," *Journal of Magnetism and Magnetic Materials*, vol. 256, no. 1-3, pp. 63–68, 2003.

[31] G. A. El-Shobaky, A. M. El-Shabiny, A. A. Ramadan, and A. M. Dessouki, "Effect of gamma irradiation on microstrain and lattice parameter of Co_3O_4 loaded on Al_2O_3," *Radiation Physics and Chemistry*, vol. 30, no. 4, pp. 233–236, 1987.

11

Theoretical Simulations of Reactive and Nonreactive Scattering of Light Diatomic Molecules from Metal Surfaces: Past, Present, and Future

C. Díaz

Departamento de Química Módulo 13, Universidad Autónoma de Madrid, 28049 Madrid, Spain

Correspondence should be addressed to C. Díaz; cristina.diaz@uam.es

Academic Editor: Qiaohui Fan

In everyday life we are surrounded by surfaces and, therefore, by phenomena involving molecule-surface interactions. Furthermore, the processes of heterogeneous catalysis, which are governed by molecule-surface interactions, are of huge practical importance, because the production of most synthetic compounds involves catalytic processes, which explains the tremendous effort that surface science scientists have invested to understand the basic principles underlying elementary interactions between light molecules and surfaces. This effort was recognized in 2007 with the Nobel prize in chemistry awarded to Gerhard Ertl. Here we revise some of the most relevant studies performed so far in this field. We also point out the major challenges that the surface science community may face in this field in the years to come.

1. Introduction

In everyday life we are surrounded by phenomena involving molecule-surface interactions. For example, the corrosion of a coin is due to the interaction between the oxygen molecules in the air and the metal surface atoms, which cause a structural damage leaving a layer of oxidized material (rust) on the coin. Another example is the green appearance of the domes of some buildings, which is due to the oxidation of copper, material from which domes are made. Atomic and molecular interactions on surfaces also play a key role in many industrial processes, such as corrosion, friction, lubrication, oxidation, hydrogen storage, and heterogeneous catalysis. Heterogeneous catalysis is of tremendous practical importance. The production of most synthetic compounds involves catalytic processes because most of the chemical reactions relevant to chemical industries are too slow in the absence of a catalyst. Therefore, understanding the basic principles that govern the geometry and electronic structure of metal surfaces and the elemental processes occurring on them, such as molecular reactivity and molecular scattering, has been and still is one major scope in surface science. At this point, it should be noticed that the importance of

this research field was recognized in 2007 with the Nobel prize in chemistry awarded to Professor Ertl for the detailed description of the sequence of elementary molecule-surface reactions by which vast quantities of ammonia are produced [1]. Ammonia production is basic for the fertilizers industry.

Metal surfaces serve as catalysts for many chemical reactions. Some of these reactions, more efficient on a metal surface than in the gas phase, are, for example, hydrogenation of O, C, N, and S to obtain H_2O, CH_4, NH_3, and H_2S; oxidation of ammonia to nitric acid, which is a basic reaction in the production of fertilizers; methanol synthesis from CO and H_2; oxidation of ethylene to ethylene oxide, a basic reaction in the production of antifreezes; and dehydrogenation of butane to butadiene, a reaction of primary importance in the production of synthetic rubber.

A detailed knowledge, at atomic scale, of the dynamic processes that govern the molecular reactions on surfaces is essential to design and develop new and improved catalysts. In this regard, detailed theoretical studies of these kinds of processes are of most fundamental interest. Theoretical simulations are used not only to provide accurate insights into experimental results, but also to predict new trends. They can be used to decrease the number of trial and

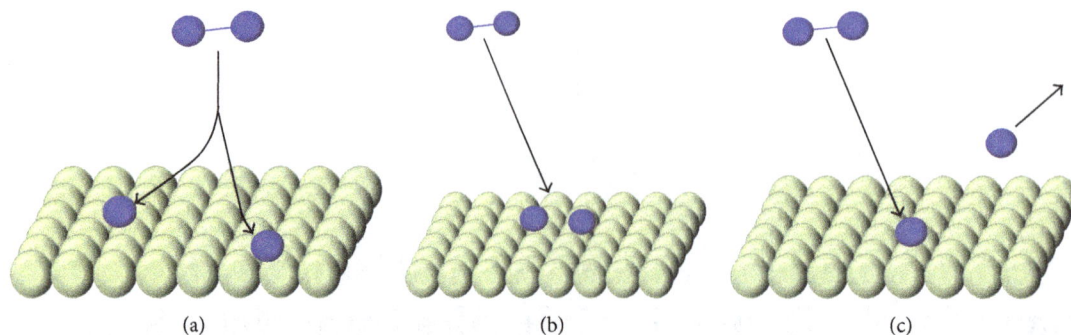

FIGURE 1: (a) dissociative adsortion mechanism; (b) molecular adsorption mechanism; and (c) abstraction mechanism.

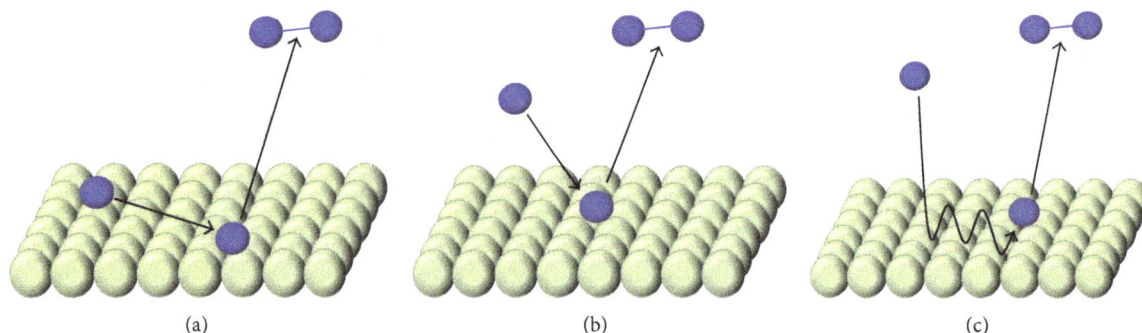

FIGURE 2: Schematic representation of the (a) Langmuir-Hinshelwood mechanism; (b) Eley-Rideal mechanism; and (c) hot-atom mechanism.

error experiments conventionally used in the design of new devices. For example, five chemical compounds could be combined in thousands of different ways to build a catalyst; the use of theoretical simulations can reduce the number of experiments needed to optimize the structure of this catalyst.

A wealth of chemical and physical processes, involving light molecules, are possible on surfaces. These processes can be divided into three main groups.

(1) Molecular adsorption (see Figure 1): it is a physical process in which a molecule coming from the vacuum (gas phase) hits a surface somehow and sticks on it. This process may induce a relatively high concentration of molecules (or atoms) at the place of contact, and, therefore, to the formation of a molecular or atomic film on the surface. Four primary physical mechanisms are associated with the molecular adsorption mechanism:

   (a) molecular physisorption: a molecule, coming from the vacuum, gets adsorbed on the surface and weakly binds to it, due to van der Waals forces;

   (b) molecular chemisorption: a molecule, coming from the vacuum, gets adsorbed on the surface and strongly binds to it, due to the formation of new chemical bonds;

   (c) dissociative chemisorption: a molecule, coming from the vacuum, breaks its bond and its atoms

get adsorbed on the surface thanks to the formation of new chemical bonds;

   (d) abstraction mechanism: a molecule, coming from the gas phase, breaks its bond; one of its atoms gets absorbed on the surface and the other one escapes back to the vacuum.

(2) Molecular desorption (see Figure 2): it is the opposite process to the molecular adsorption one. In this case a molecule previously absorbed on the surface is released. Physical mechanisms associated with molecular desorption are

   (a) Langmuir-Hinshelwood: two atoms or molecules adsorbed on a surface, in thermal equilibrium with it, meet each other and react; as a result a new molecule is formed and leaves the surface;

   (b) Eley-Rideal: an atom coming from the vacuum reacts with an atom adsorbed on the surface (in thermal equilibrium with it) forming a new molecule, which desorbs from the surface;

   (c) Hot-atom mechanism: a reaction takes place between an adsorbed atom, in thermal equilibrium with the surface, and an atom that has recently arrived from the vacuum, which is not in thermal equilibrium yet.

(3) Molecular scattering (see Figure 3): a molecule, coming from the vacuum, collides with the surface and

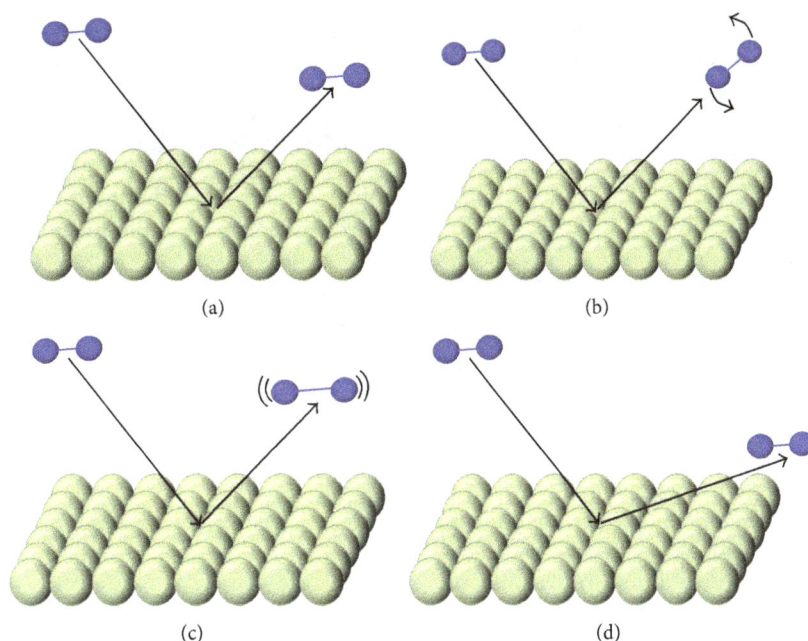

FIGURE 3: Schematic representation of the (a) elastic scattering mechanism; (b) rotational excitation mechanism; (c) vibrational excitation mechanism; and (d) diffraction mechanism.

is reflected back to it. During this physical process the molecule can transfer (gain) energy to (from) the surface. Based on this energy transfer we can distinguish the following phenomena:

(a) elastic scattering: there is not transfer of energy between the molecule and the surface during the collision;

(b) vibrationally inelastic scattering: the molecular vibrational energy increases or decreases during the collision, due to an energy transfer towards (from) the molecular vibrational motion from (toward) the surface;

(c) rotationally inelastic scattering: the molecular rotational energy increases or decreases during the collision, due to an energy transfer towards (from) the molecular rotational motion from (towards) the surface;

(d) diffraction: the parallel (to the surface) momentum of the molecule changes by discrete quantities, due to the periodicity of the surface; as a result the angular distribution of the scattered molecule presents a discrete peaks distribution.

In this review we will focus mainly on three of these mechanisms: (i) dissociative chemisorption [2, 3]; (ii) molecular scattering; and (iii) diffraction [4–6].

2. Experimental Techniques in Brief

Molecule-surface interactions experiments are, in general, carried out in ultrahigh vacuum systems (see [7] and references therein). These systems are formed by a number of connected chambers, which can be individually pumped. They contain a number of skimmers and orifices that are used to create a supersonic molecular beam (SMB) from an initial pulsed nozzle source, at high stagnation pressure. These SMBs are well defined by its stream velocity and the width of its velocity distribution. The average translational energy of the molecular beam can be varied by heating or cooling the nozzle.

SMB experiments can be used to measure the dissociative adsorption probability of a molecule-surface system using the so-called King and Wells' method [8]. In this method, the partial pressure of the molecular gas is monitored as a function of time using a quadrupole mass spectrometer (QMS). At the beginning of the measurement the supersonic beam of molecules is produced in a secondary chamber equipped with a shutter, which can intercept the beam. While the shutter intercepts the molecular beam, the QMS, located in the main chamber, measures the background signal P_0. Once the shutter is turned down, the beam enters the main chamber increasing the signal measured by the QMS to P_1. In the main chamber a second shutter intercepts the beam preventing it from striking the surface target. When this second shutter is removed from the path of the beam, the molecules can hit and stick on the surface. Whenever adsorption takes place the QMS signal decreases as $P(t)$. Thus, the relative decrease of the QMS signal gives us the absolute dissociation probability

$$R(t) = \frac{P_1 - P(t)}{P_1 - P_0}. \tag{1}$$

The dissociative adsorption probability can be also measured using temperature-programmed desorption (TPD) techniques (see, e.g., [9]). In this case, the molecules, previously adsorbed on a surface, are desorbed by heating the surface and simultaneously are detected by a mass spectrometer. In these kinds of experiments, the adsorption probabilities are obtained from coverage versus exposure measurements. Thus, the adsorption probability is determined directly as the ratio of the number of molecules that adsorb to the number of molecules that strike the surface. The number of the adsorbed molecules is proportional to the area under the TPD trace. The reliability of the TPD technique lies in the validity of the detailed balance principle [10], which assumes that the dissociation probability can be measured by looking at the reverse process, the molecular desorption.

Associative desorption measurements are also performed using time-of-flight (TOF) techniques [11], which allow obtaining molecular-state-specific probabilities. In general, TOF experiments are performed on a two-chamber apparatus. One of the chambers contains the crystal into which the molecules are adsorbed after permeation through the bulk. The molecules desorbing from the surface are then probed by laser ionization detection in the second chamber. And the TOF distributions are obtained by recording the time of flight of the photoions to a multichannel plate detector. To determine quantum state distributions, the relative signal intensity for each quantum state desorbed from the surface is compared with the corresponding signal obtained with an effusive Knudsen source of molecules.

TOF experiments are also used to measure rotational and vibrational inelastic scattering probabilities, in combination with stimulated Raman pumping (SRP) [12] and resonance enhanced multiphoton ionization (REMPI) techniques. SRP can be used to excite vibrationally the initial molecular beam, thus selecting the initial vibrational state. For this aim, the initial molecular beam is crossed with two focused laser beams, which are chosen in such a way that the frequency difference between them matches the vibration of the molecule, allowing an efficient excitation of the molecules from the ground state to an excited state. The rotational and vibrational populations as well as the quadrupole alignment of the scattered molecules can be determined using REMPI [13, 14]. In applying this technique, the probe laser beam is focused on the molecular beam few mm in front of the surface where it ionizes the molecules, and the ions are collected and detected with a microchannel electron multiplier plate.

SMB techniques can also be used to measure diffractive scattering probabilities [15]. In these kinds of experiments the monochromatic character of the molecular beams plays a crucial role; spread-velocity beams do not allow observing diffraction peaks. To measure diffraction from SMB experiments two types of apparatus are commonly used: (i) in the fixed-angle setups, the angle between the incident and the reflected beam is fixed; that is, $\Theta_i + \Theta_f = $ const, and the angular distributions of the diffracted particles are measured by rotating continuously the crystal and thus the incident angle Θ_i varies during data acquisition; (ii) in the rotary setups, the detector is able to rotate around the crystal from a given incidence angle. Therefore, the diffraction beam can be measured in a large region of the reciprocal space. The

advantage of this latter setup is the possibility of determining absolute diffraction probabilities.

3. Molecule-Surface Dynamics Simulations

Since the early 90s the interactions of light molecules with surfaces have received more and more attention from theoretical surface scientists, essentially due to the development of multidimensional quantum dynamics methods [2, 3, 16, 17] and the development of interpolation methods able to build flexible potential energy surfaces (see [18] and references therein).

Molecule-surface interactions have been usually described within the static surface Born-Oppenheimer approximation (SS-BOA). The validity of the SS approximation is strongly supported by the mass mismatch between the atoms of the molecule and the metal atoms of the surface whereas the BOA is supported by the velocity mismatch between nuclei and electrons, the latter ones being faster by few orders of magnitude than the former ones. Within the SS-BOA we describe the motion of the nuclei on a continuum potential energy surface (PES). Thus, the first step in any adiabatic simulation is to determine the electronic structure of the system, that is, the electronic landscape on which the nuclei move.

3.1. Potential Energy Surfaces. The first PES describing the electronic structure of a molecule-surface system was published in 1932 by Lennard-Jones [19]. This PES was built based on an analytical expression. Similar PESs based on London-Eyring-Polanyi-Sato (LEPS) potentials have been quite popular since then (see, e.g., [20–26]). Other analytical PESs are based on symmetry adapted functions [27, 28].

Although, for a number of molecule-surface systems, these analytical PESs have been able to describe successfully their electronic structure, the lack of flexibility of these PESs has impelled the development of methods based on interpolation of density functional theory (DFT) data. Generally speaking, the idea behind all these interpolation methods is quite similar. The first step is always to build a DFT data set; to this aim a number of high- and low-symmetry configurations (see Figure 4) are selected. For each configuration, defined by the position of the molecule over the surface (X and Y) and its orientation (θ and ϕ), a set of r (atom-atom distance) and Z (molecule-surface distance) values are computed—see Figure 5 for coordinates definition. In a second step the interpolation is performed over this data set.

First interpolation methods were applied to the development of low dimensional PESs [29, 30]. But, nowadays, to construct a PES based on the interpolation of a DFT data set, including all the molecular degrees of freedom (DOFs), has become a routine work thanks to the development of methods such as the corrugation reducing procedure (CRP) [31], the modified Shepard (MS) method [32, 33], and the neural networks (NN) method [34].

The key idea behind the CRP method is that most of the corrugation of a molecule-surface PES is due to the atom-surface interactions and, therefore, the subtraction of this

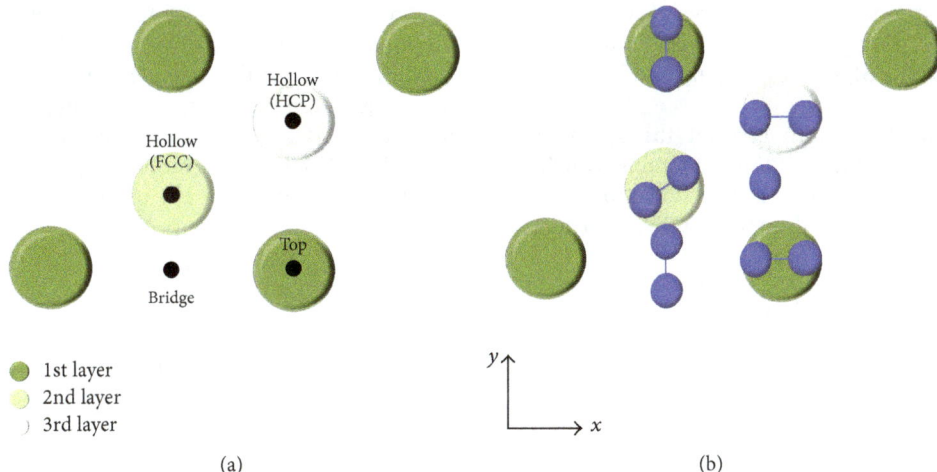

(a)   (b)

FIGURE 4: Schematic representation for a FCC(111) surface of (a) three high-symmetry surface sites; (b) some representative configurations of a diatomic molecule on the surface.

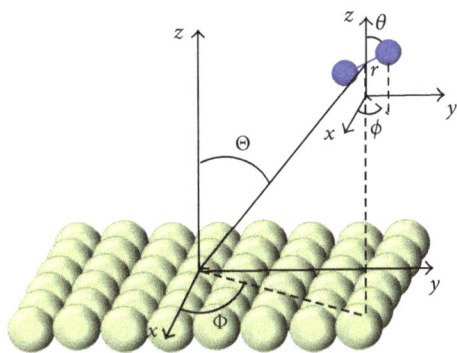

FIGURE 5: Schematic representation of the coordinate system.

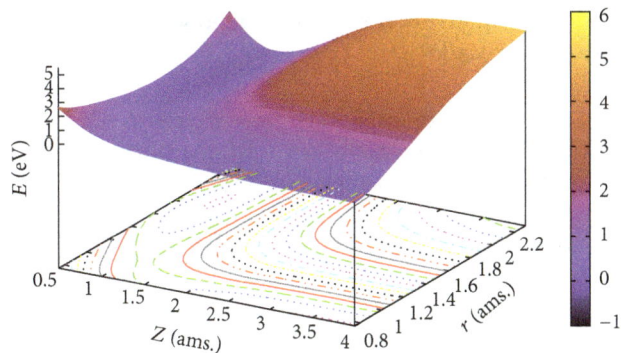

FIGURE 6: 2D cut through the H_2/Cu(111) PES obtained by applying the CRP interpolation method to set of DFT-PW91 data. This 2D cut represents a H_2 molecule dissociating on a bridge site (see Figure 4).

atomic contribution from the PES leads to a much smoother function, which can be interpolated more accurately. Thus, the 6D PES can be written as

$$V_{6D}^{CRP} = S_{6D}\left(X, Y, Z, z, \theta, \varphi\right) + \sum_{i=1}^{2} R_i\left(X_i, Y_i, Z_i\right), \quad (2)$$

where S_{6D} is the smoother function (two-body term) and R_i are the 3D PES representing the atom-surface interactions (one-body term). In the CRP method the interpolation is performed by combining analytical functions and numerical techniques. The major advantage of the CRP method is that the precision of the PES can be systematically improved by adding extra DFT data, whereas the major disadvantage is that it cannot be extended straightforwardly to describe polyatomic molecules (more than two atoms) interacting with surfaces. To date, the CRP method has been successfully used to build PESs for a wide variety of molecule-surface systems, for example H_2/Pd(111) [31], H_2/Ni(100), and H_2/Ni(110) [35], H_2/Pt(111) and H_2/Cu(100) [36], H_2/Pd(110) [37], N_2/W(110) [38], H_2/NiAl(110) [39], H_2/Pt(211) [40], H_2/Ru(0001) [41], H_2/Cu(110) [42], N_2/W(110) [43], H_2/W(100), and H_2/W(110) [44], O_2/Ag(100) [45], H_2/Cu/Ru(0001), and H_2/Pd/Ru(0001)

[46], H_2/Pd(100) [47], H_2/Cu(111) [48], and H_2/C(1 × 1)-Ti/Al(100) [49]. In Figure 6 we show, as an example, a typical 2D(r, Z) cut of the 6D PES obtained for the system H_2/Cu(111).

In the case of the MS method, initially developed to study gas phase reactions [50, 51], the 6D PES is written as a weighted series of second-order Taylor expansions:

$$V_{6D}^{MS} = \sum_{j=1}^{N_{data}} w_j\left(X_i\right) T_j\left(X_i\right), \quad (3)$$

with X_i representing the coordinates of the molecule, w_j a normalized weigh function, and T_j the second-order Taylor expansion of the potential centered on each data point. The main advantages of this method with respect to the CRP one are that (1) it can be extended straightforwardly to describe the interaction of polyatomic molecules with surfaces and (2) it requires a smaller number of ab initio points, because it focuses the interpolation on the dynamically relevant regions of the PES, which are found by means of classical dynamics calculations. The MS method has been already used to describe the energy landscape of a number of systems, such as H_2/Pt(111) [52], N_2/stepped-Ru(0001) [53], N_2/Ru(0001) [54], H_2/CO/Ru(0001) [55], H_2/Pd(111) [56], and H/H-Si(111) [57].

The NN method takes into account the symmetries underlying the system using nonfitting functions, which do not require any assumption about the functional form of the underlying problem. In this method the 6D PES is written as

$$V_{6D}^{NN} = f_2\left(W_{01}^2 + \sum_j W_{ji}^2 f_1\left(W_{oj}^1 + \sum_{i=1}^6 W_{ij}^1 X_i\right)\right), \quad (4)$$

where X_i represent the six coordinates of the H_2 molecule, f_2 and f_1 are nonlinear functions, and W_{ij} are the parameters of the representation, the so-called weights. This method has been successfully applied to study, for example, the dissociative chemisorption of H_2 on K(2×2)/Pd(100) [34], O_2 on Al(111) [58], and O_2 on Ag(111) [59]. Furthermore, this method can also be combined with the CRP method. For example, Ludwig and Vlachos [60] used a combination of these two methods to construct a PES representing the energy landscape for H_2/Pt(111).

3.2. Dynamics. Within the BOA, once the electronic potential (the PES) is known, we can solve the time-dependent nuclear Schrödinger equation:

$$\widehat{H}\Psi(\mathbf{R}, \mathbf{r}; t) = i\frac{\partial \Psi(\mathbf{R}, \mathbf{r}; t)}{\partial t}. \quad (5)$$

In this equation, $\Psi(\mathbf{R}, \mathbf{r}; t)$ represents the wave function describing the system, $\widehat{H}$ its Hamiltonian, and $\mathbf{R}$ and $\mathbf{r}$ represent the molecular center of mass and the molecular internal coordinates, respectively. In the case of a diatomic molecule, the Hamiltonian (in atomic units) can be written as

$$\widehat{H} = -\frac{1}{2M}\frac{\partial}{\partial Z^2} - \frac{1}{2M\sin^2\gamma}\left[\frac{\partial}{\partial X^2} - 2\cos\gamma\frac{\partial}{\partial X}\frac{\partial}{\partial Y} + \frac{\partial}{\partial Y^2}\right]$$
$$- \frac{1}{2\mu}\frac{\partial}{\partial r^2} + \frac{\widehat{j}^2}{2\mu r^2} + V_{6D}(X, Y, Z, r, \theta, \phi). \quad (6)$$

In this equation γ is the angle between the X and Y coordinate axis [61] ($\gamma = 90°$ if Cartesian coordinates are used). M and μ are the total and reduced mass of the molecule, respectively; $\widehat{j}$ is the rotational operator and V_{6D} the PES.

There are several methods to solve the time-dependent Schrödinger (TDS) equation. In the time-dependent wave packet (TDWP) method as implemented by Kroes et al. [17, 62] the TDS is solved by numerically exact propagation of the wave packet using a time-independent basis-set. This method is divided into three steps:

(1) the choice of the initial wave packet;

(2) the propagation of the wave packet;

(3) the asymptotic analysis.

So that, at the end of the calculation, we obtain the monoenergetic state-resolved scattering probabilities $S(v, J, n, m; E)$, with v and j being the vibrational and the rotational quantum numbers, E the incidence energy of the molecule, and (n, m) the diffraction state. From these $S(v, J, n, m; E)$ probabilities, the monoenergetic dissociative adsorption probabilities $(S(v, J; E))$ are computed as

$$S(v, J; E) = 1 - \sum_{n,m} S(v, J, n, m; E). \quad (7)$$

For further details on this method see [3, 17] and references therein.

A promising alternative method that can be also used to solve (5) is the so-called multiconfiguration time-dependent Hartree method (MCTDH) [63, 64]. The main idea behind this method is to expand the wave function in a basis-set where not only the coefficient of the expansion, but also the basis functions themselves are time-dependent. The use of time-dependent basis-sets reduces their size, decreasing the computational cost with respect to the TDWP method.

Although, in general, the nuclear motion has to be described using quantum dynamics, classical dynamics is a very useful tool to get simple physical interpretations of quantum results and experimental measurements. In the classical dynamics method a classical trajectory is obtained by integration of the classical equations of motion.

We can integrate either the Hamilton equations of motion

$$\dot{q}_i = \frac{\partial H}{\partial p_i},$$
$$\dot{p}_i = -\frac{\partial H}{\partial q_i}, \quad (8)$$

q_i and p_i being the coordinates and the conjugated momenta of the system, respectively, or the Newton equations of motion

$$M_i\frac{\partial^2}{\partial t^2}\mathbf{R_i} = -\frac{\partial}{\partial \mathbf{R_i}}V_{6D}(\mathbf{R_i}). \quad (9)$$

Disregarding the set of equations used, a classical dissociation probability is computed by averaging over internal coordinates and conjugated momenta of the molecules, which can be sampled using a standard Monte Carlo method.

In performing classical dynamics we can distinguish between pure classical and quasiclassical dynamics. In the latter, the zero point energy (ZPE) of the molecule is included in the calculation, whereas in the former the ZPE is assumed to be zero. In general, pure classical dynamics yields better results for nonactivated systems (see below for definition), and quasiclassical dynamics yields better results for activated systems (see below). For activated systems the ZPE may play a significant role due to the so-called vibrational softening, which happens whenever a molecule approaches an attractive surface. In this case, the attractive force between the atoms of the molecule and the surface becomes larger than the intramolecular force, and as a result the force constant associated with the vibrational motion is reduced, which induced

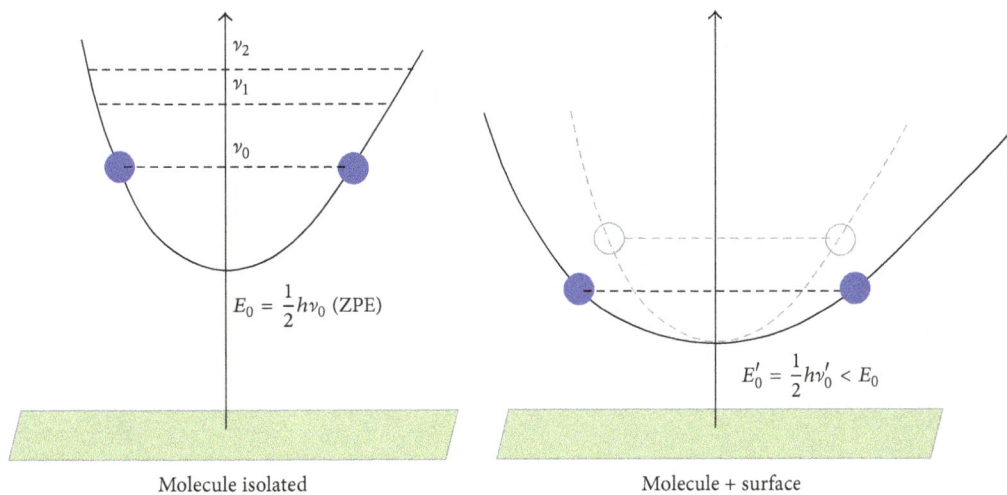

FIGURE 7: Schematic representation of the vibrational softening.

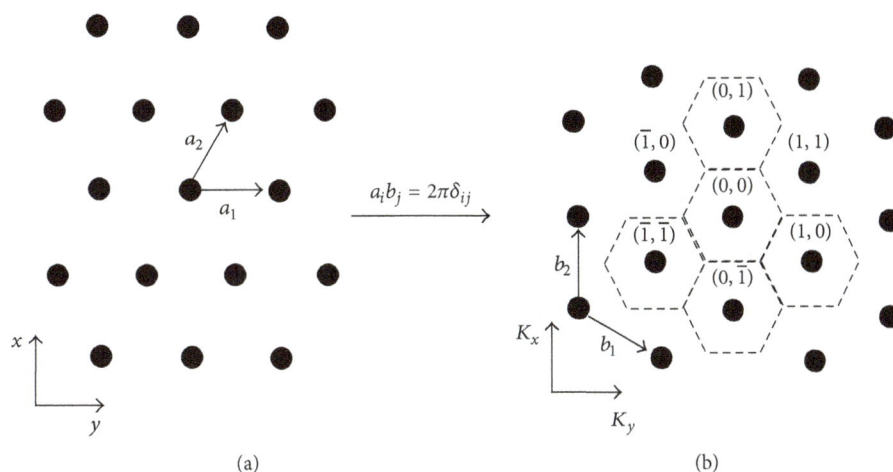

FIGURE 8: Real lattice (a) and reciprocal lattice (b) for a FCC surface. The numbers within parentheses indicate the diffraction peaks. The dashed hexagons show the Wigner-Seitz cells corresponding to some of the lattice points.

a decrease of the potential well curvature and, therefore, a decrease of the ZPE value (see Figure 7). This adiabatic energy transfer from vibration to translation is used by the molecule to overcome the reaction barrier and dissociate.

Classical dynamics can also be used to study quantum observables, such as rotational and vibrational quantum numbers [65, 66] and even diffraction probabilities [67–69]. Simulating rotational excitation using classical dynamics is possible by evaluating the closest integer that satisfied the quantum rigid rotor formula $[-1 + (1 + 4L^2/\hbar^2)^{(1/2)}]/2$, L being the classical angular momentum of the molecule. In the case of vibrational excitation, vibrational quantum numbers can be simulated by evaluating the closest integer that satisfied $S/\pi - 1/2$, S being the action variable. To simulate diffraction probabilities, the parallel momentum change is discretized by dividing the reciprocal space using as pattern the Wigner-Seitz cell associated with each lattice point (n, m) (see Figure 8). Then, the *classical diffraction* probability, for a given diffraction peak (n, m), is given by the number of trajectories

leading to a parallel momentum change contained in the Wigner-Seitz cell of the (n, m) peak, divided by the total number of trajectories.

4. A Little of History: From Low to High Dimensional Simulations

First quantum dynamics calculations of dissociative adsorption of a diatomic molecule on a metal surface were performed by Jackson and Metiu [70], for $H_2/Ni(100)$. In this study only the molecular bond (r) and the distance molecule-surface (Z) (see Figure 5) were taken into account; that is, only the vibrational and the translational motions towards the surface were properly described. Two years later a similar study was performed for $H_2/Cu(100)$ [71]. In 1992 Sheng and Zhang published the first 3-dimensional (3D) quantum study for $H_2/Ni(100)$ [72], including, in addition to r and Z, the polar rotational motion θ (see Figure 5). Similar 3D studies were performed by Mowrey [73] for $H_2(D_2)/Ni(100)$

and by Darling and Holloway [74] and Dai and Zhang [75] for dissociative adsorption of H_2 on several Cu surfaces. Other 3D simulations were performed by taking into account r, X (translational motion along the line joining two surface atoms neighbors—see Figure 5) [76]. More complex simulations including a forth degree of freedom were carried out since 1994 [77, 78]. These simulations included r, Z, θ, and ϕ [24, 79, 80], with ϕ representing the molecular azimuthal motion (see Figure 5) or r, Z, X, and Y-Y being the axis perpendicular to X (see Figure 5).

Full dimensional quantum calculations, including the 6 DOFs of the molecule, were performed for H_2/Pd(100) in 1995 [28] and later on for H_2/Cu(111) [61] and H_2/Cu(100) [81, 82]. See Section 5 for more examples.

5. Six-Dimensional Molecule-Surface Simulations

When a light diatomic molecule approaches a metal surface, at low energy (up to few eV's), the molecule can either dissociate or get reflected. Relative to the behavior of the dissociative adsorption probability as a function of the molecular incidence energy, diatomic molecule-metal surface systems are classified as activated and nonactivated systems (see Figure 9). Activated systems show a monotonous increase of the dissociative adsorption probability as a function of the incidence energy. This behavior is due to the presence of a minimum reaction barrier (MRB) in PES. At this point, it should be pointed out that the MRB is only one of the characteristics of the PES that influence the interaction between a molecule and a surface, but it is not the only one, as it is shown in the following examples. If the PES does not present a MRB (although other barriers are present in the PES), the dissociative adsorption probability exhibits a nonmonotonous behavior. The dissociation probability first decreases when the incidence energy increases, reaching a minimum, and then it increases with the incidence energy. This nonmonotonous behavior is associated with the so-called dynamic trapping [83, 84]. At low incidence energies the molecules are attracted by the attractive regions of the PES and get trapped on the surface, rebounding several times until they find a reactive path, that is, a path without a barrier, and dissociate. The trapping probability decreases when the incidence energy increases, decreasing the dissociation probability. On the contrary, the direct dissociation probability, which only depends on the dissociation barriers, increases monotonously with the incidence energy. The combination of both phenomena yields this characteristic nonmonotonous behavior (see Figure 10). In the following, we show few significant examples of both activated and nonactivated systems.

5.1. Dissociative Adsorption of H_2 on Metal Surfaces. The dissociative adsorption probability of H_2 on different metal surfaces has been widely studied during the last 20 years. This phenomenon is the first step in hydrogenation and dehydrogenation processes, which are of most importance for many industrial processes. From a purely theoretical point of view, the interest in these systems resides in the fact that

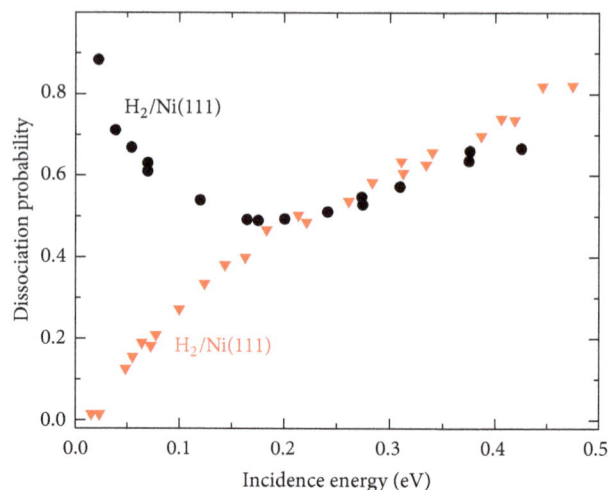

FIGURE 9: Dissociation probability of H_2 on Ni(111) (black circles) and Ni(110) (red triangles). These systems are nonactivated and activated, respectively. Experimental data are taken from [176].

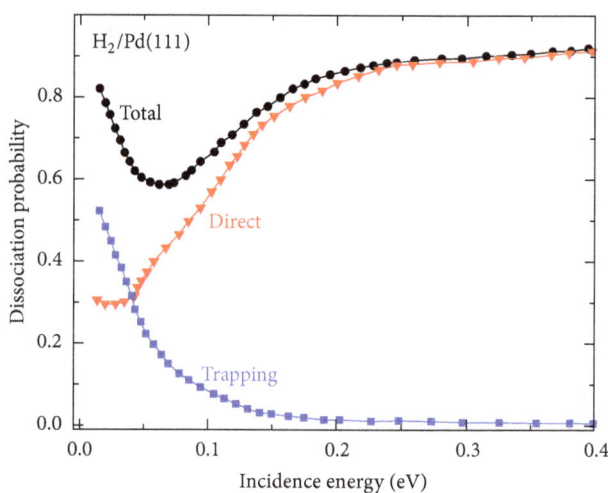

FIGURE 10: Theoretical dissociation probability versus incidence energy for H_2/Pd(111). Black line with circles: total probability; red line with triangles: direct probability; blue line with squares: trapping probability.

they are the simplest molecule-surface systems and, therefore, theoretical models and hypothesis can be more easily tested on them.

Since the development of efficient methods to build 6D PES, the number of theoretical simulations performed to study the dissociative adsorption of H_2 on metal surfaces, pure metal, alloys, and precovered-metal surfaces has experienced a significant increase. For example, in Figure 11 we show the dissociative adsorption probability of H_2, in its rovibrational ground state ($v = 0, J = 0$), for a number of activated surfaces—this is merely a sample of systems, by no means a complete list. From this figure two conclusions can be extracted: (i) although, qualitatively speaking, for all the systems the dissociative adsorption probability increases

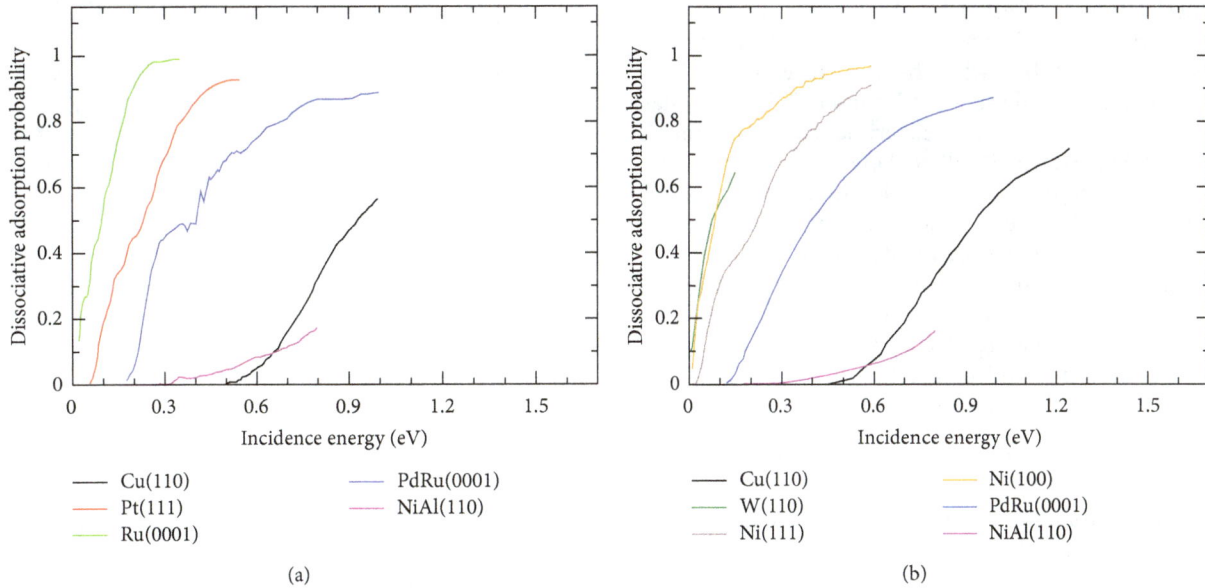

(a)   (b)

FIGURE 11: Dissociative adsorption probability of H_2, in its rovibrational ground state ($v = 0, J = 0$), as a function of the incidence energy. (a) Quantum calculations for Cu(110) (black line) [105]; Pt(111) (red line) [177]; Ru(0001) (green line) [41]; PdRu(0001) (blue line) [178]; NiAl(110) (magenta line) [179]. (b) Quasiclassical calculations for Cu(110) (black line) [42]; W(110) (green line) [44]; Ni(111) (brown line) [35]; Ni(100) (orange line) [35]; PdRu(0001) (blue line) [46]; NiAl(110) (magenta line) [39].

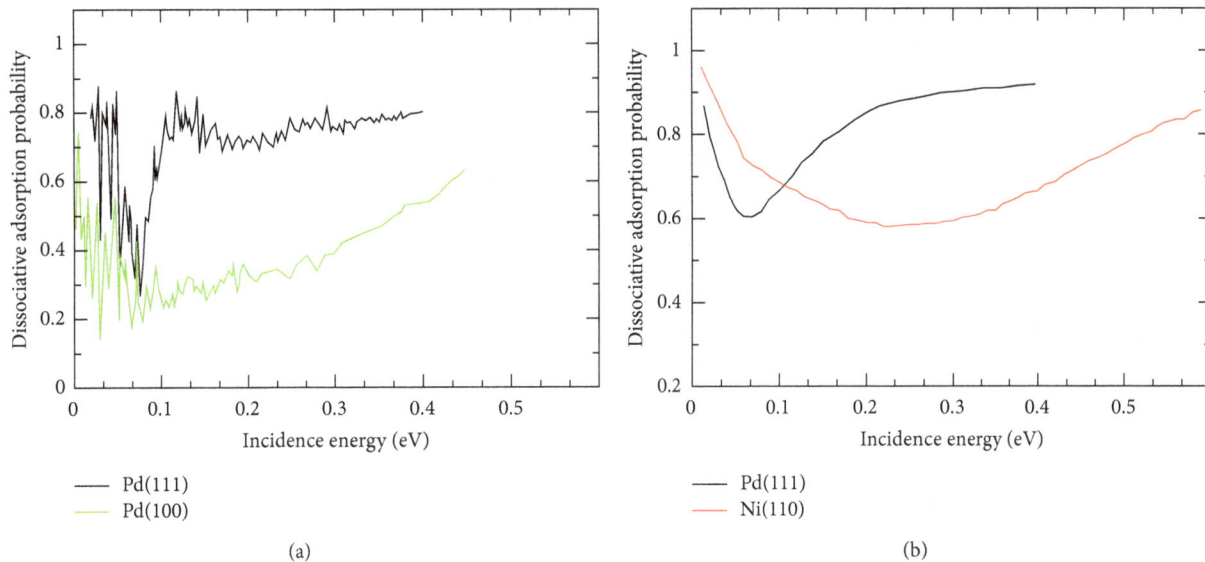

(a)   (b)

FIGURE 12: Dissociative adsorption probability of H_2, in its rovibrational ground state ($v = 0, J = 0$), as a function of the incidence energy. (a) Quantum calculations for Pd(111) (black line) [62]; Pd(100) (green line) [28]. (b) Classical calculations for Pd(111) (black line) [83]; Ni(110) (red line) [35].

monotonously when the incidence energy increases, the minimum energy leading to reaction and the slope of the curve vary from one system to another; (ii) by comparing quantum (Figure 11(a)) and classical (Figure 11(b)) calculations we can conclude that classical (quasiclassical, in fact) dynamics simulations give qualitative good results, with a smaller computational cost.

In Figure 12 we show some examples of dissociative adsorption of $H_2(v = 0, J = 0)$ on nonactivated surfaces. Once again we can see that the subtle behaviour of the

dissociative adsorption probability as a function of the incidence energy is surface-dependent and that the classical trajectory method gives pretty good results in comparison with the more sophisticate, and computational demanding time, quantum calculations.

Among the systems mentioned above, H_2/Cu has been considered for long time the prototypical activated system, which explains the huge number of experimental [85–98] and theoretical [21, 24, 42, 61, 65, 75, 78, 79, 99–113] works devoted to this system. As a result of this important theoretical effort, H_2/Cu(111) has been the first system in which theoretical simulations have been able to reproduce experimental observables with chemical accuracy [111]. To achieve this aim, beyond using a well chosen DFT functional, an appropriate comparison with experimental measurements was needed. Experimentally, dissociative adsorption probabilities are measured using SMB techniques. To know the characteristics of this beam, that is, its molecular rovibrational distribution and its energy width, is crucial to perform a meaningful comparison between theory and experiment. Thus, in order to compare with experiments, the theoretical monoenergetic state-resolved probabilities have to be convoluted using the SMB parameters. This convolution is performed through the following steps:

(1) The monoenergetic state-resolved dissociation probabilities $S(v, J; E)$, obtained from the simulations, are used to compute monoenergetic probabilities $S(T_n; E)$, which only depend on nozzle temperature and on the incidence energy as

$$S(T_n; E) = \sum_{v,J} F_B(v, J; T_n) S(v, J; E), \quad (10)$$

with $F_B(v, J; T_n)$ being the Boltzmann factor given by

$$F_B(v, J; T_n) = \frac{(2J + 1) e^{-(E_v/kT_n)} \times w(J) e^{-(E_r/0.8kT_n)}}{N}, \quad (11)$$

where N is the normalization factor and $w(J)$ is the factor characterizing the nuclear spin statistics—for example, in the case of H_2 (D_2) $w(J = 0) = (1/4)(2/3)$ and $w(J = 1) = (3/4)(1/3)$.

(2) The monoenergetic dissociation probabilities $S(T_n; E)$ are used to simulate experimental dissociation probabilities ($S(T_n)$) performing a convolution over the distribution of the molecular beam using the expression

$$S(T_n) = \frac{\int_{v=0}^{v=\infty} f(v; T_n) S(T_n; E) \, dv}{\int_{v=0}^{v=\infty} f(v; T_n) \, dv}, \quad (12)$$

where the flux weighted velocity distribution $f(v; T_n)$ is given by

$$f(v; T_n) \, dv = Cv^3 e^{[-(v-v_0)^2]/\alpha^2} dv, \quad (13)$$

and C (a constant), α (the width of the velocity distribution), and v_0 (the stream velocity) define the experimental molecular beam.

The parameters describing the molecular beam (α, v_0) may vary from one experiment to another one, which explains to a large extent the differences on the dissociation probabilities as a function of the incidence energy found experimentally for some systems. For example, experimental results on dissociative adsorption probabilities for H_2/Cu(111) measured by Rettner et al. [92] are significantly smaller than that obtained by Berger et al. [86] (see Figure 13). At this point, it should be pointed out that for long time there

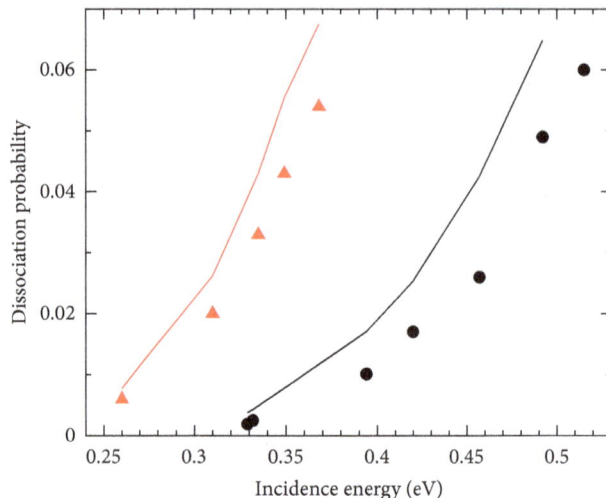

FIGURE 13: Dissociation probabilities as a function of the incidence energy for a pure H_2 molecular beam on Cu(111). Solid lines: quantum theoretical simulations from [111]. Black circles: experimental data from [92]. Red triangles: experimental data from [86].

was not a clear explanation about why, at similar translational energies, these two sets of experimental dissociation probabilities differed that much. The explanation based on the different molecular beam parameters was offered in [111] through a detailed theoretical analysis. In [111] it was shown, on one hand, that an appropriate convolution of the theoretical probabilities, using the correct experimental molecular parameters, allows simulating both experimental data sets. And, on the other hand, an appropriate convolution of the theoretical probabilities is required in order to perform adequate analysis of the experimental data. This kind of convolution is, since then, performed routinely [49, 55].

These theoretical simulations can also be used to analyze the associative desorption [114] probability, and other observables [104], such as the following.

(i) the vibrational (χ_v) and rotational (χ_J) efficacy, which are used to evaluate the effectiveness of the vibrational and rotational energy in promoting reaction. χ_v can be computed as

$$\chi_v = \frac{E_0(v = 0, J = 0) - E_0(v = 1, J = 0)}{E_v(v = 1, J = 0) - E_v(v = 0, J = 0)}, \quad (14)$$

where $E_0(v)$ is the incidence energy needed to obtain a reaction probability S equal to half of the saturation value when the molecule is initially in the vibrational state v and E_v is the vibrational energy of the molecule in the gas phase. And χ_J can be computed by the slope of the line

$$E_0(v = 0, J) = C - \chi_J \times E_J(J), \quad (15)$$

where E_J represents the rotational energy.

(ii) The associative desorption probabilities ($P_D(v, J; E)$) can be computed, invoking detailed balance, from the state-resolved dissociative adsorption probabilities as

$$P_D(v, J; E) = E e^{[-E/kT]} S(v, J; E),\qquad(16)$$

where T is the surface temperature and k the Boltzmann constant.

(iii) The average desorption energies can be computed from state-resolved dissociative adsorption probabilities as

$$\langle E(v, J)\rangle = \frac{\int S(v, J; E)\, E^2 e^{-E/kT}\, dE}{\int S(v, J; E)\, E e^{-E/kT}}.\qquad(17)$$

(iv) The quadrupole alignment parameter ($A_0^{(2)}(J)$) gives us a measurement of the reactivity of the molecule as function of its orientation, that is, as a function of m_J. $A_0^{(2)}(J)$ can be computed using the following equation:

$$A_0^{(2)}(J) = \frac{\sum_{m_J=-J}^{J} S(v, J, m_J; E)\left(3m_J^2 - J(J+1)\right)/J(J+1)}{\sum_{m_J=-J}^{J} S(v, J, m_j; E)}.\qquad(18)$$

Thus, a positive value indicates a preference for reactivity when the molecule is oriented parallel (helicopter) to the surface ($m_J = J$), whereas a negative value indicates, on the contrary, a preference for reactivity when the molecule is oriented perpendicular (cartwheel) to the surface ($m_J = 0$).

(v) The vibrational and rotational excitation probabilities of molecules scattered from surfaces.

5.2. Scattering of H_2 from Metal Surfaces. The scattering of H_2 from metal surfaces at low energy, below 1 eV, can be considered as the complementary process to dissociative adsorption—although, strictly speaking, it is the complementary process to sticking, which includes dissociative adsorption and molecular adsorption. By measuring the molecular distribution as a function of the scattering angle (Θ), and the molecular rotational and vibrational excitation upon scattering [115, 116], we can obtain extra information about the corrugation of the PES and, therefore, about the reactivity of the system. From a theoretical point of view, molecular scattering is a very useful phenomenon to test our tools. The behavior of a scattered molecule depends very much on the subtle characteristics of the PES; therefore the study of the molecular scattering procces allows us to evaluate the accuracy of PES and the accuracy of the dynamics methods.

For example, a detailed comparison between the experimental measurements and the theoretical 6D simulations, for vibrational and rotational survival probabilities and rotational excitation probabilities [65], had revealed the minor role played by nonadiabatic effects on the scattering of H_2 and D_2 from Cu(111) and accordingly on reactivity. On the other hand, TOF simulations for $H_2(v = 1, J = 3)$/Cu(111) [117] have shown that the state-of-the-art adiabatic theoretical models

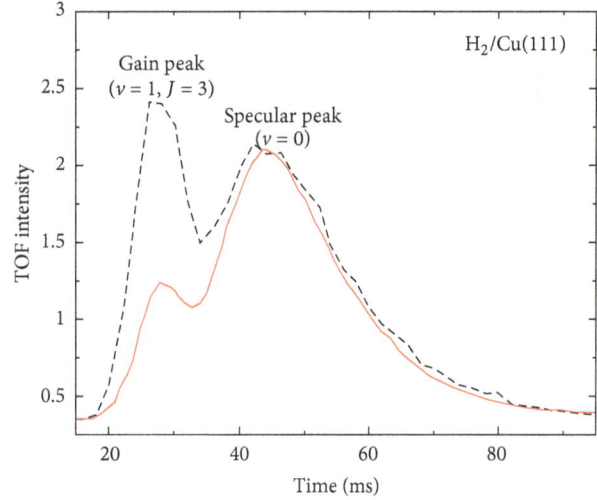

FIGURE 14: TOF spectrum of $H_2(v = 0)$ scattered from Cu(111). Black dashed line: experimental data ($T_s = 400$ k) taken from [91]; red solid line: theoretical data from [117].

underestimate the vibrational excitation. This phenomenon can be observed in Figure 14, where it can be seen that the experimental *gain peak* (at short times), due to vibrational excitation from $H_2(v = 0)$ to $H_2(v = 1, J = 3)$, is three times higher than the simulated one, whereas the agreement between theory and experiment for the *specular peak* is pretty good. In [117] it has been suggested that the discrepancy may be due to the use of the SS-BOA. As we discuss below (see Section 6), one the challenges that surface science theorists will have to face in the next future will be to go beyond this approximation.

The theoretical TOF probabilities shown in Figure 14 were computed by using the following equation:

$$
\begin{aligned}
&f(t, T_n)\\
&= c + N \times \left[\left(\frac{v_i}{v_0}\right)^4 \times \left[-\left(\frac{v_i - v_0}{\alpha}\right)^2\right]\right.\\
&\qquad\times P\left(v = v', J = J' \longrightarrow v', J'\right)\\
&\qquad+ x_t \times \left(\sum_{vJ, v'J' \neq}\left(\frac{v_i^3}{v_s v_0^4}\right)\right.\\
&\qquad\times \left(\frac{1}{x_i v_i^{-2} + x_s v_i v_s^{-3}}\right)\\
&\qquad\times \exp\left[-\left(\frac{v_i - v_0}{\alpha}\right)^2\right]\\
&\qquad\left.\left.\times w_{vJ} \times P\left(v, J \longrightarrow v', J'\right)\right)\right],
\end{aligned}
$$

$$(19)$$

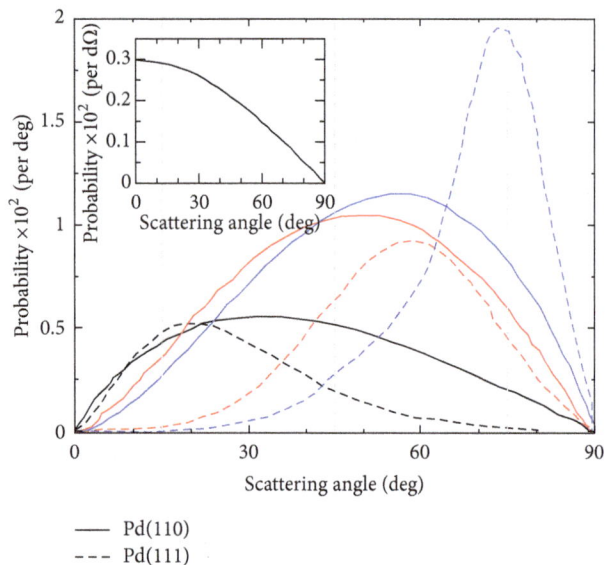

FIGURE 15: Reflection probabilities per degree as a function of the scattering angle (Θ) for H_2/Pd(110) (solid lines) and H_2/Pd(111) (dashed lines). Data are taken from [118, 119], respectively. Black lines: incidence angle, $\Theta_i = 15°$; red lines: $\Theta_i = 45°$; blue lines: $\Theta_i = 75°$. Inset: reflection probability per solid angle as a function of the scattering angle for H_2/Pd(110) ($\Theta_i = 45°$).

where $x_i(x_s)$ represents the distance travelled by the molecule from the source (surface) to the surface (detector), v_s is the velocity of the scattered molecule, $w_{v,J}$ represents the Boltzmann population of the initial molecule state (v, J) in the molecular beam divided by the population of the final state (v', J') (at the nozzle temperature used in the experiment), and $P(v, J \rightarrow v', J')$ is the transition probability from the initial state (v, J) to the final state (v', J').

The angular distributions of H_2, upon scattering from metal surfaces, have been studied, for example, for Pd(111) [118] and Pd(110) [119]. These studies revealed a quite different behavior of the angular distributions (see Figure 15). A detailed analysis of these distributions revealed that the H_2 molecules are reflected closer to Pd(110) than to the Pd(111), and therefore they are more sensitive to the surface corrugation in the case of Pd(100). This analysis also allowed establishing the signature of the dynamic trapping, a cosine-like angular distribution of the reflection probability per solid angle (see inset Figure 15). Here, it is important to remark that the same cosine-like distribution is observed in trapping-desorption experiments [120]. But the origin of trapping-desorption is totally different from the origin of dynamic trapping. The former phenomenon is due to the thermal equilibrium, that is, to the energy exchange between the surface and the molecule, which requires long interaction times (millisecond) in contrast with dynamic trapping that is a faster process (picoseconds).

5.3. Diffraction of H_2 from Metal Surfaces. The diffraction of hydrogen molecules from a metal surface [6] can be

considered as a unique technique to gauge the molecule-surface PES and the dynamics. First detailed comparison between experimental diffraction data and 6D adiabatic quantum calculations was performed on H_2/Pt(111) [121]. The agreement between the experimental diffraction peaks probabilities and the theoretical ones, for several incidence energies, was found to be remarkable—experimental results were extrapolated to 0 K surface temperature using the Debey-Wallet attenuation model (see [122] and references therein). This excellent agreement, together with the very good agreement obtained for dissociative adsorption [62], allowed discarding any significant role of nonadiabatic effects on this system. A good agreement with experimental diffraction peaks was also obtained for D_2/NiAl(110) [123, 124]. Although, in this latter case, the agreement is not that good for rotationally inelastic diffraction (RID). In this case, it has been suggested that these discrepancies may be due to inaccuracies in the PES [124], discarding surface phonons and significant nonadiabatic effects.

The systems mentioned above are activated systems, i.e., under experimental conditions most of the molecules (if not all of them) are reflected, which simplify the uptake of experimental measurements. But diffraction of nonactivated systems represents a major challenge, because under experimental conditions most of the molecules dissociate; therefore, the number of the scattered molecules is very low, and the surface gets *dirty* (covered with hydrogen) very fast. Thus, the surface temperature needs to be kept high enough to avoid hydrogen coverage, increasing phonons background and making more difficult the extrapolation of the measurements to 0 K. First measurements and subsequent simulations on these kinds of systems were carried out on H_2/Pd(111) [123]. The most remarkable result obtained from this study was a pronounced out-of-plane diffraction, showing a highly corrugated PES. This pronounced out-of-plane diffraction was considered to be associated with high reactive systems. This study revealed the importance of scanning the whole space (in-plane and out-of-plane) in order to infer trustworthy properties of the PES. Diffraction was also studied for Pd(110) [125]. In this case, the theoretical simulations showed a large number of diffraction peaks, with very low probability, which prevent the experimental observation. The presence of many diffraction peaks with low intensity was considered a signature of the dynamic trapping.

On the other hand, diffraction measurements of H_2 from Cu(111) have revealed a very low out-of-plane diffraction probability [69]. This result, together with previous results for H_2 diffraction from Pt(111), NiAl(110), Pd(111), and Pd(110), was induced to suggest that the presence of intense out-of-plane diffraction peaks in the diffraction spectrum could be considered a signature of a notable dissociative adsorption probability. However, detailed analyses of the systems H_2/Ru(0001), H_2/Cu(111), and H_2/CuRu(0001) have shown that there is not a lineal relationship between dissociative adsorption and out-of-plane diffraction [69]. In Figure 16 we can observe that H_2/Ru(0001) is the most reactive system, whereas the most intense out-of-plane diffraction is found for H_2/CuRu(0001).

(a)

(b)

(c)

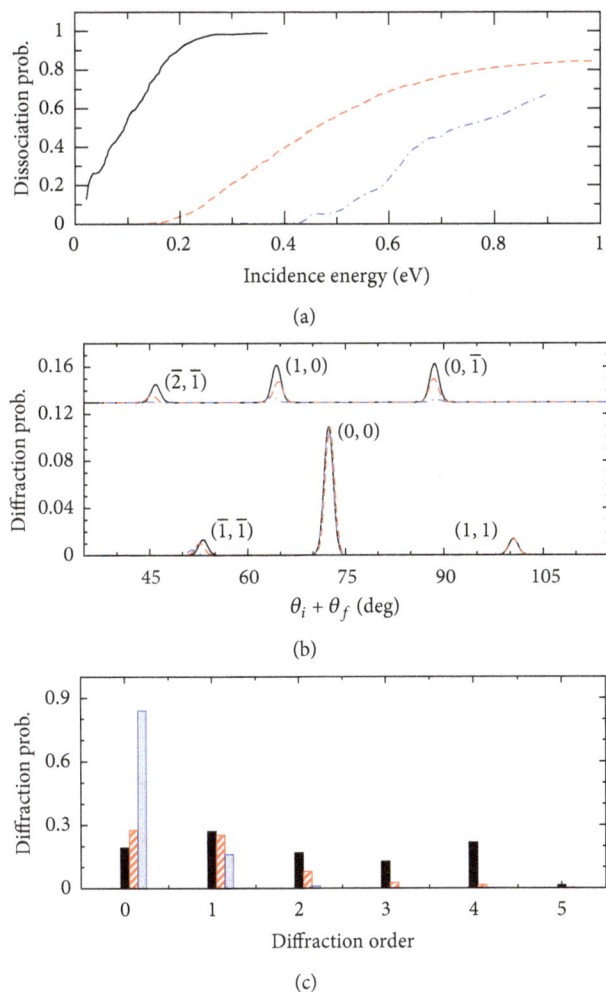

FIGURE 16: (a) dissociative adsorption probability of H_2 as a function of the incidence energy. (b) in-plane and out-of-plane (see Figure 8) diffraction spectra for an incidence energy = 0.075 eV and $\Theta_i = 30°$. Raw data have been convoluted using a Gaussian function of width $\sigma = 0.7°$, and the peaks have been normalized to the specular peak for H_2/CuRu(0001). (c) diffraction probabilities as a function of the diffraction order—defined by concentric hexagons built around the specular peak; peaks lying on the same hexagon belong to the same order (see Figure 8). Solid black bars: CuRu(0001); dashed red bar: Ru(0001); dotted bars: Cu(111).

6. Perspectives and Future Challenges

The surface static Born-Oppenheimer approximation has been essential to advance in the field of surface dynamics, but recent experiments on reactive and nonreactive scattering have questioned its applicability [126–128]. Therefore, some of the major challenges that theorists are facing in the next future are related to the development of theoretical models, which will allow including accurately electrons and phonons excitations [129, 130].

The need for such models is becoming more and more visible as surface scientists are facing more and more complex systems, for example, heavy and polyatomic molecules

interacting with surfaces, such as NO [128, 131, 132], O_2 [45, 58, 59, 133], N_2 [38, 43, 134–136], and CH_4 [137–143].

6.1. Beyond the Born-Oppenheimer Approximation.
The possible influence of nonadiabatic effects and concretely of electron-hole (e-h) pair excitations is currently under debate within the surface dynamics community. This debate comes from experiments showing unexpected results. For example, experimental results for N_2/Ru(0001) showing very low dissociative adsorption probabilities, for incidence energies well above the minimum reaction barrier [144, 145], or the isotope-dependence behavior found for vibrational deexcitation of H_2 and D_2 upon scattering from Cu(100) have been considered as a fingerprint of strong nonadiabatic effects [115, 116]. This conclusion was supported by low dimensional adiabatic calculations, which were unable to reproduce these experimental results [145, 146]. However, subsequent theoretical studies, using high (6D) dimensional adiabatic calculations, have refuted this conclusion [54, 65]. But stronger experimental indications of nonadiabatic effects, coming from experiments showing e-h pair excitation accompanying molecular chemisorption [126] and ejection of electrons accompanying scattering of highly vibrationally excited molecules, call for the development of theoretical methods able to account for these effects [128].

The first attempts to include electron-hole pair excitations in dynamics studies of molecular processes on metal surfaces date back to the 80s [147–149]. Most recently, several DFT-based approximate methods have been proposed to include nonadiabatic effects in the dynamics. Luntz et al. [146] have proposed to compute friction coefficients (η), describing electron-hole pair damping along the minimum energy barrier reaction path. This method allows performing quasiclassical nonadiabatic dynamics simulations on 2D PESs. But it cannot be extrapolated straightforwardly to multidimensional (6D) calculations. In view of the importance of taking into account the full dimensionality the system may have on the appropriate description of many molecule-surface phenomena, the applicability of this method is rather limited. Juaristi et al. [150] have proposed an approximate method to include nonadiabatic effects by keeping the multidimensionality of the problem. The main approximation of this method, from now on called LDFA (local density friction approximation), is to consider that the atoms of the molecule move independently. Using the LDFA, the nonadiabaticity is introduced in the classical equations of motion, for each atom of the molecule, through a dissipative force. Thus, the equation of motion for each atom can be written as

$$m_i \frac{d^2 r_i}{dt^2} = -\nabla_i V_{6D}\left(r_i, r_j\right) - \eta\left(r_i\right) \frac{dr_i}{dt}, \qquad (20)$$

where the second term in the right-hand side of this equation is the dissipative force experienced by each atom.

Shenvi et al. [151] have proposed to take into account nonadiabatic effects by using an independent electron surface hopping (IESM), in which the energy transfer to e-h pair excitations is described by considering hops between electronic adiabatic states. Using the IESM model, these authors have

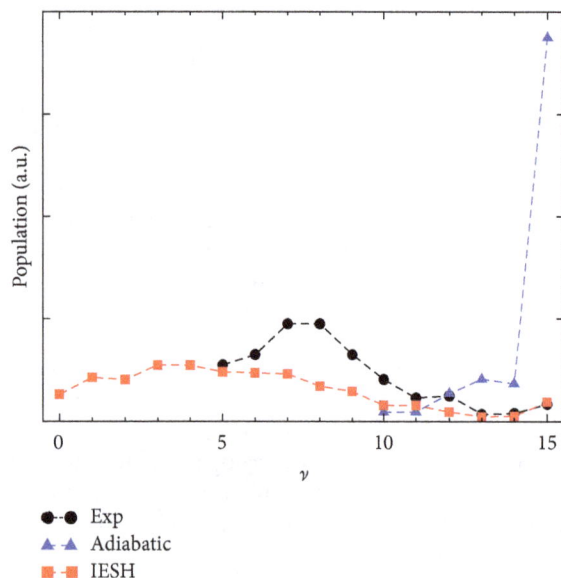

FIGURE 17: Vibrational state population distribution of NO(ν = 15)/Au(111) as a function of the vibrational quantum number (ν). Black circles: experimental results from [131]; blue triangles: adiabatic theoretical results from [132]; red squares: IESM results from [132].

been able to reproduce qualitatively the experimental vibrational deexcitation probabilities obtained for NO/Au(111) [131, 132] (see Figure 17), a system for which nonadiabatic effects have been found to be quite important [128, 152]. However, a model to accurately describe nonadiabatic effects in reactive and nonreactive scattering of molecules from metal surface is still to be developed.

6.2. Surface Temperature: The Effect of Phonons. Although the static surface approximation has yielded qualitatively good results for reactive and non-reactive scattering for many molecule-surface systems, including the effect of surface temperature may be crucial to accurately describe some systems and/or observables. For example, there are strong indications that vibrational excitation of H_2 upon scattering from Cu(111) can only be accurately described if phonons are taken into account [117].

So far, several methods have been proposed to include surface temperature, all of them within the classical dynamics framework. The simple one is the so-called surface oscillator (SO) model [153], in which a 3D harmonic oscillator is used to describe the collective motion of the surface atoms (see Figure 18). In this model, the coupling between the molecule and the surface atoms motion is described by a rigid coordinate shift of the 6D PES. Thus, the 9D PES can be written as

$$V_{9D}\left(R_A, R_B, R_S\right) = V_{6D}\left(R_A - R_S, R_B - R_S\right)$$
$$+ \frac{M}{2}\left(w_x^2 X_S^2 + w_Y^2 Y_S^2 + w_Z^2 Z_S^2\right), \quad (21)$$

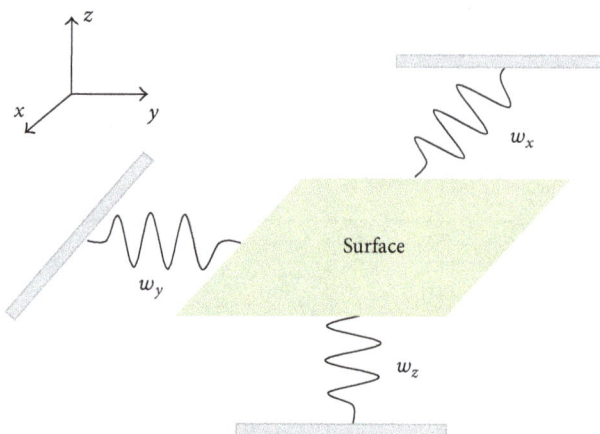

FIGURE 18: Schematic representation of the surface oscillator model.

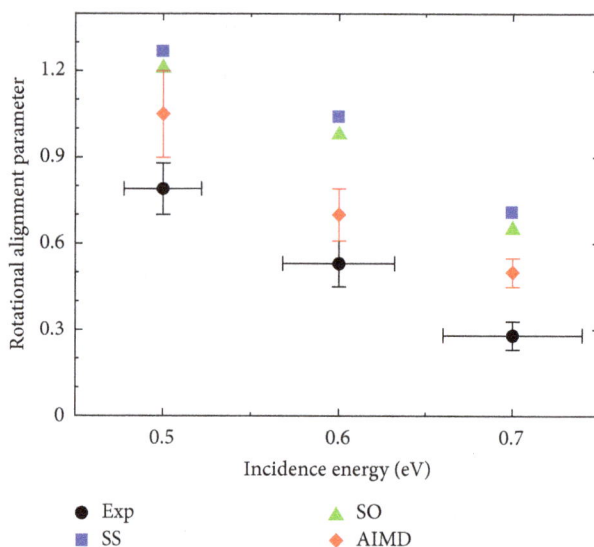

FIGURE 19: Quadrupole alignment parameter as a function of the incidence energy for H_2 on Cu(111). Black solid circles: experimental results from [93]; blue squares: static surface approximation results from [104]; green triangles: surface oscillator model results from [104]; red diamonds: AIMD results from [106].

where $w_{X,Y,Z}$ are the surface oscillation frequencies, M is the surface atoms mass, and $R_{A,B,S}$ are the vectors (X, Y, Z) defining the position of the molecule and surface atoms.

This method has been successfully used to study, for instance, the rotational excitation of H_2 upon scattering from Pd(111) [66] and the quadrupole alignment parameters for D_2/Cu(111). In the latter case, it was found that though the SO model yields better results than the static surface one, it is still too simple to obtain chemical accuracy [104]. Much better results have been obtained using *ab initio* molecular dynamics (AIMD) [106] (see Figure 19). This method allows the motion of the surface atoms in such a way that the intricate molecule-phonons coupling can be taken into account. AIMD, used for the first time to study molecule-surface reactions by Groß and Dianat [154], computes the forces on the fly which increases hugely the computational

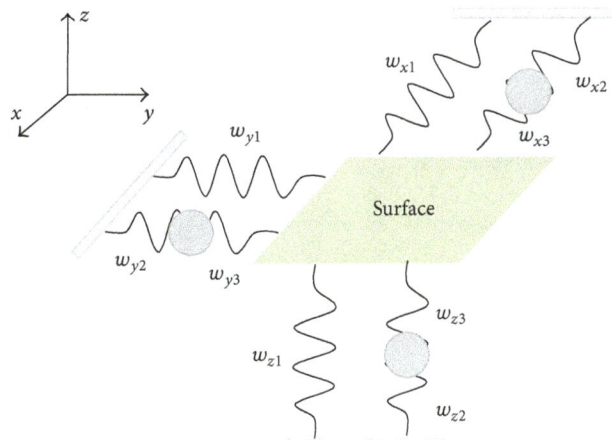

FIGURE 20: Schematic representation of the generalized Langevin oscillator model.

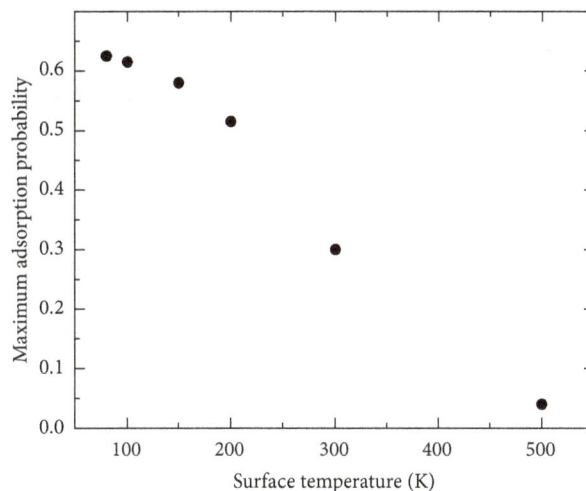

FIGURE 21: Maximum of the molecular adsorption of N_2 on Fe(110) as a function of the surface temperature. Data taken from [135].

cost, limiting the number of classical trajectories that can be computed.

An improvement over the SO model is the generalized Langevin oscillator (GLO) model [155] adapted to study molecule-surface interactions by Busnengo et al. [156] to analyze the effect of surface temperature in the dynamic trapping mediated adsorption process of H_2 on several Pd surfaces. This method includes dissipation and thermal fluctuations, thanks to *ghost* 3D oscillator (see Figure 20), which is coupled to the surface oscillator. In this model, the ghost oscillator is subject to friction and random forces. Within this framework model, the equations of motion for the 12 coordinates of the system are given by

$$\frac{d^2 R_{A,B}}{dt^2} = -\frac{1}{m_{A,B}} \nabla_{R_{A,B}} V_{6D} \left(R_A - R_S, R_B - R_s \right),$$

$$\frac{d^2 R_s}{dt^2} = -\frac{1}{m_s} \nabla_{R_S} V_{6D} \left(R_A - R_s, R_B - R_S \right) - \omega_S^2 R_S + \Lambda_{gs} U,$$

$$\frac{d^2 U}{dt^2} = -\omega_g^2 U + \Lambda_{gs} R_S + \gamma_g \frac{dU}{dt} + F_r \left(\Delta t \right).$$

$$(22)$$

Here $U(U_x, U_y, U_z)$ and ω_g are the coordinates and frequency matrix associated with the ghost oscillator, Λ_{gs} is the coupling matrix, which couples the ghost and the surface oscillators, γ_g represents the damping matrix associated with the friction force, and $F_r(\Delta t)$ is the random force.

GLO model has been used by Goikoetxea et al. [135] to study the molecular adsorption of N_2 on Fe(110), showing a strong dependence of the maximum of the molecular adsorption on the surface temperature (see Figure 21). This method has been also used to study, for example, the scattering of N_2 from W(110) [136].

Some effort to include surface phonons effects in quantum dynamics has been recently done [157]. In this case, a 7D model including the perpendicular motion of the second layer atom was proposed.

In spite of the big effort already invested by surface scientists, a method to fully take into account the complex motion associated with surface phonons is still to be developed.

6.3. *Looking for Accurate Exchange-Correlation Functionals.* Also relative to future challenges, it is worthy to mention some of the shortcomings inherent to state-of-the-art DFT simulations. Although a number of exchange-correlation functionals, such as PW91 [158], PBE [159], RPBE [160], or a specific reaction parameter approach applied to them [111] have shown to give good qualitative (and even quantitative) results, in comparison with experiment, for H_2 reactive and nonreactive scattering, other molecules present a major challenge. For example, up to now, none of the proposed functionals have been able to yield a sufficiently accurate PES to describe the dissociative adsorption of O_2 on Al(111). For this system, experimental results [161] show a very low dissociation probability at thermal energies, increasing monotonously with the incidence energy, whereas adiabatic molecular dynamics simulations show very high reactivities [162, 163] independent of the incidence energy (see Figure 22).

This failure of the standard DFT functionals to accurately describe the interaction of O_2 with Al(111), also observed for adsorption of O_2 on Si(111) [164], is due to the triplet-to-singlet spin conversion, which is not properly described by standard DFT. Contrary to most diatomic molecules, O_2 in gas phase (far from the surface) is in its triplet ground state, and when approaching the surface a transition to two oxygen atoms in their spin-singlet state occurs. Aiming to overcome this shortcoming a spin-constrained DFT approach has been proposed [58, 165]. In this model, the spin of the O_2 molecule is constrained to the Hilbert subspace, which prevents spin quenching and charge transfer before the molecule starts interacting with the surface; that is, the molecule is forced to travel in a spin-triplet configuration up to distances close to the surface. It is worth mentioning that an accurate description of triplet-to-singlet spin conversion becomes

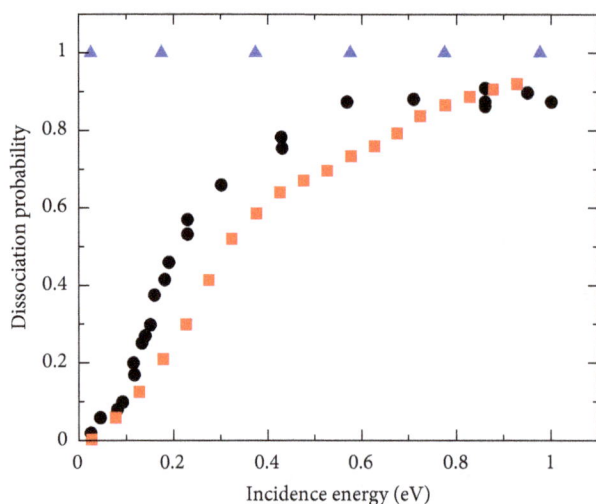

FIGURE 22: Dissociation probability as a function of the incidence energy for O_2 on Al(111). Black solid circles: experimental results from [161]; red solid squares: spin-constrained DFT results from [58]; blue triangles: adiabatic results.

crucial mostly for systems with a low density of states at the Fermi level. For example, standard DFT adiabatic dynamics for O_2/Ag(100) [45] agrees with experimental data in the absence of dissociative adsorption for energies below 1 eV.

Finally, we should mention that standard DFT does not take into account the effect of van der Waals (vdW) interactions. However these forces may play a prominent role for polyatomic molecules interacting with surfaces [166]. Attempts to include vdW forces have been made by Grimme et al. [167–169] using an empirical correction scheme and by Dion et al. [170–172] who proposed to include the vdW forces by expansion to second order in a specific quantity contained in the long range part of the correlation functional. Tkatchenko and Scheffler [173] have proposed a scheme which uses a parameter-free method to define accurately the long range vdW forces from mean-field electronic structure calculations. To conclude, it is also worth mentioning that several efficient implementations of these nonlocal density functionals are already available in commercial codes [174, 175].

Conflict of Interests

The author declares that there is no conflict of interests regarding the publication of this paper.

Acknowledgment

The author acknowledges support under MICINN Project FIS2010-25127.

References

[1] G. Ertl, "Primary steps in catalytic synthesis of ammonia," *Journal of Vacuum Science and Technology A: Vacuum, Surfaces and Films*, vol. 1, no. 2, pp. 1247–1253, 1983.

[2] A. Groß, "Reactions at surfaces studied by ab initio dynamics calculations," *Surface Science Reports*, vol. 32, no. 8, pp. 291–340, 1998.

[3] G. Kroes, "Six-dimensional quantum dynamics of dissociative chemisorption of H_2 on metal surfaces," *Progress in Surface Science*, vol. 60, no. 1, pp. 1–5, 1999.

[4] D. Farias and K. H. Rieder, "Atomic beam diffraction from solid surfaces," *Reports on Progress in Physics*, vol. 61, no. 12, p. 1575, 1998.

[5] D. Farías, H. F. Busnengo, and F. Martín, "Probing reaction dynamics at metal surfaces with H_2 diffraction," *Journal of Physics: Condensed Matter*, vol. 19, Article ID 305003, 2007.

[6] D. Farías and R. Miranda, "Diffraction of molecular hydrogen from metal surfaces," *Progress in Surface Science*, vol. 86, no. 9-10, pp. 222–254, 2011.

[7] L. Vattuone, L. Savio, F. Pirani, D. Cappelletti, M. Okada, and M. Rocca, "Interaction of rotationally aligned and of oriented molecules in gas phase and at surfaces," *Progress in Surface Science*, vol. 85, no. 1–4, pp. 92–160, 2010.

[8] D. A. King and M. G. Wells, "Molecular beam investigation of adsorption kinetics on bulk metal targets: nitrogen on tungsten," *Surface Science*, vol. 29, no. 2, pp. 454–482, 1972.

[9] C. T. Rettner, L. A. DeLouise, and D. J. Auerbach, "Effect of incidence kinetic energy and surface coverage on the dissociative chemisorption of oxygen on W(110)," *The Journal of Chemical Physics*, vol. 85, no. 2, pp. 1131–1149, 1986.

[10] G. Comsa, "Angular distribution of scattered and desorbed atoms from specular surfaces," *Journal of Chemical Physics*, vol. 48, article 3240, no. 7, 1968.

[11] H. A. Michelsen, C. T. Rettner, D. J. Auerbach, and R. N. Zare, "Effect of rotation on the translational and vibrational energy dependence of the dissociative adsorption of D_2 on Cu(111)," *Journal of Chemical Physics*, vol. 98, no. 10, pp. 8294–8307, 1993.

[12] P. Maroni, D. Papageorgopoulos, A. Ruf, R. D. Beck, and T. R. Rizzo, "Efficient stimulated Raman pumping for quantum state resolved surface reactivity measurements," *Review of Scientific Instruments*, vol. 77, no. 5, Article ID 054103, 2006.

[13] G. O. Sitz, A. C. Kummel, and R. N. Zare, "Direct inelastic scattering of N_2 from Ag(111). I. Rotational populations and alignment," *The Journal of Chemical Physics*, vol. 89, no. 4, pp. 2558–2571, 1988.

[14] E. E. Marinero, C. T. Rettner, and R. N. Zare, "Quantum-state-specific detection of molecular hydrogen by three-photon ionization," *Physical Review Letters*, vol. 48, no. 19, pp. 1323–1326, 1982.

[15] G. Scoles, *Atomics and Molecular Beam Methods*, vol. 2, Oxford University Press, New York, NY, USA, 1992.

[16] G. R. Darling and S. Holloway, "The dissociation of diatomic molecules at surfaces," *Reports on Progress in Physics*, vol. 58, no. 12, pp. 1595–1672, 1995.

[17] G. Kroes and M. F. Somers, "Six-dimensional dynamics of dissociative chemisorption of H_2 on metal surfaces," *Journal of Theoretical & Computational Chemistry*, vol. 4, no. 2, pp. 493–581, 2005.

[18] R. Diez Muino and H. F. Busnengo, Eds., *Dynamics of Gas-Surface Interactions*, Springer Series in Surface Science, 2013.

[19] J. E. Lennard-Jones, "Processes of adsorption and diffusion on solid surfaces," *Transactions of the Faraday Society*, vol. 28, pp. 333–359, 1932.

[20] L. Martin-Gondre, C. Crespos, P. Larregaray, J. C. Rayez, B. van Ootegem, and D. Conte, "Is the LEPS potential accurate

enough to investigate the dissociation of diatomic molecules on surfaces?" *Chemical Physics Letters*, vol. 471, no. 1–3, pp. 136–142, 2009.

[21] M. F. Somers, S. M. Kingma, E. Pijper, G. J. Kroes, and D. Lemoine, "Six-dimensional quantum dynamics of scattering of $(v = 0, j = 0)$ H_2 and D_2 from Cu(1 1 1): test of two LEPS potential energy surfaces," *Chemical Physics Letters*, vol. 360, no. 3-4, pp. 390–399, 2002.

[22] M. Persson, J. Strömquist, L. Bengtsson, B. Jackson, D. V. Shalashilin, and B. Hammer, "A first-principles potential energy surface for Eley-Rideal reaction dynamics of H atoms on Cu(111)," *Journal of Chemical Physics*, vol. 110, no. 4, pp. 2240–2249, 1999.

[23] S. Caratzoulas, B. Jackson, and M. Persson, "Eley-Rideal and hot-atom reaction dynamics of H(g) with H adsorbed on Cu(111)," *Journal of Chemical Physics*, vol. 107, no. 16, pp. 6420–6431, 1997.

[24] J. Dai and J. Z. H. Zhang, "Quantum adsorption dynamics of a diatomic molecule on surface: four-dimensional fixed-site model for H_2 on Cu(111)," *The Journal of Chemical Physics*, vol. 102, no. 15, pp. 6280–6289, 1995.

[25] A. Forni, G. Wiesenekker, E. J. Baerends, and G. F. Tantardini, "A dynamical study of the chemisorption of molecular hydrogen on the Cu(111) surface," *Journal of Physics: Condensed Matter*, vol. 7, no. 36, pp. 7195–7207, 1995.

[26] J. H. McCreery and G. Wolken Jr., "Atomic recombination dynamics on a solid surface: H_2+W(001)," *Journal of Chemical Physics*, vol. 64, no. 7, pp. 2845–2853, 1976.

[27] G. Wiesenekker, "An analytical six-dimensional potential energy surface for dissociation of molecular hydrogen on Cu(100)," *Journal of Chemical Physics*, vol. 104, no. 18, pp. 7344–7358, 1996.

[28] A. Gross, S. Wilke, and M. Scheffler, "Six-dimensional quantum dynamics of adsorption and desorption of H_2 at Pd(100): steering and steric effects," *Physical Review Letters*, vol. 75, no. 14, Article ID 2718, 1995.

[29] T. Ho and H. Rabitz, "A general method for constructing multidimensional molecular potential energy surfaces from ab initio calculations," *The Journal of Chemical Physics*, vol. 104, no. 7, pp. 2584–2597, 1996.

[30] D. E. Makarov and H. Metiu, "Fitting potential-energy surfaces: a search in the function space by directed genetic programming," *Journal of Chemical Physics*, vol. 108, no. 2, pp. 590–598, 1998.

[31] H. F. Busnengo, A. Salin, and W. Dong, "Representation of the 6D potential energy surface for a diatomic molecule near a solid surface," *The Journal of Chemical Physics*, vol. 112, no. 17, pp. 7641–7651, 2000.

[32] C. Crespos, M. A. Collins, E. Pijper, and G. J. Kroes, "Multidimensional potential energy surface determination by modified Shepard interpolation for a molecule-surface reaction: H_2 + Pt(1 1 1)," *Chemical Physics Letters*, vol. 376, no. 5-6, pp. 566–575, 2003.

[33] T. J. Frankcombe, M. A. Collins, and D. H. Zhang, "Modified Shepard interpolation of gas-surface potential energy surfaces with strict plane group symmetry and translational periodicity," *Journal of Chemical Physics*, vol. 137, no. 14, Article ID 144701, 2012.

[34] S. Lorenz, A. Groß, and M. Scheffler, "Representing high-dimensional potential-energy surfaces for reactions at surfaces by neural networks," *Chemical Physics Letters*, vol. 395, no. 4–6, pp. 210–215, 2004.

[35] G. Kresse, "Dissociation and sticking of H_2 on the Ni(111), (100), and (110) substrate," *Physical Review B*, vol. 62, no. 12, pp. 8295–8305, 2000.

[36] R. A. Olsen, H. F. Busnengo, A. Salin, M. F. Somers, G. J. Kroes, and E. J. Baerends, "Constructing accurate potential energy surfaces for a diatomic molecule interacting with a solid surface: H_2 + Pt(111) and H_2 + Cu(100)," *The Journal of Chemical Physics*, vol. 116, no. 9, pp. 3841–3855, 2002.

[37] M. A. Di Césare, H. F. Busnengo, W. Dong, and A. Salin, "Role of dynamic trapping in H_2 dissociation and reflection on Pd surfaces," *Journal of Chemical Physics*, vol. 118, no. 24, pp. 11226–11234, 2003.

[38] G. Volpilhac and A. Salin, "Dissociative adsorption of N2 on the W(1 0 0) surface," *Surface Science*, vol. 556, no. 2-3, pp. 129–144, 2004.

[39] P. Riviére, H. F. Busnengo, and F. Martín, "Density functional theory study of H and H_2 interacting with NiAl(110)," *The Journal of Chemical Physics*, vol. 121, p. 751, 2004.

[40] M. Luppi, D. A. McCormack, R. A. Olsen, and E. J. Baerends, "Rotational effects in the dissociative adsorption of H_2 on the Pt(211) stepped surface," *Journal of Chemical Physics*, vol. 123, no. 16, Article ID 164702, 2005.

[41] J. K. Vincent, R. A. Olsen, G. J. Kroes, M. Luppi, and E. J. Baerends, "Six-dimensional quantum dynamics of dissociative chemisorption of H_2 on Ru(0001)," *The Journal of Chemical Physics*, vol. 122, no. 4, Article ID 044701, 2005.

[42] A. Salin, "Theoretical study of hydrogen dissociative adsorption on the Cu(110) surface," *Journal of Chemical Physics*, vol. 124, no. 10, Article ID 104704, 2006.

[43] M. Alducin, R. Díez Muiño, H. F. Busnengo, and A. Salin, "Why N2 molecules with thermal energy abundantly dissociate on W(100) and not on W(110)," *Physical Review Letters*, vol. 97, no. 5, Article ID 056102, 2006.

[44] H. F. Busnengo and A. E. Martínez, "H_2 chemisorption on W(100) and W(110) surfaces," *Journal of Physical Chemistry C*, vol. 112, no. 14, pp. 5579–5588, 2008.

[45] M. Alducin, H. F. Busnengo, and R. Díez Muiño, "Dissociative dynamics of spin-triplet and spin-singlet O2 on Ag(100)," *The Journal of Chemical Physics*, vol. 129, no. 22, Article ID 224702, 2008.

[46] G. Laurent, F. Martín, and H. F. Busnengo, "Theoretical study of hydrogen dissociative adsorption on strained pseudomorphic monolayers of Cu and Pd deposited onto a Ru(0001) substrate," *Physical Chemistry Chemical Physics*, vol. 11, no. 33, pp. 7303–7311, 2009.

[47] A. Lozano, A. Gross, and H. F. Busnengo, "Adsorption dynamics of H_2 on Pd(100) from first principles," *Physical Chemistry Chemical Physics*, vol. 11, no. 27, pp. 5814–5822, 2009.

[48] C. Díaz, R. A. Olsen, H. F. Busnengo, and G. J. Kroes, "Dynamics on six-dimensional potential energy surfaces for H_2/Cu(111): corrugation reducing procedure versus modified shepard interpolation method and PW91 versus RPBE," *Journal of Physical Chemistry C*, vol. 114, no. 25, pp. 11192–11201, 2010.

[49] J. C. Chen, M. Ramos, C. Arasa et al., "Dynamics of H_2 dissociation on the 1/2 ML c(2 × 2)-Ti/Al(100) surface," *Physical Chemistry Chemical Physics*, vol. 14, no. 9, pp. 3234–3247, 2012.

[50] J. Ischtwan and M. A. Collins, "Molecular potential energy surfaces by interpolation," *The Journal of Chemical Physics*, vol. 100, no. 11, pp. 8080–8088, 1994.

[51] D. H. Zhang, M. A. Collins, and S.-Y. Lee, "First-principles theory for the H + H_2O, D_2O reactions," *Science*, vol. 290, no. 5493, pp. 961–963, 2000.

[52] C. Crespos, M. A. Collins, E. Pijper, and G. J. Kroes, "Application of the modified Shepard interpolation method to the determination of the potential energy surface for a molecule-surface reaction: H_2+Pt(111)," *The Journal of Chemical Physics*, vol. 120, no. 5, pp. 2392–2404, 2004.

[53] R. van Harrevelt, K. Honkala, J. K. Nørskov, and U. Manthe, "The reaction rate for dissociative adsorption of N_2 on stepped Ru(0001): six-dimensional quantum calculations," *Journal of Chemical Physics*, vol. 122, no. 23, Article ID 234702, 2005.

[54] C. Díaz, J. K. Vincent, G. P. Krishnamohan et al., "Multidimensional effects on dissociation of N_2 on Ru(0001)," *Physical Review Letters*, vol. 96, no. 9, Article ID 096102, 2006.

[55] I. M. N. Groot, J. C. Juanes-Marcos, C. Díaz, M. F. Somers, R. A. Olsen, and G. Kroes, "Dynamics of dissociative adsorption of hydrogen on a CO-precovered Ru(0001) surface: a comparison of theoretical and experimental results," *Physical Chemistry Chemical Physics*, vol. 12, no. 6, pp. 1331–1340, 2010.

[56] P. N. Abufager, C. Crespos, and H. F. Busnengo, "Modified Shepard interpolation method applied to trapping mediated adsorption dynamics," *Physical Chemistry Chemical Physics*, vol. 9, no. 18, pp. 2258–2265, 2007.

[57] T. J. Frankcombe and M. A. Collins, "Growing fragmented potentials for gas-surface reactions: The reaction between hydrogen atoms and hydrogen-terminated silicon (111)," *Journal of Physical Chemistry C*, vol. 116, no. 14, pp. 7793–7802, 2012.

[58] J. Behler, B. Delley, S. Lorenz, K. Reuter, and M. Scheffler, "Dissociation of O_2 at Al(111): the role of spin selection rules," *Physical Review Letters*, vol. 94, no. 3, Article ID 036104, 2005.

[59] I. Goikoetxea, J. Beltrán, J. Meyer, J. I. Juaristi, M. Alducin, and K. Reuter, "Non-adiabatic effects during the dissociative adsorption of O_2 at Ag(111)? A first-principles divide and conquer study," *New Journal of Physics*, vol. 14, Article ID 013050, 2012.

[60] J. Ludwig and D. G. Vlachos, "Ab initio molecular dynamics of hydrogen dissociation on metal surfaces using neural networks and novelty sampling," *Journal of Chemical Physics*, vol. 127, no. 15, Article ID 154716, 2007.

[61] J. Dai and J. C. Light, "Six dimensional quantum dynamics study for dissociative adsorption of H_2 on Cu(111) surface," *Journal of Chemical Physics*, vol. 107, no. 5, pp. 1676–1679, 1997.

[62] E. Pijper, G. J. Kroes, R. A. Olsen, and E. J. Baerends, "Reactive and diffractive scattering of H_2 from Pt(111) studied using a six-dimensional wave packet method," *Journal of Chemical Physics*, vol. 117, no. 12, pp. 5885–5898, 2002.

[63] C. Crespos, H.-D. Meyer, R. C. Mowrey, and G. J. Kroes, "Multiconfiguration time-dependent Hartree method applied to molecular dissociation on surfaces: H_2+Pt(111)," *Journal of Chemical Physics*, vol. 124, no. 7, Article ID 074706, 2006.

[64] M. H. Beck, A. Jäckle, G. A. Worth, and H. D. Meyer, "The multiconfiguration time-dependent Hartree (MCTDH) method: a highly efficient algorithm for propagating wavepackets," *Physics Report*, vol. 324, no. 1, pp. 1–105, 2000.

[65] A. S. Muzas, J. I. Juaristi, M. Alducin, R. D. Muio, G. J. Kroes, and C. Díaz, "Vibrational deexcitation and rotational excitation of H_2 and D_2 scattered from Cu(111): adiabatic versus non-adiabatic dynamics," *The Journal of Chemical Physics*, vol. 137, no. 6, Article ID 064707, 2012.

[66] H. F. Busnengo, W. Dong, P. Sautet, and A. Salin, "Surface temperature dependence of rotational excitation of H_2 scattered from Pd(111)," *Physical Review Letters*, vol. 87, Article ID 127601, 2001.

[67] C. Díaz, M. F. Somers, G. J. Kroes, H. F. Busnengo, A. Salin, and F. Martín, "Quantum and classical dynamics of H_2 scattering from Pd(111) at off-normal incidence," *Physical Review B*, vol. 72, Article ID 035401, 2005.

[68] C. Díaz, H. F. Busnengo, P. Rivière et al., "A classical dynamics method for H_2 diffraction from metal surfaces," *The Journal of Chemical Physics*, vol. 122, no. 15, Article ID 154706, 2005.

[69] C. Díaz, F. Martín, G. J. Kroes, M. Minniti, D. Farías, and R. Miranda, "H_2 diffraction from a strained pseudomorphic monolayer of Cu deposited on Ru (0001)," *The Journal of Physical Chemistry C*, vol. 116, p. 13671, 2012.

[70] B. Jackson and H. Metiu, "The dynamics of H_2 dissociation on Ni(100): a quantum mechanical study of a restricted two-dimensional model," *The Journal of Chemical Physics*, vol. 86, no. 2, pp. 1026–1035, 1986.

[71] M. R. Hand and S. Holloway, "A theoretical study of the dissociation of H_2/Cu," *The Journal of Chemical Physics*, vol. 91, no. 11, pp. 7209–7219, 1989.

[72] J. Sheng and J. Z. H. Zhang, "Dissociative chemisorption of H_2 on Ni surface: time-dependent quantum dynamics calculation and comparison with experiment," *The Journal of Chemical Physics*, vol. 96, no. 5, pp. 3866–3874, 1992.

[73] R. C. Mowrey, "Dissociative adsorption of H_2 using the close-coupling wave packet method," *The Journal of Chemical Physics*, vol. 99, no. 9, pp. 7049–7055, 1993.

[74] G. R. Darling and S. Holloway, "Angular and vibrational effects in the sticking and scattering of H_2," *The Journal of Chemical Physics*, vol. 97, no. 7, pp. 5182–5192, 1992.

[75] J. Dai and J. Z. H. Zhang, "Steric effect in dissociative chemisorption of hydrogen on Cu," *Surface Science*, vol. 319, no. 1-2, pp. 193–198, 1994.

[76] A. Gross, "The role of lateral surface corrugation for the quantum dynamics of dissociative adsorption and associative desorption," *The Journal of Chemical Physics*, vol. 102, no. 12, pp. 5045–5058, 1995.

[77] G. J. Kroes, G. Wiesenekker, E. J. Baerends, and R. C. Mowrey, "Competition between vibrational excitation and dissociation in collisions of H_2 with Cu(100)," *Physical Review B*, vol. 53, no. 15, pp. 10397–10401, 1996.

[78] A. Gross, B. Hammer, M. Scheffler, and W. Brenig, "High-dimensional quantum dynamics of adsorption and desorption of H_2 at Cu(111)," *Physical Review Letters*, vol. 73, no. 23, pp. 3121–3124, 1994.

[79] G. R. Darling and S. Holloway, "Rotational motion and the dissociation of H_2 on Cu(111)," *The Journal of Chemical Physics*, vol. 101, no. 4, pp. 3268–3281, 1994.

[80] R. C. Mowrey, G. J. Kroes, and E. J. Baerends, "Dissociative adsorption of H_2 on Cu(100): fixed-site calculations for impact at hollow and top sites," *Journal of Chemical Physics*, vol. 108, no. 16, pp. 6906–6915, 1998.

[81] G. J. Kroes, E. J. Baerends, and R. C. Mowrey, "Six-dimensional quantum dynamics of dissociative chemisorption of ($v = 0, j = 0$) H_2 on Cu(100)," *Physical Review Letters*, vol. 78, no. 18, pp. 3583–3586, 1997.

[82] G. J. Kroes, E. J. Baerends, and R. C. Mowrey, "Six-dimensional quantum dynamics of dissociative chemisorption of ($v = 0, j = 0$) H_2 on Cu(100)," *Physical Review Letters*, vol. 78, article 3583, 1997.

[83] H. F. Busnengo, C. Crespos, W. Dong, A. Salin, and J. C. Rayez, "Role of orientational forces in nonactivated molecular dissociation on a metal surface," *Physical Review B*, vol. 63, no. 4, Article ID 041402, 2001.

[84] C. Crespos, H. F. Busnengo, W. Dong, and A. Salin, "Analysis of H_2 dissociation dynamics on the Pd(111) surface," *Journal of Chemical Physics*, vol. 114, no. 24, pp. 10954–10962, 2001.

[85] G. Anger, A. Winkler, and K. D. Rendulic, "Adsorption and desorption kinetics in the systems H_2/Cu(111), H_2/Cu(110) and H_2/Cu(100)," *Surface Science*, vol. 220, no. 1, pp. 1–17, 1989.

[86] H. F. Berger, M. Leisch, A. Winkler, and K. D. Rendulic, "A search for vibrational contributions to the activated adsorption of H_2 on copper," *Chemical Physics Letters*, vol. 175, no. 5, pp. 425–428, 1990.

[87] H. A. Michelsen and D. J. Auerbach, "A critical examination of data on the dissociative adsorption and associative desorption of hydrogen at copper surfaces," *The Journal of Chemical Physics*, vol. 94, no. 11, pp. 7502–7520, 1991.

[88] H. A. Michelsen, C. T. Rettner, and D. J. Auerbach, "State-specific dynamics of D_2 desorption from Cu(111): the role of molecular rotational motion in activated adsorption-desorption dynamics," *Physical Review Letters*, vol. 69, no. 18, pp. 2678–2681, 1992.

[89] C. T. Rettner, D. J. Auerbach, and H. A. Michelsen, "Role of vibrational and translational energy in the activated dissociative adsorption of D_2 on Cu(111)," *Physical Review Letters*, vol. 68, no. 8, pp. 1164–1167, 1992.

[90] C. T. Rettner, D. J. Auerbach, and H. A. Michelsen, "Observation of direct vibrational excitation in collisions of H_2 and D_2 with a Cu(111) surface," *Physical Review Letters*, vol. 68, no. 16, pp. 2547–2550, 1992.

[91] C. T. Rettner, H. A. Michelsen, and D. J. Auerbach, "Determination of quantum-state-specific gas-surface energy transfer and adsorption probabilities as a function of kinetic energy," *Chemical Physics*, vol. 175, no. 1, pp. 157–169, 1993.

[92] C. T. Rettner, H. A. Michelsen, and D. J. Auerbach, "Quantum-state-specific dynamics of the dissociative adsorption and associative desorption of H_2 at a Cu(111) surface," *The Journal of Chemical Physics*, vol. 102, no. 11, pp. 4625–4641, 1995.

[93] H. Hou, S. J. Guiding, C. T. Rettner, A. M. Wodtke, and D. J. Auerbach, "The stereodynamics of a gas-surface reaction," *Science*, vol. 277, no. 5322, pp. 80–82, 1997.

[94] A. Hodgson, J. Moryl, P. Traversaro, and H. Zhao, "Energy transfer and vibrational effects in the dissociation and scattering of D_2 from Cu(111)," *Nature*, vol. 356, no. 6369, pp. 501–504, 1992.

[95] A. Hodgson, P. Samson, A. Wight, and C. Cottrell, "Rotational excitation and vibrational relaxation of H_2 (v = 1, J = 0) scattered from Cu(111)," *Physical Review Letters*, vol. 78, no. 5, pp. 963–966, 1997.

[96] M. J. Murphy and A. Hodgson, "Adsorption and desorption dynamics of H_2 and D_2 on Cu(111): the role of surface temperature and evidence for corrugation of the dissociation barrier," *Journal of Chemical Physics*, vol. 108, no. 10, pp. 4199–4211, 1998.

[97] D. Wetzig, M. Rutkowski, R. David, and H. Zacharias, "Rotational corrugation in associative desorption of D_2 from Cu(111)," *Europhysics Letters*, vol. 36, no. 1, pp. 31–36, 1996.

[98] S. J. Gulding, A. M. Wadtke, H. Hou, C. T. Rettner, H. A. Michelsen, and D. J. Auerbach, "Alignment of D_2(v, J) desorbed from Cu(111): low sensitivity of activated dissociative chemisorption to approach geometry," *The Journal of Chemical Physics*, vol. 105, p. 9702, 1996.

[99] G. R. Darling and S. Holloway, "Dissociation thresholds and the vibrational excitation process in the scattering of H_2," *Surface Science*, vol. 307-309, pp. 153–158, 1994.

[100] J. Dai and J. C. Light, "The steric effect in a full dimensional quantum dynamics simulation for the dissociative adsorption of H_2 on Cu(111)," *The Journal of Chemical Physics*, vol. 108, no. 18, pp. 7816–7820, 1998.

[101] J. Dai, J. Sheng, and J. Z. H. Zhang, "Symmetry and rotational orientation effects in dissociative adsorption of diatomic molecules on metals: H_2 and HD on Cu(111)," *The Journal of Chemical Physics*, vol. 101, no. 2, pp. 1555–1563, 1994.

[102] G. R. Darling and S. Holloway, "Translation-to-vibrational excitation in the dissociative adsorption of D_2," *The Journal of Chemical Physics*, vol. 97, no. 1, pp. 734–736, 1992.

[103] J. Sheng and J. Z. H. Zhang, "Quantum dynamics studies of adsorption and desorption of hydrogen at a Cu(111) surface," *The Journal of Chemical Physics*, vol. 99, no. 2, pp. 1373–1381, 1993.

[104] C. Díaz, R. A. Olsen, D. J. Auerbach, and G. J. Kroes, "Six-dimensional dynamics study of reactive and non reactive scattering of H_2 from Cu(111) using a chemically accurate potential energy surface," *Physical Chemistry Chemical Physics*, vol. 12, no. 24, pp. 6499–6519, 2010.

[105] G. J. Kroes, E. Pijper, and A. Salin, "Dissociative chemisorption of H_2 on the Cu(110) surface: a quantum and quasiclassical dynamical study," *Journal of Chemical Physics*, vol. 127, no. 16, Article ID 164722, 2007.

[106] F. Nattino, C. Díaz, B. Jackson, and G. Kroes, "Effect of surface motion on the rotational quadrupole alignment parameter of D_2 reacting on Cu(111)," *Physical Review Letters*, vol. 108, no. 23, Article ID 236104, 2012.

[107] P. S. Thomas, M. F. Somers, A. W. Hoekstra, and G. J. Kroes, "Chebyshev high-dimensional model representation (Chebyshev-HDMR) potentials: application to reactive scattering of H_2 from Pt(111) and Cu(111) surfaces," *Physical Chemistry Chemical Physics*, vol. 14, no. 24, pp. 8628–8643, 2012.

[108] A. Marashdeh, S. Casolo, L. Sementa, H. Zacharias, and G. J. Kroes, "Surface temperature effects on dissociative chemisorption of H_2 on Cu(100)," *Journal of Physical Chemistry C*, vol. 117, no. 17, pp. 8851–8863, 2013.

[109] L. Sementa, M. Wijzenbroek, B. J. Van Kolck et al., "Reactive scattering of H_2 from Cu(100): comparison of dynamics calculations based on the specific reaction parameter approach to density functional theory with experiment," *Journal of Chemical Physics*, vol. 138, no. 4, Article ID 044708, 2013.

[110] B. Hammer, M. Scheffler, K. W. Jacobsen, and J. K. Nørskov, "Multidimensional potential energy surface for H_2 dissociation over Cu(111)," *Physical Review Letters*, vol. 73, no. 10, pp. 1400–1403, 1994.

[111] C. Díaz, E. Pijper, R. A. Olsen, H. F. Busnengo, D. J. Auerbach, and G. J. Kroes, "Chemically accurate simulation of a prototypical surface reaction: H_2 dissociation on Cu(111)," *Science*, vol. 326, no. 5954, pp. 832–834, 2009.

[112] U. Nielsen, D. Halstead, S. Holloway, and J. K. Nørskov, "The dissociative adsorption of hydrogen: two-, three-, and four-dimensional quantum simulations," *The Journal of Chemical Physics*, vol. 93, no. 4, pp. 2879–2884, 1990.

[113] S. Nave, D. Lemoine, M. F. Somers, S. M. Kingma, and G. J. Kroes, "Six-dimensional quantum dynamics of (v = 0, j = 0)D_2 and of (v = 1, j = 0)H_2 scattering form Cu(111)," *The Journal of Chemical Physics*, vol. 122, Article ID 214709, 2005.

[114] C. Díaz, A. Perrier, and G. J. Kroes, "Associative desorption of N_2 from Ru(0 0 0 1): a computational study," *Chemical Physics Letters*, vol. 434, no. 4-6, pp. 231–236, 2007.

[115] E. Watts and G. O. Sitz, "State-to-state scattering in a reactive system: H$_2$(v=1,J=1) from Cu(111)," *Journal of Chemical Physics*, vol. 114, no. 9, pp. 4171–4179, 2001.

[116] L. C. Shackman and G. O. Sitz, "State-to-state scattering of D$_2$ from Cu(100) and Pd(111)," *The Journal of Chemical Physics*, vol. 123, no. 6, Article ID 064712, 2005.

[117] G. Kroes, C. Díaz, E. Pijper, R. A. Olsen, and D. J. Auerbach, "Apparent failure of the Born-Oppenheimer static surface model for vibrational excitation of molecular hydrogen on copper," *Proceedings of the National Academy of Sciences of the United States of America*, vol. 107, no. 49, pp. 20881–20886, 2010.

[118] C. Díaz, H. F. Busnengo, F. Martin, and A. Salin, "Angular distribution of H$_2$ molecules scattered from the Pd(111) surface," *The Journal of Chemical Physics*, vol. 118, no. 6, p. 2886, 2003.

[119] C. Díaz, F. Martín, H. F. Busnengo, and A. Salin, "Theoretical analysis of the relation between H$_2$ dissociation and reflection on Pd surfaces," *The Journal of Chemical Physics*, vol. 120, no. 1, pp. 321–328, 2004.

[120] G. Comsa and R. David, "Dynamical parameters of desorbing molecules," *Surface Science Reports*, vol. 5, no. 4, pp. 145–198, 1985.

[121] P. Nieto, E. Pijper, D. Barredo et al., "Reactive and nonreactive scattering of H$_2$ from a metal surface is electronically adiabatic," *Science*, vol. 312, no. 5770, pp. 86–89, 2006.

[122] B. Gumhalter, "Single- and multiphonon atom-surface scattering in the quantum regime," *Physics Report*, vol. 351, no. 1-2, pp. 1–159, 2001.

[123] D. Farías, C. Díaz, P. Riviére et al., "In-plane and out-of-plane diffraction of H$_2$ from metal surface," *Physical Review Letters*, vol. 93, no. 24, Article ID 246104, 2004.

[124] G. Laurent, D. Barredo, D. Farías et al., "Experimental and theoretical study of rotationally inelastic diffraction of D$_2$ from NiAl(110)," *Physical Chemistry Chemical Physics*, vol. 12, no. 43, pp. 14501–14507, 2010.

[125] D. Barredo, G. Laurent, C. Díaz et al., "Experimental evidence of dynamic trapping in the scattering of H$_2$ from Pd(110)," *The Journal of Chemical Physics*, vol. 125, no. 5, Article ID 051101, 2006.

[126] B. Gergen, H. Nienhaus, W. H. Weinberg, and E. W. McFarland, "Chemically induced electronic excitations at metal surfaces," *Science*, vol. 294, no. 5551, pp. 2521–2523, 2001.

[127] A. M. Wodtke, J. C. Tully, and D. J. Auerbach, "Electronically non-adiabatic interactions of molecules at metal surfaces: Can we trust the Born-Oppenheimer approximation for surface chemistry?" *International Reviews in Physical Chemistry*, vol. 23, no. 4, pp. 513–539, 2004.

[128] J. D. White, J. Chen, D. Matsiev, D. J. Auerbach, and A. M. Wodtke, "Conversion of large-amplitude vibration to electron excitation at a metal surface," *Nature*, vol. 433, no. 7025, pp. 503–505, 2005.

[129] G. Sitz, "Surface chemistry: approximate challenges," *Nature*, vol. 433, no. 7025, p. 470, 2005.

[130] G. Kroes, "Frontiers in surface scattering simulations," *Science*, vol. 321, no. 5890, pp. 794–797, 2008.

[131] Y. Huang, C. T. Rettner, D. J. Auerbach, and A. M. Wodtke, "Vibrational promotion of electron transfer," *Science*, vol. 290, no. 5489, pp. 111–114, 2000.

[132] N. Shenvi, S. Roy, and J. C. Tully, "Dynamical steering and electronic excitation in NO scattering from a gold surface," *Science*, vol. 326, no. 5954, pp. 829–832, 2009.

[133] J. Meyer and K. Reuter, "Electron-hole pairs during the adsorption dynamics of O$_2$ on Pd(100): Exciting or not?" *New Journal of Physics*, vol. 13, Article ID 085010, 2011.

[134] C. Díaz, J. K. Vincent, G. P. Krishnamohan et al., "Reactive and nonreactive scattering of N$_2$ from Ru(0001): a six-dimensional adiabatic study," *Journal of Chemical Physics*, vol. 125, Article ID 114706, 2006.

[135] I. Goikoetxea, M. Alducin, E. Díez-Muiño, and J. I. Juaristi, "Dissociative and non-dissociative adsorption dynamics of N$_2$ on Fe(110)," *Physical Chemistry Chemical Physics*, vol. 14, p. 7471, 2012.

[136] L. Martin-Gondre, M. Alducin, G. A. Bocan, R. Díez Muiño, and J. I. Juaristi, "Competition between electron and phonons excitation in the scattering of nitrogen atoms and moleculesb off tungsten and silver metal surface," *Physical Review Letters*, vol. 108, Article ID 096101, 2012.

[137] X. J. Shen, A. Lozano, W. Dong, H. F. Busnengo, and X. H. Yan, "Towards bond selective chemistry from first principles: methane on metal surfaces," *Physical Review Letters*, vol. 112, Article ID 046101, 2014.

[138] R. R. Smith, D. R. Killelea, D. F. DelSesto, and A. L. Utz, "Preference for vibrational over translational energy in a gas-surface reaction," *Science*, vol. 304, no. 5673, pp. 992–995, 2004.

[139] L. B. F. Juurlink, R. R. Smith, D. R. Killelea, and A. L. Utz, "Comparative study of C-H stretch and bend vibrations in methane activation on Ni(100) and Ni(111)," *Physical Review Letters*, vol. 94, Article ID 208303, 2005.

[140] R. D. Beck, P. Maroni, D. C. Papageorgopoulos, T. T. Dang, M. P. Schmid, and T. R. Rizzo, "Vibrational mode-specific reaction of methane on a nickel surface," *Science*, vol. 302, no. 5642, pp. 98–100, 2003.

[141] S. Nave and B. Jackson, "Methane dissociation on Ni(111): the role of lattice reconstruction," *Physical Review Letters*, vol. 98, no. 17, Article ID 173003, 2007.

[142] S. Nave and B. Jackson, "Vibrational mode-selective chemistry: methane dissociation on Ni(100)," *Physical Review B*, vol. 81, no. 23, Article ID 233408, 2010.

[143] A. K. Tiwari, S. Nave, and B. Jackson, "Methane dissociation on Ni(111): a new understanding of the lattice effect," *Physical Review Letters*, vol. 103, no. 25, Article ID 253201, 2009.

[144] L. Romm, G. Katz, R. Kosloff, and M. Asscher, "Vibrational and kinetic energy: molecular beam experiments and quantum mechanical calculations," *The Journal of Physical Chemistry B*, vol. 101, p. 2213, 1997.

[145] L. Diekhöner, H. Mortensen, A. Baurichter, E. Jensen, V. V. Petrunin, and A. C. Luntz, "N$_2$ dissociative adsorption on Ru(0001): the role of energy loss," *Journal of Chemical Physics*, vol. 115, no. 19, pp. 9028–9035, 2001.

[146] A. C. Luntz, M. Persson, and G. O. Sitz, "Theoretical evidence for nonadiabatic vibrational deexcitation in H$_2$ (D$_2$) state-to-state scattering from Cu(100)," *Journal of Chemical Physics*, vol. 124, no. 9, Article ID 091101, 2006.

[147] J. W. Gadzuk and H. Metiu, "Theory of electron-hole pair excitations in unimolecular processes at metal surfaces. I. X-ray edge effects," *Physical Review B*, vol. 22, no. 6, pp. 2603–2613, 1980.

[148] H. Metiu and J. W. Gadzuk, "Theory of rate processes at metal surfaces. II. the role of substrate electronic excitations," *The Journal of Chemical Physics*, vol. 74, no. 4, pp. 2641–2653, 1981.

[149] B. Gumhalter and S. G. Davison, "Effect of electronic relaxation on covalent adsorption reaction rates," *Physical Review B*, vol. 30, no. 6, pp. 3179–3190, 1984.

[150] J. I. Juaristi, M. Alducin, R. Díez-Muiño, H. F. Busnengo, and A. Salin, "Role of electron-hole pair excitations in the dissociative adsorption of diatomic molecules on metal surfaces," *Physical Review Letters*, vol. 100, no. 11, Article ID 116102, 2008.

[151] N. Shenvi, S. Roy, and J. C. Tully, "Nonadiabatic dynamics at metal surfaces: Independent-electron surface hopping," *Journal of Chemical Physics*, vol. 130, no. 17, Article ID 174107, 2009.

[152] N. H. Nahler, J. D. White, J. LaRue, D. J. Auerbach, and A. M. Wodtke, "Inverse velocity dependence of vibrationally promoted electron emission from a metal surface," *Science*, vol. 321, no. 5893, pp. 1191–1194, 2008.

[153] M. Hand and J. Harris, "Recoil effects in surface dissociation," *The Journal of Chemical Physics*, vol. 92, no. 12, pp. 7610–7617, 1990.

[154] A. Groß and A. Dianat, "Hydrogen dissociation dynamics on precovered Pd surfaces: langmuir is still right," *Physical Review Letters*, vol. 98, no. 20, Article ID 206107, 2007.

[155] J. C. Tully, "Dynamics of gas-surface interactions: 3D generalized Langevin model applied to fcc and bcc surfaces," *The Journal of Chemical Physics*, vol. 73, no. 4, pp. 1975–1985, 1980.

[156] H. F. Busnengo, M. A. Di Césare, W. Dong, and A. Salin, "Surface temperature effects in dynamic trapping mediated adsorption of light molecules on metal surfaces: H_2 on Pd(111) and Pd(110)," *Physical Review B*, vol. 72, no. 12, Article ID 125411, 2005.

[157] M. Bonfanti, M. F. Somers, C. Díaz, H. F. Busnengo, and G. J. Kroes, "7D Quantum dynamics of H_2 scattering from Cu(111): the accuracy of the phonon sudden approximation," *Zeitschrift für Physikalische Chemie*, vol. 227, pp. 1397–1420, 2013.

[158] J. P. Perdew, J. A. Chevary, S. H. Vosko et al., "Atoms, molecules, solids, and surfaces: Applications of the generalized gradient approximation for exchange and correlation," *Physical Review B*, vol. 46, no. 11, pp. 6671–6687, 1992.

[159] J. P. Perdew, K. Burke, and M. Ernzerhof, "Generalized gradient approximation made simple," *Physical Review Letters*, vol. 77, no. 18, pp. 3865–3868, 1996.

[160] B. Hammer, L. B. Hansen, and J. K. Nørskov, "Improved adsorption energetics within density-functional theory using revised Perdew-Burke-Ernzerhof functionals," *Physical Review B*, vol. 59, no. 11, pp. 7413–7421, 1999.

[161] L. Österlund, I. Zoric, and B. Kasemo, "Dissociative sticking of O_2 on Al(111)," *Physical Review B*, vol. 55, Article ID 15452, 1997.

[162] K. Honkala and K. Laasonen, "Oxygen molecule dissociation on the Al(111) surface," *Physical Review Letters*, vol. 84, no. 4, pp. 705–708, 2000.

[163] Y. Yourdshahyan, B. Razaznejad, and B. I. Lundqvist, "Adiabatic potential-energy surfaces for oxygen on Al(111)," *Physical Review B*, vol. 65, no. 7, Article ID 075416, 2002.

[164] K. Kato, T. Uda, and K. Terakura, "Backbond oxidation of the Si(001) surface: narrow channel of barrierless oxidation," *Physical Review Letters*, vol. 80, no. 9, pp. 2000–2003, 1998.

[165] J. Behler, K. Reuter, and M. Scheffler, "Nonadiabatic effects in the dissociation of oxygen molecules at the Al(111) surface," *Physical Review B—Condensed Matter and Materials Physics*, vol. 77, no. 11, Article ID 115421, 2008.

[166] P. Lazić, N. Atodiresei, V. Caciuc, R. Brako, B. Gumhalter, and S. Blügel, "Rationale for switching to nonlocal functionals in density functional theory," *Journal of Physics Condensed Matter*, vol. 24, Article ID 424215, 2012.

[167] S. Grimme, "Accurate description of van der Waals complexes by density functional theory including empirical corrections," *Journal of Computational Chemistry*, vol. 25, no. 12, pp. 1463–1473, 2004.

[168] S. Grimme, "Semiempirical GGA-type density functional constructed with a long-range dispersion correction," *Journal of Computational Chemistry*, vol. 27, no. 15, pp. 1787–1799, 2006.

[169] S. Grimme, J. Antony, S. Ehrlich, and H. Krieg, "A consistent and accurate ab initio parametrization of density functional dispersion correction (DFT-D) for the 94 elements H-Pu," *Journal of Chemical Physics*, vol. 132, no. 15, Article ID 154104, 2010.

[170] M. Dion, H. Rydberg, E. Schröder, D. C. Langreth, and B. I. Lundqvist, "Van der Waals density functional for general geometries," *Physical Review Letters*, vol. 92, no. 24, Article ID 246401, 2004.

[171] M. Dion, H. Rydberg, E. Schröder, D. C. Langreth, and B. I. Lundqvist, "Erratum: Van der Waals density functional for general geometries," *Physical Review Letters*, vol. 95, Article ID 109902, 2005.

[172] J. Klime, D. R. Bowler, and A. Michaelides, "Van der Waals density functionals applied to solids," *Physical Review B*, vol. 83, no. 19, Article ID 195131, 2011.

[173] A. Tkatchenko and M. Scheffler, "Accurate molecular van der Waals interactions from ground-state electron density and free-atom reference data," *Physical Review Letters*, vol. 102, no. 7, Article ID 073005, 2009.

[174] P. Lazić, N. Atodiresei, M. Alaei, V. Caciuc, S. Blügel, and R. Brako, "JuNoLo—Jülich nonlocal code for parallel post-processing evaluation of vdW-DF correlation energy," *Computer Physics Communications*, vol. 181, no. 2, pp. 371–379, 2010.

[175] G. Román-Pérez and J. M. Soler, "Efficient implementation of a van der Waals density functional: application to double-wall carbon nanotubes," *Physical Review Letters*, vol. 103, Article ID 096102, 2009.

[176] K. D. Rendulic, G. Anger, and A. Winkler, "Wide range nozzle beam adsorption data for the systems H_2/nickel and H_2/Pd(100)," *Surface Science*, vol. 208, no. 3, pp. 404–424, 1989.

[177] E. Pijper, M. F. Somers, G. J. Kroes et al., "Six-dimensional quantum dynamics of scattering of (v=0, j=0) H_2 from Pt(1 1 1): comparison to experiment and to classical dynamics results," *Chemical Physics Letters*, vol. 347, no. 4–6, pp. 277–284, 2001.

[178] G. Laurent, C. Díaz, H. F. Busnengo, and F. Martín, "Non-monotonic dissociative adsorption of vibrationally excited H_2 on metal surface," *Physical Review B*, vol. 81, Article ID 161404, p. 81, 2010.

[179] P. Rivière, M. F. Somers, G. J. Kroes, and F. Martín, "Quantum dynamical study of the H_2 and D_2 dissociative adsorption and diffraction from the NiAl(110) alloy surface," *Physical Review B*, vol. 73, Article ID 205417, 2006.

12

Meticulous Overview on the Controlled Release Fertilizers

Siafu Ibahati Sempeho,[1] **Hee Taik Kim,**[2] **Egid Mubofu,**[3] **and Askwar Hilonga**[1]

[1] *Department of Materials Science and Engineering, Nelson Mandela African Institution of Science and Technology,*
P.O. Box 447, Arusha, Tanzania

[2] *Department of Chemical Engineering, Hanyang University, 1271 Sa 3-dong, Sangnok-gu, Ansan-si,*
Gyeonggi-do 426-791, Republic of Korea

[3] *Chemistry Department, University of Dar es Salaam, P.O. Box 35091, Dar es Salaam, Tanzania*

Correspondence should be addressed to Askwar Hilonga; askwar.hilonga@nm-aist.ac.tz

Academic Editor: Jason Belitsky

Owing to the high demand for fertilizer formulations that will exhaust the possibilities of nutrient use efficiency (NUE), regulate fertilizer consumption, and lessen agrophysicochemical properties and environmental adverse effects instigated by conventional nutrient supply to crops, this review recapitulates controlled release fertilizers (CRFs) as a cutting-edge and safe way to supply crops' nutrients over the conventional ways. Essentially, CRFs entail fertilizer particles intercalated within excipients aiming at reducing the frequency of fertilizer application thereby abating potential adverse effects linked with conventional fertilizer use. Application of nanotechnology and materials engineering in agriculture particularly in the design of CRFs, the distinctions and classification of CRFs, and the economical, agronomical, and environmental aspects of CRFs has been revised putting into account the development and synthesis of CRFs, laboratory CRFs syntheses and testing, and both linear and sigmoid release features of CRF formulations. Methodical account on the mechanism of nutrient release centring on the empirical and mechanistic approaches of predicting nutrient release is given in view of selected mathematical models. Compositions and laboratory preparations of CRFs basing on *in situ* and graft polymerization are provided alongside the physical methods used in CRFs encapsulation, with an emphasis on the natural polymers, modified clays, and superabsorbent nanocomposite excipients.

1. Introduction

Controlled release fertilizers (CRFs) are fertilizer granules intercalated within carrier molecules commonly known as excipients to control nutrients release thereby improving nutrient supply to crops and minimize environmental, ecological, and health hazards [1]. In that sense, CRFs usage is an advanced way to supply crop's nutrients (cf. conventional ways) due to gradual pattern of nutrient release, which improves fertilizer use efficiency (FUE) [2]. In other words, depending on the thickness of the coatings within the formulation, CRFs enable nutrients to be released over an extended period leading to an increased control over the rate and pattern of release [3], consequently the excipients play a role in regulating nutrients release time and eliminate the need for constant fertilization and higher efficiency rate than conventional soluble fertilizers [1].

Occasionally the terms controlled release fertilizers (CRFs) and slow release fertilizers (SRFs) have been used interchangeably, yet they are different. Typically, the endorsed differences between slow-release and controlled-release fertilizers are not clear [4, 5]. However, the term CRF is generally applied to fertilizers in which the factors dominating the rate, pattern, and duration of release are well known and controllable during CRF preparation [5, 6]. SRFs on the other hand are characterized by the release of the nutrients at a slower rate than is usual but the rate, pattern, and duration of release are not well controlled [5, 6]; they may be strongly affected by handling conditions such as storage, transportation, and distribution in the field, or by soil conditions such as moisture content, wetting and drying, thawing and freezing, and biological activity [7–9]. Thus, while in SRFs the nutrient release pattern is fully dependent on soil and climatic conditions and it cannot be predicted (or *only* very roughly)

[10]; with CRFs, the release pattern, quantity, and time can be predicted within certain limits. For example, the classification of sulphur-coated urea (SCU) is subject to debate [5] due to a significant variation in the release patterns between different batches of fertilizer [5, 6, 11]. As a result, SCU is considered to be SRF despite being debated.

CRFs use is associated with several economic, agronomical, and environmental returns. Economically, CRFs supply nutrients to the crops for the entire season through a single application thereby saving spreading costs and reduce the demand for short-season manual labour required for topdressing operations [7]. Agronomically, CRFs usage is associated with the improvement of plant growth conditions, such as reduction of stress and specific toxicity resulting from excessive nutrient supply in the root zones. Similarly, CRFs increase the availability of nutrients due to the controlled release of nutrients into a "fixing" medium during the fixation processes in the soil as well as supplying nutrients in the forms preferred by plants; in that way the synergistic effect between nutrients in the CRFs is enhanced [7]. From the environmental perspective, CRFs improves NUE and in so doing reduces losses of surplus nutrients (over plant needs) to the environment [7]. Consequently, high levels of fertilizer accumulation in the environment are minimized, thereby lessening several environmental problems associated with conventional fertilizer use such as eutrophication which causes O_2 depletion, death of fish, unpleasant odour to the environment, and aesthetic problems [7, 12, 13].

2. Classification of CRFs

Several classifications of CRFs have been proposed. In this review, we will attempt to discuss a few of them. Based on Shaviv's grouping [6], CRFs may be classified as follows.

2.1. Organic-N-Low-Solubility Compounds. These can be subdivided into biologically decomposing compounds usually based on urea-aldehyde condensation products, such as urea-formaldehyde (UF), urea-triazone (UT), crotonylidene diurea (CDU), and chemically decomposing compounds, such as isobutylidene-diurea (IBDU). Succinctly, UF is prepared by reacting excess urea under controlled conditions of pH, temperature, U-F ratio, and reaction time. UT solution is based on the reaction of urea-ammonia-formaldehyde. CDU is prepared by reacting urea with acetaldehyde under the catalysis of an acid. IBDU is prepared by reacting liquid isobutyraldehyde with solid urea [5, 7, 10, 14].

2.2. Fertilizers in Which a Physical Barrier Controls the Release. These can be subdivided into granules coated by hydrophobic polymers or as matrices in which the soluble active material is dispersed in a continuum that restricts the dissolution of the fertilizer. The coated fertilizers can further be divided into fertilizers with organic polymer coatings that are either thermoplastic or resins and fertilizers coated with inorganic materials such as sulphur or mineral based coatings. The materials used for preparation of matrices can also be subdivided into hydrophobic materials such as polyolefins

and rubber and gel-forming polymers (hydrogels) which are hydrophilic in nature. Broadly, the use of coated fertilizers in agricultural practices is quite common as compared to the use of matrices. For instance, sulphur-coated urea (SCU) was developed at the Tennessee Valley Authority laboratories and manufactured commercially for almost 30 years [7, 15]. Its preparation is based on coating preheated urea granules with molten sulphur. The CRF alkyd-type resin-coated fertilizer (Osmocote) was first produced commercially in California in 1967. It is a copolymer of dicyclopentadiene with a glycerol ester [7]. In fact, these formulations control the rate of nutrient release offering multiple environmental, economic, and yield benefits [16]. Gel-based matrices are still being developed [17].

2.3. Inorganic Low-Solubility Compounds. This type of CRFs includes fertilizers such as metal ammonium phosphates (e.g., $MgNH_4PO_4$) and partially acidulated phosphate rocks (PAPR). Besides, the biologically and microbially decomposed N products, such as UF, are commonly referred to in the trade as slow-release fertilizers and coated or encapsulated/occluded products as controlled-release fertilizers [5, 7]. Essentially, Zhang et al.'s writings provide a deep detailed account on the subject in a much more broad sense [18].

3. Preparation of CRFs Formulations

Slowing the release of plant nutrients from fertilizers can be achieved by different methods and the resulting products are known as slow- or controlled-release fertilizers. With controlled-release fertilizers, the principal method is to cover a conventional soluble fertilizer with a protective coating (encapsulation) of a water-insoluble, semipermeable or impermeable-with-pores material. This controls water penetration and thus the rate of dissolution and ideally synchronizes nutrient release with the plants' needs. The most important manufactured materials include (i) materials releasing nutrients through either microbial decomposition of low solubility compounds, for example, organic-N low-solubility compounds, such as urea-aldehyde condensation products, or chemically decomposable compounds, for example, IBDU [5, 6]; (ii) materials releasing nutrients through a physical barrier, for example, sulphur-coated urea (SCU) [5]; (iii) materials releasing nutrients incorporated into a matrix, which itself may be coated, including gel-based matrices, which are still under development [5, 6, 17]; materials releasing nutrients in delayed form due to a small surface-to-volume ratio, for example, super-granules, briquettes, tablets, spikes, plant food sticks [5], and others [19–21].

According to Liu et al. [22], intercalation of nutrients into the excipients is normally achieved by two methods. In the first method, the compound to be loaded is added to the reaction mixture and polymerized *in situ* whereby the compound is entrapped within the gel matrix, whereas in the second method, the dry gel is allowed to swell in the compound solution and after equilibrium swelling, the gel is dried and the device is obtained. This involves

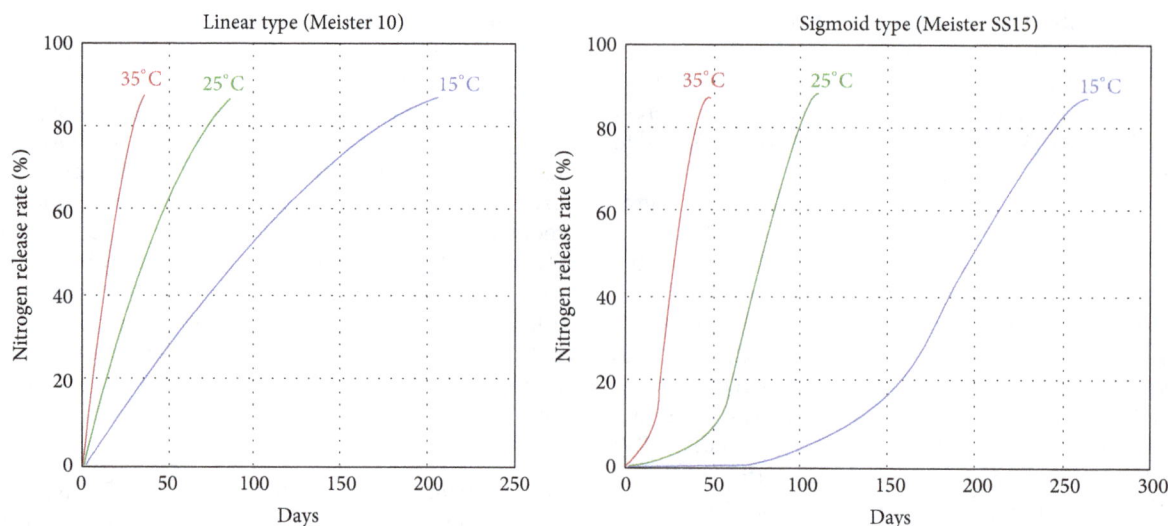

FIGURE 1: The effect of temperature on the release rate of Meister.

graft-polymerization [23–26]. The benefits and drawbacks are that for the former method, the entrapped compound may influence the polymerization process and the polymer network structure; while for the latter, the loaded compound always accumulates on the surface during the drying of the loaded hydrogel, which consequently leads to a "burst effect"; moreover, the loading amount may be low if the compound affects the water absorbency strongly.

Typical physical methods for encapsulating fertilizers include spray coating, spray drying, pan coating, and rotary disk atomization. Special equipment for these methods are rotary drum, pan or ribbon or paddle mixer, and fluidized bed [1]. The details of these methods are beyond the scope of this paper.

4. Nutrient Release in the CRFs Context

In this perspective, with regard to the European Standardization Committee (CEN), nutrient release (of course from the excipients) can be manifest by the transformation of a chemical substance or rather fertilizer nutrients into a plant-available form (e.g., dissolution, hydrolysis, degradation, etc.), whereas slow release is the release wherein the rate of a nutrient release from the fertilizer is slower than that from a fertilizer in which the nutrient is readily available for plant uptake [27]. CEN's declaration alleged that "fertilizer should be described as CRFs if at room temperature the nutrients released exceed 15% in 24 hours, or no more than 75% released in 28 days, or at least about 75% released at the stated release time" [5, 7] giving different release patterns. That is to say, CRFs that do not meet these three CEN's criteria are nonpertinent for the subject of controlled release formulations since the patterns will not comply with the standard ones, namely, linear and sigmoidal release patterns (Figure 1).

As mentioned above, release patterns can be classified into linear and sigmoidal release types [5, 28]. Examples

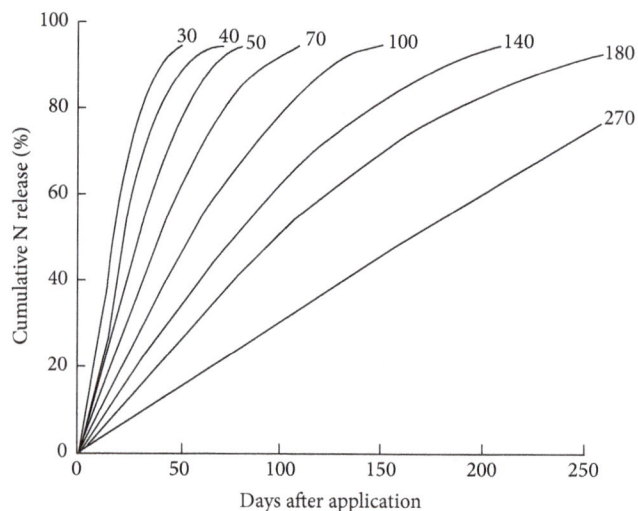

FIGURE 2: Linear release pattern.

of linear-release formulations presenting nutrient release between 30 and 270 days at 25°C for Meister formulation are given in Figure 2, whereas for sigmoidal-release formulations presenting nutrient release between 40 and 200 days at 25°C for Meister are shown in Figure 3 [5].

Actually, the characteristic features of CRFs encompass the release pattern (i.e., shape, lag, lock off); release duration; differential release between N, P, and K; effect of temperature on release; effect of the medium/environmental conditions on release [5, 6]. In most cases, the energy of activation of the release, EA_{rel}, is calculated on the basis of estimates of the rate of the release (% released per day) during the linear period obtained from the release curves [14].

As far as CEN's definition of release is concerned, the example from Meister formulation described above should comply with the criterion that at least about 75% of nitrogen should be released at the stated release time for this CRF to

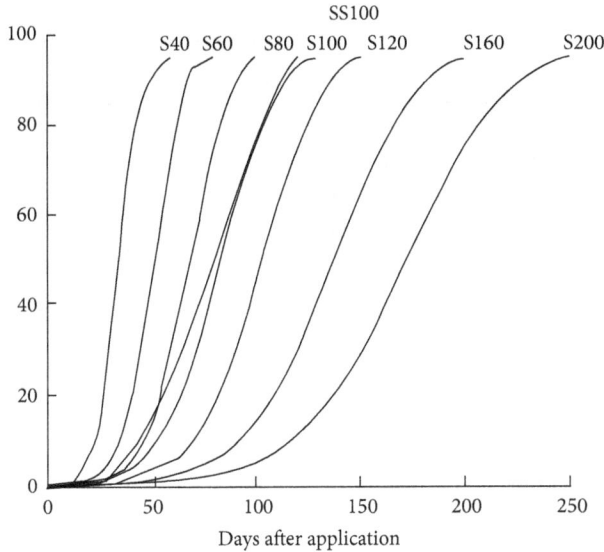

FIGURE 3: Sigmoidal release pattern.

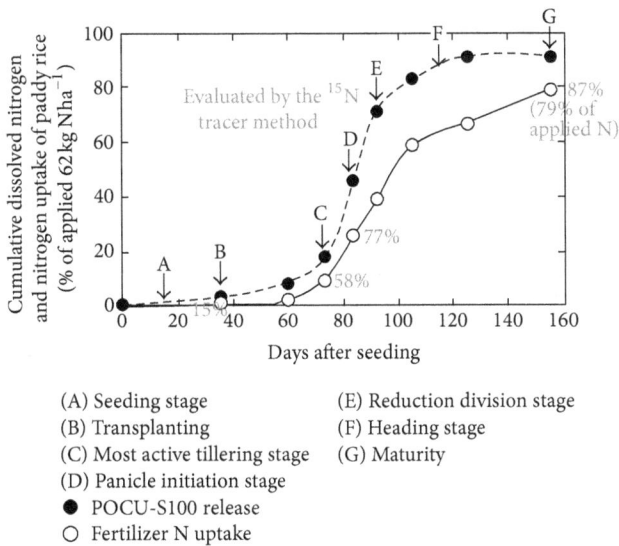

(A) Seeding stage
(B) Transplanting
(C) Most active tillering stage
(D) Panicle initiation stage
● POCU-S100 release
○ Fertilizer N uptake

(E) Reduction division stage
(F) Heading stage
(G) Maturity

FIGURE 4: Relationship between dissolved N and fertilizer derived N uptake of paddy rice.

be approved. As a matter of fact, Kanno [29] indicated that at the end of 160 days the nitrogen intake reached 79% of the applied N (Figure 4) and so conforming to CEN's conditions.

In point of fact, establishing nutrient release profiles requires data from both field testing and laboratory testing. In the laboratory, release of nutrients from the excipients is done using water and soil matrices [5, 30]. Field testing involves net bags placed in the ploughed layer of soil in the actual field [5, 30]. Industrial methods involve extract at 25°C, 40°C, and 100°C [5]. However, Du et al. [14] provide a new procedure where release characteristics are tested in three different systems, namely, (i) free water (which he termed common procedure); (ii) water saturated sand packed in columns; (iii) sand at field capacity moisture.

5. Mechanism of Nutrients Release from CRFs Formulations

Consistent experimental data with reference to release phenomena of nutrients from polymer coated CRFs are indispensably beneficial for better agronomic and environmental results [14]. Agric [31], after a period of laboratory testing of Meister CRFs, obtained the results described in Tables 1 and 2 for both linear and sigmoid patterns. The designed formulation which is marketed as Meister has its mechanism proposed by the company and the summary is given in Figure 5. The mechanism is based on three significant steps, namely, water adsorption, dissolution of urea, and leaching.

In addition to that, Guo et al. [32] proposed the mechanism of nitrogen release from urea-formaldehyde (UF) slow-release fertilizer granules based on three steps. *Step one:* the coating materials become swollen by absorbing water from the soil and so get transformed into hydrogels which contribute to increasing the orifice size of the 3D network of the coating materials so that it benefits the diffusion of the fertilizer in the core of the gel network. As a result, a layer of water between the swollen coatings and the UF granule core is formed. *Step two:* water slowly diffuses into the cross linked polymer network and dissolves the soluble part of UF; consequently the soluble part of the fertilizer gets slowly released into the soil through the swollen network with the dynamic exchange of the water in the hydrogel and the water in the soil. *Step three:* the soil microorganisms penetrate through the swollen coatings and assemble around the UF granule thereby degrading the insoluble part of nitrogen in UF granule into urea and ammonia which in turn is slowly released into the soil via dynamic exchange. Such steps have also been described as lag period, linear stage, and decay period by other researchers [14].

This mechanism can be adapted to effectively explain the release behaviour in other CRF formulations. Different mathematical mechanistic models based on empirical and mechanistic approaches plus empirical and semiempirical models have been proposed for prediction of the nutrient release using chemophysical parameters as will be discussed in the coming sections. Nevertheless, most mechanisms reveal that nutrients release from CRFs is mainly controlled by diffusion mechanism with respect to temperature, thickness of the coating material, type of nutrient, and the presence or absence of the relevant soil microorganisms.

6. Predicting Nutrient Release from CRFs

Profoundly, a number of empirical and semiempirical mechanistic mathematical models have been put forward in order to provide realistic theoretical assumptions connected to the patterns of nutrients release mechanisms based on the nature and the properties of the delivery systems (DS) [7], and in that case, release models have been used as tools for improving the CRFs' design methodology leaving behind conceivable breakthroughs in assessing prospective hazards such as leaching or volatilization losses and effects such as "bursting" or "tailing effect" [7–9]. Such conceptual approaches include the diffusion model, zero order

TABLE 1: Linear release pattern.

Brand name	80% release (days at 20°C)	80% release (days at 25°C)	Japanese brand name	N content (%)
MEISTER-5	50	30	LP-30	42
MEISTER-7	70	40	LP-40	42
MEISTER-8	80	50	LP-40	50
MEISTER-10	100	70	LP-70	42
MEISTER-15	150	100	LP-100	42
MEISTER-20	200	140	LP-140	42
MEISTER-27	270	180	LP-180	42
MEISTER-40	400	270	LP-270	42

TABLE 2: Sigmoid release pattern.

Brand name	Time lag days/release days		Japanese brand name	N content (%)
	at 20°C in soil	at 25°C in soil		
MEISTER-5	35/35	20/20	LP-S40	41
MEISTER-7	45/45	30/30	LP-S60	41
MEISTER-8	60/60	40/40	LP-S80	41
MEISTER-10	45/105	30/70	LP-S100	41
MEISTER-15	70/80	45/55	LP-SS100	41
MEISTER-20	90/90	60/60	LP-S120	41
MEISTER-27	120/120	80/80	LP-S160	41
MEISTER-40	150/150	100/100	LP-S200	41

kinetics model, first order kinetics model, Higuchi model, Korsmeyer-Peppas model, Hixson-Crowell model, Weibull model, Baker-Lonsdale model, Hoffenberg model, sequential layer model, Couarraze model, and Peppas-Sahlin model. In particular, most of the proposed release models assume that the release of nutrients from coated CRFs is either controlled by the rate of solute diffusion from the fertilizers or by the rate of water/vapour penetration into the CRF through the coating [7].

6.1. Diffusion Model. Considering a mathematical model developed for urea release from sulphur-coated granules under soil conditions [7, 33], the assumption was that urea diffuses from the granule through pores or holes caused by erosion of the coating and that the transport is influenced by temperature and soil water content; thus, diffusion occurs through the coating. This model was verified using Fick's first law as

$$\frac{dm}{dt} = -DS_k \frac{dC_k}{dx_k}, \tag{1}$$

where m is the mass of urea diffusing out of the granule, D is the effective diffusion coefficient of urea in water, S_k is the cross-sectional area through which diffusion occurs, and C_k is the urea concentration. The subscript k is the value for the internal pore coating or outside segments [7]. The predictive power of this model is certainly restricted to the fact that particle flux is directly proportional to the spatial concentration gradient. Nonetheless, it is not the spatial concentration gradient that causes particle movement, that is, particles do not push each other [34]. That is to say, particles do exhibit random motion on the molecular level and this random motion ensures that a tracer will diffuse thereby decreasing the concentration gradient [34].

Moreover, a study by Jarrell and Boersma [35] revealed that the diffusion of urea through the sulphur coating occurred in two steps represented in the following models:

$$\frac{dm_r}{dt} = \frac{DS_p}{M_o l} C_{sat} \quad \text{for } t < t_1,$$

$$\frac{dm_r}{dt} = \frac{DS_p}{M_o l} (1 - m_r) \rho, \quad \text{for } t > t_1, \tag{2}$$

where $m_r = m/M_o$, while M_o is the initial mass of urea in the granule, C_{sat} is the concentration of saturated urea solution, l is the coating thickness, ρ is the density of solid urea, and t_1 is the onset of the period of the decaying rate of release as the solution inside the granule becomes unsaturated.

Similarly, this study is also boundless for the reason that it ignores some important factors and features that are relevant to diffusion of active bioactive substances from an excipient or rather a membrane-coated granule (sphere). It is for that reason that the following Arrhenius type of model pertaining to the diffusion coefficient D was suggested [7, 33, 36]:

$$D = ATe^{(-EA_{realese}/T)}, \tag{3}$$

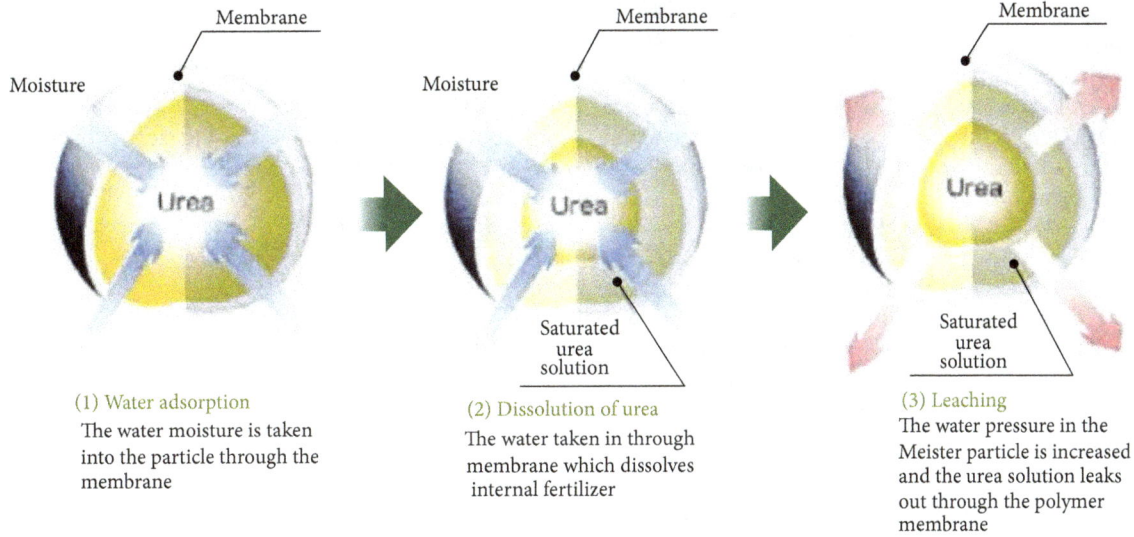

Membrane · Moisture · Urea · (1) Water adsorption — The water moisture is taken into the particle through the membrane

Membrane · Moisture · Urea · Saturated urea solution · (2) Dissolution of urea — The water taken in through membrane which dissolves internal fertilizer

Membrane · Urea · Saturated urea solution · (3) Leaching — The water pressure in the Meister particle is increased and the urea solution leaks out through the polymer membrane

FIGURE 5: Release of nitrogen from polyolefin coated urea in water at 25°C.

where T is the kelvin absolute temperature and $\mathrm{EA}_{\mathrm{release}}$ stands for the apparent energy of activation for urea diffusion from the excipients. This expression as proposed provides a conceivable explanation for the temperature dependence on the CRFs release rates. On the same side, a similar model for simulating nutrients release from the CRFs in a 1D coordinate system is known [37]; however, an additional assumption in favour of this model is that the diffusion coefficient is time dependent, thus giving the following expression:

$$D = D_0 t^n, \tag{4}$$

where t is time, D_0 is an initial value at $t = 0$, and n is an empirical constant. The time dependence of D presents a lag in the curve describing cumulative release with time (i.e., sigmoidal release pattern) which could otherwise not have been obtained by simply applying Fick's law described before [7].

6.2. Sequential Layer Model.

This model assumes that during the release of an active ingredient from the hydrophilic excipients, significant water concentration gradients are formed in the first place at the matrix/water interface and by so doing there is a creation of water imbibition into the system and as a result, and there occur dramatic physicochemical changes, namely, the exact geometry of the active substance within the excipients, axial and radial direction of the mass transport, and water diffusion coefficient on the matrix. Due to swelling of the excipients following water imbibition phenomenon, the concentration of participating species (i.e., polymer and a chemical substance) significantly changes thereby causing increased dimensions of the system. Consequently, the dissolution of the active ingredient occurs and so it diffuses out of such hydrophilic system following concentration gradients. Essentially, the amount of water available for dissolution is directly proportional to the diffusion coefficient of the active substance within the excipients. In that view, dissolution rate

constant, k_{diss}, of the active ingredient-excipient system can be computed and is given as

$$M_{pt} = M_{po} - k_{\mathrm{diss}} A_t t, \tag{5}$$

where M_{pt} and M_{po} are the dry polymer matrix mass at time t and $t = 0$, respectively; A_t is the surface area of the device at time t [20, 38–45].

6.3. Hopfenberg Model.

The primary assumption in this model is that nutrients are released from the surface-eroding excipients possessing some geometries ranging from slabs, spheres, and infinite cylinders displaying heterogeneous erosion. This approach can be mathematically expressed as

$$\frac{M_t}{M_\infty} = 1 - \left(1 - \frac{k_o t}{C_o a}\right)^n, \tag{6}$$

where M_t is the concentration of the chemical substance dissolved in time t, M_∞ is the total matrix (chemical-excipient) concentration dissolved when the system is exhausted, k_o is the erosion rate constant, C_o is the initial concentration of chemical substance/fertilizer in the matrix, and a_o is the initial radius for a sphere of cylinder or the half-thickness for a slab. The value of n is 1, 2, and 3 for a slab, cylinder, and sphere, respectively [20, 38–45].

6.4. Weibull Model.

As far as CRFs formulation is concerned, this model accounts for the release of nutrient molecules from the erodible matrix formulations with an assumption that factors influencing the overall release rate are exclusively mass dependent, while other factors stand to be time dependent [38]. The model depicts that a plot of logarithm of the amount of nutrient molecules dissolved in an excipient' solution versus the logarithm of time will be linear and it is mathematically given as

$$\mathrm{Log}\left[-\ln\left(1 - m\right)\right] = b \log\left(t - T_t\right) - \log a, \tag{7}$$

where a relates to the time scale of the process corresponding to the ordinate $[-\ln(1 - m)] = 1$. T_t refers to the lag time before the onset of the release process, t is time after release phenomena, b is the shape parameter corresponding to the ordinate value $(1/a)$ when time $t = 1$, and m relates to the fraction of the active ingredient in the excipient' solution at time t [20, 38–45].

Despite limitations associated with this model including the inability to sufficiently characterize the release kinetics of the nutrient molecules and the limited use for establishing *in vivo/in vitro* correlation, the model is known to be grander in the fact that the release half-life can easily be calculated and also the errors associated with it are only single figures, that is, minimum. In fact, the number of single figure errors is known to be higher than other models [38].

6.5. Korsmeyer-Peppas Model. Based on the CRFs context, this semiempirical model is effective in the determination of the concentration of nutrient molecules released from the excipients' membranes. Theoretically, the simple expression allied to this model is given as

$$f_t = at^n, \tag{8}$$

where a refers to a constant incorporating structural and geometric characteristics of the given active substance, n is the release exponent indicative of the release mechanism, and t is fractional release of active substance $[M_t/M_\infty]$ described in (6) above [40].

6.6. Higuchi Model. Predominantly, this model explicates the release of water soluble and low soluble nutrient substances merged into the semisolid or solid excipients molecules; it has been lengthily applied in the diffusion matrix formulations. The assumption core to this model stipulates that initial concentration of the nutrient molecules incorporated into the matrix is much higher than the solubility of the former [41]. Another assumption states that the diffusion of the nutrient molecules with excipients takes place only in one dimension such that the edge effect is negligible. The third one depicts that nutrient particles are much smaller than the system thickness. Also, matrix swelling and dissolution are negligible and so diffusivity of an active nutrient substance is constant; the last assumption is that perfect sink conditions are always attained in the release environment [41].

Considerably, the assumptions underlying this model reveal that there are two systems which may be considered when formulating mathematical expression for the release systems. Such systems are as follows: (i) when the nutrient molecules are dispersed in a homogeneous uniform matrix, which of course acts as diffusional mechanism and (ii) when they are incorporated into the planar heterogeneous matrix where their concentration in the matrix is lower than their solubility such that release process occurs through pores in the excipients by penetrative leaching out [44]. Conceptually, Figure 6 can be used to formulate a language to express this model.

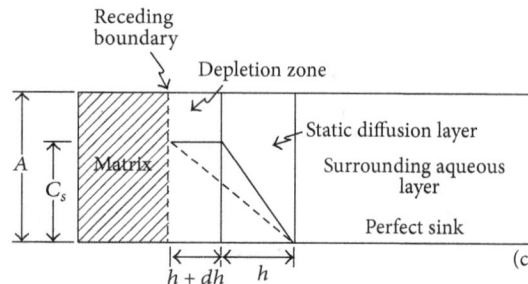

FIGURE 6: Conceptual Higuchi model.

According to Fick's first law,

$$\frac{dM}{Sdt} = \frac{dQ}{dt} = \frac{DC_s}{h}. \tag{9}$$

At this instant, when the nutrients molecules are dispersed within the homogeneous excipient matrix, the borderline indicated by the dashed vertical line (Figure 6) moves to the left by an infinitesimal distance (dh) and the infinitesimal amount (dQ) of the nutrients released because of this shift is given as

$$dQ = Adh - \frac{1}{2}C_s dh. \tag{10}$$

When (10) is substituted in (9), (11) is obtained and is given as

$$\frac{DC_s}{h} = \left(A - \frac{1}{2}C_s\right)\frac{dh}{dt}. \tag{11}$$

Based on Narender's derivation steps [44], it is possible to follow Higuchi's steps for derivation as follows:

$$2A - \frac{C_s}{2DC_s}\int hdh = \int dt,$$

$$t = (2A - C_s)\frac{h^2}{4DC_s} + C. \tag{12}$$

Integrating "C" when $h = 0$, (13) is obtained as

$$h = \left(\frac{4DC_s t}{2A - C_s}\right)^{1/2}. \tag{13}$$

Recall (10) as follows:

$$dQ = Adh - \frac{1}{2}C_s dh. \tag{14}$$

This equation can be integrated to take the following simple form:

$$Q = hA - \frac{1}{2}hC_s. \tag{15}$$

Substituting (15) into (14), we obtain the Higuchi equation for the homogeneous nutrient-excipients matrices as follows:

$$Q = \left[D\left(2A - C_s\right)C_s t\right]^{1/2}. \tag{16}$$

This can be further simplified to take a form of

$$f_t = Q = A\sqrt{D(2C - C_s)C_s t}, \qquad (17)$$

where Q is the concentration of the given nutrient released in time t per unit area A, C is the initial concentration of the nutrient, C_s is the nutrient solubility in the matrix media, and D is the diffusivity of the nutrients molecules (diffusion coefficient) in the matrix substance [41, 44].

On the other hand, the heterogeneous nutrient-excipient matrix system takes a different form and in that way (16) is modified in order to take into account the porosity and tortuosity of the matrix. Conceptually, the mathematical expression will be

$$\frac{dQ}{dt} = \left(\frac{ADC_s}{2t}\right)^{1/2}, \qquad (18)$$

where dQ/dt is the concentration of nutrients released at time t, A is the total amount of nutrients in unit volume of matrix, D is the diffusion coefficient of the nutrients in matrix, C_s is the solubility of the nutrients in polymeric matrix, and t is the time [41, 44].

6.7. Hixson Crowell Model.

This model assumes that the release rate of a nutrient molecules contained in a polymeric excipient is limited to the dissolution rate of its particles and not by the diffusion that could take place through polymeric matrix. The assumptions underlying this model include the following: (i) dissolution occurs normally to the surface of the solute particles, (ii) agitation is uniform all over the exposed surfaces and there is no stagnation, and also (iii) the particle of solute retains its geometric shape [20, 38–45].

According to Narender [44], the radius of a given bioactive particle is given as r and surface area is thus $4\pi r^2$. For that reason, during release process the radius is reduced by dr and so the infinitesimal volume of one particle fragment lost can be differentiated to be

$$dV = 4\pi r^2 dr. \qquad (19)$$

However, infinitesimal volume of n particle fragment lost can be differentiated as

$$dV = 4N\pi r^2 dr. \qquad (20)$$

Recalling (20), the surface of n particles can be found as

$$S = 4N\pi r^2. \qquad (21)$$

From the Noyes-Whitney law, the infinitesimal change in weight is given by the equation

$$dW = kSC_s dt. \qquad (22)$$

Then, the density of the nutrient molecules in the matrix could be multiplied by the infinitesimal volume change as $\rho dV = dW$ to give the following equation:

$$\rho dV = kSC_s dt. \qquad (23)$$

Substituting (20) into (21) and (23), (24) is obtained as

$$-4\rho N\pi r^2 dr = 4N\pi r^2 KC_s dt. \qquad (24)$$

Equation (24) can be simplified further by integrating it with respect to $r = ro$, at $t = 0$ to give the following expression:

$$r = ro - \frac{kC_s t}{\rho}. \qquad (25)$$

Equally, it is possible to substitute the radius in (25) with the weight of n particles to give the following expression:

$$W^{1/3} = \left(\sqrt[3]{N\rho\frac{\pi}{6}}\right)d. \qquad (26)$$

Since the diameter d can be substituted for $2r$, then it is possible to substitute d from (25) with $2r$ from (26) to yield the Hixson-Crowell cube root equation as follows:

$$W_o^{1/3} - W_t^{1/3} = \kappa t, \qquad (27)$$

where W_o is the initial concentration of nutrient molecules in the matrix, W_t is the remaining concentration of nutrient molecules in the matrix at time t, and κ (kappa) is a constant incorporating the surface-volume relation [44].

6.8. Zero Order Kinetics Model.

This model describes the delivery system at which the concentration of nutrients released per unit time is constant. This model assumes that in the course of dissolution process the area does not change and no equilibrium conditions are obtained. In that case, this model has been useful in the release of bioactive species/nutrients from the matrix that do not disaggregate and release the nutrients slowly from the excipients. Mathematically, the model can be expressed as

$$Q_o - Q_t = K_o t, \qquad (28)$$

where Q_t is the amount of nutrients dissolved in time t, Q_o is the initial amount of nutrients in the solution (most times, $Q_o = 0$), and K_o is the zero order release constant expressed in units of concentration/time. According to Mahat [42], the zero order release kinetics account for various different mass transport phenomena such as diffusion of water and bioactive species and the swelling and degradation of the excipients.

6.9. First Order Kinetic Model.

This model is applied in the release kinetics to describe the absorption and elimination of the bioactive ingredients/nutrients from the excipients. This model assumes that a graph of release data versus time will be linear. Conferring to this model, the rate of bioactive species released from the excipients matrix is directly proportional to the concentration; that is to say, the release rate of nutrient molecules is concentration dependent [39]. Mathematical expression for this model is given as

$$\ln Q_t = \ln Q_0 + k_t, \qquad (29)$$

where Q_t is the concentration of nutrients yet to be released at time t, Q_0 is the concentration of nutrients yet to be released at time zero, and k_t is the first order release constant.

6.10. Baker-Lonsdale Model. This model was established from the Higuchi model in an attempt to describe the dissolution of bioactive species from spherical matrix based excipients and hence it has been quite suitable model for microcapsules or microspheres systems. In a very simplified form it is possible to express this model as

$$F_t = \frac{3}{2}\left[1 - \left(1 - \frac{M_t}{M_\infty}\right)^{2/3}\right] - \frac{M_t}{M_\infty} = k_t, \qquad (30)$$

where F_t is the fraction of bioactive species released at time t, M_t is the amount released at time t, and M_∞ is the amount released at infinite time [20, 38–45].

7. Failure Release

Experiments on the modified polymer or sulphur-coated urea granules (PSCU) conducted by Raban [46] revealed the main processes occurring during the failure release mechanism. The release process starts as water vapours penetrate through the coating. The rate of water penetration is defined by the driving force (vapour pressure gradient), the coating thickness, and features of the coating material. The water vapours condense and dissolve the fertilizer, thus causing a buildup of internal pressure inside the coated granule. The increase of internal pressure above a threshold value is likely to cause rupture of the coating (in contrast to the case of diffusion when the coating resists the pressure). The destruction of the coating leads to instantaneous release of the fertilizer.

Zaidel [47] analysed the forces involved during water penetration into a single granule and the rate of pressure buildup in it, from which it was possible to develop an expression for the time of "burst" or rupture (t_b) of a single coating (membrane):

$$t_b \cong \frac{r_0 l_0 Y}{P_h \Delta \pi M}, \qquad (31)$$

where r_0 is the granule radius, l_0 is the coating thickness, Y is the yield stress of the coating (Pa), P_h is the water permeability of the membrane (cm^2 day^{-1} Pa^{-1}), $\Delta \pi$ is the gradient of osmotic pressure across the membrane (Pa), and M is Young's module of elasticity of the coating (Pa).

8. CRFs Release Properties

Characterization of release from a given SRF/CRF is one of the most important steps in assessing the efficacy of a given fertilizer. Trenkel and IFI Association [10] provides a partial list of methods used by several manufacturers of coated fertilizers to assess the release of different SRFs/CRFs. Tests performed at temperatures ranging from 2°C to 60°C at varying sampling frequencies are reported.

Release characteristics may be attributed to both physical effects (such as reduced diffusion rates in soils, moisture and temperature fluctuations [48]) and chemical effects (pH changes, root excretion) as well as to the action of microbes on biodegradable materials (UF, sulphur coating, waxes, etc.).

This implies that a correlation between laboratory tests and release rates obtained under field conditions is required in order to achieve the highest NUE with the CRFs [7]. Release curves are the best common methods used in the characterization of nutrient release from the CRF formulation as seen in Figures 1–4 above.

Despite the release curves, several other parameters are known to be used in evaluating the properties of a particular CRF formulation such as water permeability [49] swelling ratio, and dissolution rate [50] which account for release behaviour. Others include zeta potential (ZP) and particle size [51, 52] together with morphology and thermal degradation properties [53].

9. CRFs and Biodegradability

Biodegradability means that a material has the proven capability to decompose in the most common environment where the material is disposed within 3 years through natural biological processes into nontoxic carbonaceous soil, water, carbon dioxide, or methane.

Partial Biodegradation. This relates to the minimal transformation that alters the physical characteristics of a compound while leaving the molecule largely intact. In other words, it refers to the alteration in the chemical structure of a substance, brought about by biological action, resulting in the loss of a specific property of that substance. Partial biodegradation is not necessarily a desirable property, since the intermediary metabolites formed can be more toxic than the original substrate. Therefore, mineralization is the preferred aim is such cases.

Complete Biodegradation. This occurs when the molecular cleavage is sufficiently extensive to remove biological, toxicological, chemical, and physical properties associated with the use of the original product, eventually forming carbon dioxide and water.

Readily Biodegradable. This is an arbitrary classification of chemicals which have passed certain specified screening tests for ultimate biodegradability; these tests are so stringent that it is assumed that such compounds will rapidly and completely biodegrade in aquatic environments under aerobic conditions.

Inherently Biodegradable. This is a classification of chemicals for which there is unequivocal evidence of biodegradation (partial or complete) in any test of biodegradability.

Despite the fact that an understanding of biodegradability is vital, the questions remain to be how biodegradable is the material? According to Han et al. [49], the test of biodegradability in CRF formulation is achieved by cutting CRF films into small squares such as 3 × 3 cm. Each specimen is then weighed and placed in agricultural soil (in a pot); subsequently, the pots are exposed to ambient conditions for 50 days. Variations in film morphology and disintegration time are then recorded as a test for biodegradability [49].

Similarly, terms like "environmentally friendly," "environmentally preferable," and "environmentally responsible" have been used to describe a material produced by biodegradable starting materials. In that case, one can freely use these terms interchangeably without distorting the meaning of the biodegradability concept.

10. Composition of CRF' Formulations

Basically most CRFs may contain among others the following components.

10.1. Polymer Solution. A number of polymers have been used in fertilizer coating; such polymers could be thermosetting, thermoplastic, or biodegradable ones. Some of the common thermoset polymers include urethane resin, epoxy resin, alkyd resin, unsaturated polyester resin, phenol resin, urea resin, melamine resin, phenol resin, and silicon resin. Among them, urethane resin urethane is very commonly used [1]. In addition, polyacrylamide is known to reduce soil erosion and so in this review we recommend that more studies should be conducted for its advanced use in CRFs [2, 54]. Thermoplastic resins are not very commonly used in practice because they are either not soluble in a solvent or make a very viscous solution which is not suitable for spraying; however, polyolefin is used in the art for coating the fertilizer granules. Biodegradable polymers are naturally available and so they are known to be environment friendly because they decompose in bioactive environments and degrade by the enzymatic action of microorganisms such as bacteria, fungi, and algae and their polymer chains may also be broken down by nonenzymatic processes such as chemical hydrolysis. However, both synthetic and natural polymers containing hydrolytically or enzymatically labile bonds or groups are degradable [25]. In the field of agriculture, the use of polymers is only limited by their relatively high cost, which has restricted their use mainly in light and medium soils with high sand content [55]. Therefore, in this review we have decided to concentrate on some natural biopolymers that are useful in CRF practices, for instance, natural rubber which was used by Hanafi to develop CRFs formulations [56].

10.1.1. Natural Polymers in CRFs Practices. Hitherto, natural polymers have been used to replace synthetic ones for the reason that they are inexpensive, they can control soil erosion [54], and they have low toxicity and excellent biodegradability [57]. Basically, natural polymers are more superior to the synthetic polymers owing to their highly organized macroscopic and molecular structure which in turn adds to their strength and biocompatibility [57]. There are three basic types of natural polymers widely used in the controlled release delivery systems. These are neutral, for example, hydroxypropylmethylcellulose (HPMC), cationic, for example, chitosan, and anionic polymers, for example, κ-carrageenan and sodium alginate. Several natural polymers including a few lists below have been used in the design of controlled release formulations of drugs and fertilizers as described hereunder.

Chitosan. This is a cationic polysaccharide composed of linear copolymers of glucosamine and N-acetyl glucosamine resulting from partial deacetylation of chitin obtained from crustacean shells. The natural rich sources of chitosan include chitin of invertebrates, insects, and yeasts [58]. Researches indicate that complexes formed between chitosan with bioactive compounds and other polymers are useful in modifying the release profile characteristics in different preparations [57]. In fact, it is found to provide first order release kinetics especially when particle size of less than 75 micron was used [58]. Several studies have been conducted using chitosan nanoparticles; the findings reveal that it is possible to intercalate NPK fertilizers into chitosan nanoparticles prepared by polymerizing methacrylic acid [59]. Interestingly, chitosan nanoparticles obtained showed spherical shapes and uniform sizes of approximately 78 nm [59].

The mechanism to optimize the incorporation of the N, P, and K elements into the designed chitosan nanoparticle is yet to be described [59]; this creates a gap for further research. Jamnongkan and Kaewpirom [50] reported CRF hydrogels prepared from chitosan, polyvinyl alcohol, and polyvinyl alcohol/chitosan, using glutaraldehyde as a crosslinker. The synthesized CRF hydrogels exhibited high swelling ratio [50]. Wu and Liu [60] managed to prepare chitosan-coated nitrogen, phosphorus, and potassium compound fertilizer with controlled-release and water-retention (CFCW) capacity using inversion suspension polymerization [60]. Besides, the CFCW synthesized possessed excellent water retention capacity and thus, can potentially be considered as a suitable formulation for both agricultural uses as well as for use in the arid and desert environments reclamation endeavors.

Xanthan Gum. This is a high molecular weight, water soluble, anionic-bacterial heteropolysaccharide; it is a hydrophilic polymer, biocompatible, and inert and thus it provides time-dependent release kinetics [57]. Xanthan gum (XG) is used as a rheology modifier and is derived as a result of microbial fermentation of glucose from the bacterial coat of *Xanthomonas campestris* [57]. As a matter of fact, the applications of XG in the CRFs industry are less common however, findings prove that XG matrices exhibit quite consistent higher ability to retard drug release for controlled-release formulation [61]. This calls for further investigations on its use in CRFs.

Carrageenan. This is a naturally occurring high molecular weight anionic gel-forming polysaccharide extracted from certain species of red seaweeds (Rhodophyceae) such as *Chondrus crispus*, Eucheuma, *Gigartina stellata*, and Iridaea [57]. It is made up of the repeating units of galactose and 3, 6 anhydrogalactose. Depending on the different degree of sulfation, they are classified into various types: ι-(mono-sulfate), κ-(di-sulfate), and λ-carrageenan (tri-sulfate). ι- and κ-carrageenan forms gel while highly sulphated λ-carrageenan is a thickening agent and does not form gel, which influences their release kinetics [57]. The integrated method for production of carrageenan and liquid fertilizer from fresh seaweeds is known [62]. In their work, the fresh biomasses of seaweeds *Kappaphycus alvarezii* were crushed to release sap which was then used for extraction of κ-carrageenan. The extract was found to be a superior raw material for production of liquid fertilizer after suitable treatment with additives. A novel biopolymer-based super-absorbent hydrogel synthesized after graft copolymerization of acrylic acid onto kappa-carrageenan backbones is reported to have been successfully researched [26]. Release studies revealed that Kappa-carrageenan is effective in minimizing burst release (bust effects versus tailing effects); in fact the burst release is found to depend highly on the degree of cross-linking and the mesh space available for drug diffusion [63].

Pectin. It is a methoxyester of pectic acid found in the higher plants cell walls [57, 58]. Certain fruits such as apple, quince, plum, gooseberry, grapes, cherries, and oranges also are known to contain pectin [57]. Little is known about pectin based CRFs; however, a pectin-based hydrogel used for removing Cu^{2+} and Pb^{2+} ions from water and wastewater and in the release of phosphate, potassium, and urea has been reported. The finding revealed that the pectin based hydrogel is effective in conserving water necessary for absorption by horticultural plants [64]. Non-Fickian mechanism was seen to control the release process of fertilizer nutrients from the hydrogels.

Tamarind Seed Polysaccharide (TSP). Tamarind seed polysaccharide (TSP) is a galactoxyloglucan (a monomer of mainly three sugars-galactose, xylose and glucose in a molar ratio of 1 : 2 : 3) isolated from seed kernel of *Tamarindus indica* [57]. Being a natural biopolymer TSP is nontoxic, biocompatible and cheap agro-based material for use in CRFs practices giving zero order release kinetics [58].

Mimosa pudica Seed Mucilage. Mimosa mucilage is known to act as a matrix forming agent for sustained delivery of formulations [57]. According to Kumar and Gupta, Mimosa mucilage biopolymer exhibited bioadhesion time of 10 h and more than 85% release of drug in 10 h [58]; however, its use in CRFs industry is yet to be exploited.

Leucaena leucocephala Seed Polysaccharide (LLSP). LLSP is a galactoxyloglucan hydrophilic gum isolated from seed kernel of *L. leucocephala.* In the controlled release art, LLSP has been used for controlled release of water-soluble plus water-insoluble drugs [57]. Intercalation of nitrogen fertilizer into the *L. leucocephala* residues under different moisture situation indicated that N content in soil released from residues increased with the time. Relatively higher amount of N release was observed in *L. leucocephala,* although the rate of N release was more with low N concentration residues [65]. In fact, this biopolymer is known to be a suitable natural disintegrant thereby being potentially useful in solid dispersion formulations for modifying rheological flow properties. It is also useful as suspending and emulsifying agent owing to its pseudoplastic and thixotropic flow patterns [66].

Guar Gum. This is a nonionic naturally occurring, hydrophilic polysaccharide extracted from the seeds of *Cyamopsis*

tetragonolobus and is used as binder and disintegrant [57]. It acts as the release-retarding polymer which follows a first-order release kinetic. *C. tetragonolobus* has been confirmed to be a suitable excipient for controlled release practices, although its use in CRFs is not clear and so opening door for further researches [58]. Findings revealed that increased gum concentration raises the swelling index value which is ideal for slow release kinetics [57, 67]. In addition, Alginate which is a natural polysaccharide obtained from marine brown algae and seaweeds and produced by some bacteria such as *Pseudomonas aeruginosa* or *Azotobacter vinelandii,* could be used in the same way. In a point of fact, Alginate is a hydrophilic salt of alginic acid consisting of two uronic acids, β-D-mannuronic acid (M) and α-L-glucuronic acid (G).

Terminalia catappa Gum (TC). It is a gum exudate obtained from *Terminalia catappa* Linn. It is a natural release retarding polymer. The drug release retarding behaviour of TC gum is well studied [68, 69]. Kumar et al. [69] demonstrated the excellent swelling properties of TC gum in water and its ability to sustain the release of dextromethorphan hydrobromide from matrix tablet. Therefore, tablet formulations containing TC gum as an excipient may ensure the utility of the TC gum in controlled drug delivery systems of sparingly water-soluble, low molecular weight drug substance. Nevertheless, how suitable TC gum is in CRFs is not clear.

Gellan Gum. This is a hydrophilic, high molecular weight, anionic deacetylated exocellular polysaccharide gum isolated as a fermentation product from a pure culture of *Pseudomonas elodea.* It consists of a tetrasaccharide repeating unit of one β-D-glucuronic acid, one α-L-rhamnose, and two β-D-glucose residues. On top of that, Grewia *gum* is a natural, hydrophilic polysaccharide obtained from the inner bark of the tree; *Grewia mollis* is known to hydrate on contact with water and swells to form a highly viscous dispersion making very suitable for CRFs [50].

Mucuna Gum. Mucuna gum is a biodegradable, amorphous polymer composed of mainly D-galactose along with D-mannose and D-glucose and isolated from the cotyledons of plant *Mucuna flagillepes.* Interestingly, studies show that formulations without crosslinking showed the fastest drug release [53]. This signifies that Mucuna based CRFs would exhibit similar features.

Gum Copal (GC). It is a naturally occurring hydrophobic resin isolated from the plant *Bursera bipinnata* and follow zero order release kinetics. To add more, Gum dammar (GD) which is a GC sister is also anticipated to exhibit similar release kinetics as in GC. Primarily, GD is a naturally occurring hydrophobic gum obtained from plant *Shorea wiesneri.*

Karaya Gum. It is a hydrophilic naturally occurring gum obtained from *Sterculia urens* and composed of galactose, rhamnose, and glucuronic acid. It swells in water and is thus used as release rate controlling polymers in different formulations.

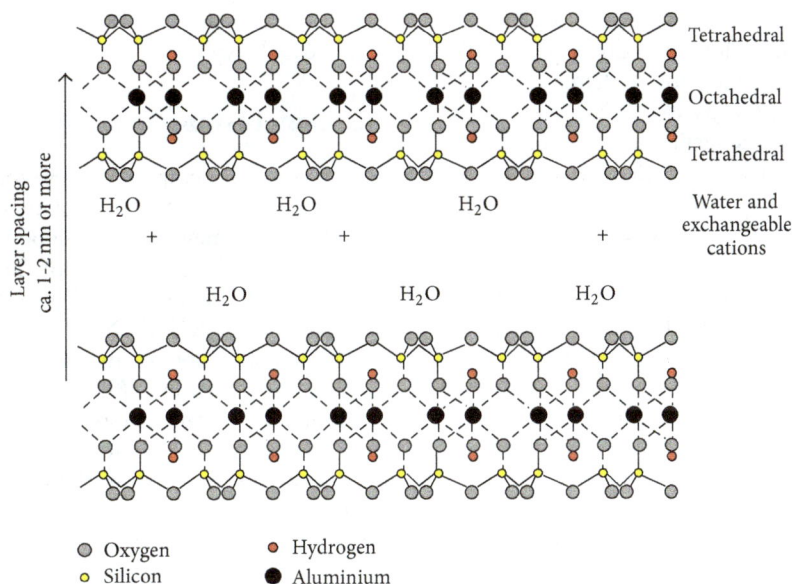

FIGURE 7: Structure of clay minerals.

Furthermore, Kumar and Gupta [58] provide an additional list of natural polymers used in controlled release systems including the following.

Rosin. a clear, pale yellow to dark amber thermoplastic resin present in oleoresins of the tree *Pinus roxburghii* and *Pinus taeda* belonging to the family Pinaceae. Rosin acts as a hydrophobic matrix forming agent for development of controlled drug delivery systems. It could be used as a binding agent and coating and matrix forming agent and so can be utilized as microencapsulating agent [70].

Gum Acacia. It is from stems of the *Acacia arabica* tree and can be used as encapsulating agent [71]. Locust bean gum provides excipient which gives sufficient mucoadhesive applications [72–74]. Other gums in a list include Khaya gum from Khaya *grandifoliola* used as binding and coating agent [75–77], Tragacanth gum used for sustained release [58], Okra gum from *Hibiscus esculentus* used in the formulation of sustained-release [78], and *Hibiscus rosa-sinesis* mucilage used to improve binding efficacy and hence acting as release-retarding agent [79, 80]. Moreover, olibanum and its resin [81], gum copal and gum damar [82], fenugreek mucilage [83], and dika nut mucilage (*Irvingia gabonensis*) are as well known to be used as binding agents in the release formulations [84].

10.2. Modified Clays. Nanoclay is the most common nanoparticle which has been used to produce CRFs. The layered clays like montmorillonite and kaolinite are made of high aspect ratio nanolayers. Large surface areas and reactivity of nanolayers are much greater than those of micrometre size materials. Also, their surfaces and interfaces provide an active substrate for physical, chemical, and biological reactions. Because of these features, nanolayers could be a suitable carrier or reservoir of fertilizers. Mechanisms which are involved in interaction between clay and organic materials depend on some factors like clay type, functional groups of organic material, and physical or chemical properties of organic material. For example basic molecules bond strongly to montmorillonite unlike anionic molecules which exhibit much weaker interactions. Similarly, benzoic acid or anionic species are adsorbed on the edge face of clay (cationic or crystal violet particles) after being adsorbed on the basal plane [1]. Basically, modification of clay can be achieved in many ways and different types of modified clay are named according to the methods followed such as pillared layered clays, organoclays, nanocomposites clays, acid and salt-induced modified clays, and thermally and mechanically induced modified clays. The use of each modified clay types given above, preparation, and their application in nano-CRFs are given by Basak et al. [85].

10.2.1. Organoclay Chemistry. Characteristically, clay minerals are natural materials with particle size <2 μm. Smectites, classified as 2 : 1 phyllosilicate clays, have a crystal lattice unit formed by one alumina octahedral sheet sandwiched between two silica tetrahedral sheets (Figure 7). The ion substitution or the site vacancies at the tetrahedral and/or octahedral sheets gives rise to a negatively charged surface. The exchangeable cations between the layers compensate the negative charge and may be easily exchanged by other metal cations, explaining the high ion exchange capacities of these minerals (70–120 meq/100 g). Due to this crystalline arrangement, smectites are able to expand and contract the interlayer while maintaining the two-dimensional crystallographic integrity. The interlayer between units contains positive cations and water molecules [86].

On the other hand, kaolinite, the main constituent of kaolin, is made up of tiny, thin, pseudohexagonal, flexible sheets of triclinic crystals with a diameter of 0.2–12 μm. The cation exchange capacity of kaolinite is considerably less than that of smectite in the order of 2–10 meq/100 g, depending on the particle size, but the rate of the exchange reaction is rapid, almost instantaneous [87]. Kaolinite adsorbs small molecular substances such as lecithin, quinolone, paraquat, and diquat. The adsorbed material can be easily removed from the particles because adsorption is limited to the surface of the particles (planes, edges), unlike the smectite where the adsorbed molecules are also bound between layers. This adsorption behaviour influences researchers to investigate kaolin as a vehicle for bioactive compounds and hence creating widespread pharmaceutical application of kaolin group of minerals as it is with smectites [88].

Properties such as colloidal particle size, crystalline structure, high specific surface area, charge, and swelling capacity confer on smectites and kaolin optimum rheological behaviour and excellent adsorption capacities for inorganic and organic substances such as drugs. In particular, the electrically charged surface of clay controls the interaction with other environmental ions, molecules, polymers, microorganisms, and particles. These processes have various technological applications such as drug delivery systems and controlled release fertilizers.

10.3. Other Components. Several other ingredients are known to compose CRFs formulations, namely, crosslinkers such as glutaraldehyde [50] and methylene-bisacrylamide (MBA) [89]; fertilizer nutrients such as urea and ammonium nitrate and initiators such as azobisisobutyronitrile (AIBN) [56], ZnO [56], and ammonium persulphate (APS) [89] have been used to create polymer before crosslinking. In addition to that, surfactants such as sodium octadecyl phosphate and sometimes a dispersion medium such as cyclohexane (which is normally used to disperse surfactant molecules) are also known in the release formulation practices [22]. In fact, different surfactants have been used in CRFs designs and the commonly used ones include nonionic surfactant molecules [90].

11. Polymer/Clay Superabsorbent Composites

According to Ekebafe et al. [23], the polymer-clay superabsorbent composites have been of great interest to researchers due to their comparative low production costs and high water absorbency. Apt superabsorbent composites by graft copolymerization reaction of acrylic acid (AA) and acrylamide (Am) on attapulgite micropowder using N, N-methylene bisacrylamide (MBA) as a crosslinker, and ammonium persulphate (APS) as an initiator in an aqueous solution are reported by the author. In fact, Ekebafe et al. describe acrylamide as a kind of nonionic monomer possessing good salt resistant performance and so a suitable raw material for superabsorbent synthesis [23]. In addition, attapulgite is described as a good substrate for superabsorbent composite

materials due to its aluminosilicate layers with reactive surface hydroxyl groups.

12. Conclusion

Regardless of being widely used, fertilizers particularly nitrogenous ones by virtue of their high nitrogen content ($\cong$46%) and somewhat low cost of production, are associated with up to 60% to 70% loss of the nitrogen being applied owing to ammonia produced through hydrolysis of say urea by soil urease ($NH_2CONH_2 + H_2O \rightarrow 2NH_3 + CO_2$). Fundamentally, due to surface runoff, leaching, and vaporization, the utilization efficiency or plant uptake of urea, for example, is generally below 50% thereby escalating fertilization expenditure per season and reducing crop productivity. Such drawbacks related to the use of nitrogenous fertilizers could be corrected by amending conventional nitrogenous fertilizers with suitable excipients in order to manufacture CRFs so as to improve FUE by plants and minimize the losses thereby reducing repeated fertilization expenditure per season and maximizing crop yields. CRFs reduce the demand for short-season manual labour obligatory during critical periods, reduce stress and specific toxicity (as a result of synchronizing nutrient release with plants' demands), increase availability of nutrients and supply of nutrient forms preferred by plants, and augment synergistic effects between nutrients and plant roots. In that view point, it is worth noting that researchers ought to design nano-CRFs by using natural excipients materials to come up with efficient, effective, reliable, and cost-effective CRFs formulations based on the prevailing resource limitations thereby minimizing food crisis and other challenges facing crop production. Essentially, scientists should anticipate mending agronomic returns through scientific novelties; the motive behind this must be geared towards researching, innovations, and commercialization of the CRF products.

Conflict of Interests

The authors declare that there is no conflict of interests regarding the publication of this paper.

Acknowledgments

The authors would like to thank the Nelson Mandela African Institution of Science and Technology (Tanzania) and the Department of Chemical Engineering, Hanyang University, (Republic of Korea) for financial support. Special thanks are due to Niconia Amos Bura for technical support in the preparation of the paper.

References

[1] Ukessays, "Controlled Release Fertilizers and Nanotechnology Traces Biology Essay," 2013, http://www.ukessays.com/essays/biology/controlled-release-fertilizers-and-nanotechnology-traces-biology-essay.php.

[2] C. V. Subbarao, G. Kartheek, and D. Sirisha, "Slow release of potash fertilizer through polymer coating," *International*

Journal of Applied Science and Engineering, vol. 11, no. 1, pp. 25–30, 2013.

[3] K. M. England, D. M. Camberato, and R. G. Lopez, *Commercial Greenhouse and Nursery Production*.

[4] AAPFCO, "Official Publication No. 50," West Lafayette, Ind, USA, 1997.

[5] M. E. Trenkel, *Slow-and Controlled-Release and Stabilized Fertilizers: An Option for Enhancing Nutrient Use Efficiency in Agriculture*, International Fertilizer Industry Association (IFA), 2010.

[6] A. Shaviv, "Controlled release fertilizers," in *Proceedings of the IFA International Workshop on Enhanced-Efficiency Fertilizers*, International Fertilizer Industry Association, Frankfurt, Germany, 2005.

[7] A. Shaviv, "Advances in controlled-release fertilizers," *Advances in Agronomy*, vol. 71, pp. 1–49, 2001.

[8] A. Shaviv, "Plant response and environmental aspects as affected by rate and pattern of nitrogen release from controlled release N fertilizers," in *Progress in Nitrogen Cycling Studies*, pp. 285–291, Springer, 1996.

[9] S. Raban, E. Zeidel, and A. Shaviv, "Release mechanisms controlled release fertilizers in practical use," in *Proceedings of the 3rd International Dahlia Greidinger Symposium on Fertilization and The Environment*, J. J. Mortwedt and A. Shaviv, Eds., pp. 287–295, 1997.

[10] M. E. Trenkel and IFI Association, *Controlled-Release and Stabilized Fertilizers in Agriculture*, vol. 11, International Fertilizer Industry Association, Paris, France, 1997.

[11] H. M. Goertz, "Technology developments in coated fertilizers," in *Proceedings of the Dahlia Greidinger Memorial International Workshop on Controlled/Slow Release Fertilizers*, Technion-Israel Institute of Technology, Haifa, Israel, 1993.

[12] A. N. Sharpley and R. G. Menzel, "The impact of soil and fertilizer phosphorus on the environment," *Advances in Agronomy*, vol. 41, pp. 297–324, 1987.

[13] R. Clark, *Marine Pollution*, Clarendon Press, Oxford, UK, 1989.

[14] C.-W. Du, J.-M. Zhou, and A. Shaviv, "Release characteristics of nutrients from polymer-coated compound controlled release fertilizers," *Journal of Polymers and the Environment*, vol. 14, no. 3, pp. 223–230, 2006.

[15] S. Landels, *Controlled-Release Fertilizers: Supply and Demand Trends in US Nonfarm Markets*, SRI International, Menlo Park, Calif, USA, 1994.

[16] IPNI, *Coated Fertilizer. Nutrient Source Specifics 2013*, 2013, http://www.ipni.net/publication/nss.nsf/0/33C6A283CC38EE26852579AF007682E3/$FILE/NSS-20%20-Coated%20Fertilizer.pdf.

[17] A. Shaviv, E. Zlotnikov, and E. Zaidel, "Mechanisms of nutrient release from controlled release fertilizers," in *Proceedings of the Dahlia Gredinger Memorial International Workshop on Controlled Release Fertilizers*, J. Hagin and J. Mortvedt, Eds., Technion, Israel Institute of Technology, Haifa, Israel, 1995.

[18] M. Zhang, Y. C. Yang, F. P. Song, and Y. X. Shi, "Study and industrialized development of coated controlled release fertilizers," *Journal of Chemical Fertilizer Industry*, vol. 32, pp. 7–12, 2005.

[19] A. Shaviv, S. Raban, and E. Zaidel, "Modeling controlled nutrient release from a population of polymer coated fertilizers: statistically based model for diffusion release," *Environmental Science & Technology*, vol. 37, no. 10, pp. 2257–2261, 2003.

[20] A. Shaviv, S. Raban, and E. Zaidel, "Modeling controlled nutrient release from polymer coated fertilizers: diffusion release from single granules," *Environmental Science & Technology*, vol. 37, no. 10, pp. 2251–2256, 2003.

[21] Y. Yang, Y. Geng, M. Zhang, J. Chen, and H. Chen, "Effects of coating properties of controlled-release fertilizers on nutrient release characteristics," *Transactions of the Chinese Society of Agricultural Engineering*, vol. 2007, no. 11, 2007.

[22] M. Liu, R. Liang, F. Zhan, Z. Liu, and A. Niu, "Preparation of superabsorbent slow release nitrogen fertilizer by inverse suspension polymerization," *Polymer International*, vol. 56, no. 6, pp. 729–737, 2007.

[23] L. Ekebafe, D. Ogbeifun, and F. Okieimen, "Polymer applications in agriculture," *Biokemistri*, vol. 23, no. 2, pp. 81–89, 2011.

[24] W. Li, L. Zhang, C. Liu, and Z. Liang, "Preparation and property of poly (acrylamide-co-acrylic acid) macromolecule slow-releasing fertilizer," *International Journal of Electrochemical Science*, vol. 7, no. 11, pp. 11470–11476, 2012.

[25] D. R. Lu, C. M. Xiao, and S. J. Xu, "Starch-based completely biodegradable polymer materials," *Express Polymer Letters*, vol. 3, no. 6, pp. 366–375, 2009.

[26] H. Hosseinzadeh, "Controlled release of diclofenac sodium from pH-responsive carrageenan-g-poly(acrylic acid) superabsorbent hydrogel," *Journal of Chemical Sciences*, vol. 122, no. 4, pp. 651–659, 2010.

[27] B. Kloth, *Aglukon Spezialdünger GmbH: Reply to the Request on Controlled-release Fertilizers*, Personal Communication, 1996.

[28] S. Shoji and A. T. Gandeza, *Controlled Release Fertilizers with Polyolefin Resin Coating*, Kanno Printing, Sendai, Japan, 1992.

[29] H. Kanno, "Use of Controlled-Release Fertilizers (CRF) for Sustainable Crop Production in Asia," Japan International Research Center for Agricultural Sciences, Ohwashi, Tsukuba, Japan, 2013, http://jircas-d.job.affrc.go.jp/Ver-1/english/files/2014/03/fanglei/2013-session-22.pdf.

[30] T. Fujita, *Reply to the Request on Controlled-Release Fertilizers*, Personal Communication, 1996.

[31] J. Agric, *Meister: Advanced Coated Fertilizer for the New Agriculture*, Jcam Agri. Co., Ltd, Tokyo, Japan, 2014, http://www.jcam-agri.co.jp/en/product/meister.html.

[32] M. Guo, M. Liu, R. Liang, and A. Niu, "Granular urea-formaldehyde slow-release fertilizer with superabsorbent and moisture preservation," *Journal of Applied Polymer Science*, vol. 99, no. 6, pp. 3230–3235, 2006.

[33] W. Jarrell and L. Boersma, "Release of urea by granules of sulfur-coated urea," *Soil Science Society of America Journal*, vol. 44, no. 2, pp. 418–422, 1980.

[34] J. Crank, *The Mathematics of Diffusion*, Clarendon Press, Oxford, UK, 2nd edition, 1975.

[35] W. Jarrell and L. Boersma, "Model for the release of urea by granules of sulfur-coated urea applied to soil," *Soil Science Society of America Journal*, vol. 43, no. 5, pp. 1044–1050, 1979.

[36] W. J. Moore, *Physical Chemistry*, Prentice Hall, Englewood Cliffs, NJ, USA, 1972.

[37] V. Glaser, P. Stajer, and J. Vidensky, "Simulace prubehu rozpousteni obalovanych prumyslovych hnojiv ve vode—II," *Chemický průmysl*, vol. 37, no. 62, pp. 353–355, 1987.

[38] M. Barzegar-Jalali, "Kinetic analysis of drug release from nanoparticles," *Journal of Pharmacy and Pharmaceutical Sciences*, vol. 11, no. 1, pp. 167–177, 2008.

[39] A. C. Salome, C. O. Godswill, and I. O. Ikechukwu, "Kinetics and mechanisms of drug release from swellable and non

swellable matrices: a review," *Research Journal of Pharmaceutical, Biological and Chemical Sciences*, vol. 4, no. 2, pp. 97–103, 2013.

[40] P. Costa and J. M. S. Lobo, "Modeling and comparison of dissolution profiles," *European Journal of Pharmaceutical Sciences*, vol. 13, no. 2, pp. 123–133, 2001.

[41] V. Dixit, Dissolution and Dissolution Models, 2014, authorSTREAM, http://www.authorstream.com/Presentation/aSGuest106867-1122372-dissolution-and-its-models/.

[42] B. S. Mahat, "Mathematical Models used in Drug Release Studies," Department of Pharmacy, Kathmandu University, Dhulikhel, Nepal, 2010, http://www.scribd.com/doc/54516124/MATHEMATICAL-MODELS-USED-IN-THE-DRUG-RELEASE-STUDIES.

[43] B. K. Nanjwade, *Fundamentals of Modified Release Formulations*, Department of Pharmaceutics, KLE University College of Pharmacy, Bangalore, India, 2013, http://www.google.com/url?sa=t&rct=j&q=&esrc=s&source=web&cd=1&cad=rja&uact=8&ved=0CCMQFjAA&url=http%3A%2F%2Fapi.ning.com%2Ffiles%2Fn5OxHQ0ab5xtv8QP-Fp3eJlZdWEcLnrQ2-yWWZlLWPoC3bly9iOzib∗Hi5lW8hWaM-D9AY5nAbRwu-Thr-inRne0FGHlzxO∗T5%2FFundamentalsofModifiedReleaseFormulations.pptx&ei=LdnFU5TMDIb-8QXQqYCoBQ&usg=AFQjCNFKcgCsWACXkE5xjyrhL9GFPONE0Q&sig2=bQ-ZXrTWW29RjXMHmyknDw.

[44] D. Narender, *Theories and Mechanisms of Dissolution Testing*, Pharmawiki, Kakatiya University, 2014, http://pharmawiki.in/ppt-theories-and-mechanisms-of-dissolution-testing/.

[45] J. Siepmann and N. A. Peppas, "Modeling of drug release from delivery systems based on hydroxypropyl methylcellulose (HPMC)," *Advanced Drug Delivery Reviews*, vol. 48, no. 2-3, pp. 139–157, 2001.

[46] S. Raban, "Release mechanisms of membrane coated fertilizers," in *Advances in Agronomy*, A. Shaviv, Ed., pp. 1–49, Faculty of Agricultural Engineering, Technion, IIT-Israel, 1994.

[47] E. Zaidel, *Models of Controlled Release of Fertilizers*, Technion-Israel Institute of Technology, Faculty of Agricultural Engineering, 1996.

[48] Southern Agricultural Insecticides, *I. SA-50 Controlled-Release Fertilizer 14-14-14: Osmocoter*, 1998, http://www.southernag.com/docs/labels_msds/contrl.pdf.

[49] X. Han, S. Chen, and X. Hu, "Controlled-release fertilizer encapsulated by starch/polyvinyl alcohol coating," *Desalination*, vol. 240, no. 1–3, pp. 21–26, 2009.

[50] T. Jamnongkan and S. Kaewpirom, "Controlled-release fertilizer based on chitosan hydrogel: phosphorus release kinetics," *Science Journal UBU*, vol. 1, pp. 43–50, 2010.

[51] NanoComposix, Zeta Potential Analysis of Nanoparticles, 2012, http://cdn.shopify.com/s/files/1/0257/8237/files/nanoComposix_Guidelines_for_Zeta_Potential_Analysis_of_Nanoparticles.pdf.

[52] E. Zeta Potential—Electrophoresis, 2013, http://www.escubed.co.uk/sites/default/files/zeta_potential_%28an011%29_elecrophoresis.pdf.

[53] P. Gill, T. T. Moghadam, and B. Ranjbar, "Differential scanning calorimetry techniques: applications in biology and nanoscience," *Journal of Biomolecular Techniques*, vol. 21, no. 4, pp. 167–193, 2010.

[54] K. N. Nwankwo, *Polyacrylamide as a Soil Stabilizer for Erosion Control*, 2001.

[55] G. K. Chatzoudis and F. Rigas, "Macroreticular hydrogel effects on dissolution rate of controlled-release fertilizers," *Journal of Agricultural and Food Chemistry*, vol. 46, no. 7, pp. 2830–2833, 1998.

[56] M. Hanafi, S. Eltaib, and M. Ahmad, "Physical and chemical characteristics of controlled release compound fertiliser," *European Polymer Journal*, vol. 36, no. 10, pp. 2081–2088, 2000.

[57] A. Singh, P. K. Sharma, and R. Malviya, "Release behavior of drugs from various natural gums and polymers," *Polimery w Medycynie*, vol. 41, no. 4, pp. 73–80, 2011.

[58] S. Kumar and S. K. Gupta, "Natural polymers, gums and mucilages as excipients in drug delivery," *Polimery w Medycynie*, vol. 42, no. 3-4, pp. 191–197, 2012.

[59] E. Corradini, M. R. de Moura, and L. H. C. Mattoso, "A preliminary study of the incorparation of NPK fertilizer into chitosan nanoparticles," *Express Polymer Letters*, vol. 4, no. 8, pp. 509–515, 2010.

[60] L. Wu and M. Liu, "Preparation and properties of chitosan-coated NPK compound fertilizer with controlled-release and water-retention," *Carbohydrate Polymers*, vol. 72, no. 2, pp. 240–247, 2008.

[61] M. M. Talukdar, I. Vinckier, P. Moldenaers, and R. Kinget, "Rheological characterization of xanthan gum and hydroxypropylmethyl cellulose with respect to controlled-release drug delivery," *Journal of Pharmaceutical Sciences*, vol. 85, no. 5, pp. 537–540, 1996.

[62] K. Eswaran, P. K. Ghosh, A. K. Siddhanta et al., "Integrated method for production of carrageenan and liquid fertilizer from fresh seaweeds," Google Patents, 2005.

[63] H. Hezaveh and I. I. Muhamad, "Controlled drug release via minimization of burst release in pH-response kappa-carrageenan/polyvinyl alcohol hydrogels," *Chemical Engineering Research and Design*, vol. 91, no. 3, pp. 508–519, 2013.

[64] M. R. Guilherme, A. V. Reis, A. T. Paulino, T. A. Moia, L. H. C. Mattoso, and E. B. Tambourgi, "Pectin-based polymer hydrogel as a carrier for release of agricultural nutrients and removal of heavy metals from wastewater," *Journal of Applied Polymer Science*, vol. 117, no. 6, pp. 3146–3154, 2010.

[65] S. K. Das, G. S. Reddy, K. L. Sharma et al., "Prediction of nitrogen availability in soil after crop residue incorporation," *Fertilizer Research*, vol. 34, no. 3, pp. 209–215, 1993.

[66] V. Pendyala, C. Baburao, and K. B. Chandrasekhar, "Studies on some physicochemical properties of *Leucaena Leucocephala* bark gum," *Journal of Advanced Pharmaceutical Technology & Research*, vol. 1, no. 2, pp. 253–259, 2010.

[67] V. N. Deshmukh, S. P. Singh, and D. M. Sakarkar, "Formulation and evaluation of sustained release metoprolol succinate tablet using hydrophilic gums as release modifiers," *International Journal of PharmTech Research*, vol. 1, no. 2, pp. 159–163, 2009.

[68] H. A. Patel, G. V. Joshi, R. R. Pawar, H. C. Bajaj, and R. V. Jasra, "Mechanical and thermal properties of polypropylene nanocomposites using organically modified indian bentonite," *Polymer Composites*, vol. 31, no. 3, pp. 399–404, 2010.

[69] S. V. Kumar, D. Sasmal, and S. C. Pal, "Rheological characterization and drug release studies of gum exudates of *Terminalia catappa* linn," *AAPS PharmSciTech*, vol. 9, no. 3, pp. 885–890, 2008.

[70] S. Kumar and S. K. Gupta, "Rosin: a naturally derived excipient in drug delivery systems," *Polimery w medycynie*, vol. 43, no. 1, pp. 45–48, 2013.

[71] E.-X. Lu, Z. Jiang, Q. Zhang, and X. Jiang, "A water-insoluble drug monolithic osmotic tablet system utilizing gum arabic as an osmotic, suspending and expanding agent," *Journal of Controlled Release*, vol. 92, no. 3, pp. 375–382, 2003.

[72] K. Malik, G. Arora, and I. Singh, "Locust bean gum as superdisintegrant—formulation and evaluation of nimesulide orodispersible tablets," *Polimery w Medycynie*, vol. 41, no. 1, pp. 17–28, 2011.

[73] M. Dionísio and A. Grenha, "Locust bean gum: exploring its potential for biopharmaceutical applications," *Journal of Pharmacy and Bioallied Sciences*, vol. 4, no. 3, pp. 175–185, 2012.

[74] C. Vijayaraghavan, S. Vasanthakumar, and A. Ramakrishnan, "In vitro and in vivo evaluation of locust bean gum and chitosan combination as a carrier for buccal drug delivery," *Die Pharmazie*, vol. 63, no. 5, pp. 342–347, 2008.

[75] O. A. Odeku and O. A. Itiola, "Characterization of khaya gum as a binder in a paracetamol tablet formulation," *Drug Development and Industrial Pharmacy*, vol. 28, no. 3, pp. 329–337, 2002.

[76] O. A. Odeku and J. T. Fell, "Evaluation of khaya gum as a directly compressible matrix system for controlled release," *Journal of Pharmacy and Pharmacology*, vol. 56, no. 11, pp. 1365–1370, 2004.

[77] O. A. Odeku and J. T. Fell, "In-vitro evaluation of khaya and albizia gums as compression coatings for drug targeting to the colon," *Journal of Pharmacy and Pharmacology*, vol. 57, no. 2, pp. 163–168, 2005.

[78] V. D. Kalu, M. A. Odeniyi, and K. T. Jaiyeoba, "Matrix properties of a new plant gum in controlled drug delivery," *Archives of Pharmacal Research*, vol. 30, no. 7, pp. 884–889, 2007.

[79] G. K. Jani and D. P. Shah, "Evaluation of mucilage of Hibiscus rosasinensis Linn as rate controlling matrix for sustained release of diclofenac," *Drug Development and Industrial Pharmacy*, vol. 34, no. 8, pp. 807–816, 2008.

[80] K. Ameena, C. Dilip, R. Saraswathi, P. N. Krishnan, C. Sankar, and S. P. Simi, "Isolation of the mucilages from *Hibiscus rosasinensis* linn. and Okra (*Abelmoschus esculentus* linn.) and studies of the binding effects of the mucilages," *Asian Pacific Journal of Tropical Medicine*, vol. 3, no. 7, pp. 539–543, 2010.

[81] K. Chowdary, P. Mohapatra, and M. Krishna, "Evaluation of olibanum and its resin as rate controlling matrix for controlled release of diclofenac," *Indian Journal of Pharmaceutical Sciences*, vol. 68, no. 4, pp. 497–500, 2006.

[82] D. Morkhade, S. Fulzele, P. Satturwar, and S. Joshi, "Gum copal and gum damar: novel matrix forming materials for sustained drug delivery," *Indian Journal of Pharmaceutical Sciences*, vol. 68, no. 1, pp. 53–58, 2006.

[83] A. Nokhodchi, H. Nazemiyeh, A. Khodaparast, T. Sorkh-Shahan, H. Valizadeh, and J. L. Ford, "An in vitro evaluation of fenugreek mucilage as a potential excipient for oral controlled-release matrix tablet," *Drug Development and Industrial Pharmacy*, vol. 34, no. 3, pp. 323–329, 2008.

[84] O. A. Odeku and B. O. Patani, "Evaluation of dika nut mucilage (Irvingia gabonensis) as binding agent in metronidazole tablet formulations," *Pharmaceutical Development and Technology*, vol. 10, no. 3, pp. 439–446, 2005.

[85] B. B. Basak, S. Pal, and S. C. Datta, "Use of modified clays for retention and supply of water and nutrients," *Current Science*, vol. 102, no. 9, pp. 1272–1278, 2012.

[86] M. E. Parolo, L. G. Fernández, I. Zajonkovsky, M. P. Sánchez, and M. Baschini, "Antibacterial activity of materials synthesized from clay minerals. Science against microbial pathogens: communicating current research and technological advances," *Formatex, Microbiology Series*, vol. 3, pp. 144–151, 2011.

[87] R. E. Grim, *Clay Mineralogy*, International Series in the Earth and Planetary Sciences, McGraw-Hill, New York, NY, USA, 1968.

[88] C. Aguzzi, P. Cerezo, C. Viseras, and C. Caramella, "Use of clays as drug delivery systems: possibilities and limitations," *Applied Clay Science*, vol. 36, no. 1–3, pp. 22–36, 2007.

[89] R. Liang, H. Yuan, G. Xi, and Q. Zhou, "Synthesis of wheat straw-g-poly(acrylic acid) superabsorbent composites and release of urea from it," *Carbohydrate Polymers*, vol. 77, no. 2, pp. 181–187, 2009.

[90] H. Talaat, M. H. Sorour, A. G. Aboulnour, and H. F. Shaalan, "Development of a multi-component fertilizing hydrogel with relevant techno-economic indicators," *American-Eurasian Journal of Agricultural & Environmental Science*, vol. 3, no. 5, pp. 764–770, 2008.

13

An Efficient One Pot Protocol to the Annulation of Face "d" of Benzazepinone Ring with Pyrazole, Isoxazole, and Pyrimidine Nucleus through the Corresponding Oxoketene Dithioacetal Derivative

Aditi Anand, Navjeet Kaur, and Dharma Kishore

Department of Chemistry, Banasthali University, Banasthali, Raj 304022, India

Correspondence should be addressed to Navjeet Kaur; nvjithaans@gmail.com

Academic Editor: Manuel Sergi

A highly facile single step approach to the annulation of face "d" of benzazepinone nucleus with pyrazole, isoxazole, and pyrimidine ring has been described. The annulation proceeded smoothly on the reaction of oxoketene dithioacetal derivative **3** with (i) NH_2–$NH_2\cdot H_2O$, (ii) $NH_2OH\cdot HCl$, (iii) acetamidine hydrochloride, (iv) guanidine nitrate, (v) urea, and (vi) thiourea which yielded the pyrazolo, isoxazolo, and pyrimido annulated analogues of benzazepinone **4–9**, respectively, in acceptable yields. The 4-ketene dithioacetal analogue of 7-fluorobenzo[b]azepine-2,5-dione (**3**) was in turn obtained from the reaction of 7-fluoro-3,4-dihydro-1*H*-benzo[b]azepine-2,5-dione (**2**) (with CS_2 + CH_3I in presence of *t*-BuOK). 7-Fluoro-3,4-dihydro-1*H*-benzo[b]azepine-2,5-dione (**2**) resulted from the acylation of *p*-fluoroaniline with succinyl chloride followed by cyclocondensation of the later with polyphosphoric acid (PPA).

1. Introduction

Koch et al. [1] have carried out a quantitative analysis of physiologically active natural products and showed that compound molecules with two or three rings were most often found in active natural products. The interest in the various facets of the chemistry and biology of small bicyclic and tricyclic (carbocyclic and heterocyclic) systems has expanded exponentially thereafter. Since then the development of small molecule libraries with potential biological activities has been a major focus of research in the area of chemical biology and medicinal chemistry. In view of this, the development of efficient methodologies to access small molecules of medicinal utility has been currently of special interest, with particular emphasis on the preparation of compound libraries from the privileged medicinal structures or from those structures akin to these.

Benzodiazepine framework in general has been recognized to belong to the family of "privileged medicinal structures," by virtue of their ability to provide ligands to a number of functionally and structurally discrete biological receptors. A derivative of benzazepine "the mirtazapine," being in close analogy to its activity, with 1,4-benzodiazepines, has emerged as an active antidepressant for the treatment of moderate to severe depression [2, 3] (Figure 1). This discovery provided optimism towards the development of other novel agents from benzazepine class of compounds to find if there were others too, to show a higher level of medicinal efficacy. This search resulted in the discovery of 7-phenyl sulfonyl-tetrahydro-3-benzazepine derivative for its use as antipsychotic agent [4] and 6-chloro-2,3,4,5-tetrahydro-3-methyl-1*H*-3-benzazepine to give relief in benign prostatic hypertrophy [5, 6].

Literature is replete with ample examples to show that the presence of an additional heterocyclic ring onto the seven membered ring of benzodiazepines exerts a profound influence on the biological activity in the resulting materials. Due to this, we considered it of interest to functionalize the benzazepine-2,5-dione system **2** with ketene dithioacetal

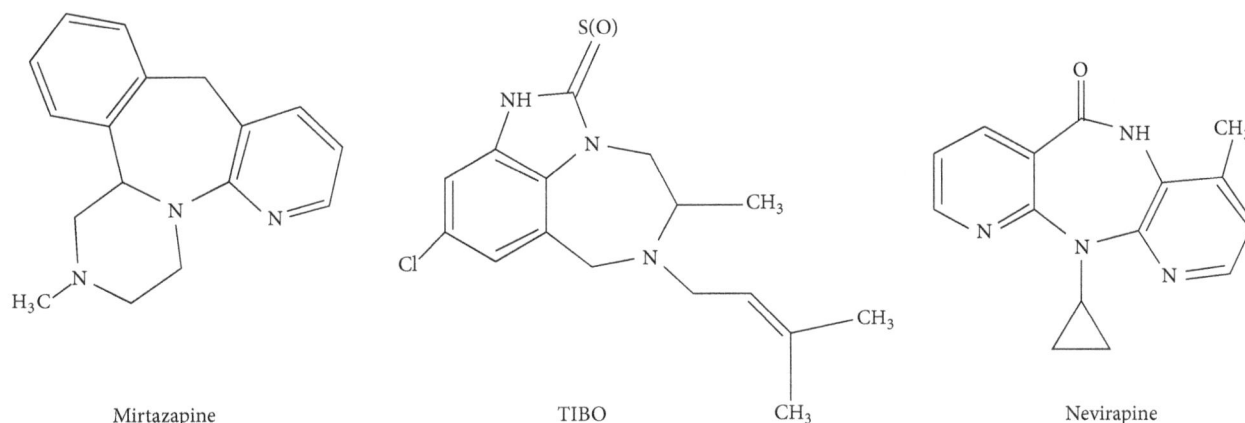

FIGURE 1

function in consideration of their amenability to the corresponding pyrazolo, isoxazolo, and pyrimido annulated analogues **4–9**, respectively, (Scheme 1) on the premise that their presence in tandem in the same molecular framework could inherit its positive impact onto the overall biological efficacy in the resulting molecules. Herein, in this communication, we report the preliminary results of our endeavour, focused in this direction.

2. Materials and Method

p-Fluoroaniline (**1**) and succinyl chloride were obtained from commercial sources and were used as obtained, from the sigma suppliers without further purification. Melting points were determined in open glass capillaries and are uncorrected. The purity of the compounds was checked by TLC on silica gel (G) plates. IR spectra were recorded on CE (Schimatzu) FTIR-9050 S. ^{1}H-NMR spectra and ^{13}C NMR spectra were recorded on Sea 400 (Bruker) using CDCl$_3$ as solvent and TMS as an internal reference. Chemical shifts are expressed in δ ppm. Mass spectra were recorded on Bosch Tech. X.

3. Experimental Section

3.1. Synthesis of 7-Fluoro-3,4-dihydro-1H-benzo[b]azepine-2,5-dione (2). *p*-Fluoroaniline (**1**) (3.60 mL, 0.03 mol) was mixed with succinyl chloride (4.92 g, 0.03 mol) in dry pyridine (20.0 mL) and the mixture was refluxed for 15 min. Cold reaction mixture was poured slowly with stirring in 150–200 mL ice cold water. The separated solid was filtered, washed with cold water, and recrystallized using methanol and water. Polyphosphoric acid (25 g) was mixed to 3.21 g (0.01 mol) of separated solid and heated at 150–160°C for 4 h (the progress of the reaction was monitored by TLC). The reaction mixture was cooled to 20°C and a concentrated aqueous solution of Na$_2$CO$_3$ was added to make it alkaline. The product was extracted with ethyl acetate (3 × 10 mL). The extract was dried over anhydrous sodium sulfate and concentrated in vacuum. The residue was purified by column chromatography on silica gel with CHCl$_3$ as an eluent to

give **2** (2.85 g, yield: 79%): m.p.: 158–160°C; IR (KBr) cm^{-1}: 3240 (N–H str.), 2990 (C–H str.), 2900, 1400 (–CH$_2$ next to C=O), 1680, 1710 (C=O str.), and 1535 (C=C str.); ^{1}H NMR (400 MHz, CDCl$_3$) δ ppm: 8.0 (1H, s, NH), 7.26–7.78 (3H, m, Ar–H), and 3.49 (4H, m, (CH$_2$)$_2$); ^{13}C NMR (400 MHz, CDCl$_3$) δ ppm: Ar–C [157.55 (CF), 120.24 (CH), 113.54 (CH), 112.44 (CH)], Ar–C [134.44 (C), 115.25 (C), azepinone], 27.6, 34.4 [(CH$_2$)$_2$ azepinone)], 176.75 (C of amide), and 183.49 (C of carbonyl); MS: *m/z* 193.17 (M$^+$); Anal. calcd./found for C$_{10}$H$_8$FNO$_2$: C 62.18/62.35, H 4.17/4.11, and N 7.25/7.48.

3.2. Synthesis of 4-(Bis(methylthio)methylene)-7-fluoro-3,4-dihydro-1H-benzo[b]azepine-2,5-dione (3). A mixture of 7-fluoro-3,4-dihydro-1*H*-benzo[b]azepine-2,5-dione (**2**) (2.82 g, 0.01 mol) and CS$_2$ (1.6 mL, 0.01 mol) was added to a well-stirred and cold suspension of *t*-BuOK (2.23 g, 0.02 mol) in dry benzene (7.0 mL) and DMF (3.0 mL) and the reaction mixture was allowed to stand for 4 h. Methyl iodide (3.3 mL, 0.02 mol) was gradually added with stirring and reaction mixture was maintained at low temperature by placing it in ice cold water bath. The reaction mixture was allowed to stand for 4 h at room temperature with occasional shaking followed by reflux on a water bath for 3 h. The mixture was poured on crushed ice and the benzene layer was separated. The aqueous portion was extracted with benzene and the combined extracts were washed with water and dried over anhydrous sodium sulfate and the solvent was removed by distillation. The product thus obtained was purified by crystallization with ethanol to give **3** (1.7 g, yield: 60%): m.p.: 155–157°C; IR (KBr) cm^{-1}: 3240 (N–H str.), 3000 (C–H str.), 2900, 1400 (–CH$_2$ next to C=O), 1640, 1685 (C=O str.), 1620 (C=C str. of α,β-unsaturated ketone), 1535 (C=C str.), and 680 (C–S str.); ^{1}H NMR (400 MHz, CDCl$_3$) δ ppm: 8.0 (1H, s, NH), 7.45–7.98 (3H, m, Ar–H), 3.53 (2H, s, CH$_2$), and 2.80 (6H, s, (CH$_3$)$_2$ of (SMe)$_2$); ^{13}C NMR (400 MHz, CDCl$_3$) δ ppm: Ar–C [164.5 (CF), 121.5 (CH), 113.1 (CH), 112.6 (CH)], Ar–C [136.8 (C), 136.0 (C), 108.7 (C), azepinone], 28.60 (CH$_2$ azepinone), 168.7 (C of amide), 187.0 (C of carbonyl), 155.3 [–C–(SMe)$_2$], and 18.0 [2C of (CH$_3$)$_2$]; MS: *m/z* 297.37 (M$^+$); Anal. calcd./found for C$_{13}$H$_{12}$FNO$_2$S$_2$: C 52.51/52.32, H 4.07/4.01, N 14.71/14.48, and S 21.57/21.38.

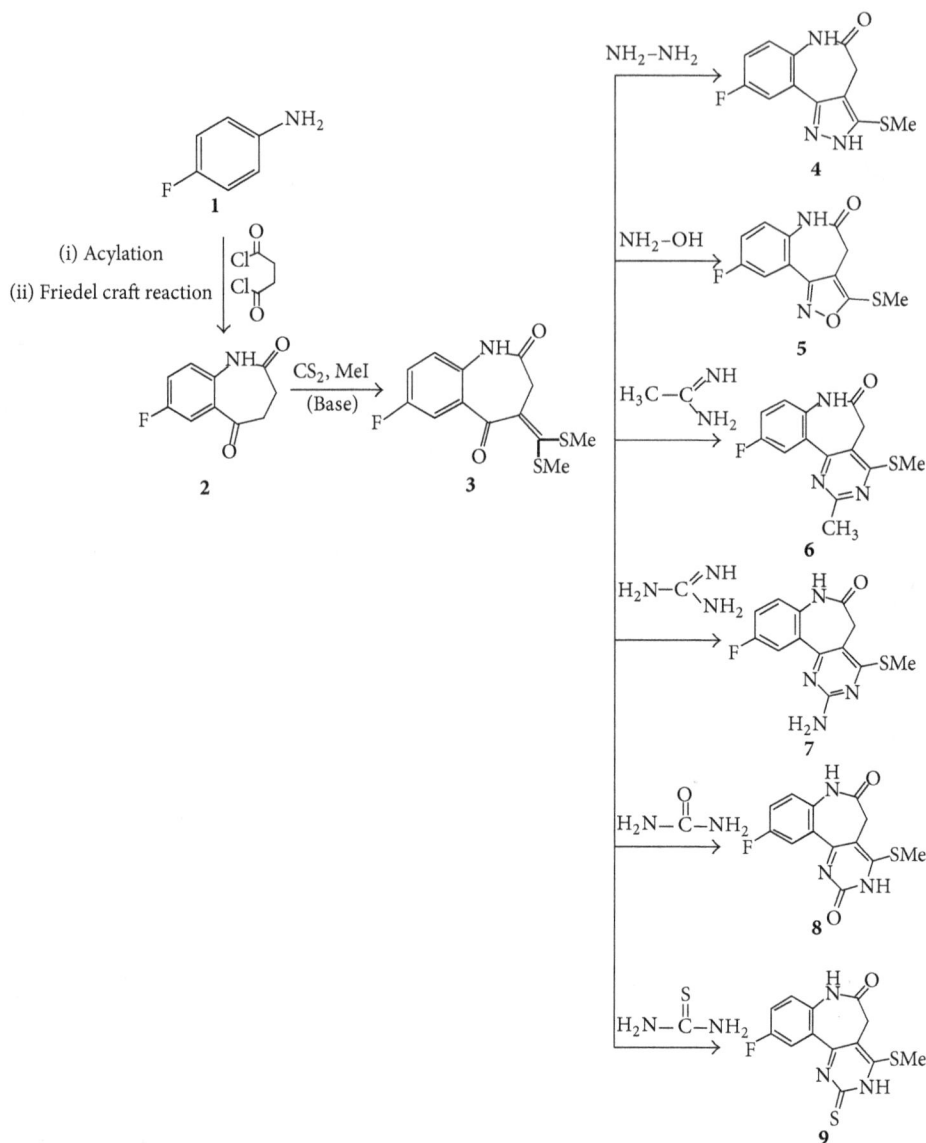

SCHEME 1

3.3. Synthesis of 9-Fluoro-3-(methylthio)-4,6-dihydrobenzo[b] pyrazolo[3,4-d]azepin-5(2H)-one (4). Hydrazine hydrate (1 mL, 0.02 mol) and 4-(bis(methylthio)methylene)-7-fluoro-3,4-dihydro-1H-benzo[b]azepine-2,5-dione (3) (0.658 g, 0.002 mol) were taken in ethanol (10.0 mL) and refluxed for 3 h. The solvent was removed and residue was dissolved in 5 mL of chloroform. On removal of the solvent, **4** was obtained (0.38 g, yield: 58%): m.p.: 230–232°C; IR (KBr) cm^{-1}: 3390 (N–H str.), 3210 (N–H str. of pyrazole ring), 3090 (C–H str.), 2990, 1400 (–CH$_2$ next to C=O), 1600 (C=O str.), 1590 (C=C str.), 1515 (C=N str.), and 690 (C–S str.); ^{1}H NMR (400 MHz, CDCl$_3$) δ ppm: 12.82 (IH, s, NH of pyrazole ring), 8.03 (1H, s, NH), 7.59–8.51 (3H, m, Ar–H), 3.72 (2H, s, CH$_2$), and 2.70 (3H, s, CH$_3$ of SMe group); ^{13}C-NMR (400 MHz, CDCl$_3$) δ ppm: Ar–C [168.83 (CF), 121.01 (CH), 114.51 (CH), 114.49 (CH)], Ar–C [143.29 (C), 131.82 (C), 123.11 (C), 97.33 (C), azepinone), 172.62 (C of amide), 39.60 (CH$_2$,

azepinone), 130.57 (pyrazole, –CSMe), and 17.62 (C of CH$_3$); MS: *m/z* 263.301 (M$^+$); Anal. calcd./found for C$_{12}$H$_{10}$FN$_3$OS: C 54.74/54.52, H 3.83/3.89, N 15.96/15.36, and S 12.18/12.36.

3.4. Synthesis of 9-Fluoro-3-(methylthio)-4H-benzo[b]isoxazolo[3,4-d]azepin-5(6H)-one (5). Hydroxylamine hydrochloride (0.34 g, 0.005 mol) was added to sodium methoxide (0.27 g, 0.005 mol) in absolute methanol (15 mL) and mixture was stirred for 10 min. 4-(bis (methylthio)methylene)-7-Fluoro-3,4-dihydro-1H-benzo[b]azepine-2,5-dione (3) (1.54 g, 0.005 mol) was added and the mixture was refluxed for 5 h. Excess of methanol was removed under reduced pressure and the mixture was poured into ice cold water. The solid thus separated was filtered, washed with diethyl ether, and dried. Recrystallization with ethanol gave the analytically pure product **5** (1.08 g, yield: 70%): m.p.: 233–235°C; IR (KBr) cm^{-1}: 3300 (N–H str.), 3010 (C–H str.), 2990, 1400

(–CH$_2$ next to C=O), 1600 (C=O str.), 1580 (C=C str.), 1510 (C=N str.), 900 (C-O-N str.), and 695 (C-S str.); ^{1}H NMR (400 MHz, CDCl$_3$) δ ppm: 8.11 (1H, s, NH), 7.25–8.34 (3H, m, Ar-H), 3.59 (2H, s, CH$_2$), and 2.82 (3H, s, CH$_3$); ^{13}C NMR (400 MHz, CDCl$_3$) δ ppm: Ar-C [163.69 (CF), 123.11 (CH), 114.60 (CH), 114.41 (CH)], Ar-C [143.11 (C), 139.13 (C), 130.62 (C), 104.65 (C), azepinone], 172.62 (C of amide), 39.22 (CH$_2$ azepinone), and 138.10 (isoxazole C, –CSMe), 17.61 (C of CH$_3$); MS: m/z 264.281 (M$^+$); Anal. calcd./found for C$_{12}$H$_9$FN$_2$O$_2$S: C 54.54/54.24, H 3.43/3.50, N 10.60/10.36, and S 12.13/12.32.

3.5. Synthesis of 10-Fluoro-2methyl-4-(methylthio)-5H-benzo[b]pyrimido[4,5-d]azepin-6(7H)-one (6). To 4-(bis (methylthio)methylene)-7-fluoro-3,4-dihydro-1H-benzo[b] azepine-2,5-dione (3) (1.48 g, 0.005 mol) in EtOH (5.0 mL) was added acetamidine hydrochloride (0.54 g, 0.005 mol) and triethylamine (0.50 mL, 0.005 mol). The solution was heated under reflux for 45 h and concentrated. The residue was extracted with ethyl acetate and washed with water. The organic layer was extracted with chloroform, dried over anhydrous magnesium sulfate, and evaporated. The residue was purified by column chromatography by eluting with hexane: ethyl acetate to give brown powder. The solid was recrystallized with ethanol to give 6 (0.98 g, yield: 66%): m.p.: 229–231°C; IR (KBr) cm^{-1}: 3300 (N-H str.), 3000 (C-H str.), 2990 (CH$_3$ str.), 2980, 1400 (–CH$_2$ next to C=O), 1600 (C=O str.), 1570 (C=C str.), 1510 (C=N str.), and 680 (C-S str.); ^{1}H NMR (400 MHz, CDCl$_3$) δ ppm: 8.15 (1H, s, NH), 7.54–8.37 (3H, m, Ar-H), 3.97 (2H, s, CH$_2$), 2.93 (3H, s, CH$_3$ of SMe), and 2.71 (3H, s, CH$_3$); ^{13}C NMR (400 MHz, CDCl$_3$) δ ppm: Ar-C [163.75 (CF), 123.11 (CH), 114.62 (CH), 114.44 (CH)], Ar-C [157.72 (C), 143.22 (C), 130.62 (C), 115.02 (C), azepinone], 39.22 (CH$_2$ azepinone), 170.50 (C of amide), 170.41 (pyrimidine C, –CCH$_3$), 172.56 (pyrimidine C, –CSMe), 24.27 (C of CH$_3$, pyrimidine), and 17.01 (C of CH$_3$, –SMe); MS: m/z 289.331 (M$^+$); Anal. calcd./found for C$_{14}$H$_{12}$FN$_3$OS: C 58.12/58.31, H 4.18/4.12, N 14.52/14.24, and S 11.08/11.26.

3.6. Synthesis of 2-Amino-10-fluoro-4-(methylthio)-5H-benzo[b]pyrimido[4,5-d]azepin-6-(7H)one (7). To 4-(bis (methylthio)methylene)-7-fluoro-3,4-dihydro-1H-benzo[b] azepine-2,5-dione (3) (1.48 g, 0.005 mol) in EtOH (5.0 mL) were added guanidine nitrate (0.61 g, 0.005 mol) and triethylamine (0.50 mL, 0.005 mol). The solution was heated under reflux for 45 h and concentrated. The residue was extracted with ethyl acetate and washed with water. The organic layer was extracted with chloroform, dried over anhydrous magnesium sulfate, and evaporated. The residue was purified by column chromatography by eluting with hexane: ethyl acetate to give brown powder. The solid was recrystallized with ethanol to give 7 (0.96 g, yield: 65%): m.p. 260–262°C; IR (KBr) cm^{-1}: 3380 (N-H str.), 3270 (NH$_2$ str.), 3000 (C-H str.), 2975, 1400 (–CH$_2$ next to C=O), 1650 (C=O str.), 1540 (C=C str.), 1510 (C=N str.), and 680 (C-S str.); ^{1}H NMR (400 MHz, CDCl$_3$) δ ppm: 8.13 (1H, s, NH), 7.58–8.83 (3H, m, Ar-H), 6.59 (2H, s, NH$_2$), 3.97 (2H, s,

CH$_2$), and 2.82 (3H, s, CH$_3$); ^{13}C NMR (400 MHz, CDCl$_3$) δ ppm: Ar-C [163.73 (CF), 123.14 (CH), 114.62 (CH), 114.44 (CH)], Ar-C [158.62 (C), 143.33 (C), 130.65 (C), 115.09 (C), azepinone], 39.24 (CH$_2$ azepinone), 170.50 (C of amide), 166.44 (pyrimidine C, –CNH$_2$), 172.54 (pyrimidine C, –CSMe), and 17.03 (C of CH$_3$); MS: m/z 290.32 (M$^+$); Anal. calcd./found for C$_{13}$H$_{11}$FN$_4$OS: C 53.78/53.59, H 3.82/3.76, N 19.30/19.11, and S 11.04/11.25.

3.7. Synthesis of 10-Fluoro-4-(methylthio)-5,7-dihydro-2H-benzo[b]pyrimido[4,5-d]azepin-2,6-(3H)-dione (8). To a mixture of urea (0.12 g, 0.002 mol) and sodium ethoxide (0.14 g, 0.002 mol) in ethanol (10.0 mL), 4-(bis(methylthi-o)methylene)-7-fluoro-3,4-dihydro-1H-benzo [b]azepine-2, 5-dione (3) (0.658 g, 0.002 mol) was added. The reaction mixture was refluxed for 12 h. The solvent was removed by distillation and the residue was treated with glacial acetic acid (10.0 mL) just enough to dissolve sodium salt of pyrimidine and the mixture was refluxed for 15 min. It was poured on crushed ice and the precipitate obtained was purified by recrystallization from ethanol to give 8 (0.41 g, yield: 62%): m.p. 225–228°C; IR (KBr) cm^{-1}: 3380 (N-H str.), 3000 (C-H str.), 2980, 1400 (–CH$_2$ next to C=O), 1670, 1710 (C=O str.), 1570 (C=C str.), 1500 (C=N str.), and 685 (C-S str.); ^{1}H NMR (400 MHz, CDCl$_3$) δ ppm: 8.04 (1H, s, NH), 7.34–8.45 (3H, m, Ar-H), 3.72 (2H, s, CH$_2$), 2.90 (3H, s, CH$_3$), and 2.0 (1H, s, NH of pyrimidine ring); ^{13}C NMR (400 MHz, CDCl$_3$) δ ppm: Ar-C [163.79 (CF), 122.08 (CH), 114.52 (CH), 114.42 (CH)], Ar-C [155.82 (C), 141.42 (C), 144.24 (C), 115.02 (C), azepinone], 39.21 (CH$_2$, azepinone), 171.47 (C of amide), 170.42 (pyrimidine C, –CSMe), 166.4 (C of carbonyl), and 17.06 (C of CH$_3$); MS: m/z 291.30 (M$^+$); Anal. calcd./found for C$_{13}$H$_{10}$FN$_3$O$_2$S: C 53.60/53.38, H 3.46/3.41, N 14.42/14.13, and S 11.01/11.28.

3.8. Synthesis of 10-Fluoro-4-(methylthio)-2-thioxo-5,7-dihydro-2H-benzo[b]pyrimido[4,5-d]azepin-2,6-(3H)-one (9). To a mixture of thiourea (0.12 g, 0.002 mol) and sodium ethoxide (0.14 g, 0.002 mol) in ethanol (10.0 mL), 4-(bis(methylthio)methylene)-7-fluoro-3,4-dihydro-1H-benzo[b]azepine-2,5-dione (3) (0.658 g, 0.002 mol) was added and the reaction mixture was refluxed for 12 h. The solvent was removed by distillation and the residue was treated with glacial acetic acid (10.0 mL) just enough to dissolve sodium salt of pyrimidine and the mixture was refluxed for 15 min. It was poured on crushed ice and the precipitate obtained was purified by recrystallization from ethanol to give 9 (0.44 g, yield: 67%): m.p.: 275–277°C; IR (KBr) cm^{-1}: 3370 (N-H str.), 2990 (C-H str.), 2980, 1400 (–CH$_2$ next to C=O), 1680 (C=O str.), 1570 (C=C str.), 1510 (C=N str.), 780 (C=S str.), and 690 (C-S str.); ^{1}H NMR (400 MHz, CDCl$_3$) δ ppm: 8.04 (1H, s, NH), 7.34–8.44 (3H, m, Ar-H), 3.90 (2H, s, CH$_2$), 2.92 (3H, s, CH$_3$), and 2.0 (1H, s, NH of pyrimidine ring); ^{13}C NMR (400 MHz, CDCl$_3$) δ ppm: Ar-C [163.58 (CF), 122.08 (CH), 114.52 (CH), 114.42 (CH)], Ar-C [156.62 (C), 143.22 (C), 136.63 (C), 115.02 (C), azepinone], 39.20 (CH$_2$ azepinone), 171.23 (C of amide), 170.42 (pyrimidine C, –CSMe), 17.07 (C of CH$_3$), and 180.4 (pyrimidine C, $-C=S$); MS: m/z 307.37 (M^+); Anal. calcd./found for $C_{13}H_{10}FN_3OS_2$: C 50.80/50.64, H 3.28/3.21, N 13.67/13.38, and S 20.86/20.44.

4. Results and Discussion

The synthetic importance of oxoketene dithioacetals specially the dimethyl thioacetal in the construction of a variety of novel fused heterocyclic systems encouraged us to explore its potential in the annulation of face "d" of 7-fluoro-benzazepine-2,5-dione (2) with such pharmacophoric scaffolds as pyrazole, isoxazole, and pyrimidine which have been accredited in the literature with a proven record of their bioactive profiles. In consideration of the easy accessibility of the corresponding ketene dimethyl acetals from the base catalyzed reaction of CS_2 and CH_3I with compounds containing an active methylene group, we applied this strategy on 2 to append this functionality onto its 4-position to form 3. The versatility of 3 in allowing a facile annulation of its face "d" with the above bioactive pharmacophores was exploited in its reaction with (i) hydrazine hydrate [7], (ii) hydroxylamine hydrochloride [8–10], (iii) acetamidine hydrochloride [11], (iv) guanidine nitrate [11], (v) urea [12], and (vi) thiourea [12] to generate 4–9, respectively, in acceptable yields.

The structures of the compounds 2–9 were established on the basis of their microanalysis, IR, ^{1}H NMR, ^{13}C NMR, and MS data which corroborated strongly to the structures assigned to these molecules. The formation of compound 2 from 1 was ascertained by the appearance of two carbonyls in the IR spectrum of 2 (at 1680 and 1710 cm^{-1}). The presence of a multiplet (due to two overlapping triplets of two CH_2's) at δ 3.49 in its NMR spectrum corroborated strongly its formation from 1. The appearance of a carbonyl group peak at 1640 cm^{-1} for the α,β-unsaturated ketone in the IR spectrum of 3 and a singlet of 2H for the CH_2 group at δ 3.53 in its NMR spectrum indicated clearly its formation from 2. The most diagnostic evidence which substantiated their formation was the disappearance of one of the carbonyl group peaks of 3 in the IR spectrum of 4–9 which confirmed its involvement in cyclocondensation reactions. The ^{1}H NMR spectrum (400 MHz, in CDCl$_3$) of 3, 5, 6, and 7 showed the presence of an NH peak at δ, 8.0, 8.11, 8.15, and 8.13, respectively, (for the azepinone NH) whereas compound 4 exhibited two NH peaks at δ, 12.82 (pyrazole NH) and δ, 8.03 (for azepinone NH). Apart from this, compounds 8 and 9 also exhibited two NH peaks at δ, 8.04 (for azepinone NH) and δ, 2.0 (pyrimidine NH), all of which exchanged with D_2O.

5. Conclusion

In summary, the unprecedented potential of oxoketene dithioacetals in synthesis was exploited to provide an easy access to face "d" pyrazolo, isoxazolo, and pyrimido annulated analogues of benzazepinone 4–9, respectively, from 4-ketene dimethyl thioacetal substituted derivative of 7-fluorobenzo-(b)-azepine-2,5-dione (3). The process is characterized by mild reaction condition and easy work-up procedure.

Conflict of Interests

The authors declare that there is no conflict of interests regarding the publication of this paper.

Acknowledgments

Authors are thankful to Punjab University, Chandigarh, for providing the spectral data of the compounds. Authors are also thankful to Department of Science and Technology (DST), New Delhi, for financial support provided to Banasthali Center for Education and Research in Basic Sciences under their CURIE (Consolidation of University Research for Innovation and Excellence in Women Universities) programme.

References

[1] M. A. Koch, A. Schuffenhauer, M. Scheck et al., "Charting biologically relevant chemical space: a structural classification of natural products (SCONP)," *Proceedings of the National Academy of Sciences of the United States of America*, vol. 102, no. 48, pp. 17272–17277, 2005.

[2] D. D. C. Wan, D. Kundhur, and R. W. J. Lam, "Mirtazapine for treatment-resistant depression: a preliminary report," *Journal of Psychiatry and Neuroscience*, vol. 28, no. 1, pp. 55–59, 2003.

[3] O. L. Munk and F. Smith, "PET kinetics of radiolabeled antidepressant, [N-methyl-^{11}C]mirtazapine, in the human brain," *EJNMMI Research*, vol. 1, article 36, 2011.

[4] H. R. Howard, "7-Phenylsulfonyl-tetrahydro-3-benzazepine derivatives as antipsychotic agents," *Informa Healthcare*, vol. 15, pp. 1811–1815, 2005.

[5] J. P. Hieble, "Method of treating benign prostatic hypertrophy," Patent 4755507, 2010.

[6] N. Dhingra and D. Bhagwat, "Benign prostatic hyperplasia: an overview of existing treatment," *Indian Journal of Pharmacology*, vol. 43, no. 1, pp. 6–12, 2011.

[7] W. W. Wardakhan and N. A. Louca, "Synthesis of novel pyrazole, coumarin and pyridazine derivatives evaluated as potential antimicrobial and antifungal agents," *Journal of the Chilean Chemical Society*, vol. 52, no. 2, pp. 1145–1149, 2007.

[8] P. K. Mahata, U. K. S. Kumar, V. Sriram, H. Ila, and H. Junjappa, "1-Bis(methoxy)-4-bis(methylthio)-3-buten-2-one: useful three carbon synthon for synthesis of five and six membered heterocycles with masked (or unmasked) aldehyde functionality," *Tetrahedron*, vol. 59, no. 15, pp. 2631–2639, 2003.

[9] J. Kaffy, R. Pontikis, D. Carrez, A. Croisy, C. Monneret, and J. Florent, "Isoxazole-type derivatives related to combretastatin A-4, synthesis and biological evaluation," *Bioorganic and Medicinal Chemistry*, vol. 14, no. 12, pp. 4067–4077, 2006.

[10] M. V. D. Tresa and P. Melo, "Recent advances on the synthesis and reactivity of isoxazoles," *Current Organic Chemistry*, vol. 9, no. 10, pp. 925–958, 2005.

[11] M. A. Ebraheem, K. M. L. Rai, N. U. Kudva.n, and A. S. Bahjat, "Synthesis of new polysubstituted (pyrazoles, pyrimidines and quinolines) five and six-membered heterocycles: reaction of α,α-dioxoketene dithioacetals with nucleophiles," *Tetrahedron Letters*, vol. 51, no. 27, pp. 3486–3492, 2010.

[12] M. M. Abdelhalim, M. M.T. El-Saidi, S. T. Rabie, and G. A. Elmegeed, "Synthesis of novel steroidal heterocyclic derivatives as antibacterial agents," *Steroids*, vol. 72, no. 5, pp. 459–465, 2007.

14

Speeds of Sound and Excess Molar Volume for Binary Mixture of 1,4-Dioxane with 1-Heptanol at Five Temperatures

Anil Kumar Koneti[1] and Srinivasu Chintalapati[2]

[1] Department of Physics, Sri Vani School of Engineering, Chevuturu, Andhra Pradesh 521 229, India
[2] Department of Physics, Andhra Loyola College, Vijayawada, Andhra Pradesh 520 008, India

Correspondence should be addressed to Anil Kumar Koneti; anilkumarkoneti@gmail.com

Academic Editor: Sandrine Bouquillon

Speed of sound and density data for dilute liquid solutions of cyclic ether 1,4-dioxane with 1-heptanol was obtained using the Anton-Paar DSA 5000 at five temperatures T = (298.15, 303.15, 308.15, 313.15, and 318.15) K at atmospheric pressure. The excess parameters were calculated from experimental data and fitted with a Redlich-Kister polynomial function and concluded the presence of weak molecular interactions.

1. Introduction

Cyclic ethers are considered some of the most important chemicals in the industry. Particularly, branched ethers (such as 2-methoxy-2-methylpropane or MTBE, 2-ethoxy-2-methyl-propane or ETBE, 2,2′-oxybis[propane] or DIPE, and 2-methoxy-2-methylbutane or TAME) have extensively been used as oxygenates in gasoline production. Cyclic ethers, in turn (e.g., 1,3-dioxolane, 1,4-dioxane, 1,3,5-trioxane, tetrahydrofuran, and tetrahydropyran), are frequently used as solvents in chemical and electrochemical processes, likewise as basic reagents (i.e., monomer) for ring-opening polymerization, and for the production of other chemical intermediaries. 1-Heptanol is often utilized in cardiac electrophysiology experiments to block gap junctions and increase axial resistance between myocytes. Increasing axial resistance will decrease conduction velocity and increase the heart's condition to reentrant excitation and sustained arrhythmias. It has a pleasant smell and is employed in cosmetics for its fragrance.

Speeds of sound and deviations in isentropic compressibilities have been previously reported by the author in their earlier studies on 1,4-dioxane + 1-butanol [1] at five temperatures T = (298.15, 303.15, 308.15, 313.15 and 318.15) K. In the continuation of investigation on excess thermodynamics functions of cyclic ethers in polar and non-polar solvents, authors are reporting the experimental values, deviations in isentropic compressibility ($\Delta\kappa_s$), excess molar volumes (V_m^E), excess free length (L_f^E), excess acoustic impedance (Z^E), and excess sound velocity (U^E) for the binary system 1,4-dioxane + 1-heptanol. The parameters are estimated using standard equations that are reported by many authors [2–5]. These excess parameters are discussed in the focus of intermolecular interactions present in the mixture at T = (298.15, 303.15, 308.15, 313.15, and 318.15) K using Anton-Paar. At the end for the best fit of the result Redlich-Kister coefficients and their related standard deviations are enclosed.

2. Experimental Procedure

In the present study the chemicals 1,4-dioxane and 1-heptanol are purchased from Sigma Aldrich chemical company, their mass fraction purities is >0.998. To begin with for the measurements purpose all prepared solutions were done at the same particular temperature, also changing the temperature the measurements were repeated.

Speeds of sound, U, densities, ρ, of the pure compounds, and their mixtures were obtained with an Anton-Paar DSA-5000 vibrating tube densimeter and sound analyzer. All controls, adjustments, and checks were done using manufacturer's software installed in the device. A computer connected

TABLE 1: Experimental speed of sound and density of pure liquids with literature values at T = (298.15, 303.15, 308.15, 313.15, and 318.15) K.

System	T (K)	U (ms^{-1})		ρ (kgm^{-3})	
		Observed	Literature	Observed	Literature
1,4-Dioxane	298.15	1344.3	1344.8 [6]	1026.8	1027.9 [6, 7]
	303.15	1322.3		1021.2	1022.5 [8, 9]
	308.15	1300.5		1015.5	1016.8 [10]
	313.15	1278.6	1279.8 [6]	1011.8	1011.3 [8, 9]
	318.15	1257.3		1004.1	1005.1 [11]
1-Heptanol	298.15	1327.9	1327.5 [12]	819.7	818.7 [12, 13]
	303.15	1310.9	1310.1 [14]	816.2	815.1 [13]
	308.15	1294.1	1293.8 [15]	812.6	812.3 [15]
	313.15	1277.3		809.0	
	318.15	1260.7		805.4	

to the U tube densimeter enabled us to read the raw data from the device memory and to perform the consequent evaluation. The temperature was automatically kept constant within ±0.01 K. The precision of the speed of sound and density measurements is ±0.1 m·s^{-1} and ±3 × 10^{-5} g·cm^{-3}, respectively. The uncertainty of the speed of sound measurements is ±1 m·s^{-1} while the uncertainty for the density is ±10^{-5} g·cm^{-3}. Calibration of the apparatus was carried out with air and degassed double-distilled water. Experimental values of the speed of sound of the pure compounds at five temperatures, along with literature values at temperature range of 298.15 to 318.15 K, are reported in Table 1. Mole fractions of these samples were determined by measuring the mass of each component with a precision balance Sartorius, model CP 225D, ±0.01 mg.

3. Results and Discussion

The derived excess parameters such as $\Delta\kappa_s$, $V_m{}^E$, $L_f{}^E$, Z^E, and U^E at above five temperature are summarized in Tables 2(a) and 2(b). From the tables, it is observed that, the experimental speed of sound values of the binary liquid mixture decreases up to the mole fraction of 0.4090 (at 298.15, 303.15 K), 0.5244 (at 308.15, 313.15, 318.15 K) and then increase with increasing mole fraction of 1,4-dioxane, whereas the density increases with increasing mole fraction. This indicates that there is a dipole-induced dipole interaction between component molecules [3, 16].

The experimental data measured values of speed of sound (U) and density (ρ) various thermoacoustical parameters:

isentropic compressibility:

$$\kappa_s = \frac{1}{\rho U^2}, \tag{1}$$

molar volume:

$$\overline{V} = \frac{\overline{M}}{\rho}, \tag{2}$$

where $\overline{M} = M_1 X_1 + M_2 X_2$,

intermolecular free length:

$$L_f = K \left(\kappa_s\right)^{1/2}, \tag{3}$$

where K is Jacobson's constant, temperature dependent,

specific acoustic impedance:

$$Z = U\rho. \tag{4}$$

The strength of interaction between the component molecules of binary liquid system is well reflected in the excess functions from ideality. The excess thermodynamic properties such as $\Delta\kappa_s$, $V_m{}^E$, $L_f{}^E$, Z^E, and U^E have been calculated using the following equation:

$$Y^E = Y_{\text{mix}} - \left(x_1 y_1 + x_2 y_2\right), \tag{5}$$

where x_1 and x_2 are mole fractions of 1,4-dioxane and 1-heptanol, respectively.

Further, the excess parameters were fitted to Redlich-Kister polynomial equation to estimate the adjustable parameters:

$$Y^E = x_1 x_2 \sum_{i=0}^{n} a_i \left(1 - 2x\right)^i. \tag{6}$$

Using least-squares regression method, the a_i coefficients are obtained by fitting the above equation to the experimental values. The optimum number of coefficients are ascertained from an examination of the variation in standard deviation $\sigma(Y)$ equation as

$$\sigma\left(Y\right) = \left[\frac{\sum\left(Y_{\text{exp}\,t} - Y_{\text{calc}}\right)^2}{N - n}\right]^{1/2}, \tag{7}$$

where N is the number of data points and n is the degree of fitting (i.e. number of coefficients) these values are reported in Table 3.

Generally, the values of the excess functions $\Delta\kappa_s$, $V_m{}^E$, and $L_f{}^E$ depend upon several physical and chemical contributions [3, 17]. The physical contribution depends mainly on two factors, namely,

TABLE 2: (a) Experimental values (U, ρ) and excess thermodynamic parameters ($\Delta\kappa_s$, V_m^E, L_f^E, Z^E, and u^E) for 1,4-dioxane + 1-heptanol system at 298.15, 303.15, and 308.15 K. (b) Experimental values (U, ρ) and excess thermodynamic parameters ($\Delta\kappa_s$, V_m^E, L_f^E, Z^E, and u^E) for 1,4-dioxane + 1-heptanol system at 313.15 and 318.15 K.

(a)

x_1	U ms^{-1}	ρ kgm^{-3}	$\Delta K_s * 10^{-10}$ m^{2}N^{-1}	$V_m^E * 10^{-4}$ m^3 mol^{-1}	$L_f^E * 10^{-11}$ m	$Z^E * 10^6$ kgm^{-2}s^{-1}	U^E ms^{-1}
				298.15 K			
0.0000	1327.9	819.7	0.0000	0.0000	0.0000	0.0000	0.0000
0.1617	1318.1	838.8	0.1906	0.0206	0.0805	−0.0300	−12.4113
0.2921	1313.3	857.3	0.2919	0.0320	0.1247	−0.0480	−19.4349
0.4090	1311.7	876.2	0.3398	0.0385	0.1469	−0.0584	−22.8951
0.5244	1312.4	897.4	0.3510	0.0412	0.1538	−0.0635	−23.9247
0.6359	1315.8	920.9	0.3291	0.0398	0.1463	−0.0627	−22.8495
0.7083	1318.3	938.2	0.2966	0.0363	0.1331	−0.0584	−21.0012
0.7947	1323.6	961.1	0.2369	0.0292	0.1077	−0.0485	−17.3789
0.8687	1329.3	982.7	0.1666	0.0204	0.0767	−0.0353	−12.7616
0.9388	1336.6	1005.9	0.0837	0.0101	0.0390	−0.0183	−6.7714
1.0000	1344.3	1026.8	0.0000	0.0000	0.0000	0.0000	0.0000
				303.15 K			
0.0000	1310.9	816.2	0.0000	0.0000	0.0000	0.0000	0.0000
0.1617	1301.6	835.0	0.1861	0.0206	0.0781	−0.0284	−11.1370
0.2921	1296.6	853.4	0.2885	0.0321	0.1223	−0.0455	−17.7549
0.4090	1294.3	872.1	0.3400	0.0386	0.1457	−0.0558	−21.2511
0.5244	1294.3	893.1	0.3549	0.0413	0.1539	−0.0608	−22.4960
0.6359	1296.8	916.3	0.3349	0.0398	0.1472	−0.0602	−21.6492
0.7083	1298.9	933.4	0.3024	0.0363	0.1341	−0.0561	−19.9310
0.7947	1303.6	956.1	0.2415	0.0292	0.1085	−0.0465	−16.4669
0.8687	1308.7	977.5	0.1694	0.0205	0.0770	−0.0338	−12.0406
0.9388	1315.4	1000.5	0.0848	0.0102	0.0390	−0.0175	−6.3496
1.0000	1322.3	1021.2	0.0000	0.0000	0.0000	0.0000	0.0000
				308.15 K			
0.0000	1294.1	812.6	0.0000	0.0000	0.0000	0.0000	0.0000
0.1617	1284.4	831.2	0.1894	0.0206	0.0787	−0.0273	−10.6121
0.2921	1279.1	849.4	0.2949	0.0321	0.1237	−0.0439	−17.0206
0.4090	1276.2	868.0	0.3477	0.0386	0.1473	−0.0538	−20.3393
0.5244	1275.9	888.7	0.3629	0.0414	0.1555	−0.0586	−21.4749
0.6359	1277.8	911.7	0.3424	0.0399	0.1486	−0.0580	−20.6255
0.7083	1279.5	928.6	0.3090	0.0364	0.1354	−0.0540	−18.9585
0.7947	1283.6	951.1	0.2463	0.0293	0.1093	−0.0447	−15.6073
0.8687	1288.1	972.3	0.1723	0.0205	0.0773	−0.0324	−11.3437
0.9388	1294.4	995.1	0.0857	0.0102	0.0390	−0.0167	−5.9248
1.0000	1300.5	1015.5	0.0000	0.0000	0.0000	0.0000	0.0000

(b)

x_1	U ms^{-1}	ρ kgm^{-3}	$\Delta K_s * 10^{-10}$ m^{2}N^{-1}	$V_m^E * 10^{-4}$ m^3 mol^{-1}	$L_f^E * 10^{-11}$ m	$Z^E * 10^6$ kgm^{-2}s^{-1}	U^E ms^{-1}
				313.15 K			
0.0000	1277.3	809.0	0.0000	0.0000	0.0000	0.0000	0.0000
0.1617	1267.4	827.4	0.1930	0.0206	0.0792	−0.0263	−10.1247
0.2921	1261.8	845.4	0.2969	0.0322	0.1232	−0.0420	−15.8967
0.4090	1259.0	863.8	0.3495	0.0387	0.1465	−0.0513	−18.9159

(b) Continued.

x_1	U ms⁻¹	ρ kgm⁻³	$\Delta K_s * 10^{-10}$ m²N⁻¹	$V_m^E * 10^{-4}$ m³ mol⁻¹	$L_f^E * 10^{-11}$ m	$Z^E * 10^6$ kgm⁻²s⁻¹	U^E ms⁻¹
0.5244	1257.8	884.3	0.3651	0.0414	0.1547	−0.0559	−19.9470
0.6359	1259.0	907.1	0.3445	0.0400	0.1478	−0.0553	−19.1057
0.7083	1261.0	923.8	0.3106	0.0365	0.1345	−0.0514	−17.5155
0.7947	1263.8	946.1	0.2474	0.0293	0.1084	−0.0425	−14.3816
0.8687	1267.7	967.1	0.1731	0.0205	0.0767	−0.0307	−10.4522
0.9388	1273.3	989.6	0.0863	0.0101	0.0387	−0.0158	−5.4790
1.0000	1278.6	1009.8	0.0000	0.0000	0.0000	0.0000	0.0000
				318.15 K			
0.0000	1260.7	805.4	0.0000	0.0000	0.0000	0.0000	0.0000
0.1617	1250.4	823.6	0.1980	0.0207	0.0805	−0.0254	−9.7832
0.2921	1244.5	841.4	0.3027	0.0323	0.1243	−0.0404	−15.1454
0.4090	1241.0	859.6	0.3599	0.0388	0.1491	−0.0497	−18.3128
0.5244	1239.5	879.9	0.3777	0.0415	0.1581	−0.0542	−19.4234
0.6359	1240.2	902.4	0.3539	0.0401	0.1500	−0.0533	−18.3277
0.7083	1241.8	919.0	0.3161	0.0366	0.1352	−0.0493	−16.5045
0.7947	1244.8	941.0	0.2489	0.0294	0.1077	−0.0405	−13.2710
0.8687	1248.3	961.8	0.1732	0.0206	0.0758	−0.0292	−9.5740
0.9388	1252.4	984.2	0.0869	0.0102	0.0384	−0.0151	−5.0789
1.0000	1257.3	1004.1	0.0000	0.0000	0.0000	0.0000	0.0000

TABLE 3: Coefficients (a_i) and standard deviation (σ) for 1,4-dioxane + 1-heptanol.

Property	T (K)	a_0	a_1	a_2	a_3	a_4	σ (10^2)
$\Delta\kappa_s$ (10^{-10} m²N⁻¹)	298.15	1.4061	−0.0223	0.1258	−0.0355	−0.1509	0.1961
	303.15	1.4189	−0.0816	0.0851	−0.0024	−0.1395	0.1805
	308.15	1.4507	−0.0821	0.0879	−0.0061	−0.1734	0.2665
	313.15	1.4594	−0.0886	0.0780	0.0403	−0.1142	0.3415
	318.15	1.5097	−0.1010	−0.0949	0.1229	0.1178	0.0559
V_m^E (m³ mol⁻¹)	298.15	0.1639	−0.0278	0.0103	0.0165	−0.0149	0.0158
	303.15	0.1643	−0.0268	0.0100	0.0139	−0.0151	0.0191
	308.15	0.1645	−0.0278	0.0099	0.0156	−0.0154	0.0183
	313.15	0.1648	−0.0275	0.0115	0.0146	−0.0194	0.0169
	318.15	0.1651	−0.0282	0.0110	0.0163	−0.0173	0.0189
L_f^E (10^{-11} m)	298.15	0.6143	−0.0463	0.0645	−0.0173	−0.0628	0.0829
	303.15	0.6135	−0.0686	0.0502	−0.0041	−0.0619	0.0756
	308.15	0.6201	−0.0676	0.0510	−0.0055	−0.0753	0.1129
	313.15	0.6169	−0.0683	0.0466	0.0117	−0.0528	0.1452
	318.15	0.6306	−0.0726	−0.0252	0.0482	0.0448	0.0227
Z^E (10^6 kgm⁻²s⁻¹)	298.15	−0.2516	0.0606	−0.0368	0.0014	0.0260	0.0290
	303.15	−0.2409	0.0620	−0.0324	−0.0020	0.0249	0.0265
	308.15	−0.2321	0.0592	−0.0314	−0.0025	0.0274	0.0341
	313.15	−0.2214	0.0564	−0.0303	−0.0065	0.0260	0.0399
	318.15	−0.2148	0.0540	−0.0123	−0.0125	0.0017	0.0128
U^E (ms⁻¹)	298.15	−95.5643	6.7116	−14.1963	14.3672	7.2122	16.5048
	303.15	−89.6441	10.8671	−9.7419	10.8158	5.9228	13.5179
	308.15	−85.6186	9.4617	−9.8055	10.9174	9.7175	20.3881
	313.15	−79.5527	8.3222	−7.9463	6.7460	3.7143	24.8586
	318.15	−77.5029	8.1303	8.6723	−0.8418	−18.5150	7.7314

(a) the dispersion forces or weak dipole-dipole interaction that leads to positive values,

(b) the geometrical effect allowing the fitting of molecules of two different sizes into each other's structure resulting in negative values.

The chemical contributions include breaking up of the associates present in pure liquids, resulting in positive $\Delta\kappa_s$, V_m^E, and L_f^E. In the present mixture the graphical representations for deviation isentropic compressibilities ($\Delta\kappa_s$), excess molar volumes (V_m^E), and excess free length (L_f^E) are

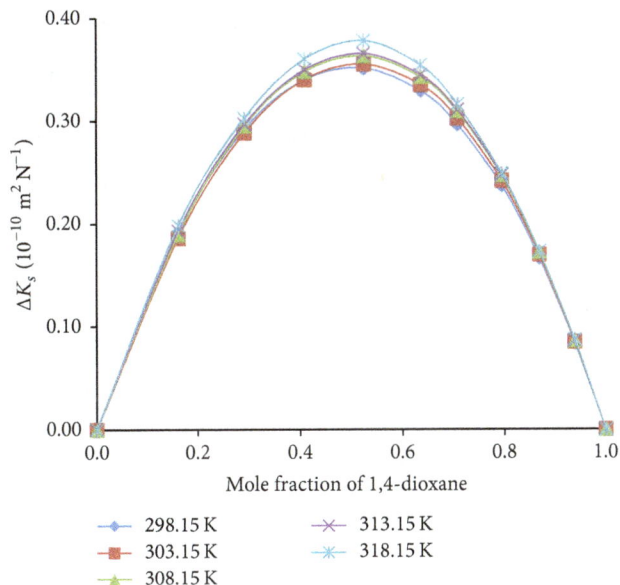

FIGURE 1: Variation of deviations in isentropic compressibility ($\Delta\kappa_s$) with mole fraction of 1,4-dioxane + 1-heptanol system at different temperatures.

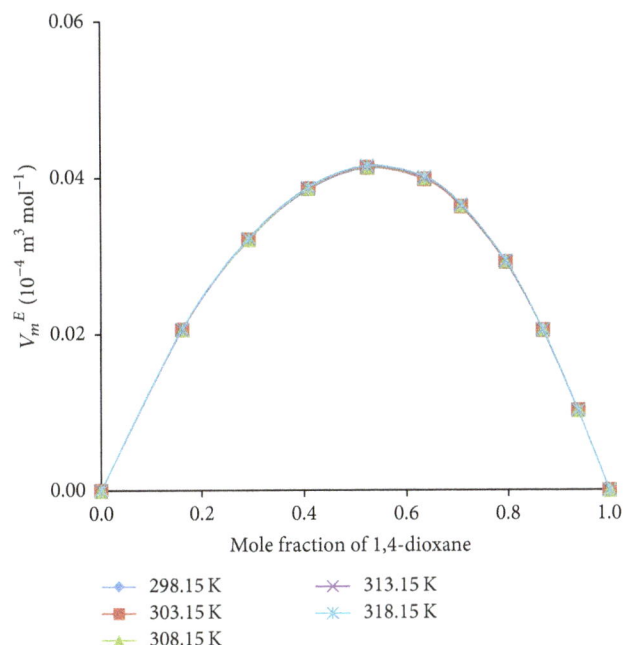

FIGURE 3: Variation of excess free length (L_f^E) with mole fraction of 1,4-dioxane + 1-heptanol system at different temperatures.

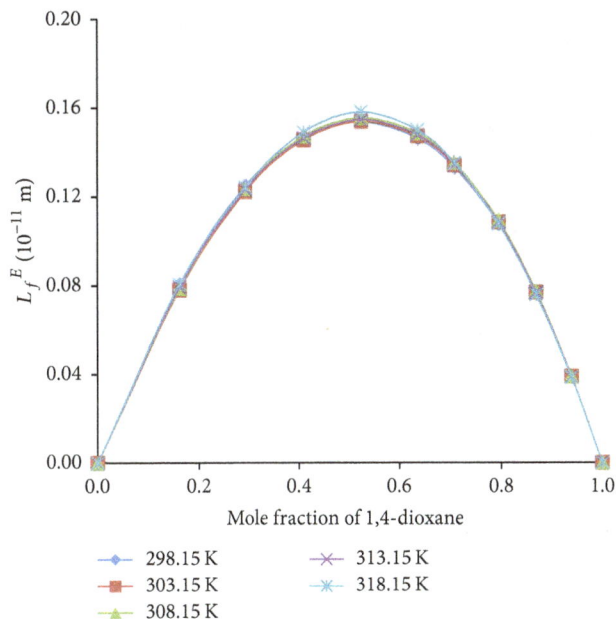

FIGURE 2: Variation of excess molar volumes (V_m^E) with mole fraction of 1,4-dioxane + 1-heptanol system at different temperatures.

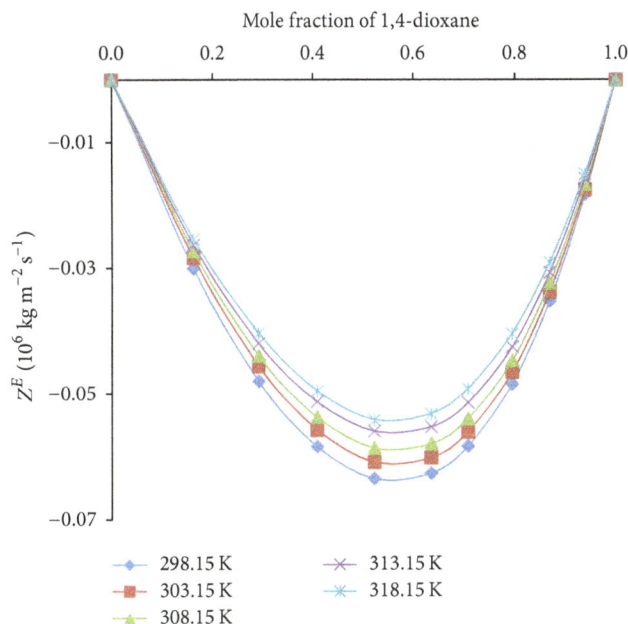

FIGURE 4: Variation of excess acoustic impedance (Z^E) with mole fraction of 1,4-dioxane + 1-heptanol system at different temperatures.

positive, presented in Figures 1, 2, and 3. The positive values reveal that there are present weak interactions in the mixture.

Figure 4 shows that the excess acoustic impedance becomes more negative at 0.5244 mole fraction of the binary mixture. The negative values of Z^E [1, 18] indicate that the breaking of hydrogen bond in 1-heptanol leads to weak packing of the structure. Also, Figure 5 shows U^E values exhibiting negative deviations over the entire composition

range of 1,4-dioxane in the mixture at all five temperatures studied. The negative deviations of Z^E and U^E suggest that dispersion forces are operative in the system. Similar reports are made by Ali et al. [19]. Further, Bahadur et al. [20] and Sumathi and Govindarajan [3]. According to their reports the negative values of U^E indicates the decrease in the strength of interaction between the molecules in the mixture. In the present work the negative deviations of Z^E and U^E suggests

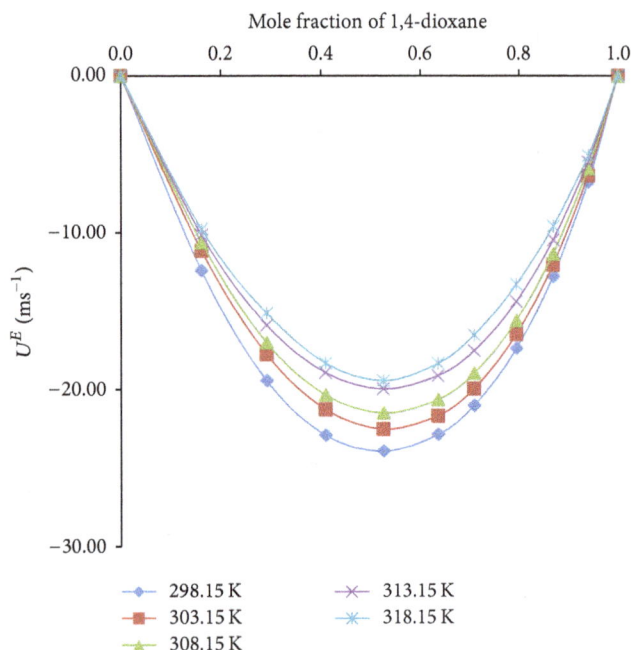

FIGURE 5: Variation of excess ultrasonic velocity (u^E) with mole fraction of 1,4-dioxane + 1-heptanol system at different temperatures.

the presence of weak interaction dispersive forces between the unlike molecules.

4. Conclusion

Speed of sound and density for binary mixture that consist of 1,4-dioxane with 1-heptanol system are measured at T = (298.15, 303.15, 308.15, 313.15, and 318.15) K using Anton-Paar. The derived acoustical parameters and their excess parameters $\Delta\kappa_s$, V_m^E, and L_f^E are positive and Z^E, U^E are negative which hint to the presence of weak dispersive forces between the component molecules in the mixture at all five temperatures studied.

Conflict of Interests

The authors declare that there is no conflict of interests regarding the publication of this paper.

References

[1] K. A. Kumar, C. Srinivasu, and K. T. S. S. Raju, "Ultrasonic velocities and excess properties of binary mixture of 1, 4-dioxane + 1-butanol at temperatures between (298.15 and 318.15) K," *Journal of Chemical, Biological and Physical Sciences C*, vol. 3, no. 4, pp. 2914–2923, 2013.

[2] A. Dan Li and W. Fang, "Thermodynamic study of binary mixtures of tricycle decane with N-methylpiperazine or triethylamine at T = 298.15–323.15 K," *Thermochimica Acta*, vol. 544, pp. 38–42, 2012.

[3] T. Sumathi and S. Govindarajan, "Molecular Interaction Studies on Some Binary Organic Liquid Mixtures at 303 K," *International Journal of Biology, Pharmacy and Allied Sciences (IJBPAS)*, vol. 1, no. 8, pp. 1153–1165, 2012.

[4] S. Thirumaran, R. Murugan, and N. Prakash, "Acoustic study of intermolecular interactions in binary liquid mixtures," *Journal of Chemical and Pharmaceutical Research*, vol. 2, no. 1, pp. 53–61, 2010.

[5] K. Anil Kumar, C. H. Srinivasu, S. K. Fakruddin, and K. Narendra, "Acoustical behavior of molecular interactions in binary liquid mixture containing 1-butanol and hexane at temperatures 298.15 K, 303.15 K and 308.15 K," *International Journal of Pharmaceutical and Chemical Sciences*, vol. 3, no. 1, pp. 10–13, 2014.

[6] A. Villares, S. Martín, M. Haro, B. Giner, and H. Artigas, "Densities and speeds of sound for binary mixtures of (1,3-dioxolane or 1,4-dioxane) with (2-methyl-1-propanol or 2-methyl-2-propanol) at the temperatures (298.15 and 313.15) K," *Journal of Chemical Thermodynamics*, vol. 36, no. 12, pp. 1027–1036, 2004.

[7] J. A. Riddick, W. B. Bunger, and T. K. Sakano, *Organic Solvents. Techniques of Chemistry*, vol. 2, Wiley/Interscience, New York, NY, USA, 4th edition, 1986.

[8] F. Corradini, G. Franchini, L. Marcheselli, L. Tassi, and G. Tosi, "Densities and excess molar volumes of 2-methoxyethanol/water binary mixtures," *Australian Journal of Chemistry*, vol. 45, no. 7, pp. 1109–1117, 1992.

[9] A. G. Oskoei, N. Safaei, and J. Ghasemi, "Densities and viscositiesfor binary and ternary mixtures of 1, 4-dioxane + 1-1-heptanol + N , N-dimethylaniline from T = (283.15 to 343.15) K," *Journal of Chemical & Engineering Data*, vol. 53, pp. 343–349, 2008.

[10] M. Das and M. N. Roy, "Studies on thermodynamic and transport properties of binary mixtures of acetonitrile with some cyclic ethers at different temperatures by volumetric, viscometric, and interferometric techniques," *Journal of Chemical & Engineering Data*, vol. 51, no. 6, pp. 2225–2232, 2006.

[11] M. A. Saleh, S. Akhtar, M. S. Ahmed, and M. H. Uddin, "Excess molar volumes and thermal expansivities of aqueous solutions of dimethylsulfoxide, tetrahydrofuran and 1,4-dioxane," *Physics and Chemistry of Liquids*, vol. 40, no. 5, pp. 621–635, 2002.

[12] E. Zorebski, M. Geppert-Rybczyńska, and B. Maciej, "Densities, speeds of sound, and isentropic compressibilities for binary mixtures of 2-ethyl-1-hexanol with 1-pentanol, 1-heptanol, or 1-nonanol at the temperature 298.15 K," *Journal of Chemical and Engineering Data*, vol. 55, no. 2, pp. 1025–1029, 2010.

[13] A. Pineiro, P. Brocos, A. Amigo, M. Pintos, and R. Bravo, "Refractive indexes of binary mixtures of tetrahydrofuran with 1-alkanols at 25°C and temperature dependence of n and ρ for the pure liquids," *Journal of Solution Chemistry*, vol. 31, no. 5, pp. 369–380, 2002.

[14] J. A. González, I. Mozo, I. g. de la Fuente, J. C. Cobos, and N. Riesco, "Thermodynamics of (1-alkanol + linear monoether) systems," *The Journal of Chemical Thermodynamics*, vol. 40, no. 10, pp. 149–152, 2008.

[15] A. Pal and R. Gaba, "Densities, excess molar volumes, speeds of sound and isothermal compressibilities for 2-(2-hexyloxyethoxy)ethanol + n-alkanol systems at temperatures between (288.15 and 308.15) K," *The Journal of Chemical Thermodynamics*, vol. 40, no. 5, pp. 750–758, 2008.

[16] R. Kumar, S. Jayakumar, and V. Kannappan, "Study of molecular interactions in binary liquid mixtures," *Indian Journal of Pure and Applied Physics*, vol. 46, no. 3, pp. 169–175, 2008.

[17] A. Ali, A. K. Nain, V. K. Sharma, and S. Ahmed, "Ultrasonic studies in binary liquid mixtures," *Indian Journal of Physics B*, vol. 75, no. 6, pp. 519–525, 2001.

[18] S. Singh, I. Vibhu, M. Gupta, and J. P. Shukla, "Excess acoustical and volumetric properties and the theoretical estimation of the excess thermodynamic functions of binary liquid mixtures," *Chinese Journal of Physics*, vol. 45, no. 4, pp. 412–424, 2007.

[19] A. Ali, Abida, S. Hyder, and A. K. Nain, "Ultrasonic, volumetric and viscometric study of molecular interactions in binary mixtures of 2,2,4-trimethyl pentane with n-hexane and cyclohexane at 308 K," *Indian Journal of Physics B*, vol. 76, no. 5, pp. 661–667, 2002.

[20] A. S. Bahadur, M. C. S. Subha, and K. C. Rao, "Ultrasonic velocity and density studies in binary liquid mixtures of N, N-dimethyl formamide with 2-methoxyethanol, 2-ethoxyethanol and 2-butoxyethanol at 308.15 K," *Journal of Pure and Applied Ultrasonics*, vol. 23, pp. 26–36, 2001.

15

New Strategy for the Cleaning of Paper Artworks:
A Smart Combination of Gels and Biosensors

Laura Micheli,[1,2] **Claudia Mazzuca,**[1] **Eleonora Cervelli,**[1] **and Antonio Palleschi**[1]

[1] Department of Chemical Sciences and Technologies, University of Rome "Tor Vergata", Via della Ricerca Scientifica, 00133 Rome, Italy
[2] Consorzio Interuniversitario Biostrutture e Biosistemi "INBB", Viale Medaglie d'Oro 305, 00136 Rome, Italy

Correspondence should be addressed to Laura Micheli; laura.micheli@uniroma2.it

Academic Editor: Ipsita Banerjee

In this work an outlook on the design and application, in the cultural heritage field, of new tools for diagnostic and cleaning use, based on biocompatible hydrogels and electrochemical sensors, is reported. The use of hydrogels is intriguing because it does not require liquid treatment that could induce damage on artworks, while electrochemical biosensors not only are easy to prepare, but also can be selective for a specific compound and therefore are suitable for monitoring the cleaning process. In the field of restoration of paper artworks, more efforts have to be done in order to know how to perform the best way for an effective restoration. Rigid Gellan gel, made up of Gellan gum and calcium acetate, was proposed as a paper cleaning treatment, and selective biosensors for substances to be removed from this gel have been obtained by choosing the appropriate enzymes to be immobilized. Using this approach, it is possible to know when the cleanup process will be completed, avoiding lengthy and sometimes unnecessary cleaning material applications.

1. Introduction

Paper is difficult to be restored, due to its fragility, its degradation process, and its multicomponent composition. The main paper component is cellulose, a polysaccharide made up of β-glucose units linked together by $\beta(1 \rightarrow 4)$ glycosidic bonds [1]. Structural changes due to ageing lead to a decrease in the stability of the material and in its strength and a change in color [2]. Critical steps, during restoring of paper material, are the cleaning of the sheets, the pH change, the optimization of the degree of humidity, and the glue removal. It is also very important that the time of the cleaning process is optimized in order to remove all pollution and degradation products and to minimize invasive treatment that can lead to irreversible damage.

In the last year, the research in cultural heritage was focalized on the study of noninvasive diagnostic tools for the determination of paper artworks degradation [3]. Different techniques promise to be useful for cultural heritage applications, due to their specific properties. The combination of innovative and noninvasive material, such as an opportune hydrogel with selective electrochemical biosensors, gives the possibility to verify the degradation conditions of the paper artworks and to clean them efficiently, monitoring the removal process of pollution and degradation products. Electrochemical biosensors, based on enzyme, have been widely used in fields such as health care, food safety, and environmental monitoring. Health care is the main area in these applications, to monitor, for example, blood glucose levels and diabetics by glucose biosensors or the reliable detection of urea at home or in the hospital on patients with renal disease. Industrial applications for biosensors include monitoring fermentation broths or food processing procedures through detecting concentrations of glucose and other fermentative end products [4, 5]. Many of these biosensors, developed in these fields, are also potentially useful for applications in cultural heritage area, for conservation or characterization of several important materials such as paper, paintings, textiles, or glass.

Gellan gel has been in use since 2003 as cleaning agent for paper artworks at the Istituto Centrale per il Restauro e la Conservazione del Patrimonio Archivistico e Librario

(ICPAL, Rome, Italy) [6]. It is made up of deacylated Gellan gum and calcium acetate. The resulting gel is very transparent, thermoreversible, and rigid; it is able to gradually release the water contained within their polymer network [7] and also to absorb the water-soluble degradation products present on paper. Finally, due to its viscoelastic properties, its application and removal are fairly simple while allowing a localized cleaning operation (Figure 1). Since it is stiff and nonsticky, it can be peeled from a surface in one piece, minimizing residues [8]. These features make this gel particularly suitable to perform safer wet cleaning treatments on paper artworks, ensuring the maintenance of the structural and aesthetic properties of paper.

The aim of this research is to achieve the effectiveness of this innovative experimental approach, through the physicochemical characterization of opportune hydrogel coupled with the monitoring electrochemical system. This system gives the possibility to verify the degradation conditions of the paper artworks and to clean it efficiently, monitoring the removal process of pollution and degradation products. Moreover, some specific examples applied on paper underline the advantage and the potentiality of the combination of these materials with biosensors with respect to the traditional old paper cleaning methodologies.

2. Gellan Gel

Innovative methods for the treatment of paper materials have been studied using materials applied in the food and biomedical areas [10, 11]. One of these is the deacylated Gellan, a high molecular weight polysaccharide, already used as gelling agent in the biomedical, pharmaceutical, and industrial fields. It is a linear anionic heteropolysaccharide produced by *Pseudomonas elodea* and consists of (1,3)-β-D-Glucose, (1,4)-β-D Glucuronic acid, (1,4)-β-D-Glucose, and (1,4)-α-L-Rhamnose repeating units [12]. In the native polymer two acyl substituents, L-glyceryl at O(2) and acetyl at O(6), are present at the 3-linked glucose and, on average, there is one glyceryl per repeating unit and one acetyl every two repeating units [13]. This polymer forms hydrogels whose sol-gel transition process is temperature dependent [14, 15]; in particular, the polysaccharide, in presence of calcium salts, forms hard and rigid gel with a slow syneresis rate; moreover, it is homogeneous, transparent, and stable to pH variations [16, 17]. The pH stability assures that the hydrogel can be safely applied to every paper sample whatever its pH value is.

The present outlook will discuss and compare the results obtained by applying the Gellan hydrogel cleaning method and the traditional cleaning technique (i.e., immersion in a deionized water bath) on different paper samples.

To assess the effectiveness and safety of the proposed cleaning method on the several samples under examination we have employed a multitechnique approach, using high-performance liquid chromatography (HPLC), Fourier transform infrared spectroscopy (FTIR), scanning electron microscopy (SEM), and pH measurements. The compatibility and cleaning ability of the Gellan gel have been assured on paper samples belonging to different centuries (from XVI to

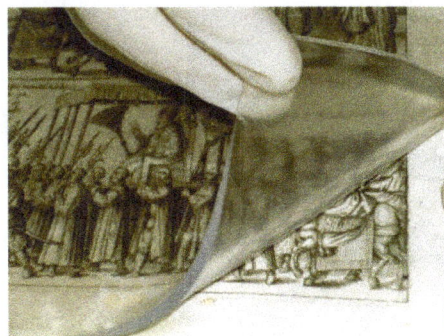

FIGURE 1: Application of Gellan gel on paper artworks.

XIX) and, therefore, characterized by different degradation condition and paper composition as obtained by Graff "C" staining experiments, HPLC measurements, and FTIR spectra [18–21]. We have assessed that Gellan gel can clean paper samples independently from their characteristics and age.

More in detail, paper samples were divided in two groups and treated, respectively, by immersion in deionized water (Figure 2(a)) and by contact with Gellan gel (Figure 2(b)). In both cases, the cleaning procedures were carried out for 60 min at room temperature. After the cleaning step, the paper samples were left to dry at room temperature. SEM images (Figure 3) showed that both samples, after immersion in water and hydrogel treatment, seemed cleaner and no swelling or fraying was present; moreover, no gel residues were visible in the images on samples analyzed after gel treatment. FTIR analysis on samples analyzed before and after treatment was indeed comparable, suggesting that no residues of gel were present (Gellan gel has characteristic IR peaks at 1602, 1408, and 1010 cm^{-1}, clearly distinguishable from those of paper samples); no detectable chemical degradation of cellulose took place as a result of the hydrogel treatment. The presence of hydrogel residues on paper samples were investigated also by performing HPLC analysis of aqueous extracts of Gellan gel before and after treatment on Whatman paper samples (data not shown). At the same time, the increase of pH values (Table 1), obtained after both cleaning procedures, indicated that acidic components involved in degradation processes were removed; anyhow, it should be noted that the decrease in acidity is slightly higher after Gellan gel cleaning procedure; this phenomenon can be explained by taking into account both the diffusive properties of molecules in the hydrogel and the intrinsic nature of it, rich in alcohol groups, capable to interact with the acidic residues on paper.

Furthermore, Gellan gel is able to remove pollution and degradation products from paper as much as the immersion in water, as shown by the comparison between the chromatograms of samples from XVIII century treated with the two methods (Figure 4). Peaks, in both cases, had almost disappeared. These results indicate that hydrogel treatment is an efficient cleaning method and does not cause change in the morphology of paper.

After investigating the cleaning ability of the Gellan gel "*per se*" (see the region between 3 and 10 minutes, where

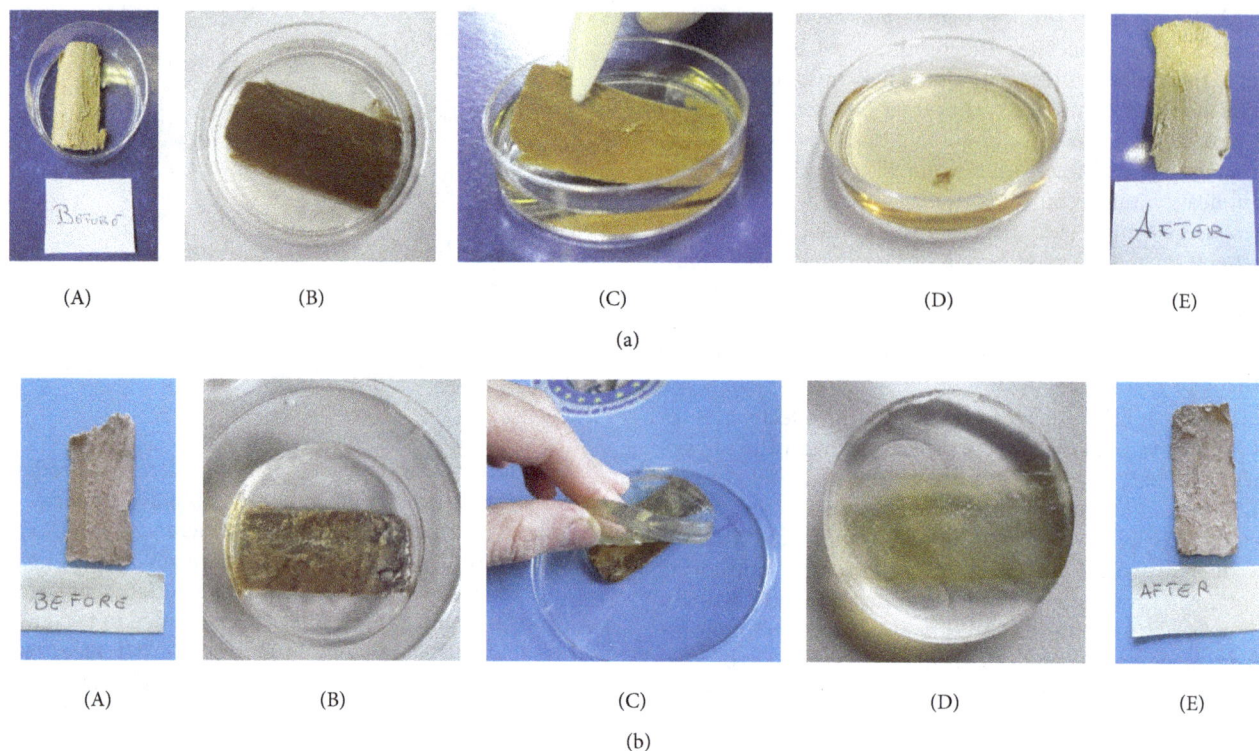

FIGURE 2: (a) Cleaning procedure with water immersion: (A) paper fragment immersed in water bath; (B) paper fragment in water; (C) removal of paper fragment from water bath after 1 h; (D) the dirt removed from paper by deionized water; (E) paper sample after water immersion. (b) Cleaning procedure using Gellan gel: (A) paper fragment before treatment with gel; (B) application of gel on paper fragment; (C) removal of gel from paper fragment in one step after 1 h; (D) the dirt removed from the paper by the yellowish color of the gel is easy to see; (E) paper sample after gel treatment.

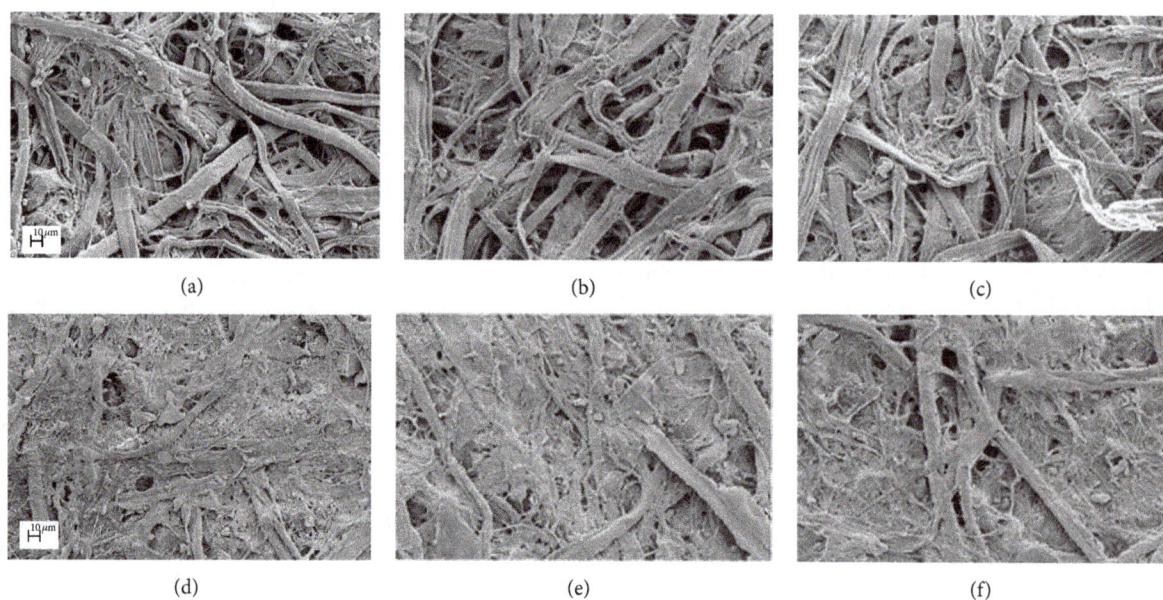

FIGURE 3: SEM images of paper samples belonging to XVII (*upper panel*) and XIX (*lower panel*) century, respectively, before cleaning (*left*), after Gellan gel treatment (*middle*), and after immersion in water (*right*) [9].

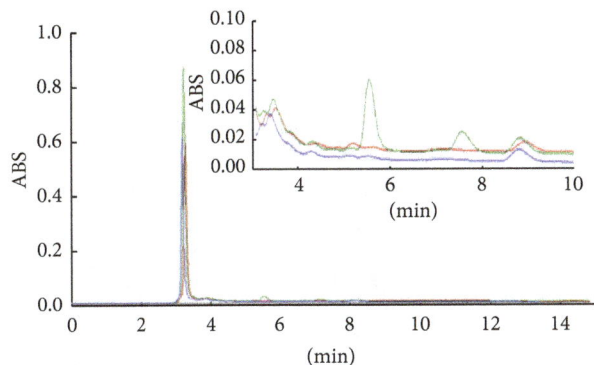

FIGURE 4: Comparison of HPLC chromatograms of XVIIII century before (green line) and after cleaning treatment with gellanGellan gel (blue) and water bath (red). Insert: detail of chromatogram, region between 3 and 10 minutes.

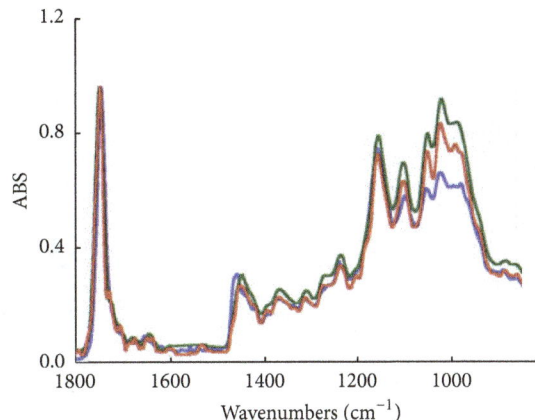

FIGURE 5: FTIR spectra of paper samples soiled with linseed oil, uncleaned (blue line), cleaned with Gellan gel (red line), and cleaned with Gellan/PLU gel (green line). Spectra are normalized to the $1740\,cm^{-1}$ peak for clarity.

TABLE 1: pH values before and after cleaning treatments [9].

	Before	After (cleaning step with gel)	After (traditional clearing step)
XVI	8.5 ± 0.1	9.4 ± 0.2	9.3 ± 0.1
XVII	7.1 ± 0.2	7.6 ± 0.1	7.9 ± 0.2
XVIII	8.6 ± 0.2	9.0 ± 0.2	8.9 ± 0.2
XIX	5.1 ± 0.1	6.4 ± 0.1	5.9 ± 0.1

the peaks of the components present on the untreated paper disappear after treatment with the gel), we also evaluated the possibility of using Gellan gel as a carrier of tuned cleaning agent. Gellan gel could be a not suitable agent to remove hydrophobic contaminants, due to its hydrophilic nature; to this end, we performed preliminary experiments by preparing a "mixed Gellan gel" (called Gellan/PLU gel); PLU gel is Gellan gel mixed with a physical hydrogel made up of α-cyclodextrin (α-CD) and of a pluronic copolymer (poly(ethylene oxide)$_{20}$-poly(propylene oxide)$_{70}$-poly(ethylene oxide)$_{20}$) (PEO_{20}-PPO_{70}-PEO_{20}) (PLU gel) [22–24]. PLU gel has amphiphilic properties and its ability to remove hydrophobic contaminants like linseed oil has been already assessed in our laboratory [25, 26].

The efficiency of cleaning of the proposed hydrogel was tested first on filter paper samples impregnated with linseed oil, as representative of fresh hydrophobic contamination; linseed oil was chosen because it is a natural substance that can mimic oily contaminants and it is widely used as a carrier for pigments in inks and as a primer for glossy paper [27, 28].

FTIR-ATR measurements (Figure 5) show that the two hydrogels and, particularly, PLU gel have partially removed the oil after 60 minutes treatment. Furthermore, in this respect, the absorbance ratio (called A.R. in the following) between two peak areas, one centered at $1740\,cm^{-1}$ due to oil [29] and the other at $1000\,cm^{-1}$ mainly due to cellulose [30], seemed to be particularly diagnostic. This ratio changes went from 0.30 for the untreated sample to 0.23 for paper cleaned with Gellan gel alone and fell to 0.19 in the case of the sample

cleaned with the Gellan/PLU gel. These data indicated that Gellan gel alone was a good cleaning agent for hydrophobic contaminants and that the addition of the PLU gel caused an increase of its cleaning ability.

To assess the efficacy of the proposed Gellan/PLU gel on real paper samples, fragments from the volume of XVIII century were soiled with linseed oil and aged as representative of old hydrophobic contamination; also in this case, the analysis of FTIR-ATR spectra indicated that A.R. decreased from 0.208 for the untreated sample to 0.102 and 0.098 for paper samples treated with Gellan gel and Gellan/PLU gel, respectively. HPLC experiments (Figure 6) confirmed these results, indicating that the Gellan/PLU gel was slightly more able to remove high molecular weight hydrophobic contaminants.

Another important issue in the paper restoration field is represented by the presence of specific contaminations on the paper sheets, difficult to remove by using the previous reported protocols. The most common example of this class of contamination is represented by the glues, often present in previous restoring actions. The removal of old glue from artworks is very important for their preservation; during ageing, in fact, the structural transformations of the glue produce a loss in compactness, yellowing, and an acidity increase, thus accelerating the mechanical fragility of the artwork itself. The use of hydrolytic enzymes, immobilized in hydrogel as agents for biocompatible cleaning, represents an interesting and advantageous alternative to traditional methods (i.e., use of water or mixed water/ethanol packs), resulting to be very important for a significant reduction in the application time of the cleaning up, due to the specific and targeted enzyme activity. Among other advantages of the hydrogel here presented, Gellan gel is suitable to be used as carrier for enzymes. In this context, some experiments were performed using the hydrogel with alpha amylase enzymes (called Enzymatic Gellan gel) on filter paper covered with starch paste for one hour (in this case, alpha amylase enzyme was used because it can hydrolyze the starch paste).

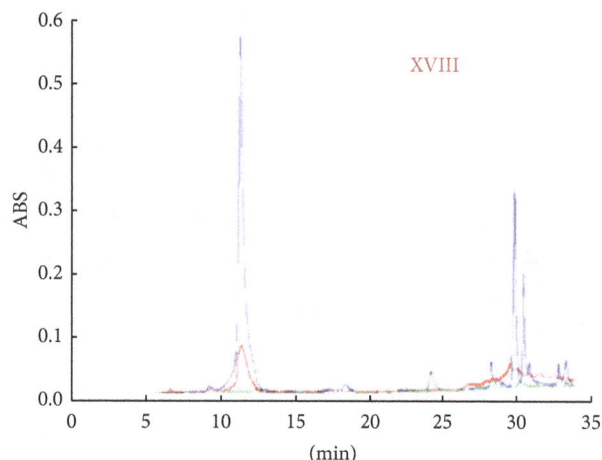

FIGURE 6: HPLC chromatograms of water extracts of paper fragments belonging XVIII century, soiled with linseed oil, aged and uncleaned (blue line), cleaned with Gellan gel (red line), and cleaned with Gellan/PLU gel (green line).

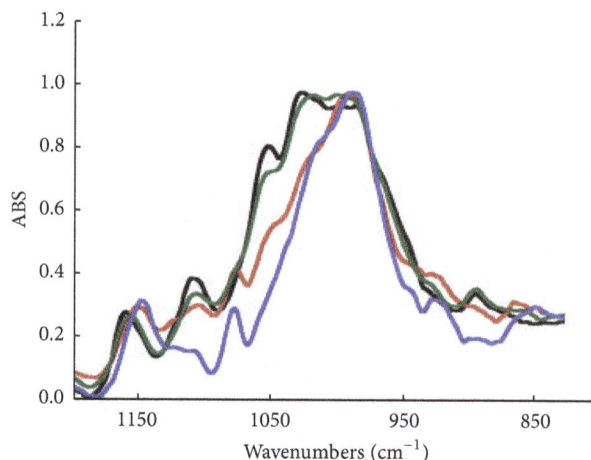

FIGURE 7: FTIR spectra of paper samples uncoated (black line), coated with starch paste (blue line), coated with starch paste after cleaning with Gellan gel (red line), and coated with starch paste and cleaned with Enzymatic Gellan gel (green line). Spectra are normalized to the maximum for clarity.

The comparison of FTIR spectra performed using Gellan gel with and without enzyme as cleaning agent on filter paper samples soiled with starch paste showed that the spectrum of paper coated with starch paste and treated with Enzymatic Gellan gel was similar to that of the uncoated paper samples, suggesting that such cleaning procedure is effective in removing starch paste. On the contrary, the spectra of paper, coated with starch paste and untreated or treated with Gellan gel, were comparable, indicating that it is not able to remove starch paste, at least after 60 minutes treatment (Figure 7) [31].

Moreover, the paper coated with starch paste showed an FTIR absorption spectrum different from that of the starting paper, due to the presence of the glue. Starch paste, in fact, showed a band at $998 \, cm^{-1}$, while cellulose has an absorption maximum at $1024 \, cm^{-1}$ [30]. The absorbance ratio between these two peaks Abs (1024)/Abs (998) was therefore diagnostic to verify the efficiency of the cleaning procedure. This ratio was 1.10 ± 0.03 for untreated paper and decreased to 0.69 ± 0.05 when paper was coated with starch paste. Treatment with the Enzymatic Gellan gel caused an increment of this ratio to 0.97 ± 0.04, very close to that of the uncoated paper; this was not the case of the starch paste coated paper cleaned with Gellan gel, as this ratio only to 0.72.

3. Sampling and Monitoring System Coupled with Electrochemical Biosensor

No data were reported in the literature that allow checking if the cleaning procedure of paper artworks was working and, simultaneously, performing a screening of the degradation state of the sample. The monitoring of the cleaning process and the identification of the degradation products or pollutant removed are very important goal required by restorers, in order to verify the degradation conditions of the paper artworks and the cleaning efficiency of the adopted procedure. A system that also combines a gel, as cleaning agent, with a selective biosensor, specific for degradation products, is therefore a very useful procedure to clean the paper artwork without damage and to estimate the extent of degradation of the paper and the time needed to remove its products and pollution. During our research study, we developed a combining tool, composed of a flow sampling plate/hydrogel/electrochemical biosensor, in order to answer to this question. In particular, the flow sampling plate was designed "*ad hoc*" and was constituted of a plate in Perspex, on which there was a serpentine with 12 channels, in order to be affixed directly on the Gellan gel during cleaning up (Figure 8). The utility of this system is to wash away the pollution and degradation products (called "material") present on the paper by using the cleaning gel and, at the same time, to remove them from the gel itself during its application, through a continuous flow of buffer in the serpentine. This "material" crosses the gel by capillarity at the contact surface between the gel and the paper sample, reaching the opposite side, where the gel is in contact with the flow sampling plate. The collected solution was analyzed by a selected electrochemical biosensor. This system was also employed to quantify specific degradation products and to know the cleanup extent.

In this outlook, we report an application to monitor the status of conservation and the cleaning process of paper material, with the quantification of the degradation products of cellulose as an example of the potentiality of this tool. The selected degradation product was the glucose, produced on paper artwork after a simulated attack by hydrolytic *cellulase*-producing fungi by use of a disposable glucose biosensor coupled with a flow sampling [32].

We chose this degradation product because it is one of the most common products of paper deterioration. Glucose is the final product of endogenous paper degradation, due to

(a)   (b)

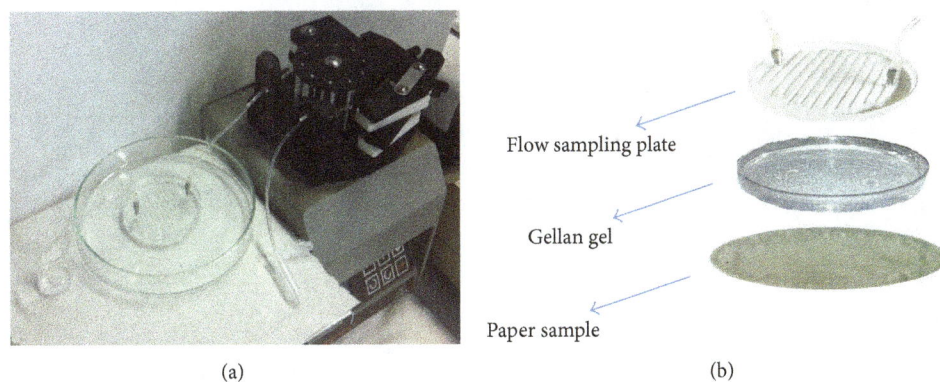

FIGURE 8: Flow sampling system applied to Gellan gel and paper sample.

the hydrolysis of cellulose (a structural component of paper-based materials) by a multicomponent enzymatic system called *cellulase*, produced by fungal species that attack and grow on the paper. In particular, *Trichoderma reesei* and *Aspergillus niger* fungi produce extracellular enzymes, called *cellulases* that metabolized cellulose sprouting mainly glucose [33, 34]. The characteristics and the sensitivity of the glucose biosensor, based on the immobilization of the *glucose oxidase* (GOx) enzyme on screen printed electrode (SPE), are well known either in the literature or in the experience of our laboratory [35], based on the following reaction scheme:

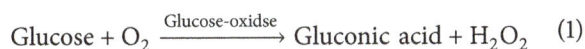

$$\text{Glucose} + O_2 \xrightarrow{\text{Glucose-oxidse}} \text{Gluconic acid} + H_2O_2 \quad (1)$$

The produced H_2O_2 is detected amperometrically using a screen printed electrode modified with Prussian Blue (*b*PB-SPE), at an applied potential of –50 mV. The advantages of Prussian blue (PB) as catalyst for the reduction of the hydrogen peroxide [36] and the screen printing technology were combined to assemble a bulk-modified screen-printed electrode. By immobilizing the glucose oxidase (GOD) onto the working electrode area, a glucose biosensor was assembled and utilized for the detection of glucose produced during the enzymatic degradation of cellulose-based materials under the action of *Aspergillus niger* and *Trichoderma reesei* cellulases.

The study of the proposed system was evaluated using a not-aged commercial filter paper as a model sample to test the proposed tool, before analyzing real paper artworks. The monitoring of the cleaning procedure was carried out using paper samples, soaked with several glucose standard solutions and dried at 37°C for 2 h. The fortified paper samples were prepared in order to estimate the sensitivity and the repeatability of the analysis and the application and the sampling time of the gel on the paper sample. The "application time" corresponds to the time necessary to remove the maximum amount of glucose from the paper by the gel, while the "sampling time" indicates the time interval between two different samples of material removed from the flow sampling system from the surface of the gel. The flow sampling plate was applied directly onto the Gellan gel during the cleaning procedure. All analytical parameters of flow system/biosensor were studied and optimized, like the flow rate (0.1 mL/min), the sampling time (10 min),

the working buffer (0.05 mM phosphate buffer + KCl 0.1 M, pH 7.4), and the applied potential for electrochemical measurement (–50 mV versus pseudoreference Ag). All collected solutions, in output from the plate, were analyzed using an electrochemical glucose biosensor (60 μL of solution was dropped on the electrode). The specific biosensor for glucose determination, based on screen printed electrodes (SPEs) produced in our laboratory, was prepared following the procedure reported elsewhere, modifying the electrode with Prussian Blue (PB) and immobilizing 0.37 U of GOx [35]. Amperometric measurements were carried out using a portable PalmSens potentiostat (Palm Instruments BV, The Netherlands) connected to a laptop computer.

Application time of Gellan gel was evaluated using a fortified paper with 10^{-3} M of glucose and treated as reported before. The current values were registered every 10 min and reported in function of the sampling time. The concentrations of glucose, removed by the gel, were extrapolated for each collected solution, using the calibration curve for glucose biosensor (LOD 2×10^{-5} M; RSD%: 2%, $n = 6$ for each concentration; linearity range: 10^{-5} M–10^{-3} M, Figure 9), obtained with standard solution absorbed by filter paper, and removed by gel after 1 h (60 μL of solution dropped on the electrode). The results obtained during the sampling study show that the maximum amount of glucose was removed by the cleaning process after 60 min and that it could be solved after 150 min. Moreover, the dilution factor of the glucose absorbed by the paper and removed from the gel was also calculated ((g glucose present in collected solution/g glucose content in the paper)∗100), showing a trend increasing with time (1 : 25 v/v after 60 min and 1 : 60 v/v after 150 min; RSD% intrasample = 6%, $n = 3$). This phenomenon was due to both the water present in the gel and the buffer flow.

4. Application of Monitoring and Cleaning Tool to Real Paper Samples

The proposed system for the monitoring of the cleaning procedure of the paper by Gellan gel was evaluated by simulation of the attack of the cellulase enzyme of paper samples produced industrially, such as newsprint paper, parchment paper, and filter paper. In addition, it was also

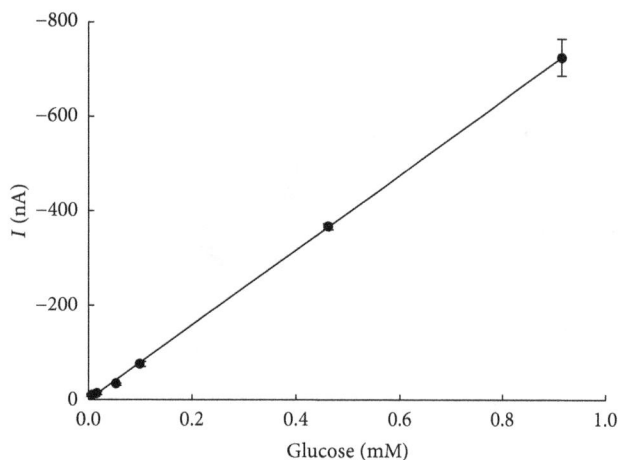

FIGURE 9: Calibration curve for glucose absorbed on filter paper and removed by gel ($y = -792645\,x + 5$; $r^2 = 0.996$). Electrochemical technique: amperometry; evaluated current after 180 s; applied potential –50 mV versus pseudoreference Ag; working buffer: 50 mM phosphate buffer + 0.1 M KCl, pH 7.4.

TABLE 2: Color of the paper samples obtained after treatment with the Graff "C".

Sample	Color after Graff "C"	Composition
Pure cellulose	Brown-red	Cotton
Filter paper	Blue	Softwood pulp
Newspaper	Brown-purple	Softwood pulp
Book (1966)	Red	?
Parchment	Brown-red	Cotton

applied to the book pages printed in 1966 in not perfect storage conditions (yellowed paper).

4.1. Paper Characterization. The composition of paper samples (old book paper (1966), newspaper, parchment paper, and filter paper) was estimated by exposure of the paper fibers to dye Graff "C" [9]. The color of the paper obtained after treatment with the Graff "C" depends on the amount of cellulose present compared to the lignin content; the different percentages of these two molecules, cellulose and lignin, depend, in turn, on the raw materials used to produce the paper such as cotton and softwood (Table 2, Figure 10).

4.2. Enzymatic Degradation of Paper and Electrochemical Measurement. The characterized paper samples (sample of 36 cm^2) underwent enzymatic degradation by commercial *cellulase* from *Aspergillus niger* (1.0 U/mg) and from *Trichoderma reesei* (6.3 U/mg). An enzyme concentration of 5 mg/mL for each cellulase or 1 : 1 w/w ratio of these two cellulases (final concentration 5 mg/mL) was brush-strokes on the papers and enzymatic reaction was carried out for 2 h at 25°C. The chosen pH (pH 6.0) for the enzymatic degradation process was a compromise between the optimal pH range of *cellulase* enzymatic activity (pH 4.5–6.5) and the normal pH values of the paper-based material in normal storage conditions (pH 6.5–8.0) [37]. The enzymatic reaction was stopped after 2 h and the proposed system (Gellan gel/flow sampling plate) was applied on the paper samples. The solution, in output of the system, was collected for 1 h. The amount of glucose produced by enzymatic reaction was monitored using the glucose biosensor, as illustrated before. Paper samples exhibit different susceptibilities towards cellulase action that can be linked to structural parameters such as crystalline and amorphous regions, since more crystalline regions in the substrate would offer stronger resistance towards enzymatic

degradation [38]. The results (Table 3) showed that the amount of glucose (mM) produced after enzymatic attack was as a function of the composition of the paper and of the percentage of cellulose present in it. For example, filter paper is composed of pure cellulose and its structure is rich in crystalline regions, not easy to be hydrolyzed by enzymes; therefore, the amount of measured glucose from filter paper is the lowest. On the contrary, newspaper is poor in cellulose, but rich in lignin and endocellulase that create amorphous regions inside of the paper, ideal substrate for fungi attack. It shows the highest amount of measured glucose.

Using a mix of the two cellulases (5 mg/mL, 1 : 1 w/w) we observed a clear effect of the synergistic action of the combined complex of enzymes. The parchment paper, under the combined action of the two cellulases, produced the highest amount of glucose.

5. Conclusion

Further efforts have to be done in order to know how to perform the best possible restoration of paper artworks. In this research work, the monitoring of simulated cleaning processes showed that the association of Gellan gel and flow sampling plate was important for following the cleanup of paper artworks [32]. During this study, the use of Gellan gel, as paper cleaning material, is useful in removing mainly hydrophilic molecules from paper artworks, without damage and also without activating anomalous long-term degradation [9]. The novelty offered by this cleaning gel, compared to the traditional one, resides in the possibility to modulate and adjust the amount of water absorbed on paper samples, according to the wet ability of each paper sample.

This hydrogel can be also modified adding surfactants or enzyme in order to remove hydrophobic molecules or glue with high yield [25, 26, 31]. Highly potential electrochemical biosensors, which have been widely employed in other scientific areas, too, may be used to know the cleanup extent of the gels used, which, in turn, are shown to be highly efficient in removing pollutants and degradation products. In fact, by choosing the appropriate enzymes to be immobilized, selective biosensors for substances to be removed from the gel have been obtained. In this way, it is possible to know when the cleanup process will be completed, avoiding lengthy and sometimes unnecessary cleaning material applications. For example, in the case of the degradation compounds and pollution of paper, the combination of glucose biosensor and gel would allow the restorers to determine when the process

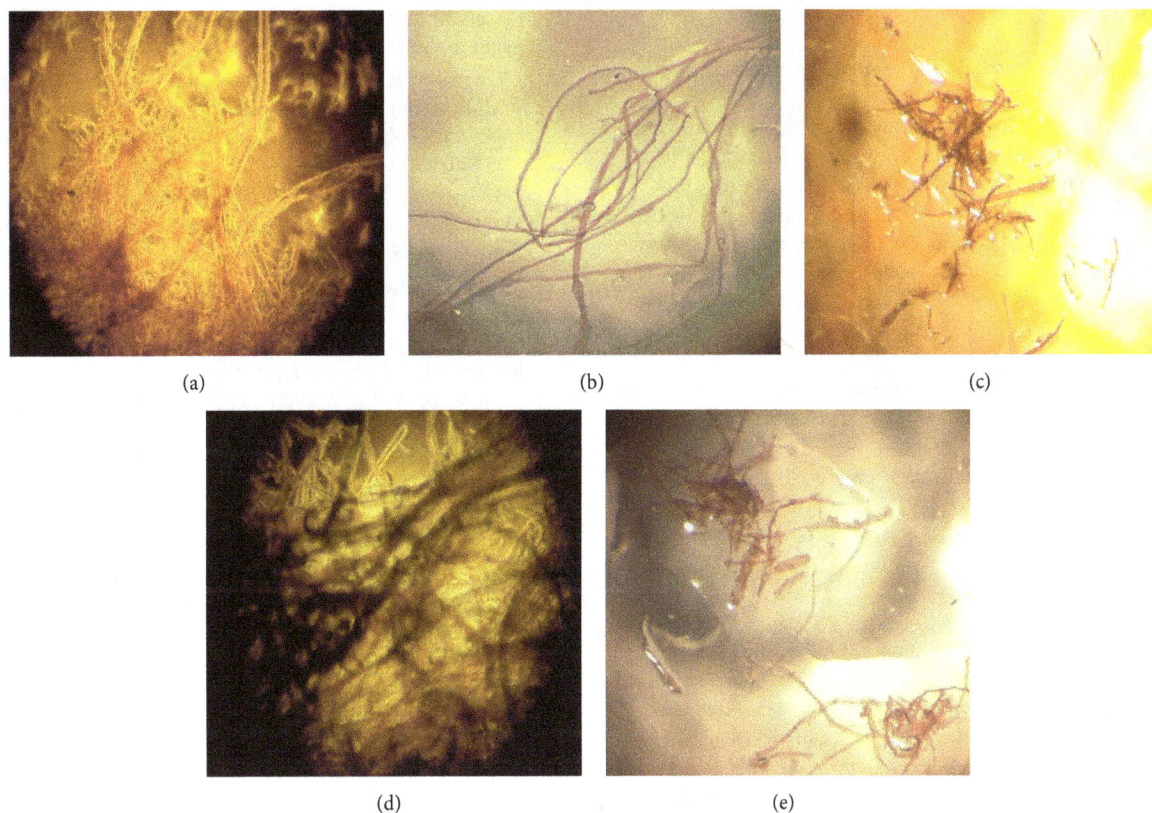

FIGURE 10: Optical microscope images of paper samples after treatment with Graf "C". (a) Pure cellulose, (b) parchment paper, (c) newspaper, (d) filter paper, and (e) book page.

TABLE 3: Concentration of glucose produced by simulated enzymatic attack.

Filter paper	
Aspergillus niger	$(3.0 \pm 0.2) \times 10^{-3}$ M
Trichoderma reesei	$(2.0 \pm 0.2) \times 10^{-3}$ M
Mix	$(5.2 \pm 0.2) \times 10^{-3}$ M
Newspaper paper	
Aspergillus niger	$(8.5 \pm 0.7) \times 10^{-3}$ M
Trichoderma reesei	$(2.8 \pm 0.2) \times 10^{-3}$ M
Mix	$(10.1 \pm 0.9) \times 10^{-3}$ M
Parchment paper	
Aspergillus niger	$(3.8 \pm 0.5) \times 10^{-3}$ M
Trichoderma reesei	$(5.7 \pm 0.7) \times 10^{-3}$ M
Mix	$(12 \pm 2) \times 10^{-3}$ M
Book (1966)	
Aspergillus niger	$(3.0 \pm 0.3) \times 10^{-3}$ M
Trichoderma	$(2.3 \pm 0.8) \times 10^{-3}$ M
Mix	$(5 \pm 1) \times 10^{-3}$ M

Analytical Chemistry and Physical Chemistry of University of Rome Tor Vergata. The results, obtained until now, are satisfactory and show the effectiveness and the potentiality of the proposed system in terms of degree of cleanup, of compatibility either with the material to be treated or with the operator, and of efficiency of monitoring the state of conservation of the material to be restored.

Conflict of Interests

The authors declare that there is no conflict of interests regarding the publication of this paper.

Authors' Contribution

Laura Micheli and Claudia Mazzuca have equally contributed to this paper.

Acknowledgments

This work was supported by the "progetto SIDE" of the Regione Lazio. The authors thank Dr. Francesco Basoli, Prof. Danila Moscone, and Dr. Raffaella Lettieri for technical support and useful discussions. They are grateful to Simonetta Iannuccelli and Silvia Sotgiu, restorers of ICPAL of Rome, for their practical support and useful discussions.

of removing this degradation product of cellulose is finished and to evaluate the cleaning efficiency [8].

In conclusion, this brief presentation summarizes an interdisciplinary research work that is being carried out together with restorers of ICPAL and the Laboratories of

References

[1] M. Copedè, *La Carta ed il Suo Degrado*, Firenze, Italy, Nardini, 1995.

[2] M. T. Doménech-Carb, "Novel analytical methods for characterization of binding media and protective coatings in artworks," *Analytica Chimica Acta*, vol. 621, pp. 109–139, 2008.

[3] J. L. Ramírez, M. A. Santana, I. Galindo-Castro, and A. Gonzalez, "The role of biotechnology in art preservation," *Trends in Biotechnology*, vol. 23, no. 12, pp. 584–588, 2005.

[4] M. I. Prodromidis and M. I. Karayannis, "Enzyme based amperometric biosensors for food," *Electroanal*, vol. 14, no. 4, pp. 241–261, 2002.

[5] K. R. Rogers, "Enzyme based amperometric biosensors for food analysis," *Biosensors and Bioelectronics*, vol. 10, no. 6-7, pp. 533–541, 1995.

[6] Quaderno Cesmar7n. 11 (2012), S. Iannuccelli e S. Sotgiu eds, il Prato, Padua.

[7] W. Gibson and G. R. Sanderson, "Gellan gum," in *Thinckening and Gelling Agents for Food*, A. Imeson, Ed., Aspen Publishers, New York, NY, USA, 1999.

[8] L. Micheli, C. Mazzuca, A. Palleschi, and G. Palleschi, "Combining a hydrogel and an electrochemical biosensor to determine the extent of degradation of paper artworks," *Analytical and Bioanalytical Chemistry*, vol. 403, no. 6, pp. 1485–1489, 2012.

[9] C. Mazzuca, L. Micheli, M. Carbone et al., "Gellan hydrogel as a powerful tool in paper cleaning process: a detailed study," *Journal of Colloid and Interface Science*, vol. 416, pp. 205–211, 2014.

[10] G. Sworn, "Gellan gum," in *Handbook of Hydrocolloids*, G. O. Philips and P. A. Williams, Eds., CRC Press/Woodhead, Cambridge, UK, 2000.

[11] I. B. Bajaj, S. A. Survase, P. S. Saudagar, and R. S. Singhal, "Gellan gum: fermentative production, downstream processing and applications," *Food Technology and Biotechnology*, vol. 45, no. 4, pp. 341–354, 2007.

[12] A. Ohtsuka and T. Watanabe, "The network structure of gellan gum hydrogels based on the structural parameters by the analysis of the restricted diffusion of water," *Carbohydrate Polymers*, vol. 30, no. 2-3, pp. 135–140, 1996.

[13] P. Matricardi, C. Cencetti, R. Ria, F. Alhaique, and T. Coviello, "Preparation and characterization of novel Gellan gum hydrogels suitable for modified drug release," *Molecules*, vol. 14, no. 9, pp. 3376–3391, 2009.

[14] E. Miyoshi, T. Takaya, and K. Nishinari, "Effects of salts on the gel-sol transition of gellan gum by differential scanning calorimetry and thermal scanning rheology," *Thermochimica Acta*, vol. 267, pp. 269–287, 1995.

[15] S. Ikeda, Y. Nitta, T. Temsiripong, R. Pongsawatmanit, and K. Nishinari, "Atomic force microscopy studies on cation-induced network formation of gellan," *Food Hydrocolloids*, vol. 18, no. 5, pp. 727–735, 2004.

[16] E. R. Morris, K. Nishinari, and M. Rinaudo, "Gelation of gellan—a review," *Food Hydrocolloids*, vol. 28, no. 2, pp. 373–411, 2012.

[17] A. Casoli, E. Cervelli, P. Cremonesi et al., "La verifica del metodo," in *Collana Quaderno Cesmar7*, S. Iannuccelli and S. Sotgiu, Eds., vol. 11, pp. 37–57, Padova, Italy, 2012.

[18] J. Vodopivec and M. Cernic, "Non destructive characterization of paper as a support of a gouache collection," in *Proceedings of the 9th International conference on NDT of Art*, Jerusalem, Israel, May 2008.

[19] P. Calvini and A. Gorassini, "FTIR—deconvolution spectra of paper documents," *Restaurator*, vol. 23, no. 1, pp. 48–66, 2002.

[20] J. Łojewska, P. Miśkowiec, T. Łojewski, and L. M. Proniewicz, "Cellulose oxidative and hydrolytic degradation: in situ FTIR approach," *Polymer Degradation and Stability*, vol. 88, no. 3, pp. 512–520, 2005.

[21] J. Lojewska, A. Lubanska, T. Lojewski, P. Miscowiec, and L. M. Proniewicz, "Kinetic approch to degradation of paper," *e-Preservation Science*, vol. 2, p. 1, 2005.

[22] J. Li, X. Ni, and K. W. Leong, "Injectable drug-delivery systems based on supramolecular hydrogels formed by poly(ethylene oxide)s and α-cyclodextrin," *Journal of Biomedical Materials Research A*, vol. 65, no. 2, pp. 196–202, 2003.

[23] J. Li, X. Li, X. Ni, X. Wang, H. Li, and K. W. Leong, "Self-assembled supramolecular hydrogels formed by biodegradable PEO-PHB-PEO triblock copolymers and α-cyclodextrin for controlled drug delivery," *Biomaterials*, vol. 27, no. 22, pp. 4132–4140, 2006.

[24] J. Li and X. J. Loh, "Cyclodextrin-based supramolecular architectures: syntheses, structures, and applications for drug and gene delivery," *Advanced Drug Delivery Reviews*, vol. 60, no. 9, pp. 1000–1017, 2008.

[25] C. Mazzuca, L. Micheli, F. Marini et al., "Rheoreversible hydrogels in paper restoration processes: a versatile tool," *Chemistry Central Journal*, vol. 8, no. 1, pp. 1–11, 2014.

[26] C. Mazzuca, G. Bocchinfuso, I. Cacciotti, L. Micheli, G. Palleschi, and A. Palleschi, "Versatile hydrogels: an efficient way to clean paper artworks," *RSC Advances*, vol. 3, no. 45, pp. 22896–22899, 2013.

[27] F. Fiorani and G. Pace, "I disegni di Étienne Du Pérac per i "Vestigi dell'antichità di Roma". Le prime carte traslucide," *Mitteilungen des Kunsthistorisches Institut in Florenz*, vol. 2, no. 2-3, pp. 240–251, 2008.

[28] A. Gambaro, R. Ganzerla, M. Fantin, E. Cappelletto, R. Piazza, and W. R. L. Cairns, "Study of 19th century inks from archives in the Palazzo Ducale (Venice, Italy) using various analytical techniques," *Microchemical Journal*, vol. 91, no. 2, pp. 202–208, 2009.

[29] M. Lazzari and O. Chiantore, "Drying and oxidative degradation of linseed oil," *Polymer Degradation and Stability*, vol. 65, no. 2, pp. 303–313, 1999.

[30] V. Librando, Z. Minniti, and S. Lorusso, "Ancient and modern paper characterization by FTIR and Micro-Raman spectroscopy," *Conservation Science in Cultural Heritage*, vol. 11, no. 1, pp. 249–268, 2011.

[31] C. Mazzuca, L. Micheli, E. Cervelli et al., "Development of a new efficient strategy to remove starch paste from paper samples," under review to *ACS Applied Materials & Interfaces*.

[32] L. Micheli, C. Mazzuca, D. Moscone, A. Palleschi, and G. Palleschi, "Electrochemical technique for Cultural heritage. An example of application of monitoring cleaning process of paper art work," in *Facciamo Germogliare le Nostre Idee—Quaderno della III Scuola di Chimica Ambientale e dei beni Culturali*, E. Papa, M. Grazia Perrone, A. Piazzalunga, and B. Varese, Eds., pp. 77–81, Società Chimica Italiana, 2014.

[33] M. Strlic, J. Kolar, and S. Scholten, "Paper and durability," in *Ageing and stabilization of paper*, M. Strlic and J. Kolar, Eds., Ljubljana National and University Library, Ljubljana, Slovenia, 2005.

[34] T. M. Wood, "Fungal cellulases," *Biochemical Society Transactions*, vol. 20, no. 1, pp. 46–53, 1992.

[35] M. Mahosenaho, F. Caprio, L. Micheli, A. M. Sesay, G. Palleschi, and V. Virtanen, "A disposable biosensor for the determination of alpha-amylase in human saliva," *Microchimica Acta*, vol. 170, no. 3, pp. 243–249, 2010.

[36] A. A. Karyakin and E. E. Karyakina, "Electroanalytical applications of Prussian Blue and its analogs," *Russian Chemical Bulletin*, vol. 50, no. 10, pp. 1811–1817, 2001.

[37] M. Itävaara, M. Siika-aho, and L. Viikari, "Enzymatic degradation of cellulose-based materials," *Journal of Environmental Polymer Degradation*, vol. 7, no. 2, pp. 67–73, 1999.

[38] S. Lorusso, F. Prestileo, L. Gregari, and M. E. Pifferi, "Tecnica, tecnologia, scienza nel settore dei beni culturali," *Scienza e Tecnica*, vol. 340, pp. 1–17, 1999.

16

Selenium (Se) Regulates Seedling Growth in Wheat under Drought Stress

Fahim Nawaz,[1] Muhammad Yasin Ashraf,[2] Rashid Ahmad,[1] Ejaz Ahmad Waraich,[1] and Rana Nauman Shabbir[1]

[1] *Department of Crop Physiology, University of Agriculture, Faisalabad 38040, Pakistan*
[2] *Nuclear Institute for Agriculture and Biology (NIAB), P.O. Box No. 128, Faisalabad, Pakistan*

Correspondence should be addressed to Fahim Nawaz; fahim5382@gmail.com

Academic Editor: Wen-Chi Hou

Selenium (Se) is an essential micronutrient with a range of physiological and antioxidative properties. Reports regarding effect of Se application on plants growth and development are not consistent. The identification of effective Se dose and application method is crucial for better understanding of Se translocation within crop plants under drought stress. The present study aimed at investigating the role of Se supplementation in improving the drought tolerance potential of wheat at early growth stages. Two wheat genotypes (Kohistan-97 and Pasban-90) were grown in plastic pots (8 × 12 cm) in green/wire-house experiments. Results demonstrated that the growth and biomass of seedlings increased at high Se foliar concentrations and decreased at low and high Se fertigation levels. The seedlings exhibited the highest values for plant height stress tolerance index (PHSI), root length stress tolerance index (RLSI), dry matter stress tolerance index (DMSI), and fresh matter stress tolerance indices (FMSI) at Se fertigation level of 7.35 μM, whereas Se foliar treatment of 7.06 μM resulted in maximum values for these indices. The seedlings foliarly sprayed with Se maintained higher DMSI and FMSI than those fertigated with Se which suggests that Se foliar spray is more effective than Se fertigation for improving drought tolerance.

1. Introduction

Drought stress has emerged as the single most critical threat to world food security. It seriously limits agricultural productivity, especially in areas where rainfall is limiting or unreliable, so improving yield under limited water conditions has become a crucial target for arid and semiarid regions of the world [1–3]. Exposure to drought stress poses serious challenges for the survival of plants, because it results in impaired germination and seedling growth [4] and affects many growth variables of the plant [5, 6], thus reducing fitness and harvestable yield of plants [3].

The physiological and antioxidant properties of selenium (Se) have increased the curiosity of many biologists in recent past. Although it does not take part in various vital metabolic processes in plants, it may help to reduce the damage under physiological stresses [7, 8]. Recently, Se has been reported to counteract the detrimental effects of various environmental stresses such as heavy metals [9], UV-B [10, 11], excess water [12], salt [13], cold [14], high temperature [15], senescence [16], and desiccation [17]. However, reports on the role of Se in plants under water stress conditions are scanty. It may regulate water status [18] and increase biomass production [19] by the activation of antioxidant apparatus of water stressed plants [20, 21].

Numerous strategies, namely, seed dressing/coating [22], seed soaking [19], soil application [23], and foliar spray [24], have been used to supply Se in plants. However, the simplicity and practicability of soil and foliar application make them widely accepted among these methods. Several studies confirmed the positive role of soil Se fertilization in various crops/plants such as rice [24, 25], maize [26], wild barley [27], and soybean [28]. The foliar Se application has been reported to significantly promote growth in vegetables such as onion bulbs and leaves [29], carrot roots and leaves

[30], radish flowers and leaves [31], garlic bulbs [32], and cereals like wheat [33].

The uptake and accumulation of Se within a narrow range are beneficial for plants [34] and are determined by the plants ability to absorb and metabolize Se. It is well documented that increase in acidity, iron oxides/hydroxides, and organic matter and high clay content of soil decrease the bioavailability of Se to plants [35, 36]. The soil moisture also affects the availability of Se to plants as it is more available under low precipitation conditions [6]. Moreover, actively growing tissues usually contain large amounts of Se [37] and accumulation is higher in shoot and leaf than in root tissues [38]. Therefore, Se fertigation and foliar spray are much more viable and effective approaches than soil application to increase Se translocation within plants. This study was conducted with the hypothesis that Se supply mitigates adverse effects of water stress in wheat seedlings.

2. Materials and Methods

2.1. Seed Material and Experimental Design. Two pot experiments were conducted in wire/green house to determine appropriate rates of Se application as fertigation and foliar spray, effective in improving the drought tolerance and biomass in wheat plants subjected to water stress at seedling stage. The seeds of two recommended spring wheat genotypes, that is, Kohistan-97 and Pasban-90, categorized as drought tolerant and sensitive, respectively, in our earlier reports [39], were used for the study. The seeds were obtained from Ayyub Agricultural Research Institute (AARI), Faisalabad (Pakistan). The experiments were laid out in the completely randomized design (CRD) with three repeats. Each repeat consisted of a 10-seedling pot in each experiment. Twenty randomly selected seeds of each genotype were sterilized with 5% sodium hypochlorite solution for five minutes and later air-dried to their original moisture level before conducting experiments.

2.2. Drought Stress Treatments. In each experiment, ten seeds of each genotype were sown at 100% field capacity (100% FC) in plastic pots (8 dia × 12 length cm) containing 430 g of sterilized, washed, fine river sand. One set of pots (control) was watered regularly while water stress was imposed in the other set of pots by stopping water application after seedling emergence. Amount of water evaporated was calculated daily and control plants were rewatered accordingly. The pots were placed under controlled temperature at 25°C, 16 h day length, 200 μmol m^{-2} s^{-1} photosynthetic active radiation (PAR), and 75–80% relative humidity for four weeks in growth chamber (Sanyo-Gallenkamp, UK). After four weeks, five plants were harvested randomly from each pot for the calculation of physiological indices. The seedlings were placed in an oven at 65°C for 72 hours to record seedling dry weight. Both experiments were repeated thrice and the data presented is the mean of values obtained in three experiments.

2.3. Selenium (Se) Treatments. The Se fertigation doses of 3.68 (0.25 mg L^{-1}), 7.35 (0.50 mg L^{-1}), 11.03 (0.75 mg L^{-1}),

and 14.70 μM (1.00 mg L^{-1}) and Se foliar treatments of 1.76 (0.12 mg L^{-1}), 3.53 (0.24 mg L^{-1}), 5.29 (0.36 mg L^{-1}), and 7.06 μM (0.48 mg L^{-1}) were developed by dissolving Na$_2$SeO$_4$ (Sigma-Aldrich, USA) in distilled water. The seedlings were fertigated and foliarly sprayed with Se after three days of exposure to stress.

The following formulae as described by [40] were used for the calculation of plant height stress tolerance index (PHSI), root length stress tolerance index (RLSI), and fresh and dry matter stress tolerance indices (FMSI, DMSI).

PHSI (%) = [plant height of stressed plant/plant height of control plant] × 100;

DMSI (%) = [dry matter of stressed plant/dry matter of control plant] × 100;

RLSI (%) = [root length of stressed plant/root length of control plant] × 100;

SFSI (%) = [shoot fresh weights of stressed plants/shoot fresh weights of control plants] × 100;

RFSI (%) = [root fresh weights of stressed plants/root fresh weights of control plants] × 100.

2.4. Statistical Analysis. All the recoded data in different experiments during this study were analyzed statistically using analysis of variance technique and STATISTICA Computer Program was used for this purpose. Tukey test at 5% probability level was used to compare means.

3. Results

The highly significant effect ($P < 0.01$) of Se fertigation and foliar spray was recorded on physiological indices of both wheat genotypes (Kohistan-97 and Pasban-90). The maximum PHSI value (86%) was noted in seedlings fertigated with 7.35 μM Se, whereas low PHSI values were recorded at high (11.03 μM and 14.70 μM) or low levels (3.68 μM) of Se fertigation (Figure 1). A gradual increase in PHSI was observed by increasing Se foliar spray levels. The application of Se at 7.06 μM gave the maximum value (88%) for this index, while the minimum value (63%) was recorded in seedlings sprayed with water which was statistically at par with the value (64%) obtained for the lowest Se treatment of 1.76 μM (Figure 1). Nonsignificant differences were recorded between genotypes in both Se supply methods (Se fertigation and Se foliar spray) for PHSI (Figure 1).

The highest value for RLSI (159%) was recorded in plants supplied with 7.35 μM Se. A nonsignificant difference ($P > 0.05$) was observed between high Se treatments, that is, 11.03 μM and 14.70 μM, which had RLSI values of 150% and 152%, respectively, whereas the lowest value for RLSI (141%) was recorded in seedlings fertigated with 3.68 μM Se (Figure 2). The plants foliarly sprayed with 7.06 μM Se maintained maximum value (121%) for RLSI, which was 15% higher as compared to water sprayed seedlings (103%). The foliar application of Se at 3.53 μM and 5.29 μM increased RLSI by 8% and 12%, respectively, as compared to water sprayed

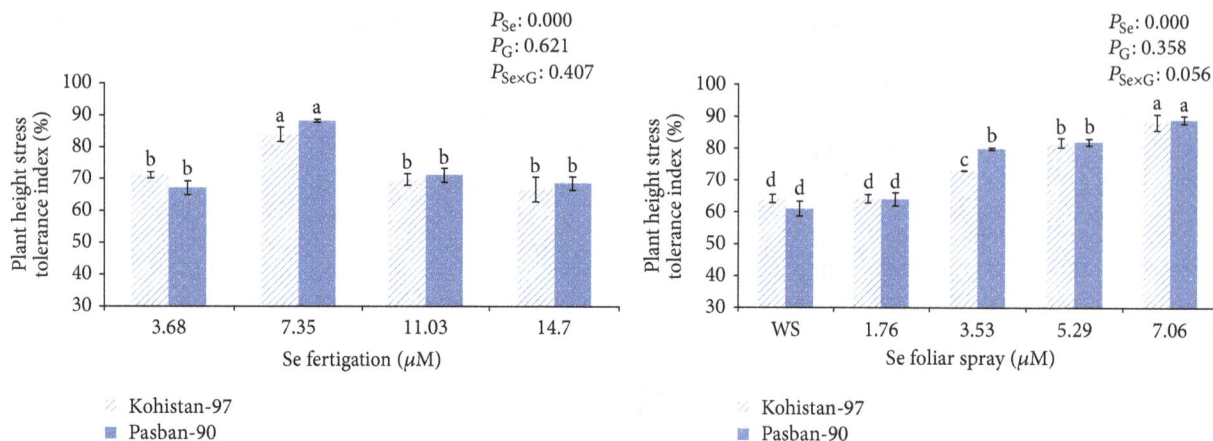

FIGURE 1: Effect of Se fertigation and Se foliar spray on plant height stress tolerance index (PHSI) of wheat seedlings. Fertigation treatments include fertigation with 3.68 (0.25 mg L^{-1}), 7.35 (0.50 mg L^{-1}), 11.03 (0.75 mg L^{-1}), and 14.70 μM (1.00 mg L^{-1}), whereas foliar treatments include foliar spray with water (WS) and Se foliar spray with 1.76 (0.12 mg L^{-1}), 3.53 (0.24 mg L^{-1}), 5.29 (0.36 mg L^{-1}), and 7.06 μM (0.48 mg L^{-1}). Values are mean $\pm$ standard error. P_{Se}: Se effects; P_{G}: genotype effects; $P_{Se \times G}$: interaction effects of Se and genotypes.

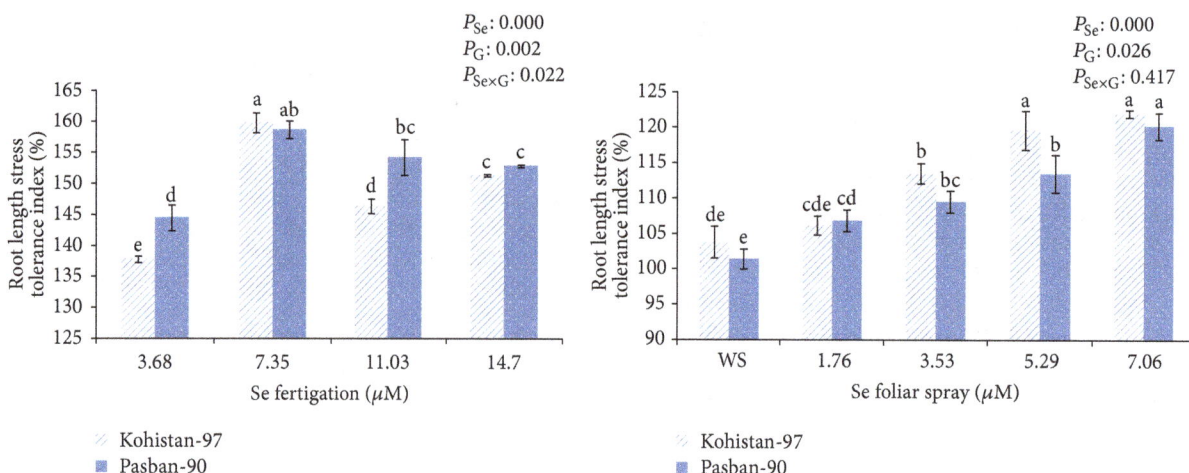

FIGURE 2: Effect of Se fertigation and Se foliar spray on root length stress tolerance index (RLSI) of wheat seedlings. Fertigation treatments include fertigation with 3.68 (0.25 mg L^{-1}), 7.35 (0.50 mg L^{-1}), 11.03 (0.75 mg L^{-1}), and 14.70 μM (1.00 mg L^{-1}), whereas foliar treatments include foliar spray with water (WS) and Se foliar spray with 1.76 (0.12 mg L^{-1}), 3.53 (0.24 mg L^{-1}), 5.29 (0.36 mg L^{-1}), and 7.06 μM (0.48 mg L^{-1}). Values are mean $\pm$ standard error. P_{Se}: Se effects; P_{G}: genotype effects; $P_{Se \times G}$: interaction effects of Se and genotypes.

seedlings (Figure 2). The results showed that Se fertigated plants maintained 31% higher RLSI than those foliarly applied with Se.

The fertigation of seedlings with 7.35 μM and 11.03 μM Se gave the maximum value (72%) for DMSI, whereas at Se fertigation level of 3.68 μM the minimum value (67%) was recorded (Figure 3). The seedlings applied with foliar Se treatment of 7.06 μM showed an increase of 85% and had maximum value (89%) as compared to plants sprayed with water exhibiting minimum value (48%) for DMSI. The foliar spray of plants with Se resulted in 33% higher DMSI than Se fertigation (Figure 3). Wheat genotype Kohistan-97 showed significantly higher RLSI and DMSI than Pasban-90 in both Se supply methods, that is, Se fertigation and foliar spray (Figures 2 and 3).

The plants fertigated with 7.35 μM Se exhibited the highest values for SFSI (53%) and RFSI (80%). The higher Se

fertigation levels of 11.03 μM and 14.70 μM also increased SFSI and RFSI and were statistically related to each other for these indices (Figures 4 and 5). The lowest dose of Se fertigation (3.68 μM) resulted in minimum SFSI and RFSI, that is, 44% and 63%, respectively. In seedlings foliarly applied with Se, the foliar Se treatment of 7.06 μM gave the maximum values for SFSI (85%) and RFSI (80%) (Figure 4). It was observed that Se foliar spray was more effective (60%) than Se fertigation in enhancing SFSI of the seedlings (Figure 4), whereas a nonsignificant difference was observed between Se supply methods for RFSI (Figure 5).

4. Discussion

The results indicate that water stress adversely influences wheat growth due to poor germination and seedling establishment. It was observed that fertigation and foliar spray are

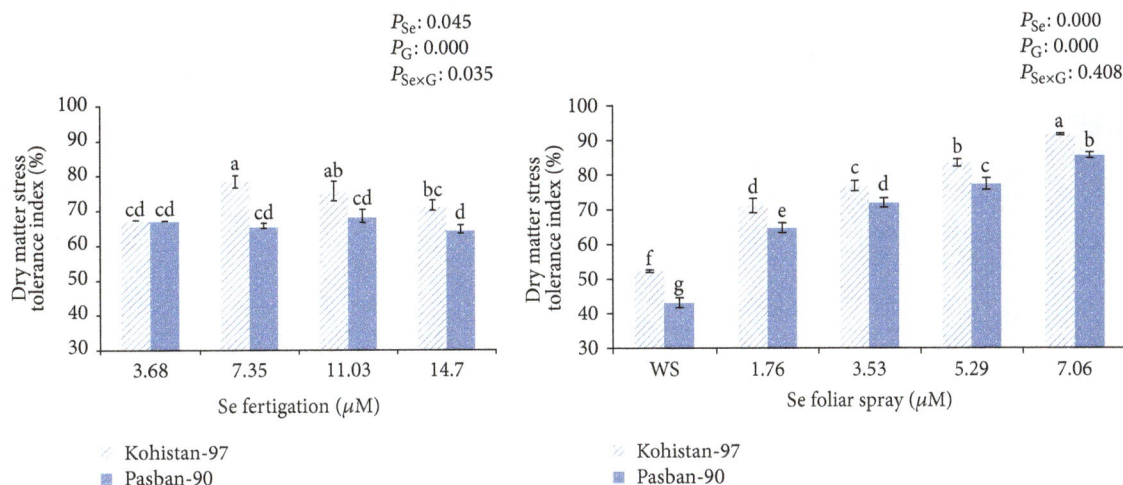

FIGURE 3: Effect of Se fertigation and Se foliar spray on dry matter stress tolerance index (DMSI) of wheat seedlings. Fertigation treatments include fertigation with 3.68 (0.25 mg L^{-1}), 7.35 (0.50 mg L^{-1}), 11.03 (0.75 mg L^{-1}), and 14.70 μM (1.00 mg L^{-1}), whereas foliar treatments include foliar spray with water (WS) and Se foliar spray with 1.76 (0.12 mg L^{-1}), 3.53 (0.24 mg L^{-1}), 5.29 (0.36 mg L^{-1}), and 7.06 μM (0.48 mg L^{-1}). Values are mean ± standard error. P_{Se}: Se effects; P_G: genotype effects; $P_{Se \times G}$: interaction effects of Se and genotypes.

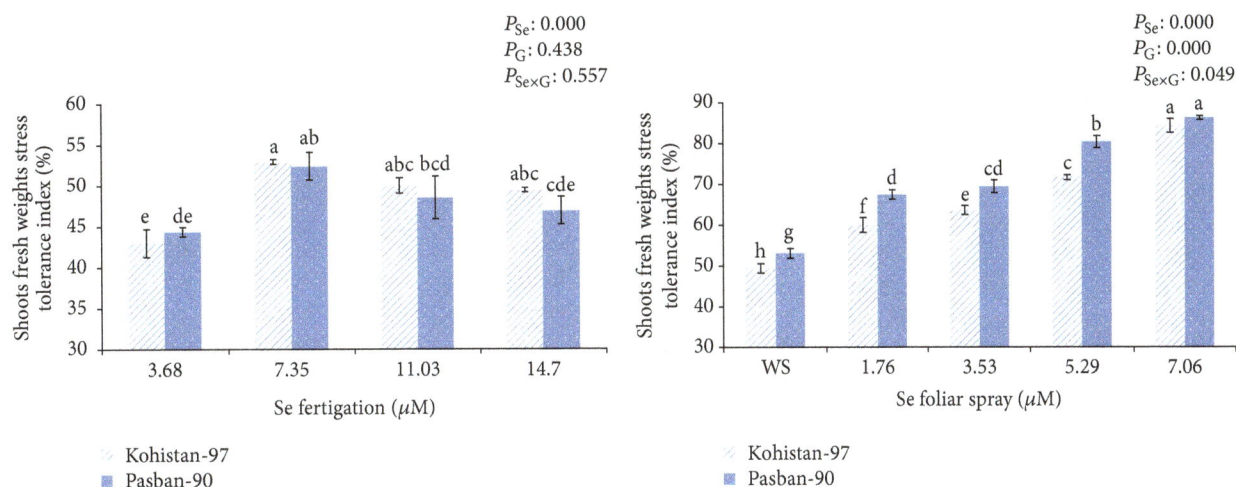

FIGURE 4: Effect of Se fertigation and Se foliar spray on shoots fresh weights stress tolerance index (SFSI) of wheat seedlings. Fertigation treatments include fertigation with 3.68 (0.25 mg L^{-1}), 7.35 (0.50 mg L^{-1}), 11.03 (0.75 mg L^{-1}), and 14.70 μM (1.00 mg L^{-1}), whereas foliar treatments include foliar spray with water (WS) and Se foliar spray with 1.76 (0.12 mg L^{-1}), 3.53 (0.24 mg L^{-1}), 5.29 (0.36 mg L^{-1}), and 7.06 μM (0.48 mg L^{-1}). Values are mean ± standard error. P_{Se}: Se effects; P_G: genotype effects; $P_{Se \times G}$: interaction effects of Se and genotypes.

efficient, viable, and effective approaches for the application of fertilizers and improving fertilizer use efficiency [41]. An increase in physiological indices with Se supply confirmed the hypothesis that Se plays a positive role in improving drought tolerance of wheat seedlings.

The increase in PHSI specifies Se role in regulation of water status of moisture stressed seedlings [42] and activation of plant hormones responsible for cell expansion and enlargement [43]. The maximum PHSI observed by Se fertigation treatment of 7.35 μM (Figure 1) and Se foliar application at 7.06 μM (Figure 1) might be attributed to the Se-regulated decrease in osmotic potential that increases the water relations of water stressed seedlings [44]. The actively growing plant parts such as young leaves and seeds

accumulate large amounts of Se which affects osmoregulation in plants [37, 45].

The growth and development of plants is directly influenced by root activity [46] so it can serve as an important index of plant resistance [20]. The increase in root length is an adaptive response of wheat plants exposed to drought stress. The highest RLSI recorded in plants fertigated (Figure 2) and foliarly sprayed (Figure 2) with Se treatments of 7.35 μM and 7.06 μM, respectively, indicates the effectiveness of Se in improving plant resistance against drought stress. The radicles treated with Se are healthier and vigorous with extensive root hairs [47]. Yao et al. [20] reported an increase in root activity (growth and uptake) of water stressed seedlings by extra Se supply resulting in an increase in dry matter which

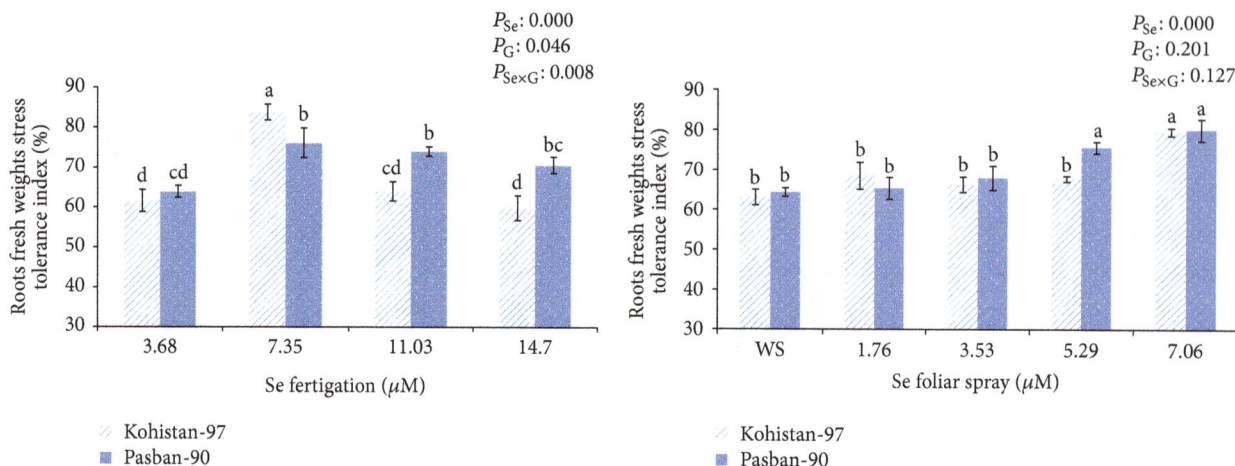

FIGURE 5: Effect of Se fertigation and Se foliar spray on roots fresh weights stress tolerance index (RFSI) of wheat seedlings. Fertigation treatments include fertigation with 3.68 (0.25 mg L^{-1}), 7.35 (0.50 mg L^{-1}), 11.03 (0.75 mg L^{-1}), and 14.70 μM (1.00 mg L^{-1}), whereas foliar treatments include foliar spray with water (WS) and Se foliar spray with 1.76 (0.12 mg L^{-1}), 3.53 (0.24 mg L^{-1}), 5.29 (0.36 mg L^{-1}), and 7.06 μM (0.48 mg L^{-1}). Values are mean ± standard error. P_{Se}: Se effects; P_G: genotype effects; $P_{Se \times G}$: interaction effects of Se and genotypes.

improved DMSI of seedlings (Figure 3). Similarly, Valadabadi et al. [48] noted a significant increase in total dry weight of rapeseed cultivars foliarly sprayed with Se under water stress conditions. The increase in root length and dry weight by Se application supports the fact that a significant relation exists between root and seedling dry weight of water stressed seedlings [49, 50].

The fertigation of seedlings by Se treatment of 7.35 μM resulted in the highest biomass accumulation (Figures 4 and 5). However, low Se fertigation doses did not significantly improve the biomass (SFSI and RFSI). These results are in line with the findings of Nawaz et al. [19] who observed that Se significantly increased growth of water stressed wheat seedlings and were of the view that Se regulates water status under drought stress. Nonsignificant effect of lower Se doses on biomass has been reported in wheat [20], perennial ryegrass, and strawberry clover [51]. The high Se fertigation dose (14.70 μM) caused a significant reduction in biomass (Figures 4 and 5). Similar results were reported by Ximénez-Embún et al. [52] in white lupine (20%) and sunflower (40%) at high Se fertilization doses. High levels of Se inhibit photosynthesis by decreasing the light energy absorbed by the antenna system and impair photosynthetic machinery of wheat [53] which results in the lower production of starch [54, 55] that may lead to a decrease in the biomass production.

The highest biomass accumulation (SFSI and RFSI) with Se foliar spray (Figures 4 and 5) may be due to the diffusion of Se ions that takes place from the surface of leaves to epidermal cells. The foliar application of Se has been reported to stimulate growth in lettuce [56], green tea [57], and potato [58]. Germ [59] observed significant increase in mass of only water stressed potato tubers supplemented with Se, whereas the mass of tubers was reduced by Se application in well-watered plants. Similar results were reported by Habibi [60] in barley. The increased Se efficiency by foliar application may be due to its direct absorption and accumulation in the plants by diffusion from the surface of leaves to epidermal cells [61]

but its high concentration can cause damage to leaf surface [62]. Therefore, concentration of solution at fertigation and foliar application of Se should be chosen with care, based on recommendations.

5. Conclusion

The identification of effective Se dose and method is crucial for better understanding of Se uptake and accumulation in water stressed crop plants. From the results of experiments, it was concluded that Se fertigation at 7.35 μM and Se foliar treatment of 7.06 μM significantly mitigated the adverse effects of drought stress in wheat seedlings. The difference among Se supply methods (Se fertigation and Se foliar spray) suggests differences in their efficiency for Se uptake in plants. It was observed that Se foliar spray was more effective than Se fertigation in increasing DMSI and biomass (SFSI and RFSI) of seedlings. However, the plants exhibited maximum RLSI by Se fertigation, whereas a nonsignificant difference was recorded between Se application methods for PHSI.

Conflict of Interests

The authors declare that there is no conflict of interests regarding the publication of this paper.

Acknowledgment

The authors are grateful to the Higher Education Commission (HEC) of Pakistan for the financial support.

References

[1] M. M. Chaves and M. M. Oliveira, "Mechanisms underlying plant resilience to water deficits: prospects for water-saving

agriculture," *Journal of Experimental Botany*, vol. 55, no. 407, pp. 2365–2384, 2004.

[2] J. J. Zhang, G. P. Hao, Z. Y. Wu et al., "Nucleotide variation in ATHK1 region of *Arabidopsis thaliana* and its association study with drought tolerance," *African Journal of Biotechnology*, vol. 7, no. 3, pp. 224–233, 2008.

[3] F. Nawaz, R. Ahmad, E. A. Waraich, M. S. Naeem, and R. N. Shabbir, "Nutrient uptake, physiological responses, and yield attributes of wheat (*Triticum aestivum* l.) exposed to early and late drought stress," *Journal of Plant Nutrition*, vol. 35, no. 6, pp. 961–974, 2012.

[4] C. M. Ashraf and S. Abu-Shakra, "Wheat seed germination under low temperature and moisture stress," *Agron Journal*, vol. 70, pp. 135–139, 1978.

[5] X. Tian and Y. Lei, "Nitric oxide treatment alleviates drought stress in wheat seedlings," *Biologia Plantarum*, vol. 50, no. 4, pp. 775–778, 2006.

[6] F. Zhao, F. J. Lopez-Bellido, C. W. Gray, W. R. Whalley, L. J. Clark, and S. P. McGrath, "Effects of soil compaction and irrigation on the concentrations of selenium and arsenic in wheat grains," *Science of the Total Environment*, vol. 372, no. 2-3, pp. 433–439, 2007.

[7] B. Hanson, G. F. Garifullina, S. D. Lindblom et al., "Selenium accumulation protects Brassica juncea from invertebrate herbivory and fungal infection," *New Phytologist*, vol. 159, no. 2, pp. 461–469, 2003.

[8] M. Seppänen, M. Turakainen, and H. Hartikainen, "Selenium effects on oxidative stress in potato," *Plant Science*, vol. 165, no. 2, pp. 311–319, 2003.

[9] M. Kumar, A. J. Bijo, R. S. Baghel, C. R. K. Reddy, and B. Jha, "Selenium and spermine alleviate cadmium induced toxicity in the red seaweed *Gracilaria dura* by regulating antioxidants and DNA methylation," *Plant Physiology and Biochemistry*, vol. 51, pp. 129–138, 2012.

[10] X. Yao, J. Chu, and C. Ba, "Antioxidant responses of wheat seedlings to exogenous selenium supply under enhanced ultraviolet-B," *Biological Trace Element Research*, vol. 136, no. 1, pp. 96–105, 2010.

[11] X. Q. Yao, J. Z. Chu, and C. J. Ba, "Responses of wheat roots to exogenous selenium supply under enhanced ultraviolet-B," *Biological Trace Element Research*, vol. 137, no. 2, pp. 244–252, 2010.

[12] C. Wang, "Water-stress mitigation by selenium in Trifolium repens L.," *Journal of Plant Nutrition and Soil Science*, vol. 174, no. 2, pp. 276–282, 2011.

[13] M. Hasanuzzaman, M. A. Hossain, and M. Fujita, "Selenium-induced up-regulation of the antioxidant defense and methylglyoxal detoxification system reduces salinity-induced damage in rapeseed seedlings," *Biological Trace Element Research*, vol. 143, no. 3, pp. 1704–1721, 2011.

[14] J. Chu, X. Yao, and Z. Zhang, "Responses of wheat seedlings to exogenous selenium supply under cold stress," *Biological Trace Element Research*, vol. 136, no. 3, pp. 355–363, 2010.

[15] M. Djanaguiraman, P. V. V. Prasad, and M. Seppanen, "Selenium protects sorghum leaves from oxidative damage under high temperature stress by enhancing antioxidant defense system," *Plant Physiology and Biochemistry*, vol. 48, no. 12, pp. 999–1007, 2010.

[16] H. Hartikainen, T. Xue, and V. Piironen, "Selenium as an antioxidant and pro-oxidant in ryegrass," *Plant and Soil*, vol. 225, no. 1-2, pp. 193–200, 2000.

[17] S. Pukacka, E. Ratajczak, and E. Kalemba, "The protective role of selenium in recalcitrant *Acer saccharium* L. seeds subjected to desiccation," *Journal of Plant Physiology*, vol. 168, no. 3, pp. 220–225, 2011.

[18] V. V. Kuznetsov, V. P. Kholodova, V. I. V. Kuznetsov, and B. A. Yagodin, "Selenium regulates the water status of plants exposed to drought," *Doklady Biological Sciences*, vol. 390, pp. 266–268, 2003.

[19] F. Nawaz, M. Y. Ashraf, R. Ahmad, and E. A. Waraich, "Selenium (Se) seed priming induced growth and biochemical changes in wheat under water deficit conditions," *Biological Trace Element Research*, vol. 151, no. 2, pp. 284–293, 2013.

[20] X. Yao, J. Chu, and G. Wang, "Effects of selenium on wheat seedlings under drought stress," *Biological Trace Element Research*, vol. 130, no. 3, pp. 283–290, 2009.

[21] M. Hasanuzzaman and M. Fujita, "Selenium pretreatment upregulates the antioxidant defense and methylglyoxal detoxification system and confers enhanced tolerance to drought stress in rapeseed seedlings," *Biological Trace Element Research*, vol. 143, no. 3, pp. 1758–1776, 2011.

[22] S. Bittman, W. T. Buckley, K. Zaychuk, and E. A. P. Brown, "Seed coating for enhancing the level of selenium in crops," USA Patent No. 6,058, 649, 2000.

[23] L. D. Temmerman, N. Waegeneers, C. Thiry, G. D. Laing, F. Tack, and A. Ruttens, "Selenium content of Belgian cultivated soils and its uptake by field crops and vegetables," *Science of the Total Environment*, vol. 468–469, pp. 77–82, 2013.

[24] P. F. Boldrin, V. Faquin, S. J. Ramos, K. V. F. Boldrin, F. W. Ávila, and L. R. G. Guilherme, "Soil and foliar application of selenium in rice biofortification," *Journal of Food Composition and Analysis*, vol. 31, pp. 238–244, 2013.

[25] Y. Wang, X. Wang, and Y. Wong, "Generation of selenium-enriched rice with enhanced grain yield, selenium content and bioavailability through fertilisation with selenite," *Food Chemistry*, vol. 141, no. 3, pp. 2385–2393, 2013.

[26] A. D. C. Chilimba, S. D. Young, C. R. Black, M. C. Meacham, J. Lammel, and M. R. Broadley, "Agronomic biofortification of maize with selenium (Se) in Malawi," *Field Crops Research*, vol. 125, pp. 118–128, 2012.

[27] J. Yan, F. Wang, H. Qin et al., "Natural Variation in Grain Selenium Concentration of Wild Barley, *Hordeum spontaneum*, Populations from Israel," *Biological Trace Element Research*, vol. 142, no. 3, pp. 773–786, 2011.

[28] F. Yang, L. Chen, Q. Hu, and G. Pan, "Effect of the application of selenium on selenium content of soybean and its products," *Biological Trace Element Research*, vol. 93, no. 1–3, pp. 249–256, 2003.

[29] E. Kápolna, K. H. Laursen, S. Husted, and E. H. Larsen, "Biofortification and isotopic labelling of Se metabolites in onions and carrots following foliar application of Se and [77]Se," *Food Chemistry*, vol. 133, no. 3, pp. 650–657, 2012.

[30] E. Kápolna, P. R. Hillestrøm, K. H. Laursen, S. Husted, and E. H. Larsen, "Effect of foliar application of selenium on its uptake and speciation in carrot," *Food Chemistry*, vol. 115, no. 4, pp. 1357–1363, 2009.

[31] K. R. Hladun, D. R. Parker, K. D. Tran, and J. T. Trumble, "Effects of selenium accumulation on phytotoxicity, herbivory, and pollination ecology in radish (*Raphanus sativus* L.)," *Environmental Pollution*, vol. 172, pp. 70–75, 2013.

[32] P. Põldma, T. Tõnutare, A. Viitak, A. Luik, and U. Moor, "Effect of selenium treatment on mineral nutrition, bulb size, and

antioxidant properties of garlic (*Allium sativum* L.)," *Journal of Agricultural and Food Chemistry*, vol. 59, no. 10, pp. 5498–5503, 2011.

[33] D. Curtin, R. Hanson, T. N. Lindley, and R. C. Butler, "Selenium concentration in wheat (*Triticum aestivum*) grain as influenced by method, rate, and timing of sodium selenate application," *New Zealand Journal of Crop and Horticultural Science*, vol. 34, no. 4, pp. 329–339, 2006.

[34] N. Terry, A. M. Zayed, M. P. de Souza, and A. S. Tarun, "Selenium in higher plants," *Annual Review of Plant Physiology and Plant Molecular Biology*, vol. 51, pp. 401–432, 2000.

[35] R. L. Mikkelsen, A. L. Page, and F.T. Bingham, "Factors affecting selenium accumulation by agricultural crops," in *Selenium in Agriculture and the Environment*, L. W. Jacobs, Ed., SSSA Special Publication Number 23, pp. 65–94, Soil Science Society of America, Madison, Wis, USA, 1989.

[36] A. Kabata-Pendias and H. Pendias, *Trace Elements in Soils and Plants*, CRC Press, Boca Raton, Fla, USA, 2nd edition, 1992.

[37] C. Kahakachchi, H. T. Boakye, P. C. Uden, and J. F. Tyson, "Chromatographic speciation of anionic and neutral selenium compounds in Se-accumulating *Brassica juncea* (Indian mustard) and in selenized yeast," *Journal of Chromatography A*, vol. 1054, no. 1-2, pp. 303–312, 2004.

[38] A. Zayed, C. M. Lytle, and N. Terry, "Accumulation and volatilization of different chemical species of selenium by plants," *Planta*, vol. 206, no. 2, pp. 284–292, 1998.

[39] M. Y. Ashraf, F. Hussain, J. Akhter, A. Gul, M. Ross, and G. Ebert, "Effect of different sources and rates of nitrogen and supra optimal level of potassium fertilization on growth, yield and nutrient uptake by sugarcane grown under saline conditions," *Pakistan Journal of Botany*, vol. 40, no. 4, pp. 1521–1531, 2008.

[40] F. Nawaz, *Wheat response to exogenous selenium supply under drought stress [Ph.D. dissertation]*, University of Agriculture, Faisalabad, Pakistan, 2014.

[41] A. Latif and M. M. Iqbal, "Fertigation techniques," in *Proceedings of the Workshop on Technologies for Sustainable Agriculture (NIAB '01)*, pp. 155–159, Faisalabad, Pakistan, 2001.

[42] M. Djanaguiraman, D. D. Devi, A. K. Shanker, J. A. Sheeba, and U. Bangarusamy, "Selenium—an antioxidative protectant in soybean during senescence," *Plant and Soil*, vol. 272, no. 1-2, pp. 77–86, 2005.

[43] K. L. Larson, "Drought injury and resistance of crop plants," in *Physiological Aspects of Dry Land Farming*, S. U. Gupta, Ed., pp. 147–162, Oxford & IBH Publishing, New Delhi, India, 1992.

[44] H. Hartikainen, "Biogeochemistry of selenium and its impact on food chain quality and human health," *Journal of Trace Elements in Medicine and Biology*, vol. 18, no. 4, pp. 309–318, 2005.

[45] P. Smrkolj, M. Osvald, J. Osvald, and V. Stibilj, "Selenium uptake and species distribution in selenium-enriched bean (*Phaseolus vulgaris* L.) seeds obtained by two different cultivations," *European Food Research and Technology*, vol. 225, no. 2, pp. 233–237, 2007.

[46] B. Z. Bai, J. Z. Jin, S. Bai, and L. P. Huang, "Improvement of TTC method determining root activity in corn," *Maize Science*, vol. 2, pp. 44–47, 1994 (Chinese).

[47] C. L. Carlson, D. I. Kaplan, and D. C. Adriano, "Effects of selenium on germination and radicle elongation of selected agronomic species," *Environmental and Experimental Botany*, vol. 29, no. 4, pp. 493–498, 1989.

[48] S. A. Valadabadi, A. H. Shiranirad, and H. A. Farahani, "Eco-physiological influences of zeolite and selenium on water deficit stress tolerance in different rapeseed cultivars," *Journal of Ecology and the Natural Environment*, vol. 2, pp. 154–159, 2010.

[49] G. Okçu, M. D. Kaya, and M. Atak, "Effects of salt and drought stresses on germination and seedling growth of pea (Pisum sativum L.)," *Turkish Journal of Agriculture and Forestry*, vol. 29, no. 4, pp. 237–242, 2005.

[50] M. Yağmur and D. Kaydan, "Alleviation of osmotic stress of water and salt in germination and seedling growth of triticale with seed priming treatments," *African Journal of Biotechnology*, vol. 7, no. 13, pp. 2156–2162, 2008.

[51] J. L. Hopper and D. R. Parker, "Plant availability of selenite and selenate as influenced by the competing ions phosphate and sulfate," *Plant and Soil*, vol. 210, no. 2, pp. 199–207, 1999.

[52] P. Ximénez-Embún, I. Alonso, Y. Madrid-Albarran, and C. Camara, "Establishment of selenium uptake and species distribution in lupine, Indian mustard, and sunflower plants," *Journal of Agricultural and Food Chemistry*, vol. 52, pp. 832–838, 2004.

[53] M. Łabanowska, M. Filek, J. Kościelniak, M. Kurdziel, E. Kuliś, and H. Hartikainen, "The effects of short-term selenium stress on Polish and Finnish wheat seedlings-EPR, enzymatic and fluorescence studies," *Journal of Plant Physiology*, vol. 169, no. 3, pp. 275–284, 2012.

[54] M. Vítová, K. Bišová, M. Hlavová, V. Zachleder, M. Rucki, and M. Cížková, "Glutathione peroxidase activity in the selenium-treated alga *Scenedesmus quadricauda*," *Aquatic Toxicology*, vol. 102, no. 1-2, pp. 87–94, 2011.

[55] Y. Wang, X. Wang, and Y. Wong, "Proteomics analysis reveals multiple regulatory mechanisms in response to selenium in rice," *Journal of Proteomics*, vol. 75, no. 6, pp. 1849–1866, 2012.

[56] T. Xue, H. Hartikainen, and V. Piironen, "Antioxidative and growth-promoting effect of selenium on senescing lettuce," *Plant and Soil*, vol. 237, no. 1, pp. 55–61, 2001.

[57] Q. Hu, G. Pan, and J. Zhu, "Effect of selenium on green tea preservation quality and amino acid composition of tea protein," *Journal of Horticultural Science and Biotechnology*, vol. 76, no. 3, pp. 344–346, 2001.

[58] M. Turakainen, H. Hartikainen, and M. M. Seppänen, "Effects of selenium treatments on potato (*Solanum tuberosum* L.) growth and concentrations of soluble sugars and starch," *Journal of Agricultural and Food Chemistry*, vol. 52, no. 17, pp. 5378–5382, 2004.

[59] M. Germ, "The response of two potato cultivars on combined effects of selenium and drought," *Acta Agriculturae Slovenica*, vol. 91, no. 1, pp. 121–137, 2008.

[60] G. Habibi, "Effect of drought stress and selenium spraying on photosynthesis and antioxidant activity of spring barley," *Acta Agriculturae Slovenica*, vol. 101, no. 1, pp. 31–39, 2013.

[61] P. Wójcik, "Uptake of mineral nutrients from foliar fertilization," *Journal of Fruit and Ornamental Plant Research*, vol. 12, pp. 201–218, 2004.

[62] H. Marschner, *Mineral Nutrition of Higher Plants*, Academic Press, London, UK, 1995.

17

Quantum Calculation for Musk Molecules Infrared Spectra towards the Understanding of Odor

Elaine Rose Maia,[1] **Daniela Regina Bazuchi Magalhães,**[1] **Dan A. Lerner,**[2] **Dorothée Berthomieu,**[2] **and Jean-Marie Bernassau**[3]

[1] *Laboratório de Estudos Estruturais Moleculares (LEEM), Instituto de Química, Universidade de Brasília (UnB), CP 4478, 70904-970 Brasília, DF, Brazil*
[2] *Institut Charles Gerhardt, UMR 5253 CNRS-UM2-ENSCM-UM1, Matériaux Avancés pour la Catalyse et la Santé, ENSCM, 8 rue de l'Ecole Normale, 34296 Montpellier Cedex 5, France*
[3] *Sanofi Montpellier, 264 rue du Professeur Blayac, 34080 Montpellier, France*

Correspondence should be addressed to Daniela Regina Bazuchi Magalhães; danielabazuchi@gmail.com

Academic Editor: Devis Di Tommaso

It is not clear so far how humans can recognize odor. One of the theories regarding structure-odor relationship is vibrational theory, which claims that odors can be recognized by their modes of vibration. In this sense, this paper brings a novel comparison made between musky and nonmusky molecules, as to check the existence of correlation between their modes on the infrared spectra and odor. For this purpose, sixteen musky odorants were chosen, as well as seven other molecules that are structurally similar to them, but with no musk odor. All of them were submitted to solid theoretical methodology (using molecular mechanics/molecular dynamics and Neglect of Diatomic Differential Overlap Austin Model 1 methods to optimize geometries) as to achieve density functional theory spectra information, with both Gradient Corrected Functional Perdew-Wang generalized-gradient approximation (GGA/PW91) and hybrid Becke, three-parameter, Lee-Yang-Parr (B3LYP) functional. For a proper analysis over spectral data, a mathematical method was designed, generating weighted averages for theoretical frequencies and computing deviations from these averages. It was then devised that musky odorants satisfied demands of the vibrational theory, while nonmusk compounds belonging either to nitro group or to acyclic group failed to fulfill the same criteria.

1. Introduction

1.1. Theories regarding Odor Prediction and Recognition. Musks are responsible for bringing sensuality and warmth to a fragrance [1], as well as adding their excellent fixative properties [2, 3]. Presently, natural products extraction involves high costs and generates small amounts of raw material, so synthetic odorants are greatly demanded. To properly design a new odorant, previous information on how odors are read is necessary. There are major recent advances on neurobiology, biophysics, biochemical fields [4–7], though it is still unclear how humans recognize odor. At this scenario, quantitative structure-activity relationship models (QSAR models) are valuable for the proposal of new potential musky smelling molecules [8–12]. Such approach allows for the reduction in the number of costly and unnecessary

chemical syntheses. Theories regarding odor recognition at the biological level have also been discussed in the literature. One of them, known as the odotope theory [13–15], states that noncovalent bonds (i.e., weak repulsive and attractive interactions) between the molecule and the olfactory receptor (OR) proteins (responsible for decoding the molecule and starting transduction) would elicit a unique response in the cerebral cortex.

A second theory regarding structure-odor relationship is the vibrational theory, which relates the molecule's vibrational modes to its odor. It was firstly proposed by Dyson in 1938, followed by Wright, in 1977, and refined by several authors [16]. It claims that ORs would be able to recognize odorant vibrational modes. In 1996 [17], Turin suggested that such reading would occur, at the protein-odorant level, through a mechanism close to inelastic electron tunneling spectroscopy (IETS), an inelastic nonoptical vibrational spectroscopy [18]. After building a specific algorithm to mimic vibrational modes at protein level, the author could state that *musks*—for instance—will present four bands near 700, 1000, 1500 (or 1750 cm^{-1} for nitromusks), and 2200 cm^{-1}, and that each band will have 400 cm^{-1} width [19]. There is much to be understood concerning odor recognition, supported on vibrational theory's ideas, as indicated by the variety of papers dealing with the subject [20–24].

1.2. Paper's Aims. In order to explore vibrational theory, musks and non-musks (see Figure 1) have been equally modeled so to come to their theoretical IR spectra. Firstly molecules have been computationally analyzed by molecular mechanics and molecular dynamics (MM/MD), in order to ascertain the conformational space available to them. After, semiempirical calculations employing Neglect of Diatomic Differential Overlap Austin Model 1 (NDDO AM1), Hamiltonian took place for geometry corrections. Then final calculations for energy optimization and IR spectra collection were set with density functional theory (DFT), using both Gradient Corrected Functional Perdew-Wang generalized-gradient approximation (GGA/PW91) and hybrid Becke, three-parameter, Lee-Yang-Parr (B3LYP) functionals, with the double numerical basis set with polarization (DNP) for GGA/PW91 and 6-311+G(d,p) for B3LYP. Methodology adopted is up-to-date with the literature and was mainly based on Lerner and coworkers paper [26]. For a complete discussion on computed harmonic vibrational frequencies and their scaling factors and accuracy, see data obtained by Scott and Radom [27].

GGA/PW91 and B3LYP theoretical spectra were compared to the experimental data (when available in free data banks as NIST and SDBS) and then were evaluated concerning the expression of the four characteristic musk bands (centered at 700, 1000, 1500 or 1750, and 2200 cm^{-1}). Now, these four bands were properly indicated using a specific algorithm built to reproduce IETS at biological level; however, the vibrational modes primarily responsible for such bands were not nominated. To clarify this question, a comparison was carried between IR theoretical data and IET data published previously, since both spectra show the same vibrational modes at the same frequencies, and it is much simpler to estimate theoretical IR spectra. Aside from the equivalence on estimating vibrational modes, there are no other comparisons that could fit in, once mode intensities are differently calculated in one technique and another, generating bands that can be either more prominent or even absent when comparing spectra. Besides, using IR spectra to predict odor or olfactory receptor affinity has failed repeatedly [28]. Finally, vibrational modes responsible for "musk bands" in the spectra are not to be taken as solely responsible for musk odor. It is well known that there are several molecular descriptors involved at major biological phenomena, such as carrying the odorant from the environment to an OR and docking odorant-OR, and they are equally relevant when considering the vibrationally assisted odor recognition mechanism.

The sixteen musks **1–16** are already in use in perfumes, so to have their activity is certified. The seven molecules **17–23** are structurally similar to the previous molecules but do not possess any musky odor; besides, nonmusks **17** and **18** are *woody*. Absence of odor is not to be taken as a different odor character but (i) as a molecule's inability to bind its OR, due to particularities of its structure, or (ii) as the lack of specific vibrational modes in their minimum quantity, precluding proper energy decrease on electron scattering and its subsequent tunneling [13, 14, 16]. So, regarding adopted criteria relying on the choice of such molecules, it addresses the main hypothesis here being tested: only musk odorants are supposed to express vibrational modes under mentioned ranges, and no other molecule is supposed to present the same vibrational pattern. All compounds have their IUPAC name and Chemical Abstracts Service (CAS) registry number listed in Table 1.

2. Methods and Methodology

2.1. Computational Methods. Calculation programs used were those available in the Materials Studio Package Program [29] and Gaussian 09 [30] for B3LYP calculations. Molecules had their conformational space studied undervacuum environment through MM/DM calculations, using the CFF99/COMPASS [31] force field implemented in the Discover Simulation program [32, 33]. Main parameters chosen for MM routine were Conjugate Gradient method [34] with the Polak-Ribiere algorithm [35], running until it achieved convergence at 1.0×10^{-4} kcal·mol^{-1} · Å^{-1}. MD trajectories were carried out at 650 K, during 1.0 ns, with a time step of 1.0 fs (therefore, collecting 1.000 frames), in the canonical ensemble. After previous scanning in which temperature and trajectory time were changed, parameters chosen showed to be effective on expressing molecules flexibility, allowing the collection of a representative subset of initial conformations.

Selected conformers from MM/DM procedure were then minimized with Vamp program [36], using NDDO AM1 [37–39], as a previous optimization step towards DFT calculation. Spin multiplicity was automatically determined, and a convergence of 0.1 kcal·mol^{-1}·Å^{-1} for the heat of formation value was set. Self-consistent field (SCF) cycles had a convergence tolerance of 5.0×10^{-7} eV·atom^{-1}.

Final geometries were minimized using the DMol^3 program [40–42] for DFT calculations. Basis set chosen was DNP [43], with GGA/PW91 [44] as exchange functional. Total energy convergence of 1.0×10^{-5} Ha was required, with a 3.7 Å global cutoff. Final cutoff of the SCF cycles was 5.0×10^{-7} eV·atom^{-1}, and the orbital cutoff quality was kept equal to 0.1 eV·atom^{-1}. The same starting structure used in GGA/PW91 calculation was employed for hybrid functional B3LYP [45] calculation with 6-311+G(d,p) basis set, now using Gaussian 09. Molden's processing program [46] was applied, along with Xming emulator [47], to achieve IR continuum spectra.

2.2. Development of a Frequency-Weighted Average and Other Useful Tools. A mathematical formula had to be developed

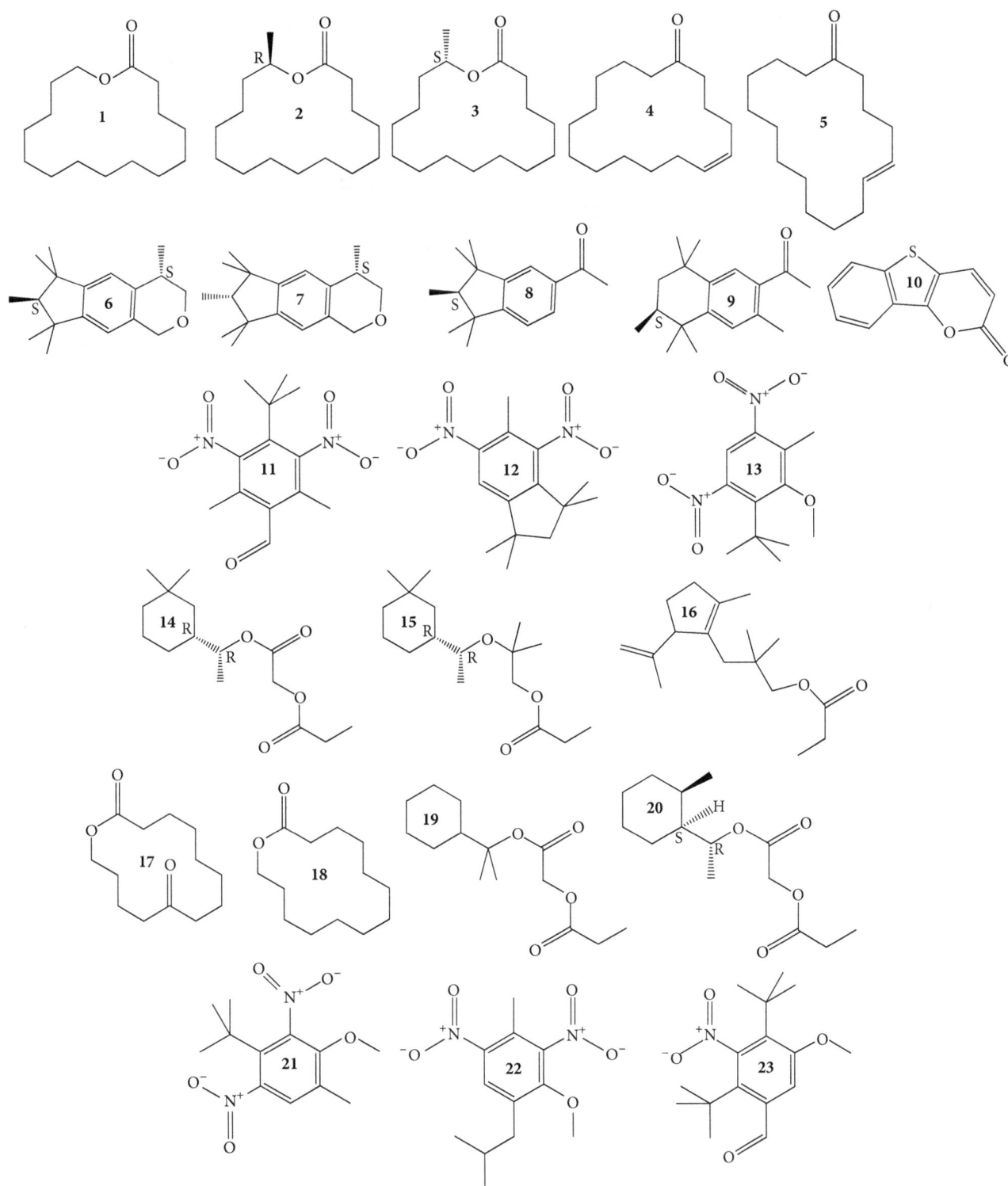

FIGURE 1: Odorants under study: macrocyclic musks (MCM): pentadecanolide (**1**), (*R*)-exaltolide (**2**), (*S*)-exaltolide (**3**), *cis*-globanone (**4**), and *trans*-globanone (**5**); polycyclic musks (PCM): (4*S*,7*S*)-galaxolide (**6**), (4*S*,7*R*)-galaxolide (**7**), indane (**8**), tetralin (**9**), and tonkene (**10**); nitromusks (NM): musk ketone (**11**), musk moskene (**12**), and musk ambrette (**13**); acyclic musks (ACM): romandolide (**14**), helvetolide (**15**), and cyclomusk (**16**); nonmusks: macrocyclic analogs (MCa) (**17**) and (**18**), acyclic analogs (Aa) (**19**) and (**20**), and nitro analogs (Na) (**21**), (**22**), and (**23**).

TABLE 1: Molecules studied.

Number	IUPAC name	CAS number
1	Oxacyclohexadecan-2-one	106-02-5
2	16(R)-Methyl-oxacyclohexadecan-2-one	69297-56-9
3	16(S)-Methyl-oxacyclohexadecan-2-one	129214-00-2
4	(5Z)-Cyclohexadec-5-en-1-one	37609-25-9
5	(5E)-Cyclohexadec-5-en-1-one	
6	(4S,7S)-Cyclopenta[g]-2-benzopyran, 1,3,4,6,7,8-hexahydro-4,6,6,7,8,8-hexamethyl	172339-62-7
7	(4S,7R)-Cyclopenta[g]-2-benzopyran, 1,3,4,6,7,8-hexahydro-4,6,6,7,8,8-hexamethyl	252332-95-9
8	1-(2,3-Dihydro-1,1,2,3,3-pentamethyl-1H-inden-5-yl)-ethanone	4755-83-3
9	1-[(6S)-5,6,7,8-Tetrahydro-3,5,5,6,8,8-hexamethyl-2-naphthalenyl]-ethanone	85549-79-7
10	2H-[1]-Benzothieno[3,2-b]pyran-2-one	5732-22-9
11	4-(1,1-Dimethylethyl)-2,6-dimethyl-3,5-dinitro-benzaldehyde	99758-50-6
12	4-(1,1-Dimethylethyl)-2,6-dimethyl-3,5-dinitro-benzaldehyde	116-66-5
13		none
14	2-(1-Oxopropoxy)-, 1-(3,3-dimethylcyclohexyl)ethyl ester	236391-76-7
15	1-Propanol, 2-[1-(3,3-dimethylcyclohexyl)ethoxy]-2-methyl-, 1-propanoate	141773-73-1
16	1-Cyclopentene-1-propanol, β,β,2-trimethyl-5-(1-methylethenyl)-, 1-propanoate	84012-64-6
17	Oxacyclotetradecane-2,10-dione	38223-27-7
18	Cyclotetradecanone	3603-99-4
19	2-(1-Oxopropoxy)-, 1-cyclohexyl-1-methylethyl ester	610770-00-8
20		none
21	2-Methoxy-1-methyl-4-(2-methyl-2-propanyl)-3,5-dinitrobenzene	99758-79-9
22	2-Methoxy-4-methyl-1-(2-methylpropyl)-3,5-dinitro-benzene	99758-78-8
23	2,4-Bis(1,1-dimethylethyl)-5-methoxy-3-nitro-benzaldehyde	107342-73-4

in this paper, in order to consider each and every mode belonging to the same 400 cm-1 range, as well as their intensities. Therefore, the use of weighted averages over theoretical data was chosen; its numerical result was called *central frequency* (cf). With this central frequency, it became possible to consider all theoretical peaks occurring in the vicinity of the diagnostic frequencies, instead of just affirming the existence of the expected band only by naming the most intense theoretical peak. Central frequencies were estimated as follows:

$$\text{Central frequency} = \frac{\sum_{i=1}^{n} \nu_i \times \text{int}_i}{\sum_{i=1}^{n} \text{int}_i}. \quad (1)$$

This mathematical approach (which simply generates a frequency-weighted average) was applied to theoretical frequencies found at $\pm 200 \text{ cm}^{-1}$ around 700, 1000, and 1500 (or 1750) cm^{-1} and resulted in central frequencies for compounds **1–23**, which were computed in Table 4 (for GGA/PW91 data) and in Table 5 (for B3LYP data). The mentioned range has been chosen so as to enable each band to complete its 400 cm^{-1} width, as established in the vibrational theory.

3. Results and Discussion

Data will be presented and discussed in three successive sections. In the first section the collection of theoretical IR spectra will be presented. In the second section they will be compared to experimental IR spectra to discuss their precision and accuracy. The third section will deal with the rationalization between musk spectra and musk odor activity. In order to establish such relation, best theoretical IR spectra will be chosen from this paper's second section.

3.1. Part 1: Computation of IR Spectra. Concerning *ab initio* methods, convolved GGA/PW91 spectra are shown in Figure 2, while B3LYP spectra are gathered in Figure 3. For GGA/PW91 spectra, carbon-oxygen single bond stretches occur on frequencies between 1018 cm^{-1} (for nitro analog **21**) and 1220 cm^{-1} (for (4S,7R)-galaxolide **7**); carbon-nitrogen single bond stretches occur under the frequency of 1200 cm^{-1} for nitro compounds; carbon-sulfur single bond stretches are observed at 1054 cm^{-1} for tonkene **10**. Carbon-carbon double bonds stretches happened at 1687 cm^{-1} for *cis*-globanone **4** and at 1706 cm^{-1} for *trans*-globanone **5**, while all other compounds expressing aromatic C=C bonds had their stretching modes between 1543 cm^{-1} (as in musk

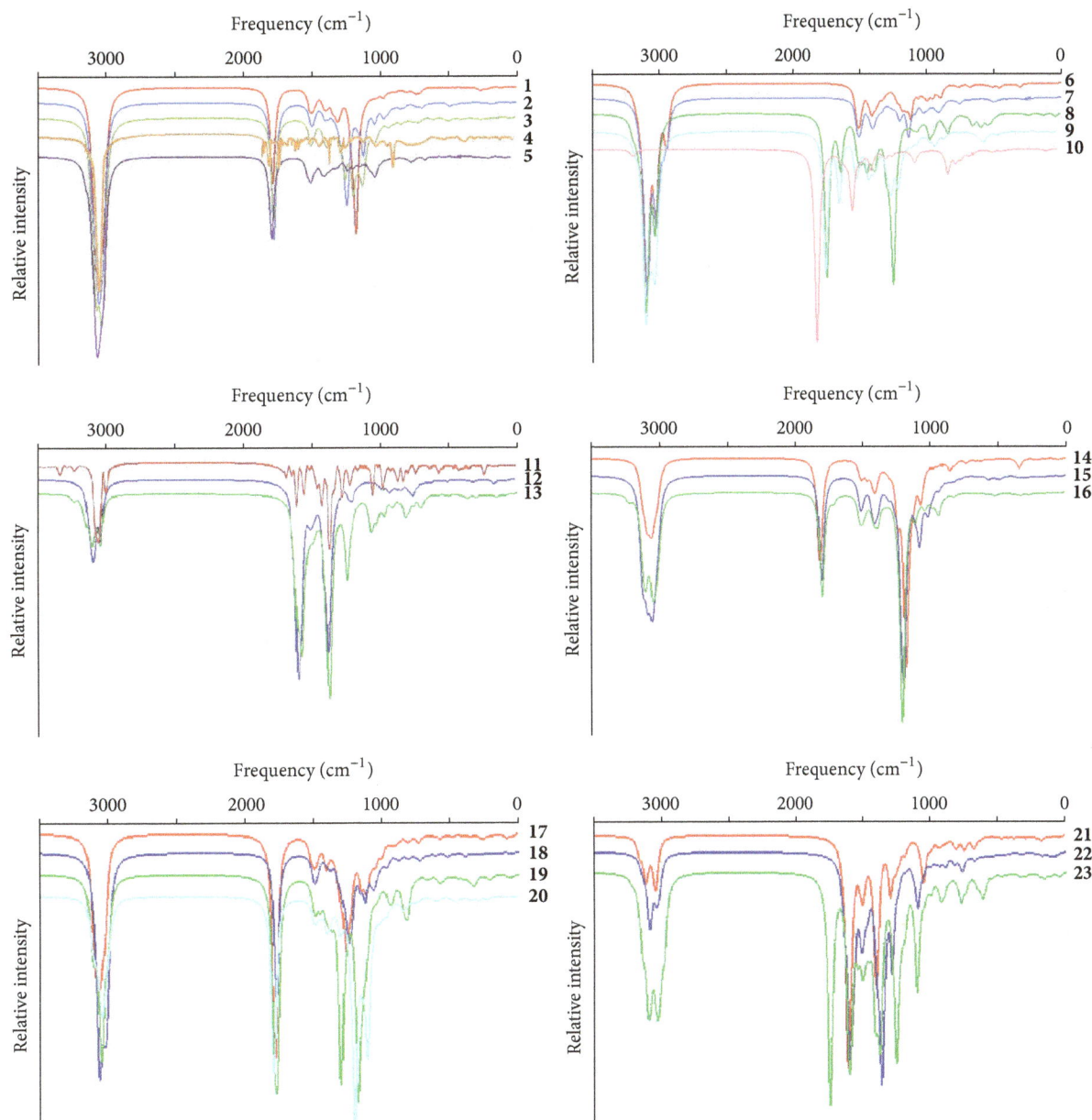

FIGURE 2: Spectra obtained with GGA/PW91 for MCM **1–5**, PCM **6–10**, NM **11–13**, ACM **14–16**, and nonmusks **17–23**.

ketone **11**) and $1618\,cm^{-1}$ (for ($4S,7R$)-galaxolide **7**). Nitro groups have presented their symmetric stretching modes near 1310–$1340\,cm^{-1}$ and their nonsymmetric stretching modes at 1520–$1560\,cm^{-1}$. For aldehydes, ketones, ethers, and esters, the carbon-oxygen double bond stretching represents an important mark for confirming spectra: here, the lowest value for such mode is found for tetralin **9**, at $1680\,cm^{-1}$, while the highest value, of $1777\,cm^{-1}$, belongs to acyclic analog **20**. Carbon-hydrogen bending modes start at very low frequencies and end at $1493\,cm^{-1}$ (for helvetolide **14**); simultaneously, the band of carbon-hydrogen stretching modes starts at $2898\,cm^{-1}$ (for ($4S,7R$)-galaxolide **7**) and finishes at $3198\,cm^{-1}$ (for nitro analog **22**).

For B3LYP spectra, the lowest frequency found for the carbon-oxygen single bond stretching was $1006\,cm^{-1}$ (for romandolide **15**), while its highest frequency was $1403\,cm^{-1}$ (for tonkene **10**); the frequency of the carbon-oxygen double bond stretching was found between $1738\,cm^{-1}$ (for musk **8**) and $1810\,cm^{-1}$ (for musk **14**). The carbon-carbon double bond stretching occurred on the frequency of 1707 and $1718\,cm^{-1}$ for compounds **4** and **5** and between 1699 and $1712\,cm^{-1}$ for compound **16**; finally, the carbon-carbon double bond stretching for aromatic compounds occurred between 1531 (for non-musk **21**) and $1650\,cm^{-1}$ (for musk **7**). The stretching for nitrogen-oxygen in nitro groups was found at 1250–$1417\,cm^{-1}$ for symmetric modes and at 1531–$1618\,cm^{-1}$

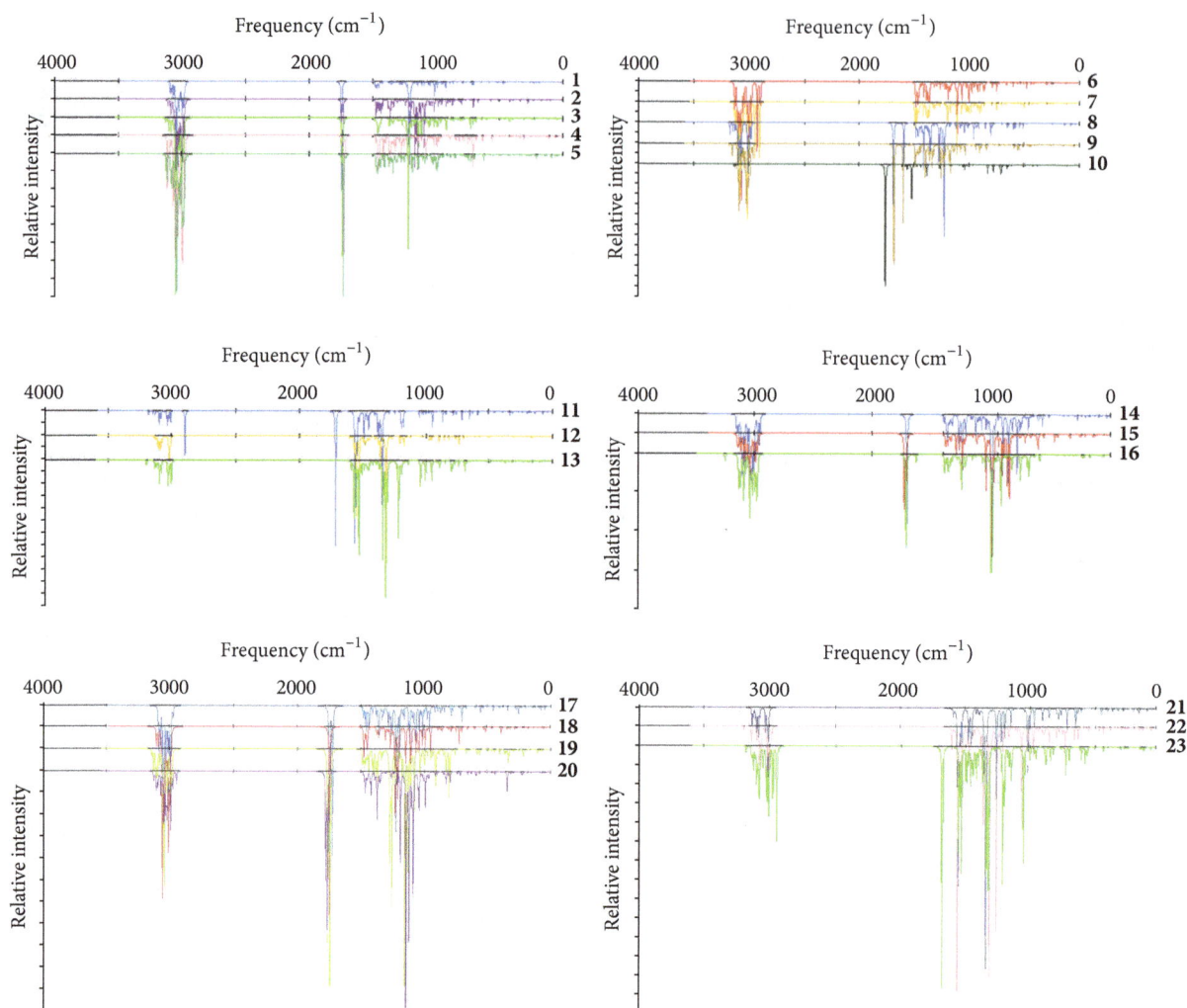

FIGURE 3: Spectra obtained at B3LYP level for MCM **1–5**, PCM **6–10**, NM **11–13**, ACM **14–16**, and nonmusks **17–23**.

for asymmetric modes. Finally, carbon-hydrogen bending modes began at very low frequencies and went up to $1540 \, cm^{-1}$, at most (for musk **13**), while carbon-hydrogen stretching occurred between $2953 \, cm^{-1}$ (for compound **6**) and $3229 \, cm^{-1}$ (for nitromusk **13**).

3.2. Part 2: Comparing GGA/PW91 and B3LYP Spectra to Experimental Data. For an overview over GGA/PW91, B3LYP, and experimental [47] IR data, Figure 4 and Table 2 were built, in which vibrational modes for the most important functional groups are organized in accordance with theoretical method adopted. Each band expresses the range in which labeled vibrational mode occurs, considering the maximum and the minimum found for all compounds. Thus, it becomes possible to state that B3LYP and GGA/PW91 results for C–H stretching modes start at 90 and $140 \, cm^{-1}$, respectively, values that are higher than the ones provided by experimental data; simultaneously, the same band ends at $+50 \, cm^{-1}$ for GGA/PW91 and at $+90 \, cm^{-1}$ for B3LYP, in comparison to the experimental data. C–H bending modes finish at $+30 \, cm^{-1}$ for GGA/PW91 and at $+80 \, cm^{-1}$ for B3LYP when compared

to experimental values. Aromatic C=C stretches occurred in an expected range for GGA/PW91, with a $+20 \, cm^{-1}$ shift for alkene compounds; B3LYP showed a continuous band for alkene and aromatic C=C stretches, which started at $+60 \, cm^{-1}$ and ended at $+30 \, cm^{-1}$ over experimental data. C–O stretching modes are found within an expected range for both GGA/PW91 and B3LYP methods; for tonkene **10**, GGA/PW91 predicted C–O mode at $1381 \, cm^{-1}$, while the same mode is found at $1403 \, cm^{-1}$ in B3LYP data. For C=O stretching modes, GGA/PW91 showed a $+20 \, cm^{-1}$ increment over lower and higher values found for experimental data, and B3LYP shifted over experimental range in $+60 \, cm^{-1}$. Finally, nitro N=O stretching modes have had both symmetric and asymmetric modes accurately and precisely estimated for GGA/PW91; for B3LYP, symmetric nitro stretching showed a broader band than that of GGA/PW91 and of experimental data, starting at a $+30 \, cm^{-1}$ shift and finishing at a $-10 \, cm^{-1}$ shift, while their asymmetric modes were correctly predicted.

Therefore, while GGA/PW91 final results were all fitted into experimental reports, considering the allowed $\pm 30 \, cm^{-1}$

TABLE 2: Collection of observed versus calculated IR mode frequencies for compounds **1–23**.

Vibrational mode[a]	Observed wavenumber (cm^{-1}) [25]	Calculated wavenumber (cm^{-1})	
		GGA/PW91	B3LYP
ν C^{b}–H	3150–2850	3198–2898	3229–2953
ν C–H (aldehyde)	2900–2800	2954–2892	2988–2939
δ C–H	up to 1465	up to 1493	up to 1540
ν C=C (alkene)	1680–1600	1706–1687	1718–1707
ν C=C (aromatic)	1600–1475	1618–1543	1650–1531
ν C=O (aldehyde)	1740–1720	1715–1693	1772–1749
ν C=O (ketone)	1725–1705	1733–1680	1784–1738
ν C=O (ester)	1750–1730	1777–1742	1810–1768
δ C–O (alcohol, ether, and ester)	1300–1000	1381, 1200–1018	1403, 1312–1006
ν^{as} nitro (R–NO$_2$)	1660–1500	1560–1520	1618–1531
ν^{s} nitro (R–NO$_2$)	1390–1260	1340–1310	1417–1250

[a]ν: stretching; ν^{as}: asymmetric stretching; ν^{s}: symmetric stretching; δ: bending.
[b]ν C(sp^2)–H, C(sp^3)–H, and C(aromatic)–H.

shifts, B3LYP results overestimated C–H stretching and bending modes, alkene C=C stretching modes, and aldehyde and ketone C=O stretching modes.

To quantify collected results, experimental IR spectrum given for each particular molecule was compared to its theoretical spectra, as shown in Table 3. In the literature—public and personal IR database—only spectra for compounds **1–5**, **8**, **10**, and **12** were found. The vibrational modes chosen as internal patterns of comparison were as follows, when occurring: (i) the two most intense peaks regarding the C–O stretching mode; (ii) the most intense peak for the C=O stretching mode; (iii) the most intense peaks for the symmetric and asymmetric C–H stretching modes; (iv) the highest wavenumber found for the C–H bending modes; (v) the highest intensity for the C=C stretching mode; and (vi) the most intense peaks for the symmetric and asymmetric N=O stretching modes. They are altogether able to express the *accuracy* relying on theoretical data, once all functional groups have had their modes listed within chosen criteria, and the whole spectrum has been covered.

Extra information comes on the so-called DIFF columns in Table 3. Consider

$$\text{DIFF}_i = \text{ExpWavenumber}_i - \text{TheoWavenumber}_i, \quad (2)$$

where the difference established between experimental and theoretical values was plotted (for both GGA/PW91 and B3LYP results), as shown in (2). In such equation, for a certain mode i, the difference (DIFF) is obtained through the subtraction of its theoretical value (TheoWavenumber) from its experimental value (ExpWavenumber). The average of the collected differences on Table 3 was estimated as the values of absolute difference were divided by the number of occurring modes, disregarding all C–H stretch modes, as follows:

$$\text{Deviation 1} = \frac{\sum_{i=1}^{n} |\text{DIFF}_i|}{n}. \quad (3)$$

This choice was made due to the fact that all C–H stretches were overestimated in more than 100 cm^{-1} for both

theoretical methods, as discussed before. This last mode, therefore, had its deviation separately estimated as follows:

$$\text{Deviation 2} = \frac{\sum_{i=1}^{n} |\text{DIFF}_i|}{n} \quad (4)$$

in which the absolute DIFF$_i$ is collected, is summed for all C–H stretching modes within the same theoretical method, and is divided by the number of occurring modes.

So finally GGA/PW91 IR results are found to be more accurate than B3LYP results, with calculated deviations that are equal to 13 cm^{-1} and 43 cm^{-1}, respectively. Nevertheless, they correspond to a 3.25% (DFT) and a 10.75% (B3LYP) shift over the 4000–0 cm^{-1} that represents the spectrum, which ensures that both methods are powerfully predictive. Simultaneously, both methods display great shift over C–H stretching modes, corresponding to an increase of 124 cm^{-1} and 145 cm^{-1}, respectively. Considering that one of this paper's aims is to compare theoretical IR data with postulates of the vibrational theory on musky odor, the last mentioned modes are irrelevant, as they are not supposed to be responsible for muskiness.

3.3. Part 3: IR Spectra and Musk Activity. Herein, both GGA/PW91 and B3LYP data will be taken into account so as to compare theoretical results with the bands described in the vibrational theory (700, 1000, 1500 or 1750, and 2200 cm^{-1}), which would assign, as supposed by the previous theory, a musky odor to these molecules.

In Tables 5 and 6 the information was gathered so as to dispose musks **1** to **23** in lines and their central frequencies (at 700, 1000 and 1500 or 1750 cm^{-1}) in columns. Besides, the compounds were gathered in groups, in accordance with their structural similarity (see first column on the left): MCM for macrocyclic musks, PCM for polycyclic musks, NM for nitromusks, and ACM for acyclic musks. Nonmusks are called MCa for macrocyclic analogs, Aa for acyclic analogs, and Na for nitro analogs. As no vibrational mode has occurred at 2200 ± 200 cm^{-1}, there is no such column.

TABLE 3: Most important wavenumbers for compounds **1**-**5**, **8**, **10**, and **12**, according to DFT, B3LYP, and experimental data.

	ν C-0	ν C-0	DIFF	DIFF	ν C=0	DIFF	ν C-H	ν C-H	DIFF	DIFF	δ C-H	DIFF	Highest ν C=C	DIFF	νˢ N=O	DIFF	νᵃˢ N=O	DIFF
1																		
GGA/PW91	1207	1215	28	33	1748	-9	3017	3034	-160	-108	1479	-18						
B3LYP	1181	1210	54	38	1786	-47	3033	3069	-176	-143	1517	-56						
Exp.	1235	1248			1739		2857	2926			1461							
2																		
GGA/PW91	1099	1231	10	7	1739	-7	2999	3027	-153	-94	1474	-14						
B3LYP	1124	1240	-15	-2	1778	-46	3012	3058	-166	-125	1512	-52						
Experimental data	1109	1238			1732		2846	2933			1460							
3																		
GGA/PW91	1121	1220	-12	18	1738	-6	2991	3037	-145	-104	1481	-21						
B3LYP	1194	1254	-85	-16	1774	-42	3029	3063	-183	-130	1510	-50						
Exp.	1109	1238			1732		2846	2933			1460							
4																		
GGA/PW91					1726	-11	2990	3051	-124	-123	1475	-15	1687					
B3LYP					1774	-59	3010	3051	-144	-123	1513	-53	1707					
Exp.					1715		2866	2928			1460							
5																		
GGA/PW91					1733	-18	2994	3039	-128	-111	1473	-13	1706					
B3LYP					1773	-58	3031	3057	-165	-129	1511	-51	1718					
Exp.					1715		2866	2928			1460							
8																		
GGA/PW91					1685	5	3005	3083	-115	-122	1483	-13	1594	11				
B3LYP					1738	-48	3033	3089	-143	-128	1518	-48	1639	-34				
Exp.					1690		2890	2961			1470		1605					
10																		
GGA/PW91					1759	-19												
B3LYP					1807	-67												
Exp.					1740													
12																		
GGA/PW91							3019	3086	-139	-105	1487	-7	1598	7	1356	4	1547	3
B3LYP							3041	3090	-161	-109	1512	-32	1610	-5	1373	-13	1599	-49
Exp.							2880	2981			1480		1605		1360		1550	

TABLE 4: Deviations estimated over differences between experimental and theoretical wavenumbers for GGA/PW91 and B3LYP IR data.

Method	Deviation 1 (cm^{-1})	Deviation 2 (cm^{-1})
GGA/PW91	13	124
B3LYP	43	145

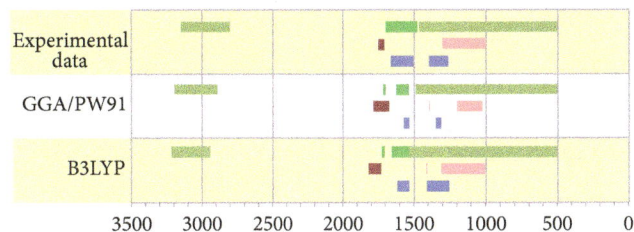

FIGURE 4: Ranges for important vibrational modes collected for compounds **1–23** through experimental data *versus* GGA/PW91 results *versus* B3LYP results. Vibrational modes come in accordance to bonding atoms and bond order: in *green*, C–H modes; in *dark green*, C=C modes; in *light pink*, C–O; in *burgundy*, C=O; and in *blue*, N=O modes.

For each musk group, the central frequencies averages (*ave*, (5)) and their standard deviations (*std*, (6)) were estimated: these values are plotted at the right side of the central frequency list. The next column on the right shows the differences between each central frequencies and the average of central frequencies (*diff*, (7)). Consider

$$\text{ave}_{\text{group}} = \frac{1}{n} \sum_{n=1}^{f} \text{cf}_n, \tag{5}$$

where cf$_n$ represents the central frequencies found for each group (MCM, PCM, etc.), having herein been defined lower and upper limits for n, so that $n = \{1, 5\}$ for MCM; $n = \{6, 10\}$ for PCM; $n = \{11, 13\}$ for NM; and $n = \{14, 16\}$ for ACM. Consider

$$\text{std}_{\text{group}} = \sqrt{\sum_{n=1}^{f} \left[\left(\text{cf}_n - \text{ave}_{\text{group}} \right)^2 \right]}, \tag{6}$$

$$\text{diff}_n = \text{cf}_n - \text{ave}_{\text{group}}. \tag{7}$$

The values found for the ave$_{\text{group}}$ and std$_{\text{group}}$ of the MCM group were the same as those found for the MCa group; those belonging to NM were repeated for Na; and data calculated for ACM were taken as parameters for Aa.

Finally, each diff$_n$ value was allowed to be lower or two times its std$_{\text{group}}$ value, as stated in

$$\text{diff}_n \leq 2 \times \text{std}_{\text{group}}. \tag{8}$$

If this happens, it is possible to say that theoretical data fit vibrational theory's demands over study band; if not, they do not agree. When repeating this process for all three bands under the spectra, values for the Total Deviations (TD) are collected. The zero input (0) means that all three bands obeyed (8). Logically, TD inputs between 1 and 3 indicate how many of the bands presented a diff$_n$ value that was higher than the std$_{\text{group}}$ doubled value.

3.4. Absence of the 2000–2400 cm^{-1} Band.

None of the previous tables bring a mode among the 2000–2400 cm^{-1} range. The 2200 cm^{-1} band was described by Turin as the carbonyl stretching mode, which only occurred at this range when working with NDDO/AM1 Hamiltonian. Through *ab initio* calculation, the mentioned mode was observed at the range of 1680–1810 cm^{-1}, and it was only taken into account for cfs estimates when dealing with the 1550–1950 cm^{-1} band for nitro compounds. Regarding structural features, compounds **6**, **7**, **12**, and **13** do not possess the carbonyl group and are real musks, while compounds **17**, **18**, **19**, **20**, and **23** contain the carbonyl group and are odorless. Therefore, carbonyl stretches cannot be taken as criteria for IR data-musky behavior profile.

3.5. GGA/PW91 cf's Plots and Musk Activity.

Table 5 (with GGA/PW91 results) presents only two molecules, nonmusks **20** and **23**, which failed to express a complete match between theoretical data and the signed bands of the vibrational theory. For nonmusk **20**, the cf value estimated near 1000 cm^{-1} was higher than its ave$_{\text{group}}$ value: this compound had both O–C$_{\text{sp}^2}$ stretching modes occurring at high intensities and frequencies below 1200 cm^{-1} (which are precisely the 1090 and 1144 cm^{-1} wavenumbers). On the other hand, musks **14–16** and nonmusk **19** have presented a stretch mode (or at least one of them) outside the 1000 ± 200 cm^{-1} range. Concerning nonmusk **23**, it can be said that it showed a lower cf value, which remained around the 700 cm^{-1} band, and a higher cf value, near the 1000 cm^{-1} band. This can be explained by the fact that, (i) referring to the 700 cm^{-1} band, **23** had an average for intensity modes that was lower than the average values for the other Na and NM compounds, and, therefore, lower *weight values* (see (4)) and lower cf; and (ii) referring to the 1000 cm^{-1} band, the O–C$_{\text{sp}^3}$ stretch mode at 1073 cm^{-1} had an intensity value that was 8.7 times higher than the second most intense mode, generating the observed overestimated cf value.

That being said and according to mathematical approach adopted in this paper, no relationship was found between IR properties and the presence of musky odor through GGA/PW91. It does not mean that GGA/PW91 calculation leads to affirm that musks and nonmusks have indistinguishable spectra. It means that, considering the weighted-averages values resource in order to simplify the adoption of all theoretical peaks under analysis, resulting central frequencies are similar among musks and nonmusks. Such observation may be due the fact that GGA/PW91 spectra had narrower bands for each vibrational mode rather than B3LYP spectra. As GGA/PW91 proved to be a more accurate method, it requires a more accurate statistical tools, so to properly deal with all theoretical modes, as to affirm (or deny) a possible difference on musks and nonmusks IR spectra.

3.6. B3LYP cf's Plots and Musk Activity.

On the other hand, Table 6 (with B3LYP data) shows that all musks from **1** to **16** have had theoretical central frequencies that fitted the frequencies demanded at 700, 1000, and 1500 (or 1750) cm^{-1}

Table 5: Calculated central frequencies for bands near 700, 1000, and 1500 cm^{-1} for **1–23** compounds for **GGA/PW91** collected data.

	Musk	Around 700	Ave, std.	Diff.	Around 1000	Ave, std.	Diff.	Around 1500	Ave, std.	Diff.	TD[*]
					Central frequency (cm^{-1})						
MCM	1	742	743	−1	1030	1049	−19	1429	1411	18	0
	2	761	12	18	1096	31	47	1404	11	−7	0
	3	748		5	1047		−2	1417		6	0
	4	734		−9	1016		−33	1405		−6	0
	5	730		−13	1057		8	1402		−9	0
PCM	6	751	733	18	1046	1020	26	1423	1480	−57	0
	7	769	34	36	1051	48	31	1424	54	−56	0
	8	678		−55	977		−43	1527		47	0
	9	731		−2	1066		46	1532		52	0
	10	736		3	959		−61	1495		15	0
NM	11	735	755	−20	1033	1019	14	1634	1589	45	0
	12	763	17	8	1002	16	−17	1567	39	−22	0
	13	766		11	1023		4	1567		−22	0
ACM	14	784	784	0	1084	1086	−2	1402	1411	−9	0
	15	762	23	−22	1079	9	−7	1401	16	−10	0
	16	807		23	1096		10	1429		18	0
MCa	17	751	743	8	1067	1049	17	1405	1411	−7	0
	18	726	12	−17	1049	31	0	1417	11	6	0
Aa	19	779	784	−6	1073	1086	−13	1411	1411	0	0
	20	757	23	−28	1105	9	19	1403	16	−8	1
Na	21	733	755	−21	1033	1019	13	1557	1589	−32	0
	22	757	17	2	1006	16	−13	1549	39	−40	0
	23	717		−38	1060		41	1655		66	2

[*]TD: total deviations.

in the vibrational theory. Simultaneously, Aa and Na showed that at least one of their cfs was too shifted from their averages. Molecule **23**, for instance, is the only nonmusk possessing an aldehyde functional group, and its C=O stretching mode is found at the 1550–1950 cm^{-1} range under study. Nevertheless, nitro musk **11** also holds an aldehyde functional group, and though its cf at 1750 cm^{-1} is indeed higher than the other cfs for the NM, it is still 70 cm^{-1} lower than **23**'s cf. Still concerning the analysis of Table 6, we can say that the acyclic analogs **19** and **20** have had their central frequencies shifted to lower values at the 1500 cm^{-1} diagnostic band, the only one in which there was no agreement between theoretical results and expectations deriving from the vibrational theory. At the 1500 ± 200 cm^{-1} range, hydrocarbon scissoring modes and alkene/aromatic C=C stretching modes are found, and **19** and **20** lack the gem-methyl at the cyclohexane that are observed in **14** and **15**; moreover, although in **16** this gem-methyl is also absent, it possesses two alkene groups, for which its stretching modes are also found at this range. Finally, macrocyclic analogs showed the same spectra data as macrocyclic musks: as compounds **18** and **19** differ from MCM in that they lack two methylene groups, it is possible to suggest that such loss leads the molecule's C–H vibrational

mode sum to be too low to be detected by its OR, as suggested elsewhere [23]. Besides, both **18** and **19** present a *woody* smell; by being smaller, they may be able to bind to a variety of different ORs, which do not reassemble the behavior of the musk, detected by possibly only one OR or just a few more [48].

4. Conclusions

This paper worked on developing a method which could state the following: (i) whether musky molecules had their IR theoretical spectra fitted into the four bands assigned by vibrational theory in 2002 Turin's paper and (ii) whether nonmusky molecules did not have their IR theoretical spectra fitted into same criteria. It was first concluded that only three bands were useful for discriminating musk from nonmusk odorants. The fourth IR band refers to a carbonyl stretching mode, and data here presented showed that this mode cannot be taken as criteria when building a muskiness profile.

Present work was carried out with two different functionals applying DFT calculation level, and both were able to accurately predict IR spectra. A mathematical strategy has been adopted in order to summarize IR theoretical

TABLE 6: Calculated central frequencies for bands near 700, 1000, and 1500 cm^{-1} for **1–23** compounds for **B3LYP** collected data.

| | Musk | Central frequency (cm^{-1}) | | | | | | | | | TD* |
		Around 700	Ave, std	Diff.	Around 1000	Ave, std.	Diff.	Around 1500	Ave, std.	Diff.	
MCM	1	754	748	6	1142	1093	48	1407	1423	−16	0
	2	768	16	20	1066	39	−27	1428	10	5	0
	3	749		1	1116		23	1421		−2	0
	4	744		−3	1099		5	1430		6	0
	5	723		−25	1044		−49	1429		6	0
PCM	6	749	722	28	1068	1019	49	1450	1476	−26	0
	7	752	51	31	1052	62	33	1450	32	−27	0
	8	692		−30	971		−48	1461		−15	0
	9	646		−75	1069		50	1499		23	0
	10	768		47	936		−83	1520		44	0
NM	11	754	766	−12	952	977	−26	1602	1597	5	0
	12	771	10	5	993	22	16	1595	5	−2	0
	13	772		7	987		9	1593		−3	0
ACM	14	722	732	−10	1101	1127	−26	1424	1435	−12	0
	15	771	35	39	1130	24	3	1431	14	−4	0
	16	703		−29	1149		22	1451		15	0
MCa	17	750	748	2	1067	1093	−26	1418	1423	−5	0
	18	726	16	−22	1072	39	−21	1432	10	9	0
Aa	19	784	732	52	1097	1127	−30	1357	1435	−78	1
	20	763	35	31	1087	24	−40	1410	14	−26	1
Na	21	742	766	−24	1015	977	38	1601	1597	4	1
	22	767	10	1	1150	22	173	1600	5	3	1
	23	702		−64	1068		91	1675		78	3

*TD: total deviations.

data and properly compare musk to nonmusk molecules. Using this approach, GGA/PW91 computed spectra showed no discriminating patterns between musks and nonmusks. B3LYP data showed theoretical frequencies and the three diagnostic bands to be matching, while nitro and macrocyclic nonmusks failed in at least one of such comparisons.

It was therefore possible to conclude that no specific mode alone is responsible for the discrimination of the musky odor. Instead, when a great amount of frequencies, expressing weak or medium intensities, were found along the 700–1700 cm^{-1} range, musk odor occurs. At this range, observed modes were CH$_2$ wag and scissor bending modes, C=C stretches, and symmetric nitro stretching modes.

This trial scanning over IR vibrational modes and their relation to musk odor perception brings new informational towards the understanding and application of vibrational theory's ideas and opens the possibility to a more extensive study, in which a broader number of molecules may be modeled so as to come to a quantitative proposal for a musky-like novel odorant, regarding their IR vibrational modes.

Conflict of Interests

The authors declare that there is no conflict of interests regarding the publication of this paper.

Acknowledgment

The authors thank Professor Inês Sabioni Resck (from Instituto de Química, Brasília, Brazil), for initial ideas and reliance.

References

[1] M. Eh, "New alicyclic musks: the fourth generation of musk odorants," *Chemistry & Biodiversity*, vol. 1, no. 12, pp. 1975–1984, 2004.

[2] P. Saiyasombati and G. B. Kasting, "Two-stage kinetic analysis of fragrance evaporation and absorption from skin," *International Journal of Cosmetic Science*, vol. 25, no. 5, pp. 235–243, 2003.

[3] G. B. Kasting and P. Saiyasombati, "A physico-chemical properties based model for estimating evaporation and absorption rates of perfumes from skin," *International Journal of Cosmetic Science*, vol. 23, no. 1, pp. 49–58, 2001.

[4] M. Shirasu, K. Yoshikawa, Y. Takai et al., "Olfactory receptor and neural pathway responsible for highly selective sensing of musk odors," *Neuron*, vol. 81, no. 1, pp. 165–178, 2014.

[5] J. Reisert and H. Zhao, "Response kinetics of olfactory receptor neurons and the implications in olfactory coding," *Journal of General Physiology*, vol. 138, no. 3, pp. 303–310, 2011.

[6] J.-P. Rospars, Y. Gu, A. Grémiaux, and P. Lucas, "Odour transduction in olfactory receptor neurons," *Chinese Journal of Physiology*, vol. 53, no. 6, pp. 364–372, 2010.

[7] B. Malnic and A. F. Mercadante, "The odorant signaling pathway," *Annual Review of Biomedical Sciences*, vol. 11, pp. T86–T94, 2009.

[8] K. Snitz, A. Yablonka, T. Weiss, I. Frumin, R. M. Khan, and N. Sobel, "Predicting odor perceptual similarity from odor structure," *PLoS Computational Biology*, vol. 9, no. 9, Article ID e1003184, 2013.

[9] B. K. Lavine, C. White, N. Mirjankar, C. M. Sundling, and C. M. Breneman, "Odor-structure relationship studies of tetralin and indan musks," *Chemical Senses*, vol. 37, no. 8, pp. 723–736, 2012.

[10] Y. Zou, H. Mouhib, W. Stahl, A. Goeke, Q. Wang, and P. Kraft, "Efficient macrocyclization by a novel oxy-oxonia-cope reaction: synthesis and olfactory properties of new macrocyclic musks," *Chemistry*, vol. 18, no. 23, pp. 7010–7015, 2012.

[11] S. Liu, "Study on the relationship between molecular odour of polycyclic musks," *Xiangliao Xiangjing Huazhuangpin*, vol. 6, pp. 24–29, 2004.

[12] S.-Y. Takane and J. B. O. Mitchell, "A structure-odor relationship study using EVA descriptors and hierarchical clustering," *Organic & Biomolecular Chemistry*, vol. 2, no. 22, pp. 3250–3255, 2004.

[13] K. J. Rossiter, "Structure-odour relationships," *Chemical Reviews*, vol. 96, no. 8, pp. 3201–3240, 1996.

[14] M. Zarzo, "The sense of smell: molecular basis of odorant recognition," *Biological Reviews*, vol. 82, no. 3, pp. 455–479, 2007.

[15] A. Kato and K. Touhara, "Mammalian olfactory receptors: pharmacology, G protein coupling and desensitization," *Cellular and Molecular Life Sciences*, vol. 66, no. 23, pp. 3743–3753, 2009.

[16] L. Turin and F. Yoshii, "Structure-odor relations: a modern perspective," in *Handbook of Olfaction and Gustation*, R. L. Doty, Ed., pp. 275–294, Marcel Dekker, New York, NY, USA, 2003.

[17] L. Turin, "A spectroscopic mechanism for primary olfactory reception," *Chemical Senses*, vol. 21, no. 6, pp. 773–791, 1996.

[18] M. A. Reed, "Inelastic electron tunneling spectroscopy," *Materials Today*, vol. 11, no. 11, pp. 46–50, 2008.

[19] L. Turin, "A method for the calculation of odor character from molecular structure," *Journal of Theoretical Biology*, vol. 216, no. 3, pp. 367–385, 2002.

[20] L. J. W. Haffenden, V. A. Yaylayan, and J. Fortin, "Investigation of vibrational theory of olfaction with variously labelled benzaldehydes," *Food Chemistry*, vol. 73, no. 1, pp. 67–72, 2001.

[21] J. C. Brookes, F. Hartoutsiou, A. P. Horsfield, and A. M. Stoneham, "Could humans recognize odor by phonon assisted tunneling?" *Physical Review Letters*, vol. 98, no. 3, Article ID 038101, 2007.

[22] M. I. Franco, L. Turin, A. Mershin, and E. M. C. Skoulakis, "Molecular vibration-sensing component in *Drosophila melanogaster* olfaction," *Proceedings of the National Academy of Sciences*, vol. 108, no. 9, pp. 3797–3802, 2011.

[23] I. A. Solov'yov, P.-Y. Chang, and K. Schulten, "Vibrationally assisted electron transfer mechanism of olfaction: myth or reality?" *Physical Chemistry Chemical Physics*, vol. 14, no. 40, pp. 13861–13871, 2012.

[24] S. Gane, D. Georganakis, K. Maniati et al., "Molecular vibration-sensing component in human olfaction," *PLoS ONE*, vol. 8, no. 1, Article ID e55780, 2013.

[25] D. L. Pavia, G. M. Lampman, G. S. Kriz, and J. R. Vyvyan, *Introduction to Spectroscopy*, Cengage Learning, São Paulo, Brazil, 2010.

[26] D. A. Lerner, D. Berthomieu, and E. R. Maia, "Modeling of the conformational flexibility and *E/Z* isomerism of thiazoximic acid and cefotaxime," *International Journal of Quantum Chemistry*, vol. 111, no. 6, pp. 1222–1238, 2011.

[27] A. P. Scott and L. Radom, "Harmonic vibrational frequencies: an evaluation of Hartree-Fock, Møller-Plesset, quadratic configuration interaction, density functional theory, and semiempirical scale factors," *The Journal of Physical Chemistry*, vol. 100, no. 41, pp. 16502–16513, 1996.

[28] S. Gabler, J. Soelter, T. Hussain, S. Sachse, and M. Schmuker, "Physicochemical vs. vibrational descriptors for prediction of odor receptor responses," *Molecular Informatics*, vol. 32, no. 9-10, pp. 855–865, 2013.

[29] Materials Studio S/W and Insight II S/W; Acelrys, San Diego, Calif, USA.

[30] M. J. Frisch, G. W. Trucks, H. B. Schlegel et al., "Gaussian 09, Revision A.02," Gaussian, Inc., Wallingford, Conn, USA, 2009.

[31] H. Sun, "Compass: an ab initio force-field optimized for condensed-phase applications—overview with details on alkane and benzene compounds," *Journal of Physical Chemistry B*, vol. 102, no. 38, pp. 7338–7364, 1998.

[32] P. Dauber-Osguthorpe, V. A. Roberts, D. J. Osguthorpe, J. Wolf, M. Genest, and A. T. Hagler, "Structure and energetics of ligand binding to proteins: *Escherichia coli* dihydrofolate reductase-trimethoprim, a drug-receptor system," *Proteins: Structure, Function and Genetics*, vol. 4, no. 1, pp. 31–47, 1988.

[33] D. H. J. Mackay, A. J. Cross, and A. T. Hagler, "The role of energy minimization in simulation strategies of biomolecular systems," in *Prediction of Protein Structure and the Principles of Protein Conformation*, D. G. Fasman, Ed., pp. 317–358, Plenum Press, New York, NY, USA, 1990.

[34] R. Fletcher and C. M. Reeves, "Function minimization by conjugate gradients," *The Computer Journal*, vol. 7, no. 2, pp. 149–154, 1964.

[35] W. H. Press, B. P. Flannery, S. A. Teukolsky, and W. T. Vetterling, *Numerical Recipes in C: the Art of Scientific Computing*, Cambridge University Press, London, UK, 1988.

[36] J. J. P. Stewart, "Mopac: a semiempirical molecular orbital program," *Journal of Computer-Aided Molecular Design*, vol. 4, no. 1, pp. 101–105, 1990.

[37] J. A. Pople, D. P. Santry, and G. A. Segal, "Approximate self-consistent molecular orbital theory. I. Invariant procedures," *The Journal of Chemical Physics*, vol. 43, no. 10, pp. S129–S135, 1965.

[38] J. A. Pople and G. A. Segal, "Approximate self-consistent molecular orbital theory. III. CNDO results for AB_2 and AB_3 systems," *The Journal of Chemical Physics*, vol. 44, no. 9, pp. 3289–3296, 1966.

[39] M. J. S. Dewar, E. G. Zoebisch, E. F. Healy, and J. J. P. Stewart, "AM1: a new general purpose quantum mechanical molecular model," *Journal of the American Chemical Society*, vol. 107, no. 13, pp. 3902–3909, 1985.

[40] B. Delley, "An all-electron numerical method for solving the local density functional for polyatomic molecules," *The Journal of Chemical Physics*, vol. 92, no. 1, pp. 508–517, 1990.

[41] B. Delley, "Fast calculation of electrostatics in crystals and large molecules," *Journal of Physical Chemistry*, vol. 100, no. 15, pp. 6107–6110, 1996.

[42] B. Delley, "DMol methodology and applications," in *Density Functional Methods in Chemistry*, J. W. Andzelm and J. K. Labanowski, Eds., pp. 101–108, Springer, New York, NY, USA, 1991.

[43] B. Delley, "Calculated electron distribution for Tetrafluoroterphthalonitrile (TFT)," *Chemical Physics*, vol. 110, no. 2-3, pp. 329–338, 1986.

[44] Y. Wang and J. P. Perdew, "Correlation hole of the spin-polarized electron gas, with exact small-wave-vector and high-density scaling," *Physical Review B*, vol. 44, no. 24, pp. 13298–13307, 1991.

[45] P. J. Stephens, F. J. Devlin, C. F. Chabalowski, and M. J. Frisch, "Ab Initio calculation of vibrational absorption and circular dichroism spectra using density functional force fields," *Journal of Physical Chemistry*, vol. 98, no. 45, pp. 11623–11627, 1994.

[46] G. Schaftenaar and J. H. Noordik, "Molden: a pre- and post-processing program for molecular and electronic structures," *Journal of Computer-Aided Molecular Design*, vol. 14, no. 2, pp. 123–134, 2000.

[47] Xming, "Xming X server for Windows," 2012, http://www .straightrunning.com/XmingNotes/.

[48] K. Nara, L. R. Saraiva, X. Ye, and L. B. Buck, "A large-scale analysis of odor coding in the olfactory epithelium," *Journal of Neuroscience*, vol. 31, no. 25, pp. 9179–9191, 2011.

18

Optical Breakdown in Liquid Suspensions and Its Analytical Applications

Tatiana Kovalchuk-Kogan, Valery Bulatov, and Israel Schechter

Schulich Department of Chemistry, Technion, Israel Institute of Technology, 32000 Haifa, Israel

Correspondence should be addressed to Israel Schechter; israel@techunix.technion.ac.il

Academic Editor: Armando Zarrelli

Micro- and nanoparticles persist in all environmental aquatic systems and their identification and quantification are of considerable importance. Therefore, the application of Laser-induced breakdown to aquatic particles is of interest. Since direct application of this method to water samples is difficult, further understanding of the breakdown is needed. We describe several optical techniques for investigation of laser breakdown in water, including Mach-Zehnder interferometry, shadow, and Schlieren diagnostic. They allow for studying the time dependent structure and physical properties of the breakdown at high temporal and spatial resolutions. Monitoring the formation of microbubbles, their expansion, and the evolution of the associated shockwaves are described. The new understanding is that the plasma column in liquids has a discrete nature, which lasts up to 100 ns. Controlling the generation of nanoparticles in the irradiated liquids is discussed. It is shown that multivariate analysis of laser-induced breakdown spectroscopy allows for differentiation between various groups of suspended particulates.

This review is dedicated to the memory of Gregory Toker, a devoted scientist who had major contributions to the field of optical breakdown and its investigation using interferometric techniques, who passed away during the preparation of this review

1. Liquid Suspensions

The effect of micro- and nanoparticles on migration of pollutants is a major environmental concern [1]. Micro- and nanoparticles are present in all aquatic systems [1–4]. The characteristic particle sizes are between 1 nm and 10 μm and they possess a high surface to mass ratio. Sorption of contaminants on their surfaces can result in severe pollution. In addition, micro- and nanoparticles may provide a medium for microbial growth. The common waterborne particulates include the following:

(a) mineral particles: clay, silica, hydroxides, and metallic salts;

(b) organic particles: humic and fulvic acids stemming from the decomposition of vegetable and animal matter;

(c) biological particles: microorganisms such as pollens, bacteria, plankton, algae, and viruses.

Besides the environmental effects, waterborne particles are often unwanted contaminants in industrial processes, since they often reduce the product quality [5]. Therefore the development of fast and simple methods for characterization of particles in water is of considerable importance. There are several known methods for detection of particles in water [5] but laser-induced breakdown detection (LIBD) is one of the most promising experimental techniques for direct quantification of aquatic colloids.

LIBD has the potential to provide online analysis of particles in water. Therefore understanding the breakdown mechanism in water is of considerable importance. In principle, LIBD is based on generation of discharge events induced by a laser at colloidal particles [6]. In this method, a high power laser pulse is focused in a sample, thus generating a local plasma or laser spark. The emission from the atoms and ions in the plasma is collected by an optic fiber and guided to a detector. LIBD involves the collection and processing of the spectral signature resulting from the plasma and the analytes

[7]. The generic nature of the breakdown has been utilized as a robust and conceptually simple method for elemental characterization of gases [8], solids [9], and aerosols [7, 10, 11].

When suspensions are concerned, LIBD is capable of detecting very small particulates, even at low concentrations [12–17]. However, the currently available methods suffer from considerable difficulties associated with quantification of particles in water [5, 13]. In this environment, particles of diameter smaller than 100 nm can be detected only at relatively high concentrations.

2. Laser Breakdown in Liquids

The interaction of high intensity light with condensed matter leads to breakdown. The origin of breakdown in air has been investigated [18], as well as the characteristics of the breakdown process itself [19]. The process requires high power densities of about 10^8 W cm^{-2}. Such power densities can be produced by focusing a pulsed laser beam of a few mJ and several ns pulse duration [12, 20]. As a result, the electrons are released from atoms and molecules by a multiphoton absorption avalanche mechanism. The initially released electrons are accelerated in the electric field of the laser pulse. Electron avalanche is produced by the electrons which knock out other electrons from the atoms and molecules. During the plasma generation process, the plasma heats up by several thousand Kelvins and its volume expands by creating a shockwave. The plasma is eventually cooled down by emission of light and other processes. Figure 1 represents the generation of plasma in liquids.

The breakdown energy threshold is the minimum energy density required for plasma generation and depends on the state of aggregation of the medium. It is highest for gases, lower for liquids, and lowest for solids [21]. Often the medium density (liquids and solids) can be modified for improved LIBD. For example, when the laser energy is not sufficient for inducing breakdown in pure water, addition of particles can solve the problem.

3. Visualization Techniques for Optical Breakdown

First experiments on visualizing the region of laser breakdown have been performed in gaseous atmosphere [22, 23]. Studying the optical breakdown has been carried out in the microsecond and partially in nanosecond ranges, by applying coherent and noncoherent techniques [24–28]. The morphology of breakdown on surfaces was also investigated in detail, using several technologies [29, 30]. Breakdown in water in picosecond and femtosecond temporal ranges has also been investigated [20, 31, 32]. However, studying the discrete character of breakdown events requires better spatial resolution than was done in the past. The following three methods are especially suited for visualization of breakdown processes in water: shadowgraphy, Schlieren, and Mach-Zehnder interferometry.

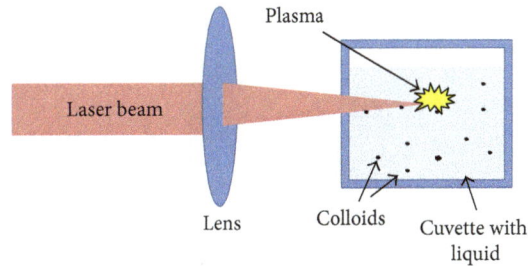

Figure 1: Schematic illustration of breakdown generation in liquid.

3.1. Shadowgraphy. Shadowgraphy is an optical technique that reveals nonuniformities in transparent media such as air, water, or glass [33–42]. Regular vision and optical inspection methods are not sensitive to temperature differences, to variations in gas composition, or to shockwaves in air. However, these disturbances refract light, so they can cast shadows. A point (or collimated) light source is used for direct illumination of the transparent object under study. The phase of the diagnostic wave can be changed by warm air or other events that cause nonhomogeneity in the medium. This phase change causes shadows on the screen.

Applications of shadowgraphy in science and technology are very broad. It is used in aerospace engineering to visualize the flow about high-speed aircrafts and missiles, as well as in combustion research, ballistics, and explosions and in testing of glass quality. A shadowgram (the result of applying the shadowgraphic technique) is not a focused image; rather it is a mere shadow. In a shadowgram, the differences in light intensities are proportional to the second spatial derivative of the refractive index field in the transparent medium under study.

3.2. Schlieren Technique. The basic optical Schlieren system uses light from a collimated source. Variations in refractive index caused by density gradients in the fluid distort the collimated light beam. This distortion creates a spatial variation in the intensity of the light, which can be visualized directly by using optical diaphragms [33–42].

In Schlieren photography, the collimated light is focused on a knife-edge obstacle (the so-called Foucault knife) placed in the focal plane, so that the source image is projected onto the edge of the knife. If the sample contains no density gradients or other optical inhomogeneities, the light is blocked by that obstacle. Otherwise, a part of the light is scattered and passes above the barrier and forms an image on the detector.

The result is a set of lighter and darker patches corresponding to positive and negative density gradients in the direction normal to the knife-edge. This system measures the first derivative of density in the direction perpendicular to the knife-edge. Schlieren method provides a high contrast image.

An experimental setup representing the application of Schlieren technique to breakdown in liquids is shown in Figure 2. In this setup, the optical collimation system sharply focused the breakdown region onto the imaging detector.

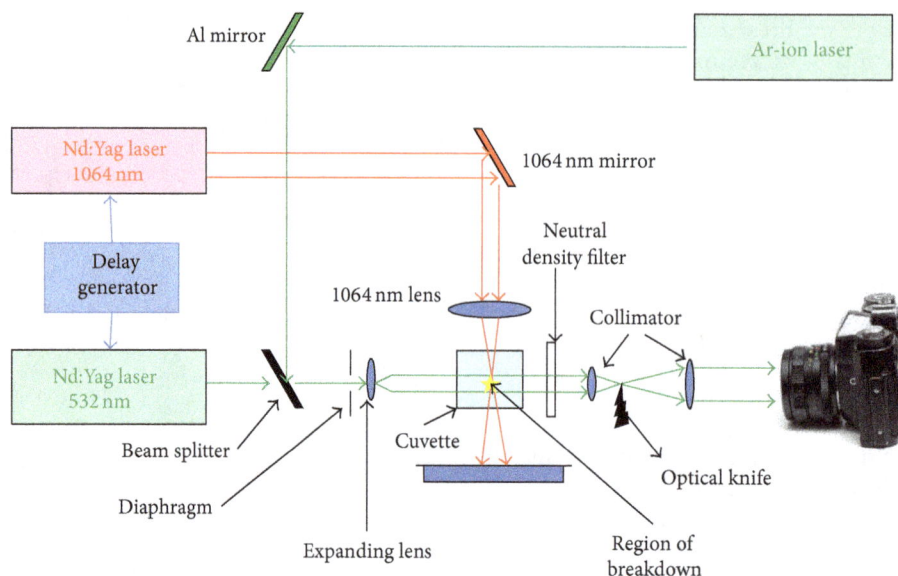

FIGURE 2: Experimental setup for Schlieren technique and its application to breakdown in liquid suspensions.

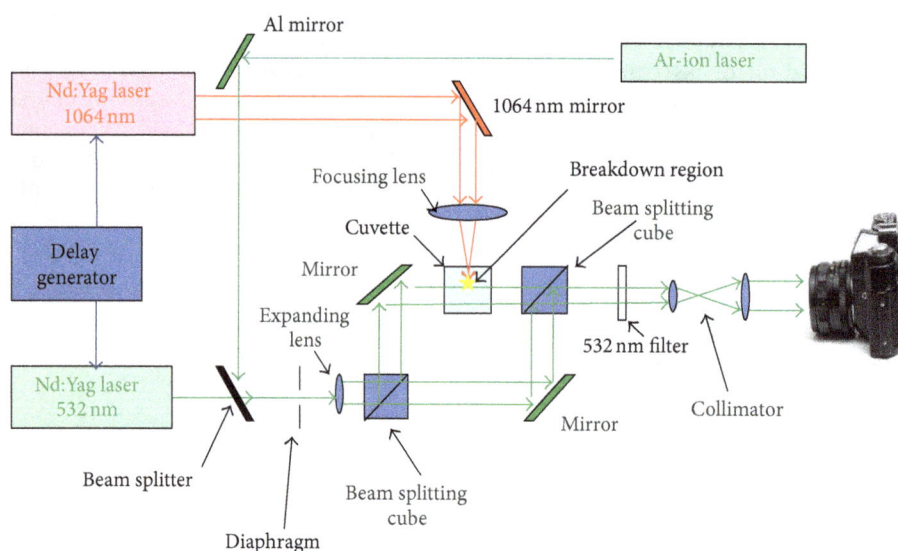

FIGURE 3: Experimental setup for application of Mach-Zehnder interferometry to breakdown in liquids.

In this setup the plasma was initiated by a Nd:YAG laser ($\lambda = 1064$ nm, 6 ns), which was focused into a quartz-windowed cuvette, where the liquid under study was located. A second laser (2nd harmonic of a Nd:YAG) was used for probing the cuvette in a direction perpendicular to the first laser beam. The probing laser was fired at controllable time delay after the first laser pulse, such that time-resolved information can be achieved. An argon-ion laser was used for alignment and adjustment of the optical components.

3.3. *Mach-Zehnder Interferometry.* Mach-Zehnder interferometer is a device used to determine the phase shift caused by a small sample which is placed in the path of one of two collimated beams (thus having plane wave fronts)

from a coherent light source [33–42]. A collimated beam is split by a half-silvered mirror. The two resulting beams (the "sample beam" and the "reference beam") are each reflected by a mirror. The two beams then pass a second half-silvered mirror. Reflection at the surface of a medium with a higher refractive index causes a phase shift. Mach-Zehnder interferometer has two important features. One is that the two paths are widely separated and are traversed only once; the other is that the region of localization of the fringes can be made to coincide with the test object, so that an extended source of high intensity can be used. The experimental setup presenting the application of Mach-Zehnder interferometry to breakdown in liquids is shown in Figure 3.

In all the above methods (interferometry, Schlieren, and shadowgraphy), the image formation relies on refractive index changes. However there are some important differences between these techniques.

(1) Interferometry has a larger number of optical elements. Since it is based on differential phase measurements, it is sensitive to alignment. Schlieren technique has fewer optical components and is less sensitive. Shadowgraphy, being the simplest configuration, is the least sensitive to factors like alignment, vibrations, and other extraneous effects.

(2) Interferograms are clear and useful in experiments with low density gradients. However the intensity contrast in Schlieren and shadowgraphy may not be large enough to provide a vivid picture. In high gradient experiments, both Schlieren and shadowgraphy yield clear and interpretable images. Interferograms are corrupted by refraction errors.

(3) In unsteady flow fields, Schlieren and shadowgraphy track temporal changes in temperature and concentration in the form of light intensity variations. These are less obvious in interferograms where information is localized at fringes.

(4) The number of fringes formed in water is large, since the sensitivity of refractive index to density is large in such media. Interferograms recorded in such experiments are affected by refraction errors. Schlieren and shadowgraphy have an advantage in this respect. However, interferograms can provide information about physical data of the experiment which cannot be obtained from Schlieren and shadowgraphy.

(5) In experiments with shockwave formation, interferography has an advantage since it can provide much sharper borders of the shockwave front.

(6) Shadowgraphy and Schlieren have the advantages of low cost and simplicity of apparatus and ease of real-time qualitative interpretation. However, Mach-Zehnder interferometry has the advantage of providing quantitative information.

4. The Structure of Optical Breakdown in Liquids

Most studies of the phenomena of optical breakdown in water refer to the dynamics of the radiation emitted from the breakdown region and only a few address the integral hydrodynamic picture of the process. Understanding the optical breakdown mechanism in liquids, formation of laser spark columns, and their time evolution are of considerable importance to a variety of tasks. These include biomedical applications, plasma-mediated eye and biological tissues surgery [43, 44], selective cell targeting with light-absorbing nanoparticles, chemical engineering, and generation of micro/nanoparticles. It is also relevant to numerous analytical applications based on laser plasma spectroscopy [45–48]. Laser-induced breakdown mechanism in air and in solids is well established [49–54]. The hydrodynamic characteristics of laser-induced breakdown in water, as well

as in some other liquids, have also been studied [24, 50, 53–62]. In earlier studies [25, 55–57], the breakdown was analyzed at microsecond resolution, which was not sufficient for revealing the early stages of the breakdown mechanism. Better spatial and temporal resolution down to the ns range allowed a better insight into the breakdown phenomenon [63–65]; however, the discrete structure of the laser spark column was not observed. Microsperical shockwaves in liquids were investigated [24] and numerical simulations have also been carried out. Such simulations and experimental data on plasma dynamics [64, 66] and shockwave formation [28, 64, 67] explained the cavitation dynamics and bubbles arising in water after breakdown and lasting longer than 1 μs [65, 68]. It has been suggested that plasma formation in water is initiated by electrons generated by multiphoton absorption in the liquid itself or in persistent impurities [64, 69, 70].

In almost all published articles in the last few decades the laser spark column in water was considered as a continuous object. The discrete structure and dynamics of laser spark columns in water have been only recently revealed by applying shadowgraphy and Schlieren diagnostic techniques [71, 72]. These high temporal and spatial resolution techniques allowed, for the first time, determining the discrete structure of the breakdown in its early stages. The laser spark in water was interferometrically imaged at its earlier moments of time. The application of Mach-Zehnder interferometry provided both high spatial and temporal resolutions, as was needed for studying the discrete microscopic structures of a laser spark column.

The structure of the laser spark depends on the laser beam characteristics, on the liquid (including its contaminants), and on time. We shall first address the structure of the laser spark columns in water and their dependence upon the nature of persistent impurities. These microscopic solid particulates initiate the plasma formation. In principle, double distilled water (DDW) should not contain any particles. In practice, the distillation process is not free of contamination and any exposure to air contributes to considerable additional contamination. The actual particulate concentration in DDW slightly exposed to air is ca. $2 \cdot 10^4 \, \text{cm}^{-3}$. This concentration is at least two orders of magnitude lower than that usually found in tap water. Filtration of tap water through 0.45 μm pore-size filter reduces the particulate concentration to ca. $1 \cdot 10^5 \, \text{cm}^{-3}$ and a 0.22 μm pore-size filter can further reduce the concentration to ca. $3 \cdot 10^4 \, \text{cm}^{-3}$. Therefore, tap water, DDW, and filtered water were all good media for investigating the breakdown threshold in the presence of impurities.

Time-resolved breakdown in tap water can be imaged using Schlieren technique and some representative results are shown in Figure 4. One can see that shortly (a few ns) after the laser pulse the laser spark column is filled with microplasma balls of tens of μm in size. The avalanche ionization is probably responsible for the observed luminous balls over the particulates [20]. In a few nanoseconds the particulates are evaporated and the thermal explosion of the vapors creates microbubbles, as the surrounding water is forced out. These microbubbles and the associated shockwaves evolve in time as shown in this figure. These images demonstrate the

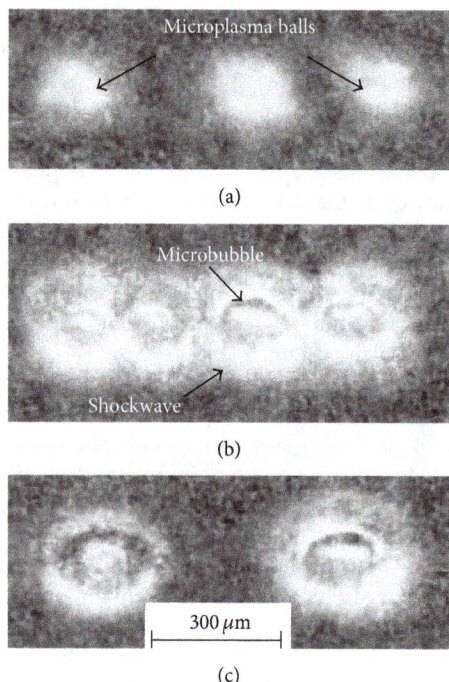

(a)

(b)

300 μm

(c)

Figure 4: Examples of dark field Schlieren pictures of a part of the laser breakdown area in tap water. Laser pulse energy: 74 mJ. Time delay of the diagnostic light: (a) 1 ns, (b) 14 ns, and (c) 102 ns. Reprinted from [71], with permission from Elsevier.

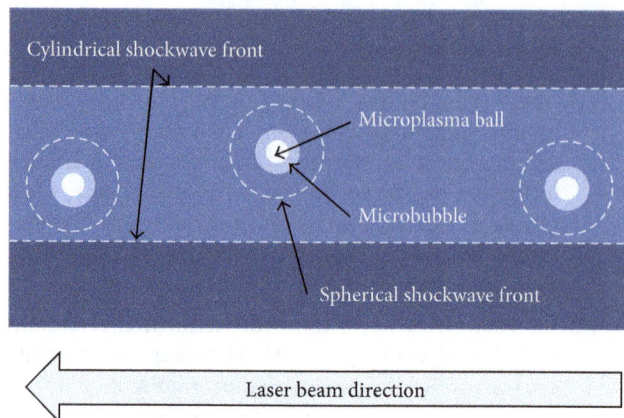

Laser beam direction

Figure 5: Schematic structure of a laser spark column. Reprinted from [72], with permission from Elsevier.

capabilities of Schlieren technique to provide time-resolved images of breakdown events in water.

Similar temporal resolutions (in the low ns range) can also be achieved using Mach-Zehnder interferometry. It provides more detailed information on laser sparks in liquids, with spatial resolution of ≤10 μm. The shifts of the interference fringes and their sign allow calculating the local changes of the refraction index in the examined liquids. For example, a region compressed by a shockwave has positive refraction index change. This can be deduced from the corresponding sign of the fringe shifts. A region warmed by the laser radiation has negative refractive index change. Positive and negative changes of the refraction index lead to fringe shifts in opposite directions, perpendicular to the fringes.

The basic features of the breakdown in liquids are schematically presented in Figure 5. Numerous plasma events are formed in the vicinity of the focal point and along the laser beam. All of them together form the so-called laser spark column. It consists of microplasma balls, microbubbles, and spherical shockwaves of well-defined front (which were clearly presented in the Schlieren picture in Figure 4).

During the laser radiation time, a warmed channel is formed in the liquid along the laser beam. In some liquids, such as alcohols, this channel results in a cylindrical shockwave propagating apart from the laser beam. In water, a different kind of cylindrical shockwave evolves at longer times: the spherical shockwaves expand and interfere with each other. These interferences result in a cylindrical pattern, called interfering shockwave cylinder.

We shall now address the various time domains in the breakdown life. Clearly, the discrete structure of laser spark columns is formed during a few ns after the heating pulse. The laser spark evolution is conveniently divided into three major time intervals: the first stage ($\tau_d < 50$ ns), the second time stage (50–200 ns), and the third stage (>200 ns). These are discussed in the following.

The first 50 ns of the spark generation is presented in Figure 6. At the end of the laser pulse (of ca. 5 ns) the spark column is full of luminous microplasma balls of ca. 100 μm diameter. They arise due to avalanche ionization in the vapors of ablated material and successive heating of the plasma due to the bremsstrahlung effect. The vapors move as a piston working on the surrounding liquid and generate a powerful microspherical shockwave in it. The microplasma balls last up to ~1 μs. Since the plasmas are generated at the surface of particles, the initial sizes and characteristics of the microplasma balls within the plasma column depend on the nature of the suspended particulates and on their concentration. It can be seen that the size of the microplasma balls monotonically increases in the first ca. 500 ns, in both water and alcohols [71].

In both tested liquids, the microspherical shockwaves are getting clearly visible in about 20 ns. However, while in water the size of the spherical shockwave fronts is remarkably larger than the size of microplasma balls, in ethanol the size of the spherical shockwaves only slightly exceeds that of the luminous microplasma balls. Also the speed of the shockwaves (visible as coaxial rings) is larger in water (measured at the same time). This can be attributed to the much lower compressibility of water ($4.4 \cdot 10^{-10}$ Pa^{-1}) compared to ethanol ($11.0 \cdot 10^{-10}$ Pa^{-1}) [73].

Note that no shockwaves are observed in butanol at time delay of 10.5 ns. At this early time the spherical shockwaves have been just formed and have not yet detached from the microplasma balls.

The concentration of the induced plasma balls depends on the tested liquid. It is of the orders of $60 \cdot 10^4$ cm^{-3} in water and $2 \cdot 10^4$ cm^{-3} in ethanol. These concentrations can be compared to the concentration of particles in these

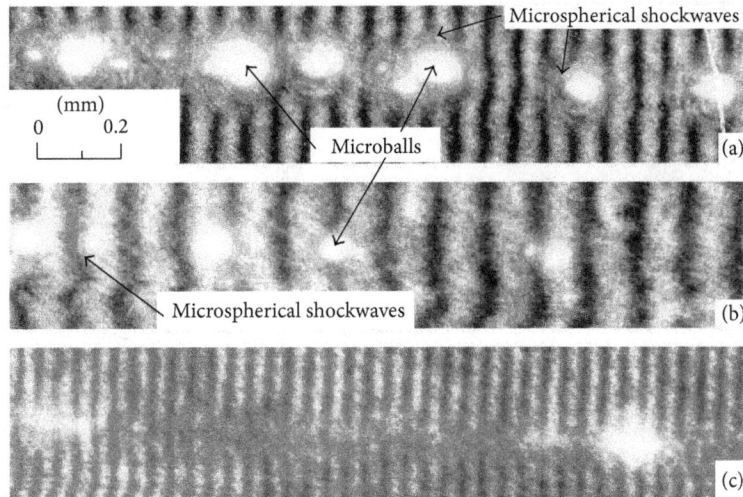

FIGURE 6: Fragments of interferograms of laser spark columns at the first time stage: $\tau_d < 50$ ns: (a) in water, time delay $\tau_d = 17$ ns; (b) in ethanol, $\tau_d = 22$ ns; (c) in butanol, $\tau_d = 10.5$ ns. The laser pulse energy was 70 mJ. The laser beam was focused from right to left. The focal spot coincides with the picture's left margin. Reprinted from [72], with permission from Elsevier.

liquids. In tap water only about 50% of the particles ignited breakdown processes, while in ethanol almost all particles were associated with plasma events. The differences might be attributed to the differences in size distribution and in the chemical nature of the suspended particulates. Nevertheless, the large percentage of particles involved in the breakdown mechanism supports the model of the discrete nature of plasma generation in liquids.

The second time rage in the spark's life starts at 50 ns and lasts till 200 ns. It is characterized by larger microbubble sizes and spherical shockwaves. This process is illustrated in Figure 7, where both the large microbubbles and the associated spherical shockwaves are clearly imaged. The microbubble sizes are larger than those of the microplasma balls.

At this time domain, the characteristic size of the spherical shockwaves is comparable to the average distance between suspended particles. Therefore, an interfering shockwave cylinder is formed, as a result of overlapping neighboring spherical shockwaves. This is manifested in a mixture of interferometric fringes of cylindrical symmetry. Note that, in the past, this cylindrical shockwave was considered as a continuous object, since the old experimental setups were not able to resolve the discrete structure of the laser spark column [50, 59–62]. It is now clear that when the spherical shockwaves emerge in a cylindrical pattern (50–90 ns), it is still a shockwave (Mach number >1). Only at later times does it become a strong acoustic wave.

As the spark evolves beyond 200 ns, the size of the microbubbles becomes remarkably larger than the diameter of the microplasma balls. The evolution of microbubbles and spherical shockwaves in this range is shown in Figure 7. Also the interfering shockwave cylinder becomes a cylindrical acoustic wave, which is clearly observed and its radius is much increased.

It is at this stage that the plasmas change from discrete nature to continuous one. The cylindrical shockwaves are merged into spherical shockwaves. This happens when the numerous microsperical shockwaves expand and combine into a single cylindrical shockwave. The transformation that the spark undergoes from discrete nature in the very beginning to a continuous event at longer times is a recent concept. Cylindrical shockwaves were well recognized [25, 74]; however, the experimental setups used in the past did not allow for observing this transition. The old measurements were carried out either at too long times (of ca. 6 μs), when only the continuous spark could be observed, or at short times but using experimental setups that did not allow for observation of the discrete events. Therefore only the new interferometric measurements were able to reveal the mechanism of the formation of the cylindrical shockwaves.

The above described spark generation mechanism is valid under most experimental conditions, where the laser focusing is moderate and the liquid contains a reasonable concentration of particulates. This might not always be the case. Under very sharp focusing conditions and at very low particulate concentrations, the mechanism might be different. Under such sharp focusing conditions, spherical shockwaves are expected and not cylindrical. If the particulate concentration is very small such that not even a single particle is present in the focal volume, the mechanism of the breakdown might be different. This was probably the case when much shorter plasma luminescence lifetimes were observed [66, 75]. However, it was found that, in most commercially available liquids of analytical grade, there are a few particles present in the focal volume of a singlet of $f = 50$–100 mm, and at least 1-2 breakdown events can be observed.

5. Breakdown Dynamics in Suspensions

5.1. Temporal Resolution. Interferometric imaging methods allow for detailed investigation of the time dependent evolution of the laser-induced microbubbles and the associated

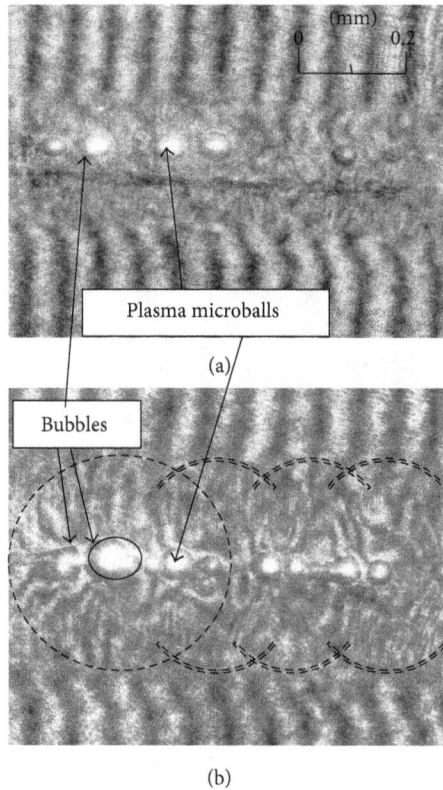

(a)

Plasma microballs

Bubbles

(b)

FIGURE 7: Fragments of interferograms of laser spark columns in ethanol. The laser pulse energy is 70 mJ. The laser beam is focused from right to left. The focal spot coincides with the picture's left margin. (a) Delay of 152 ns (second time range); (b) delay of 333 ns (third time range). Dotted circles mark spherical shockwave fronts. Reprinted from [72], with permission from Elsevier.

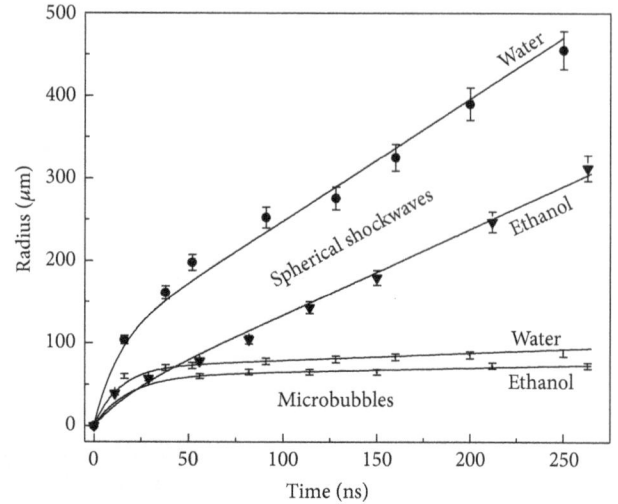

FIGURE 8: Experimental dependence of microbubbles and associated spherical shockwave radii in water and in ethanol as a function of time; energy of the laser pulse was 65 mJ. Reprinted from [72], with permission from Elsevier.

shockwaves. For example, the radii of the microbubbles and of the surrounding spherical shockwaves in water and ethanol as a function of time are presented in Figure 8. The results support the following interpretation of the breakdown dynamics. First, the suspended particulates in the irradiated liquid absorb the laser pulse. The surfaces of these inclusion particles begin to evaporate and at the same time the expanding motion of the vapors is strongly restricted by the surrounding water. This leads to high pressures, which support the ionization avalanche. Multiphoton absorption and ionization cause the appearance of the first electrons [76–78]. The expanding vapors act as a spherical piston, which effectively compresses a layer of liquid and generates a spherical shockwave in this medium. This shockwave starts its expansion with supersonic velocity together with the microbubble walls. After a certain time the shockwave outdistances the microbubble wall, leaving it behind.

This qualitative interpretation is also supported by a quantitative description and modeling of the process: the expansion of the plasma is often considered as a thermal explosion, approximated by the known dynamics of a point blast [76, 79]. This model describes the situation in which

a large amount of energy is liberated in a small volume during a short time interval:

$$R_c = \left(\frac{E_i}{\rho}\right)^{1/5} \tau^{2/5}, \tag{1}$$

where R_c is the radius of the microbubble, E_i is the laser energy absorbed by the particle during the laser pulse, ρ is water density, and τ is time.

The energy E_i is proportional to the energy of the laser beam E_{lp} and depends on d_i, the size of the particle, and d_F, the diameter of a focal spot:

$$E_i \approx \left(\frac{d_i}{d_F}\right)^2 E_{lp}. \tag{2}$$

Equations (1) and (2) provide

$$R_c \approx \left(\frac{d_i}{d_F}\right)^{2/5} \left(\frac{E_{lp}}{\rho}\right)^{1/5} \tau^{2/5}. \tag{3}$$

In this model the microbubble size depends on the laser radiation and on the size of the particle which has induced the plasma. Evaluating dR_c/dt shows that the microbubbles expand at supersonic velocity only during the laser radiation time. Afterwards, the expansion rapidly slows down and after ca. 20 ns they start moving with subsonic velocities. This is clearly confirmed in Figure 8: up to ca. 10 ns the microbubble and the shockwave are moving together at supersonic speeds. After that the shockwave outdistances the microbubble.

The Mach numbers of the spherical shockwave fronts are shown in Figure 9 for two liquids. The graphs indicate that the expanding velocities of the spherical shockwaves and of the microbubbles in water are remarkably higher than in ethanol [73]. This is attributed to the higher compressibility of ethanol and to the lower speed of sound in this liquid (1.16 km/s in

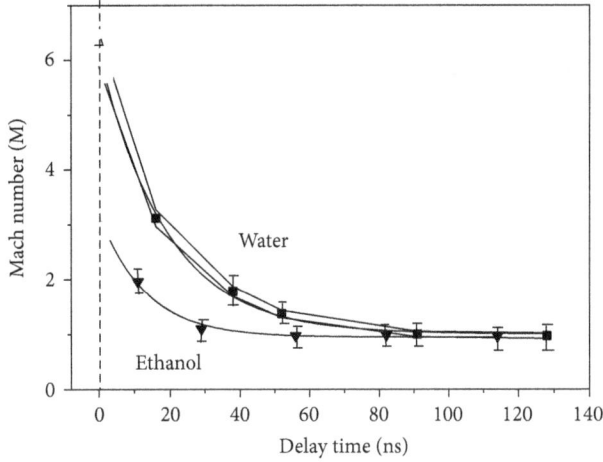

FIGURE 9: Mach number of spherical shockwaves as a function of time. Reprinted from [72], with permission from Elsevier.

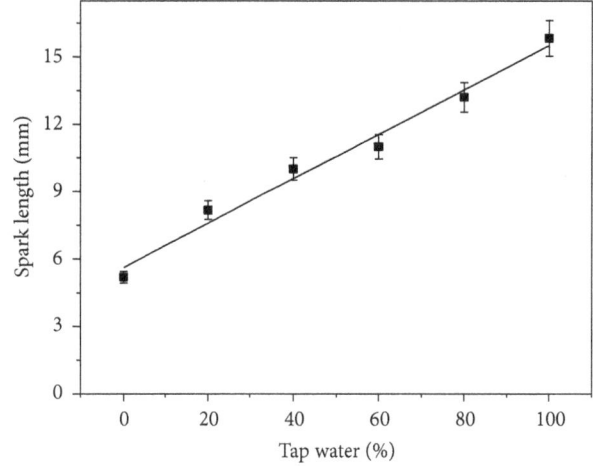

FIGURE 10: Experimental dependence of laser spark length on concentration of tap water in mixtures of tap/pure water at fixed laser pulse energy of 70 mJ.

ethanol and 1.46 km/s in water). The lower compressibility of water may also lead to higher pressures at the shockwave front.

Similar investigations were carried out for laser plasmas in continuous media [64]. Comparison to these data indicates that the shockwaves around a particle are much more intensive than those from a plasma spark column.

5.2. *Microbubble Radius as a Function of Energy.* According to (1) the microbubble radius depends on the energy of the laser pulse absorbed by the particle. This model of plasma generation in water suspensions cannot be directly confirmed, since in this medium the laser pulse simultaneously generates numerous microbubbles. However, the statistical validity of this model can still be tested using shadowgraphic measurements obtained at the same delay time and at different laser energies. The results have shown that at each energy there is a distribution of microbubble radii. Higher pulse energy generates a larger concentration of microbubbles in the focal volume and these are of larger average radii.

Using the relation $R_c \sim E_p^{1/5}$ and summing over a measured set of microbubbles provide

$$\frac{E_N}{E_M} = \frac{\sum_{n=1}^{N} (R_{c,n})^5}{\sum_{m=1}^{M} (R_{c,m})^5}, \tag{4}$$

where N and M are the numbers of microbubbles of radii $R_{c,n}$ and $R_{c,m}$, formed when the absorbed energies are E_N and E_M. This measured ratio was ≈2.4. This result correlated well with the ratio of the actual absorbed energies E_N/E_M, which was ≈2.4.

5.3. *Plasma Column Length.* A significant difference between the plasma generated in liquids and that generated in air is its structure: in air, a single plasma spot is observed, while in liquids numerous tiny plasma spots are generated. These appear in a column extending in the axis direction on both sides of the focal point. The actual length of this column is a function of laser pulse energy.

Theoretical investigations [28, 31] have indicated that the length of the plasma column, z, increases with the laser irradiance according to

$$z = z_R (\beta - 1)^{1/2}, \tag{5}$$

where β is the ratio of the peak power of the laser pulse to the breakdown threshold and z_R is Rayleigh's length. Quite interesting, the experimental z values do not quantitatively agree with those calculated in this model. The experimentally obtained results remarkably exceeded the values calculated using (5).

In water suspensions, the length of the laser spark column depends on the concentration of particles and on their size distribution and it increases with the particulate concentration. This effect was demonstrated by measuring the spark column length in pure water (DDW), in tap water, and in a series of mixtures of pure and tap water. The particulate concentration was $2.5 \cdot 10^4$ cm^{-3} in the DDW and $1.2 \cdot 10^6$ cm^{-3} in the tap water. The results are shown in Figure 10, for fixed laser pulse energy of 70 mJ.

5.4. *Breakdown Threshold in Suspensions.* Generation of plasma requires laser pulse energies over a specific threshold. Usually, this threshold is defined statistically, since plasma generation close to the threshold has a specific probability [78]. This definition of the threshold is problematic, since it requires numerous measurements and statistical evaluation of the results. A new definition of the breakdown threshold has been recently suggested [71].

Experimental results showed that the plasma column length depends on the amount of energy absorbed by the plasma, as presented in Figure 11. The intercept of the corresponding regression line with the x-axis indicates an energy value which generates a plasma column of zero length. In tap water this value is 13 mJ, which can be considered as

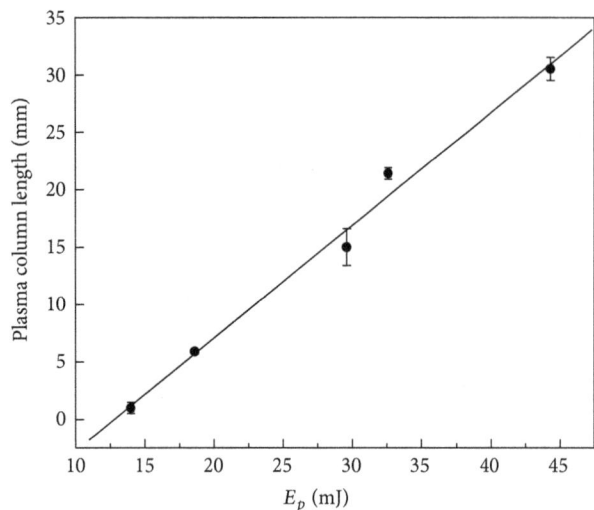

FIGURE 11: Length of the plasma column as a function of the absorbed energy.

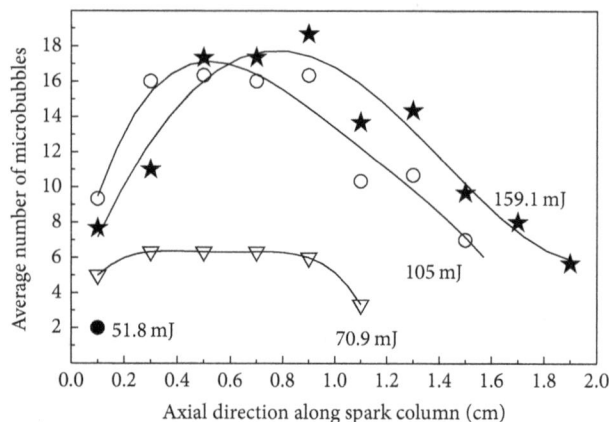

FIGURE 12: Average number of microbubbles as a function of location along the spark column for various laser pulse energies. Each point represents an interval of 2 mm. Laser beam direction was from right to left. Reprinted from [72], with permission from Elsevier.

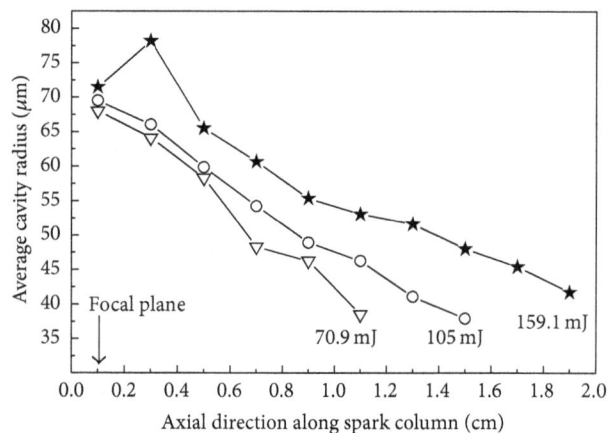

FIGURE 13: Average microbubble radius as a function of location along the spark column, for various laser pulse energies. Each point represents an interval of 2 mm. Laser beam direction was from right to left. Reprinted from [71], with permission from Elsevier.

the breakdown threshold. This definition of the threshold is readily obtained and does not depend upon statistical assumptions.

Measurements of the experimental breakdown threshold in pure water (DDW) resulted in a value of 15.8 mJ. The threshold in tap water was 5.0 mJ [71]. The breakdown threshold defined by extrapolation to zero plasma column length is higher by the factor of ~3 in pure water. Other results, using a different breakdown threshold definition, indicated a factor of 5 for ns laser pulses [80].

5.5. The Number of Microbubbles and Their Size along the Laser Spark Column. The number of microbubbles and their size along the laser spark column can also be analyzed from the plasma shadowgraphs. The average number of microbubbles as a function of the location along the laser spark column is shown in Figure 12. In all studied cases the microbubble concentration had a maximum, which was found ahead of the optical focus. The absolute number of microbubbles increased with the laser energy. However, it is interesting that the location of the maximal microbubbles concentration moved towards the laser source, as the energy increases. This was attributed to the shielding of the laser light by the plasma and to effects related to modifications in the matrix optical properties at high laser powers.

While the microbubbles concentration possesses a maximum, their average radius monotonically increases along the laser irradiation axis and reaches a maximum value close to the optical focal plane (Figure 13). The average microbubble radius in each interval along the plasma column strongly depends on the laser pulse energy.

5.6. Spatial Distribution of the Absorbed Laser Energy along the Spark Column. Elemental chemical analysis based on plasma emission depends on absorption of laser radiation. Integrated spectral data collect considerable noise, mainly

from the peripheral locations where only little laser absorption takes place. Therefore, optimization of spectral measurements requires information on the spatial distribution of the absorbed laser energy along the spark column.

The measurement of microbubble sizes and concentrations can be used as a diagnostic tool for characterizing the spatial distribution of the absorbed laser energy along the laser spark column. This requires the assumption that the absorbed energy is proportional to the sum of the values $R_c{}^5$ along the laser spark. The results (shown in Figure 14) indicate that the maximal absorption of the laser beam takes place close to the focal point. As expected, at lower energies the absorbed energy monotonically decreases in the axial direction. However, at the highest energy (of 160 mJ) a local maximum was observed just before the focal point. This was explained in terms of the plasma shielding effect. Another possible explanation was that at high energies the focus moves

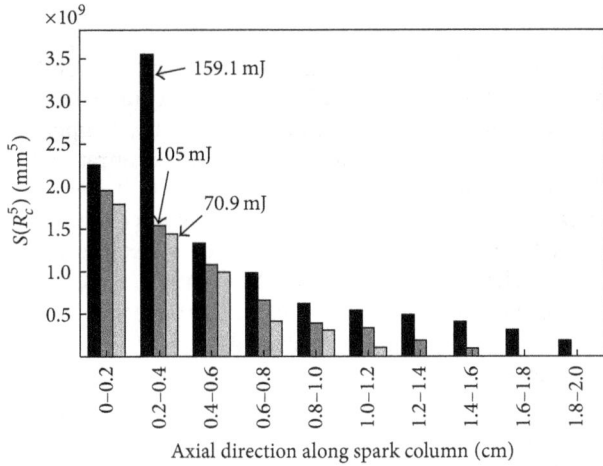

FIGURE 14: Axial distribution of the absorbed energy along the laser spark column for different laser pulse energies. The laser beam direction was from right to left. Reprinted from [71], with permission from Elsevier.

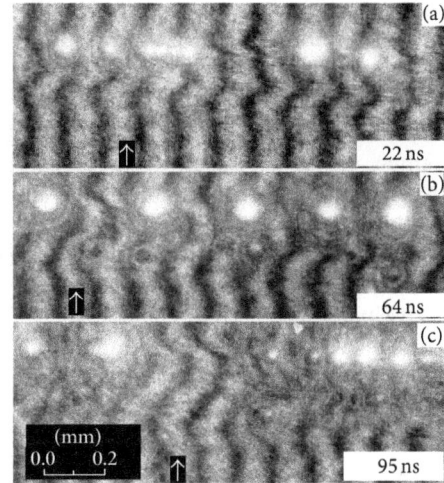

FIGURE 15: Fragments of interferograms measured at different time delays, illustrating the dynamics of a warmed channel in ethanol. The output of the laser pulse was 65 mJ. The radiation goes from right to left. The arrows indicate the planes where the radial distributions of the refraction index were calculated. Reprinted from [71], with permission from Elsevier.

towards the heating laser source due to nonlinear effects such as self-focusing.

5.7. Warmed Channels in Liquid Suspensions.
The laser radiation might form warmed channels in liquid suspensions. It has to be taken into account especially in spectral measurements performed at high repetition rates. Interference imaging allows for measurement of this effect in the focal volume region.

The temperature change, ΔT, was estimated using the interference measurements. The change in the refractive index was approximated using the following expression: $\Delta n = \Delta k \lambda / dc$, where Δk is the experimentally measured fringe shift in the warmed region, λ is the wavelength of the diagnostic radiation, and dc is the diameter of the warmed region. This way, the temperature change was estimated as

$$\Delta T = \frac{\Delta n}{(\partial n / \partial T)_\rho}. \qquad (6)$$

For water, the constant $(\partial n / \partial T)_\rho = -0.8 \cdot 10^{-4}$ [81]. Therefore, the corresponding temperature change in water is $\Delta T \approx 50$ K. This value was in agreement with the result obtained from conservation of energy consideration. It was found that the warmed channels in water last more than 1 μs and then they are deactivated by the process of thermal conductivity.

A different mechanism was observed in alcohols. In this case the warmed channels generate cylindrical shockwaves, expanding in radial direction. The dynamics of such cylindrical shockwaves in alcohols are interesting and are described in the following.

5.8. Dynamics of Cylindrical Shockwaves in the Alcohols.
The dynamics of the cylindrical shockwaves generated in alcohols were obtained from interferograms of the warmed regions. Examples of the interferograms obtained from ethanol suspensions and the associated cylindrical shockwaves at several

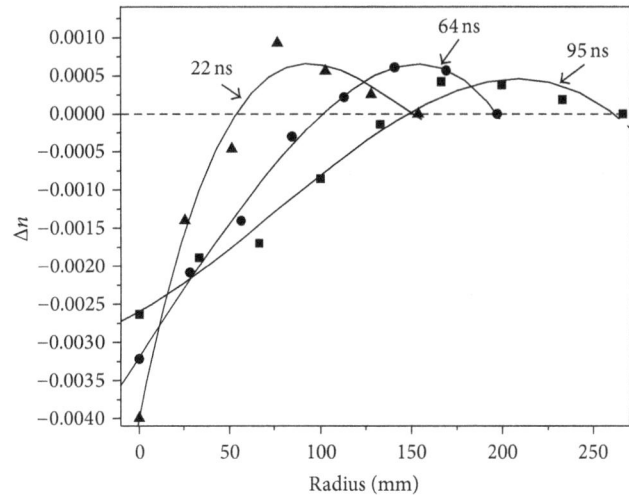

FIGURE 16: The calculated radial distribution of the refraction index of ethanol in the warmed region at three time delays. The data characterize the cylindrical shockwaves. Reprinted from [72], with permission from Elsevier.

delay times are shown in Figure 15. Such data were used for calculating the corresponding changes in the refractive index in the warmed channel and in the associated cylindrical shockwave.

The radial distribution of the refraction index change in ethanol is presented in Figure 16. In general, warming of the liquid is associated with reducing the index of refraction, while compression in the shockwave increases the index of refraction. The maximum negative change of the refractive index is on the axis of the heating beam ($r = 0$), where the temperature achieves a maximal value. The maxima in this figure indicate the locations of the higher compression, namely, the shockwave fronts. As time evolves, the maxima are found at larger radii. The results also indicate that the quantity of the thermal energy absorbed in the focal volume is sufficient for generating the cylindrical shockwave. It moves in radial direction with the speed of ~1.5 km s^{-1}. This way, the actual temperature change at any time delay and location could also be estimated. For example, the temperature change in ethanol, at 22 ns and at $r = 0$, was $\Delta T \sim 10°$K.

No cylindrical shockwaves were observed in water suspensions. Estimations have shown that the energy absorbed in water is dissipated by thermal conductivity. Full explanation of this difference between water and alcohols is not yet understood. A possible explanation was suggested, based on the differences in the coefficient of volumetric thermal expansion of these liquids.

The cylindrical shockwaves were observed in many alcohols. The dynamics are, in principle, very similar: starting their motion with supersonic speeds the cylindrical shockwaves quickly decelerate and expand asymptotically approaching the speed of sound.

6. Analytical Applications of Laser Breakdown in Water Suspensions

6.1. LIBS Method. Laser-induced breakdown spectroscopy (LIBS) is a multielemental analytical method based on time-resolved measurement of atomic emission lines from laser plasma generated at a sample surface or inside. The most appealing quality of LIBS is the potential of remote and in situ applications. LIBS offers the ability to transfer the laser energy through an optical fiber to generate plasma and transport the resulting emission back through the fiber to a remote spectrometer, while keeping the sample free from contamination and the analyst isolated from potentially harmful environments.

The advantages of LIBS for direct spectrochemical analysis include the following: (a) no or little sample preparation needed; (b) versatile sampling of all media; (c) very small amount of sample vaporized; (d) ability to analyze extremely hard materials that can hardly be digested or dissolved, such as ceramics and superconductors; (e) local analysis in microregions offering a spatial resolving power of about 1–100 μm; (f) possibility of simultaneous multielement analysis.

Good analytical performance requires high signal to noise ratio. At the very beginning of the plasma formation a continuum light emission takes place; this tends to overshadow the emission lines from the excited species in the plasma. Thus, the LIBS spectral acquisition has to be delayed until the plasma cools down. This may improve the signal to background ratio. The delay time (t_d) is the time between the laser pulse and the starting of the spectral acquisition. The gate width (t_w) is the time period over which the plasma light is collected [6]. By using a time-resolved detection system to control t_d and t_w, the intense initial continuum emission at the early stage of the plasma formation can be gated off, and the signal to noise ratio can be improved dramatically. It was found that gate width longer than 50 μs was unnecessary because most of the analytical emission lines had completely decayed [82].

LIBS performance is often affected by environmental factors, instrumental parameters, the chemical and physical properties of the sample, and the homogeneity and the ambient states of the sample. Many instrumental parameters have been optimized, including air humidity, temperature and pressure, laser pulse energy [83, 84], pulse duration [85], pulse repetition rate [86], laser wavelength [87], delay time and gate width [58–89], and the number of laser shots [90].

The LIBS spectra contain large and complex information about the elemental composition of the sample, which can be used for qualitative as well as quantitative analysis of many elements. In a matter of microseconds, tens of thousands of data points can be collected. Therefore, the use of a suitable chemometric data analysis method to efficiently and accurately analyze the complex LIBS spectra is of considerable importance. Several relevant algorithms have been developed [91–93]. In most applications, LIBS spectra are analyzed using univariate approach (based on single emission lines), which causes the loss of useful information [94, 95]. However, also multivariate analysis methods have been applied to LIBS [96–100]. The results indicated that using several emission lines in combination provides improved performance [97, 101].

6.2. LIBS of Water Suspensions. Direct application of LIBS to analysis of various materials in bulk liquids results in very low spectral intensities [102, 103]. This might be surprising, since LIBS works well for aerosols in air [11, 104–106] or on solid particulate samples [107–113]. Even analysis of individual aerosol particles in air was successfully accomplished [106]. The main difference in water is that this medium, in contrast to air, suppresses the plasma expansion, which results in reduced sensitivities. The plasma emission in water has a very short lifetime and consists of a superposition of the elemental lines over a wide and intense continuum emission.

In order to overcome these problems, several alternative methods were developed. For example, a coaxial nozzle was used in order to enhance the spectral intensities obtained from colloidal iron. The nozzle generated a constant flow and the stream of water was irradiated by a laser pulse. This method resulted in much higher LIBS intensities than in liquid bulk and allowed for the determination of colloidal iron in the low ppm concentration range. Another approach was the collection of the suspended particles on a filter for subsequent LIBS analysis. This way, analysis of biological materials such as pollen, bacteria, molds, fungi, and viruses in water could be carried out [114, 115]. Similarly, boron, uranium, and thorium [116, 117] were also analyzed in water and results in the low ppm range were obtained. Filter assisted LIBS analysis was also developed for analysis of single biological microparticles such as pollens [118]. The measurements showed that bioaerosols could be distinguished from other common aerosols, for example, dust. Also *E. coli* bacteria could be identified on paper filters using LIBS [119]. Samples were prepared by aspirating a solution of bacteria on a filter paper and drying for an hour. The results showed that the samples could be characterized by their profile of spectral intensity and varieties of trace mineral elements. These trace elements are divalent cations, which maintain the cohesion

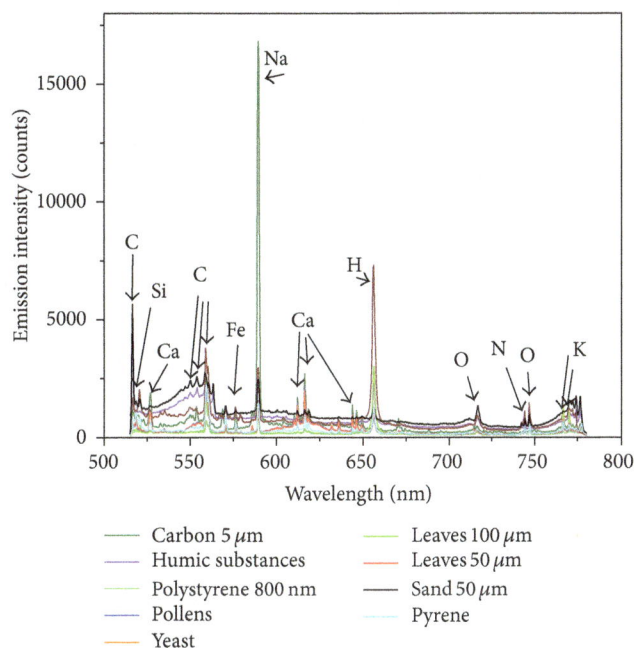

FIGURE 17: LIBS raw spectra of a set of suspension materials. The gate delay was 2 μs and gate width was 10 μs. The laser pulse energy was 230 mJ.

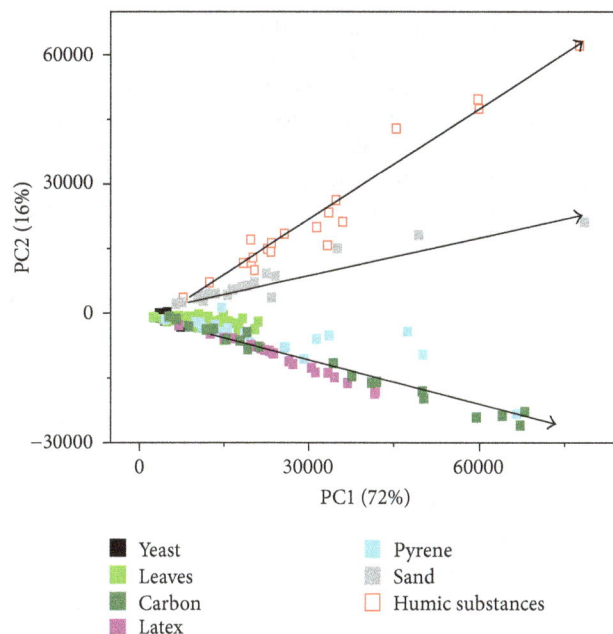

FIGURE 18: Principal component analysis (PCA) score plot of raw LIBS spectra.

of proteins in the outer bacterial membrane. It comes out that bacteria may be identified based on these trace mineral elements.

The main problem in LIBS analysis of suspended materials is explained by the following example. Figure 17 shows full LIBS spectra of a series of suspension materials. These include organic, inorganic, and biological particles, which represent particles naturally persistent in water. The tested materials were as follows:

(a) organic matter of biological origin: leaves (50 μm) and pollens (*Acacia baileyana*);

(b) organic matter of nonbiological origin: carbon (5 μm), pyrene, polystyrene (800 nm), and humic compounds;

(c) biological matter: yeast (*S. cerevisiae*, representing eukaryotic microorganisms);

(d) inorganic matter: sand (50–100 μm).

The LIBS spectra of these materials contain multiple emission lines of many elements. Many of them contain similar elemental constituents: N, Na, K, C, Ca, Si, O, and H. Inspection of these spectra shows that simple analysis is not sufficient for distinction between the various materials. The conclusion is that LIBS analysis of suspended materials requires a statistical approach. Therefore, multivariate analysis was applied to these spectra, as described in the following.

6.3. Application of Multivariate Analysis to LIBS Spectra. Multivariate analysis (MVA) is an excellent tool for summarizing data of a large number of variables. Essentially MVA

defines new variables that capture most of the variability associated with the collected data. Principal component analysis (PCA) is a basic MVA method that helps in understanding the common variations within a set of variables.

The simplest PCA approach to LIBS data includes its application to the raw spectra. A scores plot model was constructed, based on four components. This model clustered materials of similar composition, as can be seen in Figure 18. Each point represents a single laser shot. The first two principal components, PC1 and PC2, accounted for 72% and 16% of the spectral variability, respectively, while the remaining variability was described by higher PCs. These results confirmed that most of the sample set could be classified by this model.

Figure 18 shows three vectors that include all organic materials of nonbiological origin (pyrene, carbon, and polystyrene), sand, and humic substances. Organic materials of biological origin (leaves and yeast) were concentrated in one location on the plot. These materials have similar elemental composition of atoms such as N, Na, K, C, Ca, O, and H. Organic materials of nonbiological source have a high concentration of carbon and combined into one vector on the PCA score plot. Sand differs from the other materials because of its high Si concentration and thus it formed its own vector. Humic substances also differ from the other materials in their basic composition. This model allowed for easy detection of these particles in water.

Due to laser power fluctuations and due to the particulate nature of the suspended materials, LIBS spectra suffer from instabilities. Therefore, analysis based on the raw spectra might not be optimal for detection of particles. Many articles suggested utilizing peak intensity ratios in LIBS analysis [120–123]. For example, classification of explosive-containing residues was successfully achieved using MVA based on the emission intensity ratios of O, H, C, Ca, and N lines. It was found that most explosives have higher O and N content relative to C and N. Thus, the atomic emission ratios could differentiate explosive organics from nonexplosive organics. Other ratios were useful in separating inorganic and organic samples. These findings indicate that identifying the correct peak ratios can improve the classification of suspended materials.

Concerning the above suspended materials, good results were obtained using PCA analysis of the discrete peak intensity ratios of C/H, Na/K, and Na/Ca. The thus obtained score plots are shown in Figure 19. The model was based on four components, where each point represented a single laser shot. This model grouped materials of similar composition. The first two principal components, PC1 and PC2, accounted for 77% and 17% of the spectral variability, respectively (94% together). These results indicated that the model based on these particular peak ratios performed better than the one based on the raw spectra. The model clearly identified four groups of materials: biological matter, humic substances, inorganic matter, and organic matter of biological and non-biological origin. As before, sand was well separated from the other materials on the score plot. This was attributed to the fact that this material has higher K and Na content relative to C, H, and Ca.

A separate group of all organic materials was expected, since they have a high concentration of carbon and their C/H peak intensity ratios are similar. However, the results showed that yeast was well separated from other materials. While all cells contain the same elements, the ratios differ between eukaryotic cells, which do not have a rigid cell wall, and plant cells, which have chloroplasts and vacuoles [124–126]. Hence, the C/H ratio of yeast differs from that of leaves. Therefore, this model was useful for separating particles of plant cells from eukaryotic cells, which are more harmful to humans. Actually, the combination of both models provided a very useful tool for analysis of suspended materials.

6.4. Application of Optical Breakdown in Liquids to Generation of Suspended Nanoparticles. Optical breakdown in liquids attracts much attention as a new technique of material processing [127]. In particular, generation of suspended nanoparticles is of considerable interest [128]. For example, the unique properties of gold and silver nanoparticles make them attractive in nanophotonics [129] and in biological labeling and sensing [130, 131]. Metallic nanostructures can be utilized in the development of sensors for volatile organic compounds (VOCs) and biomolecules.

Interferometric methods have the potential to elucidate the dynamics of the generation of suspended nanoparticles. The dynamic information can be used for optimizing the ablation process and for controlling the size and shape of the particles, as well as improving the yield.

6.4.1. Visualization of Suspended Nanoparticles Generation in Water. Visualization was carried out using time-resolved Mach-Zehnder interferometry, as previously described. The

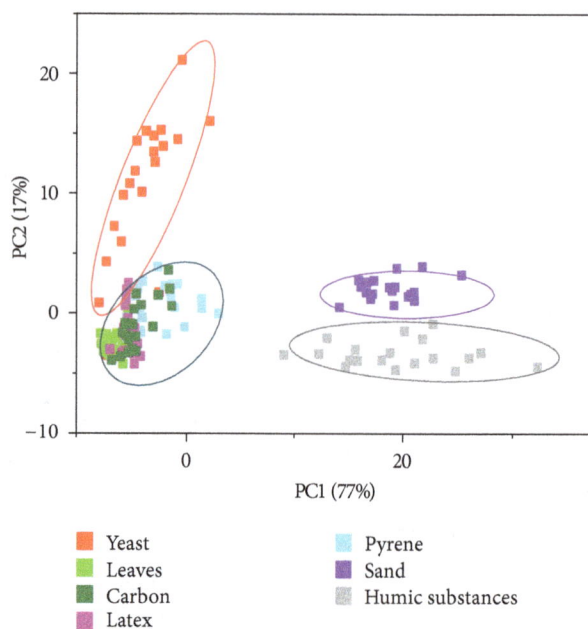

FIGURE 19: Principal component analysis (PCA) score plot of element peak ratios C/H, Na/K, and Na/Ca.

visualized process was initiated by irradiation of metal targets in double distilled water by a Nd:YAG laser ($\lambda = 1064$ nm, 6 ns). The laser was focused by singlets of different focal lengths.

The time-resolved interferograms obtained from irradiation of silver and gold targets are shown in Figures 20 and 21. These figures illustrate the different stages of the expansion dynamics of the high temperature plasma generated on gold and silver targets. The expanding plasma is associated with a strong shockwave in the surrounding liquid under the conditions of the well-developed vaporization. The surface plasma expands half-spherically and accelerates a layer of the surrounding liquid out of the target. In the early stages of plasma expansion, the associated shockwaves move in the water at supersonic velocities in opposite direction from the metal target. It seems that in these cases the dynamics can be approximated by the model of point explosion, with expansion into half-sphere.

At time delays of $\tau_d \leq 40$ ns the shockwave front propagates together with the plasma and moves with supersonic velocity. At longer delays, the front continues its motion with supersonic velocities and outdistances the contact boundary between the plasma and water. After this, the plasma rapidly decelerates its motion. Approximately this moment of time is illustrated by the interferograms in Figures 20(b) and 21(b). The detached spherical shockwave rapidly diminishes its speed and moves at moderate supersonic speeds in opposite direction from the target. For time delays longer than ~80 ns the shockwave asymptotically moves at the sonic speed in water (ca. 1.46 km sec^{-1}).

Clear differences are observed between gold and silver targets. In silver, the front of the shockwave is much wider

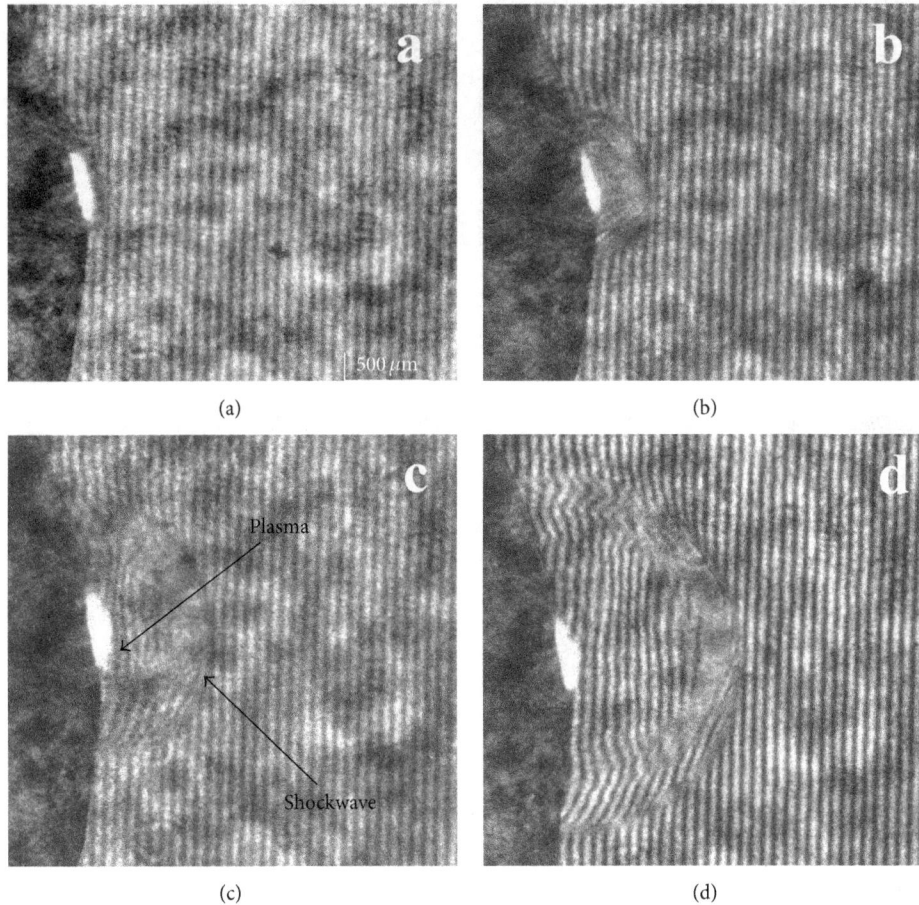

FIGURE 20: Interferograms measured at several time delays after the laser pulse: (a) 23 ns, (b) 163 ns, (c) 380 ns, and (d) 614 ns. The target was gold surface under water. The laser was focused using a lens of f = 77 mm and the power was 75 mJ (at target).

than in gold, which might indicate removal of more material from the bulk.

The time-resolved interferograms allow for estimating the shockwave radius as a function of time. This information can be converted (under certain reasonable assumptions) to expansion velocity as a function of time. The results are shown in Figure 22. Clearly, at the very beginning of the shockwave expansion, the velocities are higher for gold targets than for silver. After ca. 8 ns the order is reversed and at long times both shockwaves attenuate and reach the asymptotic M = 1 velocity.

6.4.2. Effect of Surface Energy Density on Particle Generation.

The energy density at a gold surface target was varied by changing its distance from the focusing lens (of f = 77 mm). The laser pulse energy on the target was unchanged (75 mJ). The absorption spectra of thus generated water suspensions were measured. These spectra were measured since the absorption spectra of nanoparticles depend on their size and shape [80, 132–134]. The results are shown in Figure 23.

The results indicated that surface energy density plays an important role. The highest absorbance was observed at the highest surface energy density, which means that this parameter directly affects the particle generation yield.

Moreover, the wavelength of maximum absorbance of gold nanoparticles in suspension indicates their morphology. In this particular case, the maximum absorbance peak was detected at 528 nm, which indicates that the mean size of the nanoparticles was ~35 nm. The wavelength of maximum absorbance is not affected by the surface energy density, which means that this parameter cannot modify the mean size of the nanoparticles.

6.4.3. Effect of Laser Pulse Duration on Particle Generation.

The laser pulse-width controls the temporal distribution of the energy on the solid surface; thus, it can influence the particles production [134]. Evaluation of this effect was examined by using two different pulse lengths (at constant total pulse energy and under the same geometrical conditions). The short pulses were of 6 ns and the long pulses were on 120 μs. The absorption spectra of the suspensions obtained under these conditions were measured. The results for gold target are shown in Figure 24.

The first observation was that in both cases the wavelength of maximum absorbance is the same (528 nm). Therefore, the conclusion was that the mean size of the particles is not affected by the pulse duration (ca. 35 nm in both cases). However, the production yield was much different.

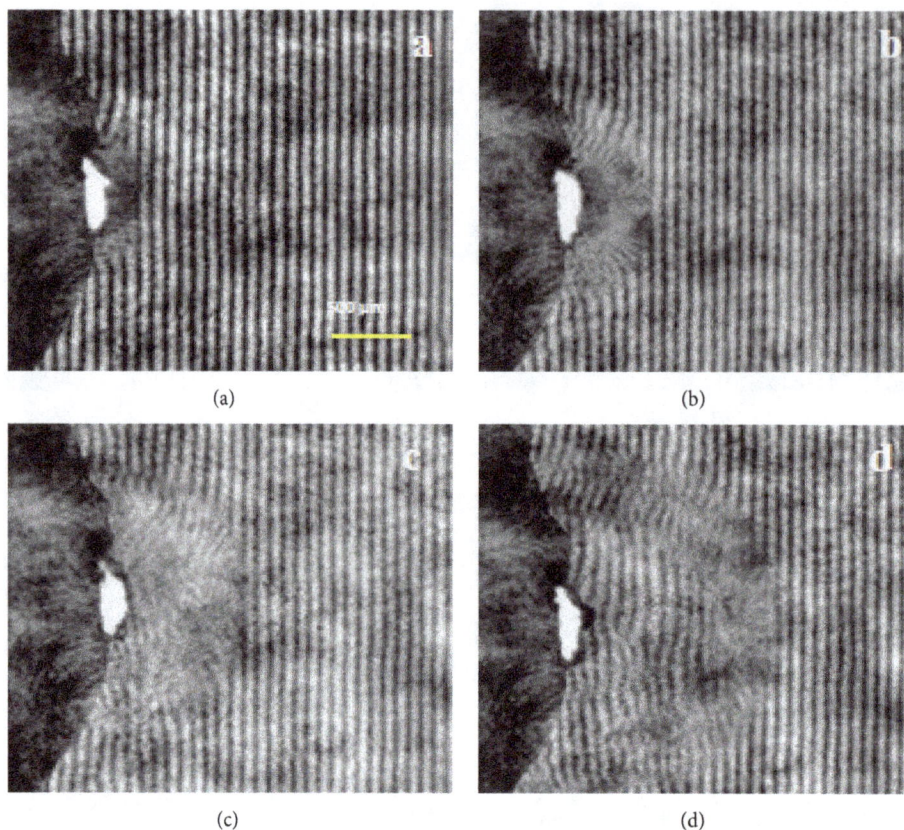

FIGURE 21: Interferograms measured at several time delays after the laser pulse: (a) 25 ns, (b) 173 ns, (c) 352 ns, and (d) 600 ns. The target was silver surface under water. The laser was focused using a lens of $f = 77$ mm and the power was 75 mJ (at target).

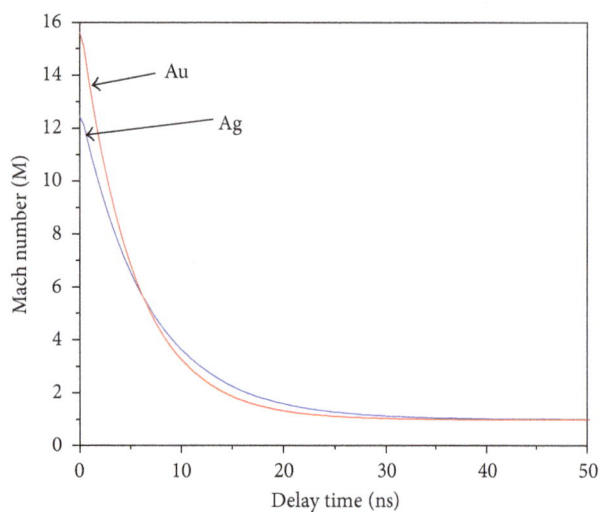

FIGURE 22: Shockwaves Mach numbers in gold and silver suspensions as a function of time; laser energy on target was 75 mJ, lens $f = 77$ mm.

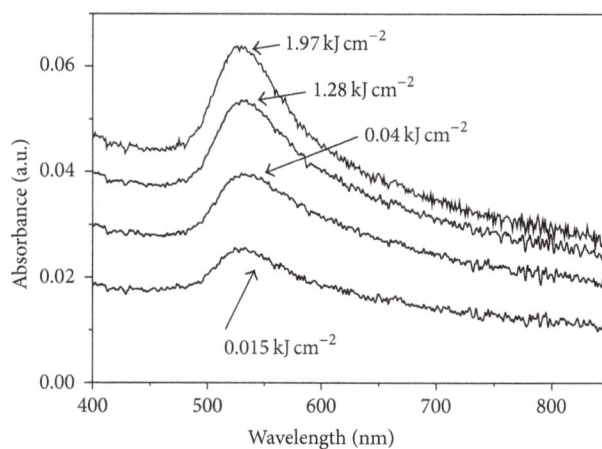

FIGURE 23: Absorbance spectra of gold suspensions at several surface energy densities. The laser pulse energy on the target was 75 mJ.

Note that 60 min of irradiation using the long pulses was needed in order to reach the same absorbance as obtained by 3 min of short pulse irradiation. Longer pulses decrease the production yields. Another observation was that the particles size distribution obtained using long pulses was narrower than that obtained using the short pulses. This can be attributed to the fact that at the long pulses the energy flux is lower, which allows for local equilibration, thus preventing extreme events that might result in very small or very large particles.

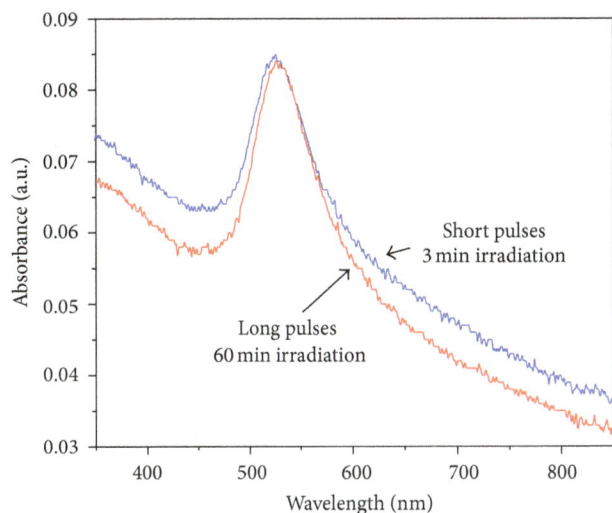

FIGURE 24: Absorbance spectra of gold nanoparticles in water, generated by short (6 ns) and long (120 μs) laser pulses. Note that much longer irradiation time (60 min) was needed using the long pulses, in order to obtain an absorbance similar to that obtained from short pulses.

The above results indicate that pulsed laser ablation of metals (e.g., silver and gold) in water enables synthesis of a variety of highly pure nanoparticles that can be integrated as functional components into sensors and different materials. The size and shape of the generated nanoparticles were found to be affected by illumination parameters (focusing lens, pulse energy, and duration).

7. Conclusions and Future Directions

Breakdown in colloidal suspensions is of considerable interest, both due to its scientific importance and due to its potential applications in chemical analysis of liquids. Since the atmosphere is rich in a large variety of aerosols at high concentrations, it comes out that all liquids exposed to air are actually suspensions. Therefore, breakdown in liquids should be almost always treated as breakdown in suspensions. The main consequence is that these breakdown events are of discrete nature and they are mainly governed by the nature of particulates.

The discrete nature of the breakdown and the crucial effects of the persistent particulates were not always understood. This was only recently proven, when high temporal and spatial measurements were performed using interferometric techniques. These measurements provided insight into the microstructure of the breakdown and into its dynamic processes, starting from the first few nanoseconds and up to the millisecond range. The differences in the behavior of various liquids could be explained in terms of the differences in their physical properties. The origin of the cylindrical shockwaves observed in some liquids and the way they are generated from merging of spherical individual events was thus revealed. Understanding the nature of the breakdown in liquids allowed for a new definition of their energy

threshold. While old definitions were based on statistical measurements, recent studies suggested a definition based on simple measurements of the spark column lengths.

Understanding breakdown in liquids is also of considerable practical value. It can contribute to better analytical applications of this phenomenon. The limitations of the related analytical method in liquids are now clear and possible solutions have been suggested. While the performance of LIBS in liquids is still inferior, application of multivariate analysis methods has the potential to succeed in classifying suspended materials.

Besides the important application to LIBS analysis, optical breakdown in liquids can also be utilized for generation of suspended nanoparticles. This process is of importance in development of new sensors and various functional components. It was found that the size and shape of the suspended nanoparticles can be controlled by the laser irradiation characteristics, such as pulse energy, pulse duration, and focusing conditions.

It seems that further investigations are still needed. Clearly, temporal and spatial information on multipulse events is of great potential. So far, the powerful interferometric methods were only applied for single-pulse events, while it is well known that multiple pulses provide better LIBS performance in liquids. Note that although the signals are commonly integrated over many pulses, these experiments are still called single-pulse measurements, because all breakdowns were individual events. In contrast, in multipulse experiments pulses are temporally and spatially correlated. In a multipulse event in water suspension, the plasmas are generated in the microenvironment created by the previous pulses. In these cases, the shockwave dynamics described above are no longer valid. Interferometric visualization of the plasma evolution in multipulse events is expected to provide insight into the breakdown processes, which might result in better LIBS performance.

An additional promising research direction is the detailed understanding of the effects of the chemical properties of individual inclusion particulates upon the induced breakdown. Currently, only integrated results are obtained and one can speculate that some of the observed signal fluctuations are attributed to variations in the properties of the particulates. Therefore, understanding of these effects might result in better signal to noise ratios.

Breakdown on solid surfaces in water was also shown to be instrumental for fabrication of nanoparticles. So far, single-pulse measurements were performed and it has been shown that the properties of the suspended nanoparticles can be controlled by the irradiation parameters. One can envision new multipulse experiments, where the breakdowns are carried out in the microenvironment of the previous events. Under such experimental conditions, the dynamics are expected to be different and will probably result in different characteristics of the nanoparticles. Consider, for example, a sequence of pulses, controlled in time and phase, hitting the same surface site. This might provide much better control over the ablation processes and the resulting nanoparticle characteristics.

Another research of considerable interest is the optical breakdown in microsuspensions of various chemical compositions, for their fragmentation into nanoparticles. It has been shown that breakdown in suspensions is of discrete nature and it occurs at the inclusions. The fate of these inclusions has not been investigated yet. One can expect that the result would be an explosion of the microparticles and their fragmentation into nanoparticles. The resulting size distribution and the parameters that allow controlling the outcome of such microexplosions are still to be investigated.

Conflict of Interests

The authors declare that there is no conflict of interests regarding the publication of this paper.

References

[1] S. Yariv and H. Gross, *Geochemistry of Colloid Systems for Earth Scientists*, Springer, Berlin, Germany, 1979.

[2] D. Shaw, *Introduction to Colloid and Surface Chemistry*, Butterworth-Heinemann, London, UK, 1980.

[3] T. Wagner, T. Bundschuh, R. Schick, T. Schwartz, and R. Köster, "Investigation of colloidal water content with laser-induced breakdown detection during drinking water purification," *Acta Hydrochimica et Hydrobiologica*, vol. 30, no. 5-6, pp. 266–274, 2003.

[4] J. N. Ryan and M. Elimelech, "Colloid mobilization and transport in groundwater," *Colloids and Surfaces A: Physicochemical and Engineering Aspects*, vol. 107, pp. 1–56, 1996.

[5] M. Plaschke, T. Schäfer, T. Bundschuh et al., "Size characterization of bentonite colloids by different methods," *Analytical Chemistry*, vol. 73, no. 17, pp. 4338–4347, 2001.

[6] D. Cremers and L. Radziemski, *Handbook of Laser Induced Breakdown Spectroscopy*, Wiley, New York, NY, USA, 2006.

[7] D. Mukherjee, A. Rai, and M. R. Zachariah, "Quantitative laser-induced breakdown spectroscopy for aerosols via internal calibration: application to the oxidative coating of aluminum nanoparticles," *Journal of Aerosol Science*, vol. 37, no. 6, pp. 677–695, 2006.

[8] Ş. Yalçin, D. R. Crosley, G. P. Smith, and G. W. Faris, "Influence of ambient conditions on the laser air spark," *Applied Physics B: Lasers and Optics*, vol. 68, no. 1, pp. 121–130, 1999.

[9] D. E. Kim, K. J. Yoo, H. K. Park, K. J. Oh, and D. W. Kim, "Quantitative analysis of aluminum impurities in zinc alloy by laser-induced breakdown spectroscopy," *Applied Spectroscopy*, vol. 51, no. 1, pp. 22–29, 1997.

[10] M. Martin and M.-D. Cheng, "Detection of chromium aerosol using time-resolved laser-induced plasma spectroscopy," *Applied Spectroscopy*, vol. 54, no. 9, pp. 1279–1285, 2000.

[11] I. Schechter, "Direct aerosol analysis by time resolved laser plasma spectroscopy—improvement by single shot measurements," *Journal of Analytical Science and Technology*, vol. 8, pp. 779–786, 1995.

[12] T. Bundschuh, T. U. Wagner, and R. Köster, "Laser-induced Breakdown Detection (LIBD) for the highly sensitive quantification of aquatic colloids. Part I: Principle of LIBD and mathematical model," *Particle and Particle Systems Characterization*, vol. 22, no. 3, pp. 172–180, 2005.

[13] H. G. Barth and R. B. Flippen, "Particle size analysis," *Analytical Chemistry*, vol. 67, no. 12, pp. 257–272, 1995.

[14] F. J. Scherbaum, R. Knopp, and J. I. Kim, "Counting of particles in aqueous solutions by laser-induced photoacoustic breakdown detection," *Applied Physics B*, vol. 63, no. 3, pp. 299–306, 1996.

[15] C. Walther, S. Büchner, M. Filella, and V. Chanudet, "Probing particle size distributions in natural surface waters from 15 nm to 2 μm by a combination of LIBD and single-particle counting," *Journal of Colloid and Interface Science*, vol. 301, no. 2, pp. 532–537, 2006.

[16] T. Bundschuh, R. Knopp, and J. I. Kim, "Laser-induced breakdown detection (LIBD) of aquatic colloids with different laser systems," *Colloids and Surfaces A: Physicochemical and Engineering Aspects*, vol. 177, no. 1, pp. 47–55, 2001.

[17] R. Kaegi, T. Wagner, B. Hetzer, B. Sinnet, G. Tzvetkov, and M. Boller, "Size, number and chemical composition of nanosized particles in drinking water determined by analytical microscopy and LIBD," *Water Research*, vol. 42, no. 10-11, pp. 2778–2786, 2008.

[18] E. Damon and R. Thomlinson, "Observation of ionization of gases by a ruby laser," *Applied Optics*, vol. 2, no. 5, pp. 546–547, 1963.

[19] G. Weyl, *Physics of Laser-Induced Breakdown*, Marcel Dekker, New York, NY, USA, 1989.

[20] Y. P. Raizer, *Laser-Induced Discharge Phenomena*, Consultants Bureau, New York, NY, USA, 1977.

[21] J. R. Bettis, "Correlation among the laser-induced breakdown thresholds in solids, liquids, and gases," *Applied Optics*, vol. 31, no. 18, p. 3448, 1992.

[22] A. Alcock, C. DeMichelis, K. Hamal, and B. Tozer, "Expansion mechanism in a laser-produced spark," *Physical Review Letters*, vol. 20, no. 20, pp. 1095–1097, 1968.

[23] P. Chan and C. Lee, "Holographic schlieren investigation of laser-induced plasmas," *Physics Letters A*, vol. 62, no. 1, pp. 33–35, 1977.

[24] C. E. Bell and J. A. Landt, "Laser-induced high-pressure shock waves in water," *Applied Physics Letters*, vol. 10, no. 2, p. 46, 1967.

[25] M. P. Felix and A. T. Ellis, "Laser–induced liquid breakdown-a step-by-step account," *Applied Physics Letters*, vol. 19, no. 11, p. 484, 1971.

[26] W. Lauterborn and K. J. Ebeling, "High-speed holography of laser-induced breakdown in liquids," *Applied Physics Letters*, vol. 31, no. 10, pp. 663–664, 1977.

[27] F. Docchio, P. Regondi, M. R. C. Capon, and J. Mellerio, "Study of the temporal and spatial dynamics of plasmas induced in liquids by nanosecond Nd:YAG laser pulses 1: analysis of the plasma starting times," *Applied Optics*, vol. 27, no. 17, pp. 3661–3668, 1988.

[28] A. Vogel, K. Nahen, D. Theisen, and J. Noack, "Plasma formation in water by picosecond and nanosecond Nd:YAG laser pulses—part I: optical breakdown at threshold and superthreshold irradiance," *IEEE Journal on Selected Topics in Quantum Electronics*, vol. 2, no. 4, pp. 847–859, 1996.

[29] V. Bulatov, R. Krasniker, and I. Schechter, "Study of matrix effects in laser plasma spectroscopy by combined multifiber spatial and temporal resolutions," *Analytical Chemistry*, vol. 70, no. 24, pp. 5302–5311, 1998.

[30] I. Schechter and V. Bulatov, "Plasma morphology," in *LIBS—Fundamentals and Applications*, A. Miziolek, V. Palleschi, and I. Schechter, Eds., Cambridge University Press, Cambridge, Mass, USA, 2007.

[31] E. N. Glezer, C. B. Schaffer, N. Nishimura, and E. Mazur, "Minimally disruptive laser-induced breakdown in water," *Optics Letters*, vol. 22, no. 23, pp. 1817–1819, 1997.

[32] B. Zysset, J. G. Fujimoto, and T. F. Deutsch, "Time-resolved measurements of picosecond optical breakdown," *Applied Physics B Photophysics and Laser Chemistry*, vol. 48, no. 2, pp. 139–147, 1989.

[33] C. Vest, *Holographic Interferometry*, John Wiley & Sons, 1979.

[34] J. N. Butters, *Holography and Its Technology*, Peter Peregrinus, 1971.

[35] P. Hariharan, *Optical Holography*, Cambridge University Press, Cambridge, UK, 1984.

[36] P. Hariharan, *Basics of Interferometry*, Academic Press, New York, NY, USA, 1992.

[37] P. Hariharan, *Optical Interferometry*, Academic Press, 1985.

[38] T. Kreis, *Holographic Interferometry Principles and Methods*, Akademie Verlag, Berlin, Germany, 1996.

[39] R. Collier, C. Burckhardt, and L. Lin, *Optical Holography*, Academic Press, 1971.

[40] L. Bergmann and C. Schaefer, *Optics of Waves and Particles*, Walter de Gruyter, 1999.

[41] E. Hecht, *Optics*, Addison-Wesley, 1998.

[42] M. Born and E. Wolf, *Principles of Optics*, Cambridge University Press, Cambridge, UK, 1999.

[43] V. Venugopalan, A. Guerra III, K. Nahen, and A. Vogel, "Role of laser-induced plasma formation in pulsed cellular microsurgery and micromanipulation," *Physical Review Letters*, vol. 88, Article ID 078103, 2002.

[44] A. Vogel and V. Venugopalan, "Mechanisms of pulsed laser ablation of biological tissues," *Chemical Reviews*, vol. 103, no. 2, pp. 577–644, 2003.

[45] C. R. Phipps, *Laser Ablation and Its Applications*, Springer, Berlin, Germany, 2007.

[46] R. E. Russo, X. Mao, H. Liu, J. Gonzalez, and S. S. Mao, "Laser ablation in analytical chemistry—a review," *Talanta*, vol. 57, no. 3, pp. 425–451, 2002.

[47] V. Bulatov, R. Krasniker, and I. Schechter, "Converting spatial to pseudotemporal resolution in laser plasma analysis by simultaneous multifiber spectroscopy," *Analytical Chemistry*, vol. 72, no. 13, pp. 2987–2994, 2000.

[48] B. Dolgin, Y. Chen, V. Bulatov, and I. Schechter, "Use of LIBS for rapid characterization of parchment," *Analytical and Bioanalytical Chemistry*, vol. 386, no. 5, pp. 1535–1541, 2006.

[49] I. Schechter, "Laser induced plasma spectroscopy: a review of recent advances," *Reviews in Analytical Chemistry*, vol. 16, no. 3, pp. 173–298, 1997.

[50] A. Miziolek, V. Palleschi, and I. Schechter, Eds., *LIBS—Fundamentals and Applications*, Cambridge University Press, Cambridge, UK, 2007.

[51] V. Bulatov, L. Xu, and I. Schechter, "Spectroscopis imaging of laser-induced plasma," *Analytical Chemistry*, vol. 68, no. 17, pp. 2966–2973, 1996.

[52] R. Krasniker, V. Bulatov, and I. Schechter, "Study of matrix effects in laser plasma spectroscopy by shock wave propagation," *Spectrochimica Acta*, vol. 56, no. 6, pp. 609–618, 2001.

[53] V. Bulatov, A. Khalmanov, and I. Schechter, "Study of the morphology of a laser-produced aerosol plume by cavity ringdown laser absorption spectroscopy," *Analytical and Bioanalytical Chemistry*, vol. 375, no. 8, pp. 1282–1286, 2003.

[54] Y. Chen, V. Bulatov, L. Singer, J. Stricker, and I. Schechter, "Mapping and elemental fractionation of aerosols generated by laser-induced breakdown ablation," *Analytical and Bioanalytical Chemistry*, vol. 383, no. 7-8, pp. 1090–1097, 2005.

[55] G. A. Askaryan, A. M. Prokhorov, G. F. Chanturiya, and G. P. Shipulo, "The effects of a laser beam in a liquid," *Soviet Physics—JETP*, vol. 17, pp. 1463–1465, 1963.

[56] R. Brewer and K. Riekhoff, "Stimulated Brillouin scattering in liquids," *Physical Review Letters*, vol. 13, no. 11, pp. 334–336, 1964.

[57] E. F. Carome, C. E. Moeller, and N. A. Clark, "Intense ruby-laser-induced acoustic impulses in liquids," *The Journal of the Acoustical Society of America*, vol. 40, no. 6, p. 1462, 1966.

[58] L. C. Yang and V. J. Menichelli, "Detonation of insensitive high explosives by a Q–switched ruby laser," *Applied Physics Letters*, vol. 19, no. 11, p. 473, 1971.

[59] V. S. Teslenko, "Initial stage of extended laser breakdown in liquids," *IEEE Transactions on Electrical Insulation*, vol. 26, no. 6, pp. 1195–1200, 1991.

[60] F. Haruhisa, Y. Katsumi, and I. Yosio, in *Proceedings of the 6th International Conference on Conduction and Breakdown in Dielectric Liquids*, p. 247, Frontiers, Dreux, France, 1978.

[61] O. Martynenko, N. Stolovich, G. Rudinand, and S. Levchenko, in *Proceedings of the 8th ICOGER*, p. 64, Minsk, Belarus, 1981.

[62] N. Melikechi, H. Ding, O. A. Marcano, and S. Rock, "Laser-induced breakdown spectroscopy of alcohols and protein solutions," in *Proceedings of the 6th Ibero–American Conference on Optics (RIAO); 9th Latin-American Meeting on Optics, Lasers and Applications*, vol. 992 of *AIP Conference Proceedings*, pp. 1177–1182, 2008.

[63] A. Vogel, S. Busch, and U. Parlitz, "Shock wave emission and cavitation bubble generation by picosecond and nanosecond optical breakdown in water," *The Journal of the Acoustical Society of America*, vol. 100, no. 1, pp. 148–165, 1996.

[64] J. Noack and A. Vogel, "Single-shot spatially resolved characterization of laser-induced shock waves in water," *Applied Optics*, vol. 37, no. 19, pp. 4092–4099, 1998.

[65] A. Vogel, N. Linz, S. Freidank, and G. Paltauf, "Femtosecond-laser-induced nanocavitation in water: implications for optical breakdown threshold and cell surgery," *Physical Review Letters*, vol. 100, Article ID 038102, 2008.

[66] F. Docchio, "Lifetimes of plasmas induced in liquids and ocular media by single Nd:YAG laser pulses of different duration," *Europhysics Letters*, vol. 6, p. 407, 1988.

[67] A. Vogel, J. Noack, D. Theisen et al., "Energy balance of optical breakdown in water at nanosecond to femtosecond time scales," *Applied Physics B: Lasers and Optics*, vol. 68, no. 2, pp. 271–280, 1999.

[68] E.-A. Brujan and A. Vogel, "Stress wave emission and cavitation bubble dynamics by nanosecond optical breakdown in a tissue phantom," *Journal of Fluid Mechanics*, vol. 558, pp. 281–308, 2006.

[69] Y. Yasojima, "Experimental studies on breakdown probability in liquids by nanoseconds laser," *Kinki Daigaku Kogakubu Kenkyu Hokoku*, vol. 41, p. 65, 2007.

[70] Y. Yasojima, "Optical field dependence of breakdown in liquids by Q-switched lasers," in *Proceedings of the IEEE 15th International Conference on Dielectric Liquids (ICDL '05)*, vol. 67, pp. 67–70, Coimbra, Portugal, July 2005.

[71] G. Toker, V. Bulatov, T. Kovalchuk, and I. Schechter, "Microdynamics of optical breakdown in water induced by nanosecond laser pulses of 1064 nm wavelength," *Chemical Physics Letters*, vol. 471, no. 4-6, pp. 244–248, 2009.

[72] T. Kovalchuk, G. Toker, V. Bulatov, and I. Schechter, "Laser breakdown in alcohols and water induced by $\lambda = 1064$ nm nanosecond pulses," *Chemical Physics Letters*, vol. 500, no. 4–6, pp. 242–250, 2010.

[73] H. Young, R. Freedman, T. Sandin, and A. Ford, *Sears and Zemansky's University Physics*, Section 11-6, Addison-Wesley, 10th edition, 1999.

[74] A. Vogel and J. Noack, "Shock-wave energy and acoustic energy dissipation after laser-induced breakdown," in *Laser Tissue Interaction IX*, vol. 3254 of *Proceedings of SPIE*, p. 180, 1998.

[75] A. Vogel and W. Lauterborn, "Acoustic transient generation by laser–produced cavitation bubbles near solid boundaries," *The Journal of the Acoustical Society of America*, vol. 84, p. 719, 1988.

[76] L. I. Sedov, *Similarity and Dimensional Methods in Mechanics*, Mir Publishers, Moscow, Russia, 1982.

[77] P. K. Kennedy, D. X. Hammer, and B. A. Rockwell, "Laser-induced breakdown in aqueous media," *Progress in Quantum Electronics*, vol. 21, no. 3, pp. 155–248, 1997.

[78] T. Bundschuh, T. U. Wagner, and R. Köster, "Laser-induced breakdown detection (LIBD) for the highly sensitive quantification of aquatic colloids. Part I: principle of LIBD and mathematical model," *Particle & Particle Systems Characterization*, vol. 22, no. 3, pp. 172–180, 2005.

[79] Y. Zeldovich and Y. Raizer, *Physics of Shock Waves and High-Temperature Hydrodynamic Phenomena*, vol. 1, Academic Press, New York, NY, USA, 1966.

[80] J.-P. Sylvestre, A. V. Kabashin, E. Sacher, and M. Meunier, "Femtosecond laser ablation of gold in water: influence of the laser-produced plasma on the nanoparticle size distribution," *Applied Physics A*, vol. 80, no. 4, pp. 753–758, 2005.

[81] D. Lide, *Physical Constants of Organic Compounds*, CRC Press, Boca Raton, Fla, USA, 2005.

[82] A. S. Eppler, D. A. Cremers, D. D. Hickmott, M. J. Ferris, and A. C. Koskelo, "Matrix effects in the detection of Pb and Ba in soils using laser-induced breakdown spectroscopy," *Applied Spectroscopy*, vol. 50, no. 9, pp. 1175–1181, 1996.

[83] B. Charfi and M. A. Harith, "Panoramic laser-induced breakdown spectrometry of water," *Spectrochimica Acta Part B: Atomic Spectroscopy*, vol. 57, no. 7, pp. 1141–1153, 2002.

[84] M. A. Gondal, T. Hussain, and Z. H. Yamani, "Optimization of the LIBS parameters for detection of trace metals in petroleum products," *Energy Sources A*, vol. 30, no. 5, pp. 441–451, 2008.

[85] K. Y. Yamamoto, D. A. Cremers, L. E. Foster, M. P. Davies, and R. D. Harris, "Laser-induced breakdown spectroscopy analysis of solids using a long-pulse (150 ns) Q-switched Nd:YAG laser," *Applied Spectroscopy*, vol. 59, no. 9, pp. 1082–1097, 2005.

[86] R. Wisbrun, I. Schechter, R. Niessner, H. Schröder, and K. L. Kompa, "Detector for trace elemental analysis of solid environmental samples by laser plasma spectroscopy," *Analytical Chemistry*, vol. 66, no. 18, pp. 2964–2975, 1994.

[87] L. M. Cabalín and J. J. Laserna, "Experimental determination of laser induced breakdown thresholds of metals under nanosecond Q-switched laser operation," *Spectrochimica Acta Part B: Atomic Spectroscopy*, vol. 53, no. 5, pp. 723–730, 1998.

[88] L. J. Radziemski and T. R. Loree, "Laser-induced breakdown spectroscopy: time-resolved spectrochemical applications," *Plasma Chemistry and Plasma Processing*, vol. 1, no. 3, pp. 281–293, 1981.

[89] L. Dudragne, P. Adam, and J. Amouroux, "Time-resolved laser-induced breakdown spectroscopy: application for qualitative and quantitative detection of fluorine, chlorine, sulfur, and carbon in air," *Applied Spectroscopy*, vol. 52, no. 10, pp. 1321–1327, 1998.

[90] J. R. Wachter and D. A. Cremers, "Determination of uranium in solution using laser-induced breakdown spectroscopy," *Applied Spectroscopy*, vol. 41, no. 6, pp. 1042–1048, 1987.

[91] L. Xu and I. Schechter, "Wavelength selection for simultaneous spectroscopic analysis. Experimental and theoretical study," *Analytical Chemistry*, vol. 68, no. 14, pp. 2392–2400, 1996.

[92] I. Litani-Barzilai and I. Schechter, "Spectroscopic prediction of nonlinear properties by principal component regression," *Analytica Chimica Acta*, vol. 348, no. 1–3, pp. 345–356, 1997.

[93] L. Xu and I. Schechter, "A calibration method free of optimum factor number selection for automated multivariate analysis. Experimental and theoretical study," *Analytical Chemistry*, vol. 69, no. 18, pp. 3722–3730, 1997.

[94] R. Barbini, F. Colao, R. Fantoni et al., "Semi-quantitative time resolved LIBS measurements," *Applied Physics B: Lasers and Optics*, vol. 65, no. 1, pp. 101–107, 1997.

[95] L. Xu, V. Bulatov, V. V. Gridin, and I. Schechter, "Absolute analysis of particulate materials by laser-induced breakdown spectroscopy," *Analytical Chemistry*, vol. 69, no. 11, pp. 2103–2108, 1997.

[96] M. Z. Martin, N. Labbé, T. G. Rials, and S. D. Wullschleger, "Analysis of preservative-treated wood by multivariate analysis of laser-induced breakdown spectroscopy spectra," *Spectrochimica Acta Part B: Atomic Spectroscopy*, vol. 60, no. 7-8, pp. 1179–1185, 2005.

[97] N. Labbé, I. M. Swamidoss, N. André, M. Z. Martin, T. M. Young, and T. G. Rials, "Extraction of information from laser-induced breakdown spectroscopy spectral data by multivariate analysis," *Applied Optics*, vol. 47, no. 31, pp. G158–G165, 2008.

[98] M. Z. Martin, N. Labbé, N. André et al., "High resolution applications of laser-induced breakdown spectroscopy for environmental and forensic applications," *Spectrochimica Acta Part B: Atomic Spectroscopy*, vol. 62, no. 12, pp. 1426–1432, 2007.

[99] B. Bousquet, J.-B. Sirven, and L. Canioni, "Towards quantitative laser-induced breakdown spectroscopy analysis of soil samples," *Spectrochimica Acta Part B: Atomic Spectroscopy*, vol. 62, no. 12, pp. 1582–1589, 2007.

[100] R. Gaudiuso, M. Dell'Aglio, O. de Pascale, G. S. Senesi, and A. de Giacomo, "Laser induced breakdown spectroscopy for elemental analysis in environmental, cultural heritage and space applications: a review of methods and results," *Sensors*, vol. 10, no. 8, pp. 7434–7468, 2010.

[101] I. Schechter, R. Wisbrun, R. Nießner, H. Schröder, and K. L. Kompa, "Signal processing algorithm for simultaneous multi-element analysis by laser-produced plasma spectroscopy," in *International Society for Optical Engineering*, vol. 2093 of *Proceedings of SPIE*, 1994, pp. 310–321.

[102] R. Knopp, F. J. Scherbaum, and J. I. Kim, "Laser Induced Breakdown Spectroscopy (LIBS) as an analytical tool for the detection of metal ions in aqueous solutions," *Fresenius' Journal of Analytical Chemistry*, vol. 355, no. 1, pp. 16–20, 1996.

[103] Y. Ito, O. Ueki, and S. Nakamura, "Determination of colloidal iron in water by laser-induced breakdown spectroscopy," *Analytica Chimica Acta*, vol. 299, no. 3, pp. 401–405, 1995.

[104] M. Adamson, A. Padmanabhan, G. J. Godfrey, and S. J. Rehse, "Laser-induced breakdown spectroscopy at a water/gas interface: a study of bath gas-dependent molecular species," *Spectrochimica Acta Part B: Atomic Spectroscopy*, vol. 62, no. 12, pp. 1348–1360, 2007.

[105] L. Xu, V. Bulatov, V. V. Gridin, and I. Schechter, "Absolute analysis of particulate materials by laser-induced breakdown spectroscopy," *Analytical Chemistry*, vol. 69, no. 11, pp. 2103–2108, 1997.

[106] J. E. Carranza, B. T. Fisher, G. D. Yoder, and D. W. Hahn, "On-line analysis of ambient air aerosols using laser-induced breakdown spectroscopy," *Spectrochimica Acta Part B: Atomic Spectroscopy*, vol. 56, no. 6, pp. 851–864, 2001.

[107] K. Y. Yamamoto, D. A. Cremers, M. J. Ferris, and L. E. Foster, "Detection of metals in the environment using a portable laser-induced breakdown spectroscopy instrument," *Applied Spectroscopy*, vol. 50, no. 2, pp. 222–233, 1996.

[108] J.-B. Sirven, B. Bousquet, L. Canioni, and L. Sarger, "Laser-induced breakdown spectroscopy of composite samples: comparison of advanced chemometrics methods," *Analytical Chemistry*, vol. 78, no. 5, pp. 1462–1469, 2006.

[109] F. C. de Lucia Jr., J. L. Gottfried, C. A. Munson, and A. W. Miziolek, "Multivariate analysis of standoff laser-induced breakdown spectroscopy spectra for classification of explosive-containing residues," *Applied Optics*, vol. 47, no. 31, pp. G112–G122, 2008.

[110] R. Wisbrun, I. Schechter, R. Nießner, and H. Schröder, "Laser-induced breakdown spectroscopy for detection of heavy metals in environmental samples," in *International Conference on Monitoring of Toxic Chemicals and Biomarkers*, vol. 1716 of *Proceedings of SPIE*, March 1993.

[111] I. Schechter, R. Wisbrun, R. Nießner, H. Schröder, and K. L. Kompa, "Real-time detection of hazardous elements in sand and soils," in *Substance Detection Systems*, vol. 2092 of *Proceedings of SPIE*, pp. 174–185, International Society of Optical Engineering, March 1994.

[112] I. Schechter, V. Bulatov, L. Xu, V. V. Gridin, and R. Krasniker, "Improved laser-induced-breakdown analysis of environmental samples by plasma imaging," *Trends in Optics and Photonics*, vol. 36, pp. 178–180, 2000.

[113] H. Schröder, I. Schechter, R. Wisbrun, and R. Nießner, "Detection of heavy metals in environmental samples using laser spark analysis," in *Excimer Lasers: The Tools, Fundamentals of Their Interactions with Matter, Fields of Applications*, L. D. Laude, Ed., pp. 269–287, Kluwer Academic Publishers, Boston, Mass, USA, 1994.

[114] J. L. Gottfried, R. S. Harmon, F. C. de Lucia Jr., and A. W. Miziolek, "Multivariate analysis of laser-induced breakdown spectroscopy chemical signatures for geomaterial classification," *Spectrochimica Acta—Part B Atomic Spectroscopy*, vol. 64, no. 10, pp. 1009–1019, 2009.

[115] A. C. Samuels, F. C. DeLucia Jr., K. L. McNesby, and A. W. Miziolek, "Laser-induced breakdown spectroscopy of bacterial spores, molds, pollens, and protein: initial studies of discrimination potential," *Applied Optics*, vol. 42, no. 30, pp. 6205–6209, 2003.

[116] A. Sarkar, D. Alamelu, and S. K. Aggarwal, "Determination of thorium and uranium in solution by laser-induced breakdown spectrometry," *Applied Optics*, vol. 47, no. 31, pp. G58–G64, 2008.

[117] A. Sarkar, S. K. Aggarwal, K. Sasibhusan, and D. Alamelu, "Determination of sub-ppm levels of boron in ground water samples by laser induced breakdown spectroscopy," *Microchimica Acta*, vol. 168, no. 1-2, pp. 65–69, 2010.

[118] A. R. Boyain-Goitia, D. C. S. Beddows, B. C. Griffiths, and H. H. Telle, "Single-pollen analysis by laser-induced breakdown spectroscopy and raman microscopy," *Applied Optics*, vol. 42, no. 30, pp. 6119–6132, 2003.

[119] M. Yao, J. Lin, M. Liu, Q. Li, and Z. Lei, in *Proceedings of the 3rd International Conference on Biomedical Engineering and Informatics (BMEI '10)*, Yantai, China, October 2010.

[120] C. A. Porcelli, F. H. Gutierrez Boem, and R. S. Lavado, "The K/Na and Ca/Na ratios and rapeseed yield, under soil salinity or sodicity," *Plant and Soil*, vol. 175, no. 2, pp. 251–255, 1995.

[121] M. A. Khan, M. U. Shirazi, S. M. Mujtaba et al., "Role of proline, K/NA ratio and chlorophyll content in salt tolerance of wheat (*Triticum aestivum* L.)," *Pakistan Journal of Botany*, vol. 41, no. 2, pp. 633–638, 2009.

[122] F. C. De Lucia Jr., J. L. Gottfried, C. A. Munson, and A. W. Miziolek, "Multivariate analysis of standoff laser-induced breakdown spectroscopy spectra for classification of explosive-containing residues," *Applied Optics*, vol. 47, no. 31, pp. G112–G122, 2008.

[123] A. Kumar, F.-Y. Yueh, J. P. Singh, and S. Burgess, "Characterization of malignant tissue cells by laser-induced breakdown spectroscopy," *Applied Optics*, vol. 43, no. 28, pp. 5399–5403, 2004.

[124] T. E. Fox, E. G. H. M. van den Heuvel, C. A. Atherton et al., "Biovailability of selenium from fish, yeast and selenate: a comparative study in humans using stable isotopes," *European Journal of Clinical Nutrition*, vol. 58, no. 2, pp. 343–349, 2004.

[125] K. Okamoto and K. Fuwa, "Preparation and certification of tea leaves reference material," *Fresenius' Zeitschrift für Analytische Chemie*, vol. 326, no. 7, pp. 622–626, 1987.

[126] R. Giulian, C. E. I. dos Santos, S. de Moraes Shubeita, L. M. da Silva, J. F. Dias, and M. L. Yoneama, "Elemental characterization of commercial mate tea leaves (*Ilex paraguariensis* A. St.-Hil.) before and after hot water infusion using ion beam techniques," *Journal of Agricultural and Food Chemistry*, vol. 55, no. 3, pp. 741–746, 2007.

[127] S. Barcikowski, F. Devesa, and K. Moldenhauer, "Impact and structure of literature on nanoparticle generation by laser ablation in liquids," *Journal of Nanoparticle Research*, vol. 11, no. 8, pp. 1883–1893, 2009.

[128] G. W. Yang, "Laser ablation in liquids: applications in the synthesis of nanocrystals," *Progress in Materials Science*, vol. 52, no. 4, pp. 648–698, 2007.

[129] B. Khlebtsov, V. Zharov, A. Melnikov, V. Tuchin, and N. Khlebtsov, "Optical amplification of photothermal therapy with gold nanoparticles and nanoclusters," *Nanotechnology*, vol. 17, no. 20, pp. 5167–5179, 2006.

[130] T. A. Taton, C. A. Mirkin, and R. L. Letsinger, "Scanometric DNA array detection with nanoparticle probes," *Science*, vol. 289, no. 5485, pp. 1757–1760, 2000.

[131] G. Raschke, S. Kowarik, T. Franzl et al., "Biomolecular recognition based on single gold nanoparticle light scattering," *Nano Letters*, vol. 3, no. 7, pp. 935–938, 2003.

[132] S. Link and M. A. El-Sayed, "Size and temperature dependence of the plasmon absorption of colloidal gold nanoparticles," *The Journal of Physical Chemistry B*, vol. 103, no. 21, pp. 4212–4217, 1999.

[133] A. V. Kabashin and M. Meunier, "Femtosecond laser ablation in aqueous solutions: a novel method to synthesize non-toxic metal colloids with controllable size," *Journal of Physics: Conference Series*, vol. 59, article 354, 2007.

[134] B. C. Stuart, M. D. Feit, S. Herman, A. M. Rubenchik, B. W. Shore, and M. D. Perry, "Optical ablation by high-power short-pulse lasers," *Journal of the Optical Society of America B: Optical Physics*, vol. 13, no. 2, pp. 459–468, 1996.

19

Antioxidants Activity and Color Evaluation of Date Fruit of Selected Cultivars Commercially Available in the United States

Fahad Mohammed Al-Jasass,[1] Muhammad Siddiq,[2] and Dalbir S. Sogi[3]

[1]*King Abdulaziz City for Science & Technology, Life Science and Environment Research Institute,*
National Center for Agricultural Technology, P.O. Box 6086, Riyadh 11442, Saudi Arabia
[2]*Food Science Consulting, Windsor, ON, Canada N9H 2M4*
[3]*Department of Food Science and Technology, Guru Nanak Dev University, Amritsar, India*

Correspondence should be addressed to Fahad Mohammed Al-Jasass; aljasass@kacst.edu.sa

Academic Editor: C.-Y. Oliver Chen

Dates (*Phoenix dactylifera* L.) are nutrient-rich fruit consumed throughout the world, either directly or in several food products. Six commercially available date cultivars in the US were analyzed for total phenolics, antioxidant activity using ABTS, DPPH, FRAP, and ORAC assays, and instrumental color. Total phenolics content varied from 33 to 125 mg GAE/100 g dry weight, with the highest in Barni (Saudi Arabia). Antioxidant values as determined by the ABTS in Deglet Nour (Algeria), Deglet Nour (California), Deglet Noor (Tunisia), Shahia (Tunisia), Barni (Saudi Arabia), and Khudri (Saudi Arabia) were 1300, 1047, 796, 452, 776, and 341 μmol TE/g dry weight, respectively. Antioxidative properties as measure by DPPH, FRAP, and ORAC varied from 3.27 to 3.54, 3.29 to 5.22, and 189 to 243 μmol TE/g dry basis, respectively. Fruit and pulp color of Deglet Nour (Algeria) was lighter whereas pulp of Barni (Saudi Arabia) was the darkest. Antioxidant values varied with different techniques used and also followed a different pattern than that of phenolics content.

1. Introduction

The date palm (*Phoenix dactylifera* L.) is grown in over 30 countries. Nutrient-rich dates are relished for their sweet, succulent, and exotic flavor. In recent years, dates have found acceptance among consumers in North America and European countries. Beside fresh consumption, this fruit is also processed into a wide variety of value-added products, such as dry dates, date paste, date syrup, date juice concentrate, date jam, date butter, date bars, date chutney, date relish, and date pickles, whereas date oil and date coffee are some of the by-products produced from date seeds [1, 2].

Phytochemicals are naturally produced, nonnutritive, and bioactive compounds which are synthesized by plants for protection against external stresses and attack by pathogenic microorganisms [3]. Phytochemicals are reported to have various biological effects, such as antimutagenic, anticarcinogenic, antioxidant, antimicrobial, and anti-inflammatory [4].

These compounds can be divided into several classes—phenolics, alkaloids, steroids, terpenes, and saponins. Phenolic compounds are characterized as having potent antioxidants and free radical scavengers, which can act as hydrogen donors, reducing agents, metal chelators, and singlet oxygen quenchers [5]. Phenolic compounds are active antioxidants playing an important role in neutralization of free radicals and decomposition of peroxides [6].

Date fruit has been shown to possess strong antioxidant activity among twenty-eight fruits commonly consumed in China [6]. Studies on various date fruit cultivars demonstrated a linear relationship between antioxidant activity and phenolic content [7]. Dates serve as a good source of natural antioxidants and could potentially be considered as functional food or ingredient [8, 9]. Date fruit lowers the incidence of cancers, especially pancreatic cancer due to antitumor activity or antimutagenic properties, and boosts immune system [10–13]. Consumption of dates may also be beneficial in glycemic and lipid control in diabetic patients [14–16].

The objective of this study was to evaluate the total phenolics, antioxidant activity, and instrumental color of six commercially available date fruit cultivars in the United States.

2. Materials and Methods

2.1. Materials. Six commercially available date fruits, that is, Deglet Nour cultivated in California (DNC), Deglet Nour imported from Tunisia (DNT), Deglet Nour imported from Algeria (DNA), Khudri imported from Saudi Arabia (KSA), Barni imported from Saudi Arabia (BSA), and Shahia imported from Tunisia (SHT), were purchased from a local source in Lansing, MI, USA. The moisture content of cultivars was variable: DNC (24.1%), DNT (26.4%), DNA (25.3%), KSA (23.78%), BSA (20.4%), and SHT (22.3%). Therefore, all results are reported on dry weight basis to overcome the variation due to moisture content.

2.2. Sample Preparation. Samples were deseeded and crushed in Waring blender (20 g of date in 40 mL of distilled water); then, 2.5 g of the sample were taken in centrifuge tubes to which 20 mL of 80% methanol was added. The samples were stirred for 1 h at 200 rpm in a water bath shaker, followed by centrifugation at 10,000 ×g for 10 min. Supernatants were collected and the pellet was reextracted twice with 10 mL 80% methanol by vortexing and centrifugation at 10,000 ×g for 5 min. Finally, all the supernatants were pooled and volume was made to 50 mL with 80% methanol.

2.3. Total Phenolics. Total phenolic contents were determined using the procedures described by Zieslin and Ben-Zaken [17]. Briefly, 0.5 mL of methanolic extract was mixed with 0.5 mL Folin-Ciocalteu reagent by annual shaking for 15–20 sec. After 3 min, 1 mL saturated sodium carbonate and 1 mL of distilled water were added. The reaction mixture was incubated in the dark at room temperature for 2 h and its absorbance was measured at 725 nm against deionized water using spectrophotometer. Results were expressed in mg gallic acid equiv. (GAE)/100 g dry weight (dw).

2.4. Antioxidant Capacity by DPPH. Analysis was carried out following the methods of Brand-Williams et al. [18]. The stock solution (24 mg DPPH/100 mL methanol) was diluted with methanol to obtain an absorbance of 1.1 at 515 nm using spectrophotometer (Milton Roy, Warminster, PA, USA). 0.6 mL of the sample extracts, blank, or Trolox solution as standard was allowed to react with 3 mL of the DPPH working solution for 20 min under dark conditions. Then, the absorbance was taken at 515 nm. Radical scavenging capacity was calculated from the absorbance of sample, blank, or Trolox solution as standard. The standard curve was prepared using Trolox as standard versus radical scavenging activity and the results were expressed in terms of micromole Trolox Equivalence (μmol TE)/g (dw).

2.5. Antioxidant Capacity by ABTS. ABTS radical cation (ABTS$^{\bullet+}$) was produced by reacting ABTS stock solution (7 mM) with potassium per sulfate (2.45 mM) in the dark at room temperature for 12–16 h. The ABTS$^{\bullet+}$ Solution was diluted with methanol (80%) to an absorbance of 1.1 at 734 nm. Diluted ABTS$^{\bullet+}$ solution (3 mL) was added to 30 μL of sample or methanol for blank or Trolox solution as standard and the absorbance reading was taken at 30°C at 1 min interval up to 6 min. The oxidation index at 6 min and then percent antioxidant activity were calculated. Antioxidant activity was plotted against the concentration of Trolox to get standard curve. The results of samples were computed as μmol TE/g dw.

2.5.1. Ferric Reducing Antioxidant Power (FRAP). The ferric reducing ability of dates extract was measured calorimetrically according to the method developed by Benzie and Strain [19]. The stock solutions included 300 mM acetate buffer (3.1 g $C_2H_3NaO_2\cdot3H_2O$ and 16 mL $C_2H_4O_2$) pH 3.6, 10 mM TPTZ solution in 40 mM HCl, and 20 mM $FeCl_3\cdot6H_2O$ solution. The fresh experimental solution was prepared by mixing 2 mL acetate buffer, 2.5 mL TPTZ solution, and 2.5 mL $FeCl_3$ solution. A 3 mL FRAP reagent was taken in test tubes and 300 μL of standard, blank, or sample was added. After 5 min, the absorbance was recorded at 593 nm. Standard curve was constructed using Trolox and the results were expressed in μmol TE/100 g dw.

2.6. Oxygen Radical Absorbance Capacity (ORAC). The ORAC assay for extracted dates samples was conducted on FLx800 Fluorescence Microplate Reader and Gen5 Data Analysis Software (BioTek Instruments Inc., Winooski, VT, USA). Exterior wells were filled with 300 μL water; then, 150 μL diluted fluorescein solution was added to all the experimental wells. For blank wells, 25 μL 75 mM sodium phosphate buffer (pH 7.4) was added. A 25 μL of Trolox dilutions (6.25–100 μM) was added to Trolox wells. A 25 μL of the sample dilutions was added to sample wells. After incubation at 37°C for 30 min, 25 μL AAPH was added to all the wells. The fluorescence was recorded for 3 h and results computed using software were expressed as μmol TE/100 g dw color measurement.

Color of whole and crushed dates was measured by Hunter Color Meter (Hunter Associates Lab, Reston, VA, USA). Instrument was calibrated using standard black and white tiles. Samples were placed in the standard cup and color values were recorded as L (0, black; 100, white), a (−a, greenness; +a, redness), and b (−b, blueness; +b, yellowness). Eight readings of each sample were taken from different sides.

2.7. Statistical Analysis. The data from three replicates were statistically analyzed using analysis of variance (ANOVA) following K. A. Gomez and A. A. Gomez [20]. The treatment means were compared using the least significant difference (LSD) at the 5% level and were used to examine multiple comparisons between means according to Waller and Duncan [21]. All statistical analysis was performed using SAS software package, version 8.0 [22].

TABLE 1: Antioxidant activity of different date fruit cultivars as determined by ABTS, DPPH, FRAP, and ORAC assays.

Cultivars	Antioxidant activity (μmol Trolox equiv./g dry weight)			
	ABTS	DPPH	FRAP	ORAC
Deglet Nour (Algeria) (DNA)	1300.50 ± 73.36^a	3.54 ± 0.85^a	5.22 ± 0.34^a	235.05 ± 9.39^a
Deglet Nour (California) (DNC)	1047.34 ± 61.20^b	3.33 ± 0.86^a	3.80 ± 0.77^b	189.11 ± 14.61^c
Deglet Nour (Tunisia) (DNT)	795.74 ± 59.36^c	3.50 ± 0.88^a	4.65 ± 0.83^{ab}	243.05 ± 19.35^a
Shahia (Tunisia) (SHT)	451.80 ± 32.19^d	3.27 ± 0.81^a	3.70 ± 0.47^b	202.49 ± 17.74^c
Barni (Saudi Arabia) (BSA)	775.97 ± 65.87^c	3.31 ± 0.80^a	4.03 ± 0.22^{ab}	224.28 ± 17.77^b
Khudri (Saudi Arabia) (KSA)	341.38 ± 71.79^d	3.35 ± 0.78^a	3.29 ± 0.41^{ab}	206 ± 15.33^b

Data is mean of three replicates ± standard error. Means with different letters in the same column are significantly different at $p \leq 0.05$.

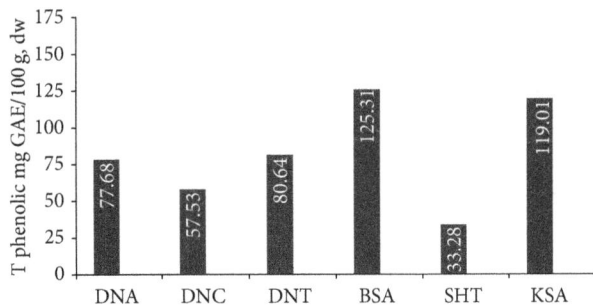

FIGURE 1: Total phenolic content of six date cultivars as gallic acid equivalent (GAE).

3. Results and Discussion

3.1. Total Phenolics.

The total phenolics contents of six cultivars varied from 33 to 125 mg GAE/100 g dw (Figure 1). The highest total phenolic content was observed in BSA whereas the lowest content was found in SHT. Wide variation was observed in the total phenolics contents among the date cultivars. ANOVA of total phenolics revealed significant differences ($p \leq 0.05$) among the various date cultivars. Dates of the Deglet Nour cultivar (DNA, DNC, and DNT) from different countries showed significant difference ($p \leq 0.05$), indicating the possible effect of location, weather, and agricultural practices. Al-Turki et al. [23] found that the range of total phenolics content in fresh dates was 225.0 to 507.0 mg GAE/100 g, which was significantly higher than that observed in the present study. Similarly, Al-Farsi et al. [24] reported the total phenolics content of 172.0 to 246.0 mg GAE/100 g fresh weight in selected Omani date varieties. In another study, Al-Farsi et al. [9] reported the phenolics content to be in the range of 217.0 to 343.0 mg GAE/100 g fresh weight. Mohamed et al. [25] reported total polyphenols content of 35.82 and 99.34 mg GAE/100 g in six varieties of dates cultivated in Sudan. Wu et al. [26] reported that dates contain relatively higher amounts of total phenolics as compared to other fresh or dried fruits in a comparative study of total phenolics in different fruits. Present results indicated lower phenolics content as compared to the previous reports which might be due to the genetic makeup, agricultural practices, and analytical procedures.

3.2. Antioxidant Activity of Date Fruit.

The results of the antioxidant assays showed that the antioxidant activity of date cultivars analyzed by ABTS, DPPH, FRAP, and ORAC varied to a large extent. Antioxidant values in terms of Trolox equivalent were low in DPPH assay than the highest observed by ABTS method. It indicated that components of the date cultivars reacted differently with chemicals involved in different antioxidant analytical protocols.

ABTS assay indicated high antioxidant activity that varied from 341 to 1300 μmol TE/g dw (Table 1). DNA showed about four-time higher antioxidant activity than KSA. Dates cultivars had more than 600 μmol TE/g dw activity except for KSA and SHT. Statistical analysis revealed that DNA had significantly higher antioxidant activity followed by DNC ($p \leq 0.05$). The difference between DNT and BSA cultivars was nonsignificant in antioxidant activity. Similar observations were noted between KSA and SHT cultivars. DPPH assay revealed the antioxidant values in the range of 3.27–3.54 μmol TE/g dw. Statistical analysis showed that the antioxidant values by DPPH method did not vary significantly ($p > 0.05$) among the six cultivars. It indicated that DPPH decolorizing reaction was not supported by date constituents and thus showed a little activity.

Antioxidant activity assayed by FRAP showed that values varied from 3.29 to 5.22 μmol TE/g dw. DNA cultivar also exhibited the highest activity while KSA showed the minimum activity. The general pattern of antioxidant activity was comparable to ABTS but values were quite low. Statistical analysis showed significant difference ($p \leq 0.05$) in FRAP values among the six cultivars.

The ORAC values of six date cultivars varied from 189 to 243 μmol TE/g with higher activity in DNA and DNT. The lowest in ORAC values were observed in DNC and SHT. Statistical analysis showed that ORAC values varied significantly among the cultivars ($p \leq 0.05$).

The ORAC values were higher than DPPH and FRAP but lower than ABTS values. The variations in antioxidant activities of different date cultivars were expected due to variation in agroclimatic conditions, variety, and country of origin. In addition, temperature, relative humidity, maturity, and processing can have a significant impact on phytochemicals profile of the date fruit [8, 27, 28]. The antioxidant values measured by different techniques did not give similar results. ABTS and FRAP analysis indicated that DNA had the highest

TABLE 2: Hunter color values of whole dates and date paste from different cultivars.

Date cultivars	Whole fruit			Ground fruit (paste)		
	L	a	b	L	a	b
Deglet Nour (Algeria) (DNA)	19.35 ± 2.59^{abc}	5.04 ± 0.93^{a}	4.54 ± 0.62^{a}	42.13 ± 1.14^{b}	3.98 ± 0.49^{c}	9.71 ± 0.47^{b}
Deglet Nour (California) (DNC)	21.06 ± 0.85^{a}	4.28 ± 1.35^{a}	4.21 ± 0.45^{ab}	47.05 ± 1.37^{a}	3.89 ± 1.43^{c}	11.50 ± 3.51^{b}
Deglet Nour (Tunisia) (DNT)	18.40 ± 0.61^{bc}	4.35 ± 0.44^{a}	3.61 ± 0.43^{b}	37.58 ± 3.68^{c}	6.15 ± 1.66^{ab}	8.63 ± 1.68^{b}
Shahia (Tunisia) (SHT)	20.19 ± 0.34^{ab}	4.51 ± 1.34^{a}	3.68 ± 0.70^{b}	45.46 ± 1.04^{ab}	4.44 ± 0.20^{bc}	10.85 ± 0.33^{b}
Barni (Saudi Arabia) (BSA)	19.26 ± 1.36^{c}	1.80 ± 0.37^{b}	2.20 ± 0.35^{c}	37.56 ± 4.30^{c}	5.85 ± 2.14^{abc}	9.35 ± 2.83^{b}
Khudri (Saudi Arabia) (KSA)	17.81 ± 0.41^{c}	4.81 ± 0.81^{a}	2.58 ± 0.09^{c}	42.84 ± 1.44^{b}	6.86 ± 1.56^{bc}	15.23 ± 3.82^{a}

Data is mean of eight replicates ± standard error. Means with different letters in the same column are significantly different at $p \leq 0.05$.

antioxidant activity while ORAC analysis showed that DNT and DNA had the highest antioxidant activity.

The antioxidant activity data is not supported by the total phenolics content as reported by some researchers. Study also indicated that the composition of the date cultivars varied significantly. Moreover, dates participate in reaction involving antioxidant estimation differently in different techniques used.

Al-Turki et al. [23] reported that the antioxidant activity measured by ABTS was higher than that measured by DPPH. Their results support the present findings where the antioxidant activity values of DPPH were lower than the ABTS values.

3.3. Date Fruit and Paste Color. Color is the important parameter affecting the consumer acceptability. It is also an important quality parameter in commercial date fruits. The results of color measurement of whole fruit dates and date paste of different cultivars, in terms of Hunter L, a, and b values, are presented in Table 2. Measuring of whole date fruits color is useful for the comparison between different date cultivars and for quality control of processed date products in the international trade. In whole date fruits, Hunter L values varied in the range of 18.44 to 21.06 among the cultivars. Higher L values indicate "lighter" color; thus, KSA had dark color whereas DNC had light color. Statistical analysis showed a significant difference ($p \leq 0.05$) in L values of whole fruit. The Hunter color L values for date pastes showed a similar trend but were in the range of 37.57–47.05. The L values of the paste increased indicating that color became lighter. Among the cultivars BSA and DNT had dark color whereas DNC had lighter color.

The Hunter color a values, of the whole dates ranged from 1.8 to 5.04. Statistical analysis showed a significant ($p \leq 0.05$) difference of "a" value in whole date. Hunter values of paste varied from 3.9 to 6.9. None of the dates or pastes having Hunter values were in the negative (−) range, which reflected the absence of any greenish tint. Hunter b values varied from 2.20 to 4.53 in whole fruits and from 8.63 to 15.23 in ground paste of date. The Hunter color b values of dates were observed to be high for DNA whole fruit while being low for BSA. Similarly, KSA cultivars had the highest a value whereas DNT had the lowest. Measuring the color of date fruit paste is useful in comparing different dates. Overall, the difference in color is mainly due to the genetic variability [27].

Hasnaoui et al. [29] evaluated the fruit color for twenty-seven of date palm cultivars collected from Moroccan oases at Tamar stage (fully ripened) and found the L values in the range of 12.12 to 38.93, a values in the range of 1.35 to 15.29, and b in the range of 0.86 to 35.12.

4. Conclusion

Characterization of date fruit for antioxidant properties can expand this fruit's consumption, as consumers are increasingly looking for healthy food. Barni and Khudri cultivars had relatively higher phenolics content than other four cultivars. With respect to assessing antioxidant capacity, the ABTS method gave higher antioxidant values than other three methods used in this study. Deglet Nour from Algeria has higher antioxidant properties than others cultivars. Light colored dates are considered better; Deglet Nour from California showed a better color with higher L values. Based on the results of this study, it can be concluded that date cultivars had medium phenolic contents and significant antioxidant activity. However, relationship between phenolic contents and the antioxidant activity of the respective date cultivars could not be established conclusively. Our work adds expanded scientific information on antioxidant analysis of different cultivars. We used different methods to measure the antioxidants capacity of dates.

Conflict of Interests

The authors declare that there is no conflict of interests regarding the publication of this paper.

Acknowledgments

This research was supported financially by the King Abdulaziz City for Science & Technology (KACST) through the Summer Grant for the Ph.D. Degree Staff. Thanks are extended to Dr. Kirk Dolan (Michigan State University) for allowing the use of his lab equipment for analyzing antioxidant activity.

References

[1] D. Huntrods, "Date profile," Agricultural Marketing Resource Center Bulletin, USDA, Washington, DC, USA, 2011.

[2] P. Vijayanand and S. G. Kulkarni, "Processing of dates into value-added products," in *Dates Production, Processing, Food and Medicinal Values*, A. Manickavasagan, M. Mohamed Essa, and E. Sukumar, Eds., pp. 255–264, CRC Press, Boca Raton, Fla, USA, 2012.

[3] Y.-L. Chew, J.-K. Goh, and Y.-Y. Lim, "Assessment of *in vitro* antioxidant capacity and polyphenolic composition of selected medicinal herbs from Leguminosae family in Peninsular Malaysia," *Food Chemistry*, vol. 116, no. 1, pp. 13–18, 2009.

[4] G.-C. Yen and C.-L. Hsieh, "Antioxidant activity of extracts from Du-zhong (*Eucommia ulmoides*) toward various lipid peroxidation models in vitro," *Journal of Agricultural and Food Chemistry*, vol. 46, no. 10, pp. 3952–3957, 1998.

[5] G. C. Yen, P. D. Duh, and C. L. Tsai, "Relationship between antioxidant activity and maturity of peanut hulls," *Journal of Agricultural and Food Chemistry*, vol. 41, no. 1, pp. 67–70, 1993.

[6] C. Guo, J. Yang, J. Wei, Y. Li, J. Xu, and Y. Jiang, "Antioxidant activities of peel, pulp and seed fractions of common fruits as determined by FRAP assay," *Nutrition Research*, vol. 23, no. 12, pp. 1719–1726, 2003.

[7] A. A. A. Allaith, "Antioxidant activity of Bahraini date palm (*Phoenix dactylifera* L.) fruit of various cultivars," *International Journal of Food Science and Technology*, vol. 43, no. 6, pp. 1033–1040, 2008.

[8] M. Al-Farsi, C. Alasalvar, M. Al-Abid, K. Al-Shoaily, M. Al-Amry, and F. Al-Rawahy, "Compositional and functional characteristics of dates, syrups, and their by-products," *Food Chemistry*, vol. 104, no. 3, pp. 943–947, 2007.

[9] M. Al-Farsi, A. Morris, and M. Baron, "Functional properties of omani dates (*Phoenix dactylifera* L.)," *Acta Horticulturae*, vol. 736, pp. 479–487, 2007.

[10] P. K. Vayalil, "Antioxidant and antimutagenic properties of aqueous extract of date fruit (*Phoenix dactylifera* L. Arecaceae)," *Journal of Agricultural and Food Chemistry*, vol. 50, no. 3, pp. 610–617, 2002.

[11] O. Ishurd and J. F. Kennedy, "The anti-cancer activity of polysaccharide prepared from Libyan dates (*Phoenix dactylifera* L.)," *Carbohydrate Polymers*, vol. 59, no. 4, pp. 531–535, 2005.

[12] A. Mansouri, G. Embarek, E. Kokkalou, and P. Kefalas, "Phenolic profile and antioxidant activity of the Algerian ripe date palm fruit (*Phoenix dactylifera*)," *Food Chemistry*, vol. 89, no. 3, pp. 411–420, 2005.

[13] J. A. Duke, *Handbook of Phytochemical of GRAS Herbs and Other Economic Plants*, 1992.

[14] S. A. A. Jassim and M. A. Naji, "In vitro evaluation of the antiviral activity of an extract of date palm (*Phoenix dactylifera* l.) pits on a pseudomonas phage," *Evidence-Based Complementary and Alternative Medicine*, vol. 7, no. 1, pp. 57–62, 2010.

[15] C. J. Miller, E. V. Dunn, and I. B. Hashim, "Glycemic index of 3 varieties of dates," *Saudi Medical Journal*, vol. 23, no. 5, pp. 536–538, 2002.

[16] C. J. Miller, E. V. Dunn, and I. B. Hashim, "The glycaemic index of dates and date/yoghurt mixed meals. Are dates 'the candy that grows on trees'?" *European Journal of Clinical Nutrition*, vol. 57, no. 3, pp. 427–430, 2003.

[17] N. Zieslin and R. Ben-Zaken, "Peroxidase activity and presence of phenolic substances in peduncles of rose flower," *Plant Physiology and Biochemistry*, vol. 31, no. 3, pp. 333–339, 1993.

[18] W. Brand-Williams, M. E. Cuvelier, and C. Berset, "Use of a free radical method to evaluate antioxidant activity," *LWT—Food Science and Technology*, vol. 28, no. 1, pp. 25–30, 1995.

[19] I. F. F. Benzie and J. J. Strain, "Ferric reducing/antioxidant power assay: direct measure of total antioxidant activity of biological fluids and modified version for simultaneous measurement of total antioxidant power and ascorbic acid concentration," *Methods in Enzymology*, vol. 299, pp. 15–27, 1999.

[20] K. A. Gomez and A. A. Gomez, *Statistical Procedures for Agricultural Research*, John Wiley & Sons, Ames, Iowa, USA, 2nd edition, 1984.

[21] SAS, *SAS for Windows, SAS User's Guide: Statistics*, Version 8.0e, SAS Institute, Cary, NC, USA, 2001.

[22] R. A. Waller and D. B. Duncan, "A Bayes rule for the symmetric multiple comparisons problem," *Journal of the American Statistical Association*, vol. 64, pp. 1484–1499, 1969.

[23] S. Al-Turki, M. A. Shahba, and C. Stushnoff, "Diversity of antioxidant properties and phenolic content of date palm (*Phoenix dactylifera* L.) fruits as affected by cultivar and location," *Journal of Food, Agriculture and Environment*, vol. 8, no. 1, pp. 253–260, 2010.

[24] M. Al-Farsi, C. Alasalvar, A. Morris, M. Baron, and F. Shahidi, "Comparison of antioxidant activity, anthocyanins, carotenoids, and phenolics of three native fresh and sun-dried date (*Phoenix dactylifera* L.) varieties grown in Oman," *Journal of Agricultural and Food Chemistry*, vol. 53, no. 19, pp. 7592–7599, 2005.

[25] R. M. A. Mohamed, A. S. M. Fageer, M. M. Eltayeb, and I. A. M. Ahmed, "Chemical composition, antioxidant capacity, and mineral extractability of Sudanese date palm (*Phoenix dactylifera* L.) fruits," *Food Science & Nutrition*, vol. 2, no. 5, pp. 478–489, 2014.

[26] X. Wu, G. R. Beecher, J. M. Holden, D. B. Haytowitz, S. E. Gebhardt, and R. L. Prior, "Lipophilic and hydrophilic antioxidant capacities of common foods in the United States," *Journal of Agricultural and Food Chemistry*, vol. 52, no. 12, pp. 4026–4037, 2004.

[27] F. Biglari, A. F. M. AlKarkhi, and A. M. Easa, "Antioxidant activity and phenolic content of various date palm (*Phoenix dactylifera*) fruits from Iran," *Food Chemistry*, vol. 107, no. 4, pp. 1636–1641, 2008.

[28] J. Gross, O. Haber, and R. Ikan, "The carotenoid pigments of the date," *Scientia Horticulturae*, vol. 20, no. 3, pp. 251–257, 1983.

[29] M. Hasnaoui, A. Elhoumaizi, A. Hakkou, B. Wathelet, and M. Sindic, "Physico-chemical characterization, classification and quality evaluation of date palm fruits of some Moroccan cultivars," *Journal of Scientific Research*, vol. 3, no. 1, pp. 139–149, 2011.

20

Conical Intersections Leading to Chemical Reactions in the Gas and Liquid Phases

Yehuda Haas

Institute of Chemistry, the Hebrew University of Jerusalem, 91904 Jerusalem, Israel

Correspondence should be addressed to Yehuda Haas; yehuda.haas@mail.huji.ac.il

Academic Editor: Rosendo Valero

The current status of the role of conical intersections (CoIns) in molecular photochemistry is reviewed with a special emphasis on the procedures used to locate them. Due to space limitations, the extensive literature of the subject is given by referring the reader to representative references, whereas the author group's work is described in detail. The basic properties of CoIns are outlined and contrasted with those of transition states in thermal reactions. Location of CoIns using the method of Longuet-Higgins sign-inverting loops is described in detail. The concept of "anchors"—valence bond structures that represent stable molecules and other stationary points on the potential energy surface—is introduced and its use in constructing loops is described. The authors' work in the field is outlined by discussing some specific examples in detail. Mathematical aspects and details are left out. The main significance of the method is that it explains a large body of photochemical reactions (for instance, ultrafast ones) and is particularly suitable for practicing chemists, using concepts such as reaction coordinates and transition states in the search.

1. Introduction

Analysis of chemical reactions is usually based on the concept of potential energy surfaces (PESs), which are derived from the Born-Oppenheimer (BO) approximation. Reactions usually start in the ground electronic state, where the reactant is found in a local minimum. Energy must be supplied to the system in order to initiate the reaction and move on to the product. In thermal reactions the energy source is heat. The reaction proceeds along a trajectory that leads adiabatically to a transition state, which is a local maximum along the reaction coordinate (RC). Subsequently, the system goes down to the product. The entire route occurs on one potential surface (PES) only, along a single coordinate; this is thus a one-dimensional (1D) process. The RC is a combination of internal degrees of freedom, usually expressed in normal coordinates. This scenario, introduced in the 1930's, appears to account well for the vast majority of thermal reactions. There are exceptions, of course: some reactions lead to electronically excited states, creating the phenomenon of chemiluminescence [1]. Recently, a new mechanism termed roaming reactions has been introduced [2]. According to this approach, some reactions start out on a certain route (for instance to bond cleavage) but at a certain point begin a roaming motion resulting in a different reaction pattern. Such developments are important but are rather limited compared to the classical transition state theory and are ignored in this work.

In photochemical reactions the energy required to launch the reaction is supplied by light (most frequently UV and visible radiation). Light absorption promotes the molecule to an electronic excited state; thus, the reaction is nonadiabatic, involving more than one PES. Most photochemical reactions end up in the ground state of the products. If, as often happens, the quantum yield of the reaction is smaller than unity part of the electronically excited molecules, return back to the original state of the reactant. Consequently, a mechanism that transfers the molecule from the initial, light induced state to the ground state necessarily exists (possibly via other excited states). As discovered experimentally, this happens even under collision-free conditions.

One ever-present process that accomplishes this task is the radiative pathway, wherein an excited molecule emits a photon (fluorescence, phosphorescence) on returning to the GS. However, many relaxation processes are measured to be much faster than the radiative one; the radiative route provides an intrinsic clock to which nonradiative transitions (NRTs) can be compared. The radiative lifetime of a molecule can be derived from the absorption spectrum of the transition, as shown by Einstein's equations. An allowed electronic transition leads to an excited state having a typical radiative lifetime of the order of 10^{-9} s. A fast nonradiative transition is of the order of 10^{-12} s, leading to a fluorescence quantum yield of 10^{-3}. Faster transitions are referred to as ultrafast and can be probed directly using femtosecond lasers. It turns out that many interesting photochemical reactions proceed at an ultrafast speed, for instance, in biological systems.

Experiment shows that a nonradiative transition between two states becomes faster as they approach each other, an effect expressed by the *energy gap law*—the probability of crossing between two electronic states, P, increases as the energy gap between their potential surfaces decreases [3, 4]. In many cases the relationship is approximately exponential; that is, $P = \exp(-\Delta E)$ where ΔE is the energy gap. The maximum rate is attained when the gap vanishes and the two PESs cross. This crossing (a nonadiabatic process) constitutes a violation of the BO approximation. Thus, in contrast with thermal reactions, the theory of photochemical reactions must seek the conditions under which nonadiabatic processes become efficient.

According to the BO approximation, the total molecular energy (and hence the Hamiltonian) is the sum of an electronic and nuclear terms. The total wave function Ψ may be written as the product of electronic (ϕ) and nuclear (χ) wave functions. In this approximation, the effect of nuclear motion on the electronic wave function is neglected (namely, $\partial\phi/\partial q = 0$ for any nuclear coordinate q). In the Born Oppenheimer (BO) approximation, the electronic PES is the potential under which the nuclei move.

Teller [5] showed that in a polyatomic molecule *two* coordinates are required to bring about a crossing. When plotted as a function of the two coordinates the PES has the shape of a double cone, the two PESs cross at the apex (where the wavefunction is degenerate); the energy surface near the crossing is called the conical intersection (CoIn). As the crossing is vital in photochemical reactions, it follows that, in contrast with thermal reactions, photochemical ones are 2D ones.

CoIns were considered for a long time to be rather rare and were largely ignored in the photochemical literature. However, they were of interest to quantum physicists and chemists, who continued to explore their properties. In 1963, Herzberg and Longuet-Higgins [6] in their study of the Jahn-Teller effect showed that if the electronic wave function was carried through a closed loop around the apex, it changed its sign (underwent a 2π phase shift). This came to be known as the sign-change theorem or the geometric phase effect. (Obviously, the nuclear wave-function underwent the same shift, so that the total wave-function did not change.) It was later extended by Longuet-Higgins [7] to any system and numerically demonstrated by using ab-initio electronic structure calculations for the LiNaK system [8]. Mead and Truhlar [9] discussed a related but different approach using vector potentials. Nowadays, there is a consensus that CoIns are quite common and wide-spread. Claims that "In the last

two decades, it became clear that state crossings play a central role for internal conversion" are commonly made [10]; see also [11].

The two coordinates [12, 13] that lift the degeneracy are defined by two vectors, the *derivative coupling* one which depends on the coupling between the two states ($H_{\text{coupling}} \sim \langle \phi_i | \partial / \partial q | \phi_j \rangle$) and the *energy difference gradient* one, called the tuning vector ($H_{\text{tuning}} = \partial |E_i - E_j| / \partial q$). The first is nonsymmetric and leads, in a 1D space, to avoided crossing; the second tunes the energy difference to achieve crossing and is a symmetric one. Their combined effect violates the BO approximation—leading to a nonadiabatic process. The proof that the touching is a CoIn (and not, for instance, a Renner-Teller intersection) is by testing the sign-change upon carrying the electronic wave function in a complete loop around the intersection point. Only a true CoIn will lead to a sign change of the electronic wave function.

In the 1980s interest in CoIns grew, as experiments were extended to the picosecond and subpicosecond time regime. Many ultrafast processes could be accounted for by assuming the involvement of CoIns; in fact, there is still no better explanation. The growing interest in the application of CoIns to photochemical systems led to the development of methods to find them as explained in some representative examples: [13–19]; these methods, based on the concept of vibronic coupling, were crucial for the progress in the field. These efforts were greatly helped by the development of efficient algorithms for calculating wave functions and coupling matrix elements. Computer programs are available to find the minimum energy point in a CoIn (MECI) [20–22]; in fact, the CoIn itself is a multidimensional surface so the programs actually find minimum points in a *conical intersection crossing seam* (MXS). The acronym CoIn refers to this surface, which is in fact a seam and not a single point.

In this review, the focus is on the important special case of a S_1/S_0 intersection, which, like a transition state, is located on both the ground and the excited states' PESs. In distinction with transition states, the CoIn is not a stationary point. The fast passage through a CoIn prevents its direct detection; its presence must, therefore, be experimentally realized by indirect methods only.

An interesting recent development uses some modifications of density functional theory (DFT) for studying CoIns [23, 24]. In view of the popularity and relatively easy and cheap use of DFT, these developments may become a key factor in the field.

2. LH Loops and Anchors

Current quantum chemical calculations are largely performed by advanced methods of the molecular orbital (MO) theory originally developed by Hund, Slater, and Mulliken, such as Hartree-Fock, complete active space (CAS), and multireference configuration interaction (MRCI). In the early days of quantum chemistry, valence bond (VB) theory based on the Heitler-London approximation and developed by Pauling [25] was quite popular. VB is more adapted to conventional chemical concepts, especially bonding and resonance structures, than MO (in which a chemical bond is not defined). Its basic feature is the spin-paired presentation of the chemical bond. VB structures are constructed by considering all possible spin-pairing forms in which two electrons have opposite spins. Unfortunately, VB methods are less suitable to computations than MO and are rarely used. For a review on VB and MO, [26] may be consulted.

Pross and Shaik [27] offered a simplified presentation of the TS by using the wave functions of the reactant (R) and product (P). At any time during the reaction, the system may be described by a combination of the two:

$$|R\rangle (t) = c_R (t) |R\rangle + c_P (t) |P\rangle , \qquad (1)$$

where $c_R(t = 0) = 1$, $c_P(t = 0) = 0$, $c_R(t = \infty) = 0$, and $c_P(t = \infty) = 1$.

We define an anchor as a stationary point on the PES having a specific spin-pairing scheme, independent of the nuclear configuration. An anchor may consist of a single or a group of VB structures. If a reaction is elementary, there is only a single transition state between the reactant R and product P. In the transition state region, a spin-pairing change must take place. At this nuclear configuration, the electronic wave function may be written as

$$|TS\rangle_{el} = k_R |R\rangle_{el} + k_P |P\rangle_{el} . \qquad (2)$$

If the sign of k_R is equal to that of k_P, the TS is the in-phase combination of $|R\rangle$ and $|P\rangle$; if the signs are different, TS is their out-of-phase combination (the k's are assumed positive). The reactions are named p-type and i-type, if the TS is the in-phase (sign-preserving) or out-of-phase (inverting) combinations, respectively. The opposite combination is an electronically excited state. Using the determinant form of the electronic wave functions for Hückel-type reactions, which are the most ordinary ones (others are Möbius ones (the concept of Hückel and Möbius type reactions is clearly presented in [28, 29]). For a generalization of Heilbronner's derivation for Möbius systems bearing one half twist to those bearing n half twists, see [30]), it is readily shown that in i-type reactions an even number of electron pairs are exchanged, whereas in an in-phase reactions, an odd number of electron pairs are exchanged [31, 32]. Thus, repairing of 4, 8,... pairs leads to i-type reactions, and repairing of 2, 6,... pairs leads to a p-type reactions.

An anchor, as defined above, contains stable molecules, conformers, all pairs of radicals and biradicals formed by a simple bond fission in which no spin re-pairing took place.

2.1. The Sign-Change Rule and the Construction of Loops. According to the sign-change theorem, when a wave-function of the given electronic state in the BO approximation is adiabatically transported round a closed path encircling a conical intersection, the wave-function changes its sign. A number of approaches for locating conical intersections based on the sign-change theorem are available, among them the recognition that the sign-inverting method is a variant of Berry's phase [33–36]. Generalizing on the H3 system [6, 7], a loop is constructed by using a sequence of two or three

elementary reactions (in both, three or more VB structures are required). The reactions comprising the loop must be elementary, having a single TS. According to the uniqueness theorem [37] if a loop formed by two or three elementary reactions is sign-inverting, it encircles one and only one conical intersection.

The following terminology is used: each reaction in the loop is termed a leg, which is either p-type or i-type. A three-legged loop is sign inverting if it is an i^3 or ip^2 loop, and sign-preserving otherwise. A two-legged loop is sign inverting if the two TSs differ (one is p-type, the other i-type) and sign-preserving otherwise.

Crossings between more than two states [38–40] have been discussed in the literature but are out of the scope of this paper.

3. Goal and Distinction of the LH Loop Method

There are numerous excellent procedures for finding and optimizing CoIns. One might wonder what new message is provided by the present method; it does not endeavor to present a better or more elegant optimization procedure. Its main distinction is in the beginning stage of the search. Other programs use a guess, based on intuition and chemical properties, and invest most of the effort in the optimization procedure. The need for optimization is not immediately obvious. The search for it implicitly assumes that the system will tend to reach the minimum energy of the CoIn (a multidimensional surface) prior to the actual crossing. This is not necessarily the case and is yet to be demonstrated. More important, the (educated) guess may be wrong and lead to nonphysical species. In the proposed method, the structures of reactant and product are essential for constructing the loop, so that no guess is required. At the same time, if the loop is indeed sign-inverting, as can be verified without the need of a calculation, a conical intersection is bound to be found within it. Moreover, the method is particularly useful in the case of large molecules, when several reaction paths are possible. In it, the desired reaction is chosen and a CoIn is sought specifically for that reaction.

4. Examples

Many systems were analyzed using the elementary reaction scheme. These include the butadiene-cyclobutene one [41, 42], cis-trans isomerization of ethylene and polyenes [43–45], the E-Z isomerization of the formaldiminium cation [37], the allyl radical—cyclopropyl radical ring closure reaction [44], and most recently styrene, stilbene, and the protonated Schiff base, PSB3 [24].

The basic concepts and their advantages are exemplified by a discussion of benzene photochemistry. Benzvalene is one of the products of the S_1 *photolysis* of benzene [46]; the sole observable product of the *thermal* isomerization of benzvalene is benzene [47]. Haas and Zilberg [44], in their study of the photochemical reaction, chose benzene (reactant) and benzvalene (BZ1, product) as two of the anchors. The third anchor was another benzvalene isomer, BZ2. The reasoning was that the BZ1 to BZ2 conversion is antisymmetric, and the two BZ to benzene reactions are of the same symmetry, so that the loop is sign-inverting and a conical intersection must be present.

Vanni et al. [48] discussed the same reaction. They offered a reformulation of the elementary reaction loop method and pointed out its connection with the concepts of coupling and tuning vectors used in the CoIn literature. They also criticized the work on the benzvalene to benzene reaction. Two BZ isomers were used as anchors, although a physical argument justifying this choice was not given.

It was pointed out that the selection of benzene as a third anchor cannot be correct because once two anchors (i.e., VB structures) are chosen all possible remaining anchors are unambiguously determined as combinations of the two, but no such combination can be found to represent benzene. A biradical structure, termed prebenzvalene (Figure 14 in [48]), was suggested as the third anchor. In an earlier paper [49], this structure was found computationally (CASSCF/4-31G) to be a local minimum on the S_0 surface having a barrier of 1.3 kcal/mol to form benzene and 0.1 kcal/mol to form benzvalene. These tiny activation energies were considered by others to be insufficient evidence for a real minimum [50].

Since Haas and Zilberg actually found a sign inverting loop and a CoIn in their analysis, an explanation is called for. The prebenzvalene entity is not an experimentally observable species; may it be replaced by benzene? These two molecules are each represented by two VB spin-pairing schemes. Inspection shows that one of the benzene VB structures (one Kekule structure) is transformed to one VB prebenzvalene structure by a four electron re-pairing scheme and the other by another four-electron scheme. This may be interpreted to represent a sign-inverting elementary reaction, as suggested by Palmer's results [49]. This reaction is quantitative—no other product is formed. Thus, although formally benzene cannot be the third anchor, for all practical purposes, it is. The upshot of this analysis is that both methods lead to the same result. Therefore, in establishing the connection to experiment, whether prebenzvalene is a minimum or not is inconsequential; benzene can be the third anchor in the benzvalene-benzene reaction.

5. Solvent Effect and Two-Legged Loops

Most of the theoretical work on CoIns dealt with isolated molecules; in contrast most photochemical experiments are conducted in the bulk. In this section the effect of external fields and of solvents on the properties of CoIns is discussed. The topic is introduced using an example of two-legged loops, a circumstance that is often encountered in the field.

Consider a reactant and product whose ground states have an appreciable polar character.

The VB presentations may be written:

$$|R\rangle = |R_{\text{cov 1}}\rangle + |R_{\text{ion}}\rangle,$$
$$|P\rangle = |P_{\text{cov 2}}\rangle + |P_{\text{ion}}\rangle, \tag{3}$$

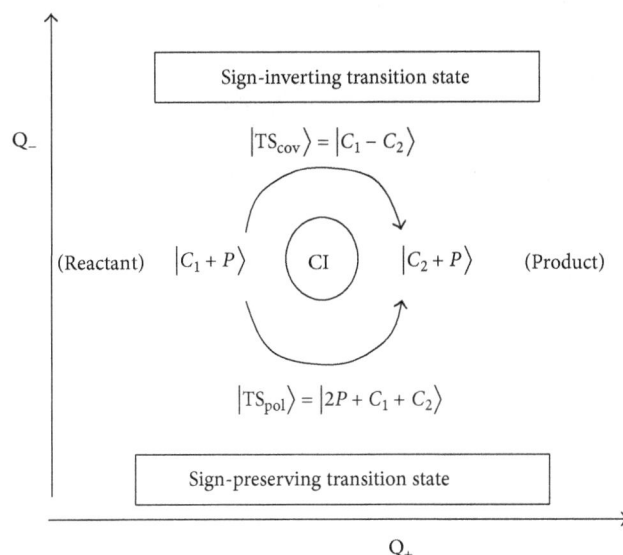

FIGURE 1: An example of two-legged sign-inverting domain, adapted from [37]. CI is a conical intersection; TS is a transition state.

(numerical coefficients and normalization are suppressed, to simplify notation).

An example is an olefin carrying electron donor and electron acceptor groups at opposite sides of the C=C double bond. The reaction is the cis-trans isomerization around the double bond. For simplicity the polar part is identical for the two molecules. Two reaction coordinates, having two TSs, can be constructed. The EWF of the transition states are the out-of-phase and in-phase combinations of the EWFs of R and P:

$$|TS_{\text{cov}}\rangle = |R\rangle - |P\rangle = |R_{\text{cov 1}}\rangle - |R_{\text{cov 2}}\rangle$$

Sign-inverting,

$$|TS_{\text{ion}}\rangle = |R\rangle + |P\rangle = 2|R_{\text{ion}}\rangle + |R_{\text{cov 1}}\rangle + |R_{\text{cov 2}}\rangle \tag{4}$$

Sign-preserving.

A two legged loop can be constructed as shown in Figure 1. Due to the symmetry properties of the two TSs, it is a sign-inverting loop and therefore contains a CoIn.

Examples for these reaction types abound, for instance, the photoisomerization of the formaldiminium cation and of many Schiff bases, important in the vision mechanism. Here it is demonstrated for the cis-trans isomerization of 1-Butyl-4-(1H-inden-1-ylidene)-1,4-dihydropyridine [51, 52] (BIDP, Scheme 1, III). This molecule is easier to handle experimentally than 4-cyclopentadienyl-1,4-dihydropyridine, (CPDHP) that was used in the calculations. The structures of the biradical and ionic transition states of CPDHP are shown in Scheme 2.

The ionic form is stabilized in this case by aromatization. The isomerization around the CC double bond can proceed along two reaction coordinates: the usual covalent one with a biradical TS (BRTS) and another one having ionic TS (ZWTS). The reaction coordinate connecting the reactant

(I) $R_1 = R_2 = H$
(II) $R_1 = n\text{-Bu}, R_2 = H$
(III) $R_1 = n\text{-Bu}, R_2 = CH_3$

(a)

(b)

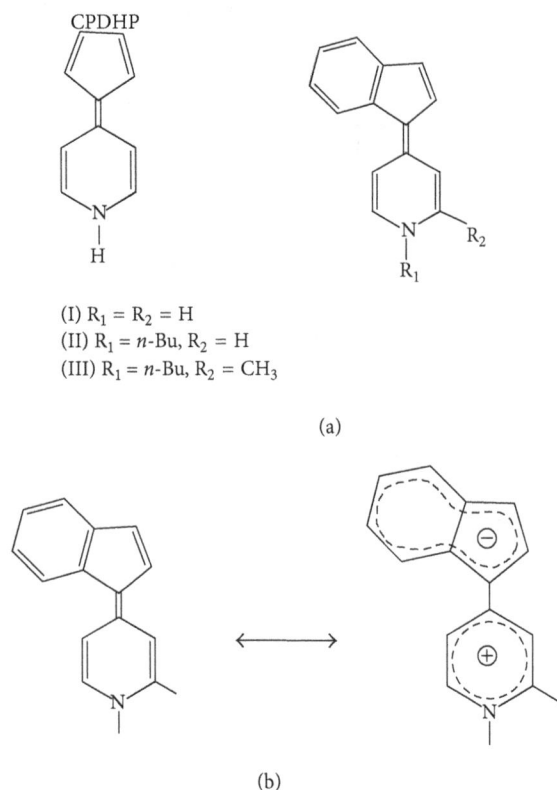

SCHEME 1: (a) The molecular structure of CPDHP (left) and the title compounds (right); II is BIDP and III is BIMDP; (b) the covalent and zwitterion VB structures of I that contribute to the electronic wave-function of the compound.

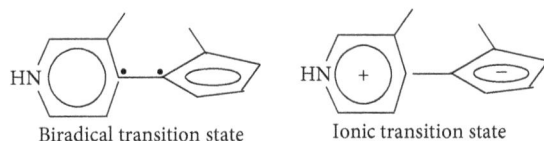

Biradical transition state   Ionic transition state

SCHEME 2: The structures of the biradical and ionic transition states of CPDHP.

with the product (Q_+) is mainly torsion, whereas the coordinate connecting the two TSs (Q_-) is mainly C=C bond stretch and aromatization. A CAS calculation confirmed the presence of a CoIn in the gas phase. The calculation showed that the ground state of CPDHP has a rather small dipole moment in the gas phase (4.3 D), whereas the ZWTS dipole moment (16.3 D) is very large. The BRTS has a dipole moment of only 2.2 D; thus, the BRTS has an essentially pure biradical nature and the ZWTS a purely ionic (charge transfer) character. In the gas phase the ZWTS lies at a higher energy than the BRTS (2.05 versus 1.95 eV above the GS). The solvent effect, approximated using the Kirkwood and Onsager model, was calculated to be remarkable. The energy of the BRTS was practically unchanged upon increasing the polarity of the solvent, whereas that of the ZWTS was considerably reduced, becoming 0.96 eV in acetonitrile. The BRTSin this solvent lies in fact on the excited state, and

the CoIn becomes an avoided crossing. These results are depicted pictorially in Figure 2, taken from [53].

Further support came from ultrafast measurements of the lifetime of MRCN in different solvents. The two major VB structures of MRCN are shown in Scheme 3: An ionic one (bottom), and a covalent one (top). The lifetime of the excited state is about 300–400 fs in toluene and 900–1000 fs in MeCN, in agreement with the trend predicted by theoretical predictions [54].

6. Future Research Directions

The reaction coordinate approach is uniquely suited for studying complex photochemical transformations. In particular it could be used in systems exhibiting several reaction pathways. Prime examples are photobiology, green chemistry, and development of new materials such as molecular switches.

An important and still highly controversial example is the origin of the photostability of DNA and DNA bases, which is believed to have played crucial role in the appearance of life. Recent experiments support a direct transition from the excited state of these molecules to the ground state, circumventing any chemical change in spite of the high energy absorbed (4–6 eV) [55–57]. Most proposed mechanisms for the individual bases and their protonated derivatives involve a CoIn, but its nature is still being debated [58–60]. Ring puckering and frustrated H-atom ejection are among the proposed routes for single bases. Mechanisms suggested include among others deactivation via proton transfer, which is mediated by an excited singlet state of charge-transfer character or vibronic coupling between different pp states. A careful analysis helped by the reaction coordinate method may point out all possible routes, and the competition between them. A more challenging issue is the case of DNA itself. Many chemical processes are possible, especially considering the initial high energy content after photoexcitation. Ground state reactions such as isomerization, cyclobutane formation, and hydrogen atom transfer must be considered and are shown to be less efficient than direct S_1-S_0 transition. The latter necessarily involves a large amplitude motion which might appear as a relatively slow process. The experimental evidence indicates that it is in fact ultrafast. An extensive study of the multidimensional potential energy is probably required.

Another field is the synthesis of novel materials. Considerable efforts are invested currently in environment friendly reactions ("green chemistry"). In this case a central challenge is to replace harmful reactants or solvents with environmentally friendly ones, avoiding toxic, flammable, and/or long-term hazardous products and by-products. Solvents are believed to play a central role; thus, for instance, replacing organic solvents with water could revolutionize the chemical industry. Although some progress has been reported, the much work remains to be accomplished. Photochemistry offers many opportunities that have been so far only superficially considered in the field. However, even solution-free reactions present a challenge. An example is the preparation of fire-works and high energy density materials (HEDMs).

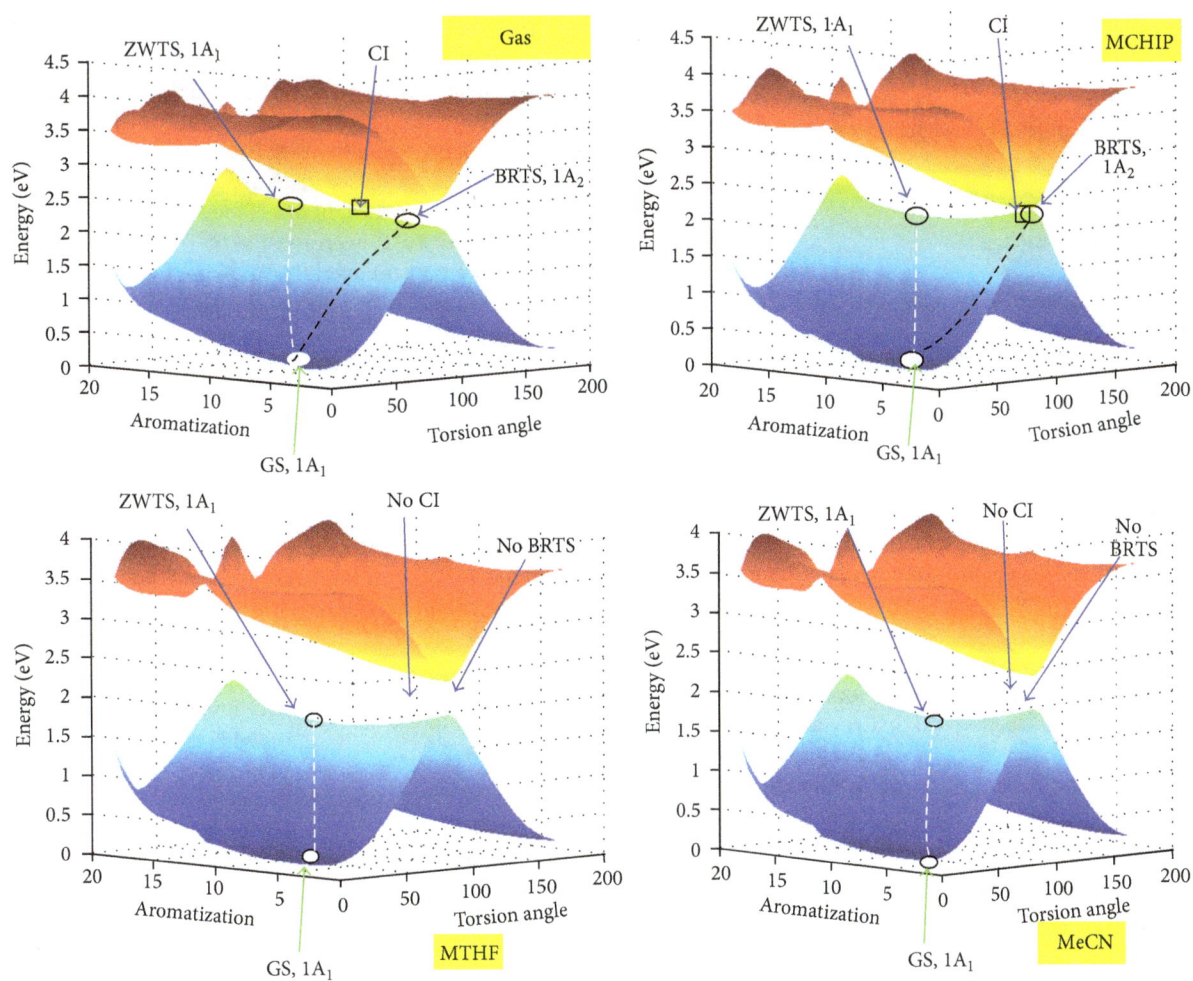

FIGURE 2: The calculated potential energy surface CPDHP in the gas phase and some solvents in the vicinity of the conical intersection. The positions of the ground state (planar) and two transition states (perpendicular) both having C_{2V} symmetry are marked in the figure along with that of the CoIn. The approximate trajectories of the two possible thermal reactions are schematically indicated by the dashed curves. In the more polar solvents (MTHF and MeCN) the BR structure lies on S_1; there is no biradical transition state and the CoIn vanishes [53].

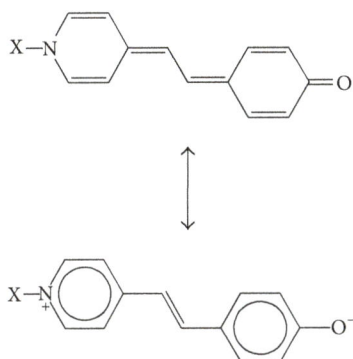

SCHEME 3: The two major VB structures of MRCN.

Effort to produce materials that will replace currently widely used chlorine containing compounds and decompose only (or mainly) to yield N_2 gas is underway [61, 62].

The preparation of all-nitrogen compounds is of the prime interest, practically and also theoretically. For instance, the cyclo-N_5^- anion has been prepared in the gas phase [63, 64] but not in solution. A photochemical mechanism involving a CoIn [65] has been proposed. A reaction in which an aryl pentazole is the starting material is of interest. The desired product (cyclo-N_5^-) can be prepared by breaking the CN bond between the phenyl and the cyclopentazole ring in aryl pentazoles (see Scheme 4 for the structure of an often used example).

However, a reaction having much lower activation energy, extrusion of N_2 to form the corresponding azide, is heavily favored thermally. A photochemical process circumventing the ground state reaction to form the less likely product is possible theoretically (resembling, in principle, the ultrafast route in the DNA bases). The reaction coordinate method, being readily adaptable to introduce solvent effects, can help in the search for conditions favoring the thermally less probable route.

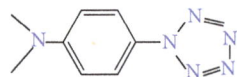

SCHEME 4: The structure of an aryl pentazole (para-dimethylamino phenyl pentazol).

7. Molecular Switches

Photochemical molecular switches are of great interest as alternatives to the present silicon-based technology. Devices based on reactions such as isomerization around an N=N double bond [66, 67] and ring opening [68] are actively being considered as they offer a substantial structural change at an ultrafast rate. Syn-anti rotational isomerism around an N=N bond and cyclohexadiene ring opening are only two examples that have been analyzed as passing through a conical intersection. For use in a real device, the molecules must be embedded in a solid matrix, organic or inorganic. The effects of such environments on photochemical reactions have not yet received extensive analysis, although some promising studies have been reported [69, 70]. This may be one of the important future developments of the field. It is very likely that treating the matrix as a cage will not correctly predict their effect. Rather, changes in electric fields, hydrogen bonding, polarization changes, and other parameters may be used to optimize the desired reaction.

8. Summary and Conclusions

The principal motivation of this study is the elucidation of photochemical reaction mechanisms. As the theory of thermal reactions is well-developed, an attractive option is the development of a model on equal (or similar) footing for photochemical ones. Thermal reactions are analyzed as 1D processes, wherein the concept of transition state plays a major role. For photochemical reactions, two dimensions are needed and the transition between 2 electronic states via a conical intersection is a crucial step. Restricting the scope to the special but important case of reactions involving only the S_1 and S_0 states, a method for locating CoIns is introduced. Some CoIn traits resemble those of TSs—both are transient species through which the system must pass during a reaction and their role can be assessed only indirectly. The transit thru both is very rapid—the system never stays in them for a measurable time; both can lead to either the product or back to the reactant. Yet, a CoIn is utterly different from a TS: in addition to requiring two dimensions for its definition, it is a $n - 2$ multidimensional surface (n is the number of nuclear degrees of freedom) and not a point on the PES; no location on this surface is a stationary point (i.e., $\partial E/\partial q \neq 0$ for any q); although it is part of the S_0 PES, it is normally not involved in thermal reactions. One might hope that the CoIn concept could provide a common means to analyze many diverse reaction schemes. A desirable (and possibly natural) tactic is to base the analysis on well-founded chemical concepts, such as the ones used in the analysis of thermal reactions. A central concept is that of a chemical bond based on the spin-pairing of two electrons.

This work is a preliminary step in that direction. It shows that a S_1/S_0 CoIn can in principle be located using ground state features only. The actual application uses VB methodology, as it provides a simple means to represent bond breaking and forming. Molecules are represented by anchors, which are VB structures or combination of them, enabling the construction of LH loops. The recent application of VB theory to the properties of transition states makes this choice a natural one [71, 72]. Nonetheless, computational implementation is usually by the CI-MO method. The sign variation of the electronic wave-function on traversing the loop signals the presence of a CoIn.

Naturally, problems arise. Some are due to the lack of an exact mathematical description of basic concepts: for instance even the exact definition of a molecule is sometimes arbitrary. The method uses spin-pairing as the practical means for presenting their wave-function. The sign convention of a reaction in which more than one VB structure is required is not always easy to establish. In particular, participation of several resonance structures may make the sign assignment inconclusive.

A basic impediment encountered in searching for avoided crossings and CoIns has been summarized by Worth and Cederbaum [73] "The complete evaluation of potential energy surfaces is an impossible task for systems comprising more than a few atoms." Thus, some simplifying assumptions have to be made. These assumptions are embodied, for instance, in the approximation made in the initial steps of the search for the CoIn.

The rationale for using reaction coordinates of elementary reactions (RCER) adopted in the present case is to minimize intuitive guess in the initial location of the CoIn. Other methods start at the geometry of a single point (for instance, the minimum energy on the GS surface) and assume that a CoIn exists in its vicinity [74]. The actual search is often initiated on the excited state at this geometry [49]. In the RCER method, it is assumed that the reactant and product VB wave-functions are known, and the third anchor is not independent on the other two—it is some combination of the VB structures that form them. However, elementary reactions can lead instantaneously from the third anchor to a more stable different chemical species (see the discussion of the benzvalene-benzene reaction). Thus, all of stable chemical species involved in the reaction can be observed experimentally. The method is uniquely appropriate for large polyatomic systems, in which several degeneracies are common.

Conflict of Interests

The author declares that there is no conflict of interests regarding the publication of this paper.

Acknowledgments

This study was supported by the Deutsche Forschungsgemeinschaft in the framework of Project no. MA 515/22-2. The author is deeply indebted to his coworkers and students, who made this work possible: S. Zilberg, B. Dick, J. Manz, W. Fuss, O. Deeb, S. Al-Falah, L. Belau, S. Cogan, X. F. Xu, B. Bazanov, A. Kahan, and U. Geiger.

References

[1] N. J. Turro, P. Lechtken, N. E. Schore, G. Schuster, H. C. Steinmetzer, and A. Yekta, "Tetramethyl-1,2-dioxetane. Experiments in chemiexcitation, chemiluminescence, photochemistry, chemical dynamics, and spectroscopy," *Accounts of Chemical Research*, vol. 7, no. 4, pp. 97–105, 1974.

[2] J. M. Bowman and A. G. Suits, "Roaming reactions: the third way," *Physics Today*, vol. 64, pp. 33–37, 2011.

[3] G. W. Robinson and R. P. Frosch, "Electronic excitation transfer and relaxation," *Journal of Chemical Physics*, vol. 38, pp. 118–120, 1963.

[4] K. Freed and J. Jortner, "Multiphonon processes in the non-radiative decay of large molecules," *The Journal of Chemical Physics*, vol. 52, pp. 6272–6291, 1970.

[5] E. Teller, "The crossing of potential surfaces," *The Journal of Physical Chemistry*, vol. 41, no. 1, pp. 109–116, 1937.

[6] G. Herzberg and H. C. Longuet-Higgins, "Intersection of potential energy surfaces in polyatomic molecules," *Discussions of The Faraday Society*, vol. 35, pp. 77–82, 1963.

[7] H. C. Longuet-Higgins, "The intersection of potential energy surfaces of polyatomic molecules," *Proceedings of the Royal Society of London A*, vol. 344, pp. 147–156, 1975.

[8] A. J. C. Varandas, J. Tennison, and J. N. Murrell, "Chercher le croisement," *Chemical Physics Letters*, vol. 61, pp. 431–434, 1979.

[9] C. A. Mead and D. G. Truhlar, "On the determination of Born-Oppenheimer nuclear motion wave functions including complications due to conical intersections and identical nuclei," *The Journal of Chemical Physics*, vol. 70, pp. 2284–2296, 1979.

[10] M. Barbatti, "Photorelaxation induced by water-chromophore electron transfer," *Journal of the American Chemical Society*, vol. 136, no. 29, pp. 10246–10249, 2014.

[11] D. G. Truhlar and C. A. Mead, "The relative likelihood of encountering conical intersections and avoided intersections on the potential energy surfaces of polyatomic molecules," *Physical Review A*, vol. 68, no. 3, Article ID 032501, pp. 32501-2, 2003.

[12] W. Domcke, D. R. Yarkony, and H. Köppel, Eds., *Conical Intersections. Theory, Computation, and Experiment*, vol. 17 of *Advanced Series in Physical Chemistry*, World Scientific, Singapore, 2011.

[13] F. Bernardi, M. Olivucci, and M. A. Robb, "Potential energy surface crossings in organic photochemistry," *Chemical Society Reviews*, vol. 25, no. 5, pp. 321–328, 1996.

[14] D. R. Yarkony, "Conical intersections: diabolical and often misunderstood," *Accounts of Chemical Research*, vol. 31, no. 8, pp. 511–518, 1998.

[15] M. Ben-Nun, J. Quenneville, and T. J. Martinez, "Ab initio multiple spawning: photochemistry from first principles quantum molecular dynamics," *The Journal of Physical Chemistry A*, vol. 104, no. 22, pp. 5161–5175, 2000.

[16] M. A. Robb and M. Olivucci, "Photochemical processes: potential energy surface topology and rationalization using VB arguments," *Journal of Photochemistry and Photobiology A: Chemistry*, vol. 144, no. 2-3, pp. 237–243, 2001.

[17] A. L. Sobolewski and W. Domcke, "Conical intersections induced by repulsive $^1\pi\sigma^*$ states in planar organic molecules: malonaldehyde , pyrrole and chlorobenzene as photochemical model systems," *Chemical Physics*, vol. 259, no. 2-3, pp. 181–191, 2000.

[18] M. G. D. Nix, A. L. Devine, R. N. Dixon, and M. N. R. Ashfold, "Observation of geometric phase effect induced photodissociation dynamics in phenol," *Chemical Physics Letters*, vol. 463, no. 4-6, pp. 305–308, 2008.

[19] W. Domcke and G. Stock, "Theory of ultrafast nonadiabatic excited-state processes and their spectroscopic detection in real time," in *Advances in Chemical Physics*, vol. 100, pp. 1–169, John Wiley & Sons, New York, NY, USA, 1997.

[20] M. J. Bearpark, A. Michael, M. A. Robb, and H. B. Schlegel, "A direct method for the location of the lowest energy point on a potential surface crossing," *Chemical Physics Letters*, vol. 223, no. 3, pp. 269–274, 1994.

[21] H. Lischka, M. Dallos, P. G. Szalay, D. R. Yarkony, and R. Shepard, "Analytic evaluation of nonadiabatic coupling terms at the MR-CI level. I. Formalism," *The Journal of Chemical Physics*, vol. 120, no. 16, pp. 7322–7329, 2004.

[22] M. Dallos, H. Lischka, R. Shepard, D. R. Yarkony, and P. G. Szalay, "Analytic evaluation of nonadiabatic coupling terms at the MR-CI level. II. Minima on the crossing seam: formaldehyde and the photodimerization of ethylene," *The Journal of Chemical Physics*, vol. 120, no. 16, pp. 7330–7339, 2004.

[23] N. Minezawa and M. S. Gordon, "Optimizing conical intersections of solvated molecules: the combined spin-flip density functional theory/effective fragment potential method," *The Journal of Chemical Physics*, vol. 137, no. 3, Article ID 034116, 2012.

[24] M. Filatov, "Assessment of density functional methods for obtaining geometries at conical intersections in organic molecules," *Journal of Chemical Theory and Computation*, vol. 9, no. 10, pp. 4526–4541, 2013.

[25] L. Pauling, *The Nature of the Chemical Bond and the Structure of Molecules and Crystals*, Cornell University Press, 1939.

[26] M. W. Schmidt and M. S. Gordon, "The construction and interpretation of MCSCF wavefunctions," *Annual Review of Physical Chemistry*, vol. 49, pp. 233–266, 1998.

[27] A. Pross and S. S. Shaik, "A qualitative valence-bond approach to organic reactivity," *Accounts of Chemical Research*, vol. 16, p. 363, 1983.

[28] H. E. Zimmerman, "The Möbius-Hückel concept in organic chemistry. Application to organic molecules and reactions," *Accounts of Chemical Research*, vol. 4, no. 8, pp. 272–280, 1971.

[29] E. Heilbronner, "Hückel molecular orbitals of Möbius-type conformations of annulenes," *Tetrahedron Letters*, vol. 5, no. 29, pp. 1923–1928, 1964.

[30] P. W. Fowler and H. S. Rzepa, "Aromaticity rules for cycles with arbitrary numbers of half-twists," *Physical Chemistry Chemical Physics*, vol. 8, no. 15, pp. 1775–1777, 2006.

[31] S. Zilberg and Y. Haas, "Locating electronic degeneracies of polyatomic molecules: a general method for nonsymmetric molecules," *The Journal of Physical Chemistry A*, vol. 107, no. 8, pp. 1222–1227, 2003.

[32] S. Zilberg and Y. Haas, "The electron-pair origin of antiaromaticity: spectroscopic manifestations," *International Journal of Quantum Chemistry*, vol. 71, no. 2, pp. 133–145, 1999.

[33] L. Salem, "Theory of photochemical reactions," *Science*, vol. 191, pp. 822–830, 1976.

[34] W. Domcke, H. Koppell, and L. S. Cederbaum, "Spectroscopic effects of comical intersections of molecular-potential energy surfaces," *Molecular Physics*, vol. 43, no. 4, pp. 851–875, 1981.

[35] M. V. Berry, "Quantal phase factors accompanying adiabatic changes," *Proceedings of the Royal Society Series A*, vol. 392, no. 1802, pp. 45–57, 1984.

[36] G. J. Atchity, S. S. Xantheas, and K. Ruedenberg, "Ptential energy surfaces near intersections," *The Journal of Chemical Physics*, vol. 95, pp. 1862–1876, 1990.

[37] Y. Haas, S. Cogan, and S. Zilberg, "The use of elementary reaction coordinates in the search for conical intersections," *International Journal of Quantum Chemistry*, vol. 102, no. 5, pp. 961–970, 2005.

[38] S. Matsika, "Three-state conical intersections in nucleic acid bases," *The Journal of Physical Chemistry A*, vol. 109, no. 33, pp. 7538–7545, 2005.

[39] A. J. C. Varandas, "Geometrical phase effect in JahnTeller systems: Twofold electronic degeneracies and beyond," *Chemical Physics Letters*, vol. 487, no. 1–3, pp. 139–146, 2010.

[40] J. A. Gámez, L. Serrano-andres, and M. Yáñez, "Two- and three-state conical intersections in the electron capture dissociation of disulfides: the importance of multireference calculations," *International Journal of Quantum Chemistry*, vol. 111, no. 13, pp. 3316–3323, 2011.

[41] P. Celani, F. Bernardi, M. Olivucci, and M. A. Robb, "Excited-state reaction pathways for *s-cis* buta-1,3-diene," *The Journal of Chemical Physics*, vol. 102, no. 14, pp. 5733–5742, 1995.

[42] B. Dick, Y. Haas, and S. Zilberg, "Locating conical intersections relevant to photochemical reactions," *Chemical Physics*, vol. 347, pp. 65–77, 2008.

[43] M. Ben-Nun and T. D. Martinez, "Photodynamics of ethylene: ab initio studies of conical intersections," *Chemical Physics*, vol. 259, pp. 237–247, 2000.

[44] Y. Haas and S. Zilberg, "Conical intersections in molecular photochemistry: the phase-change approach," in *The Role of Degenerate States in Chemistry*, vol. 124 of *Advances in Chemical Physics*, pp. 433–504, John Wiley & Sons, New York, NY, USA, 2002.

[45] S. Zilberg and Y. Haas, "The singlet-state photophysics and photochemistry of polyenes: application of the twin-state model and of the phase-change theorem," *The Journal of Physical Chemistry A*, vol. 103, pp. 2364–2374, 1999.

[46] S. A. Le, J. M. White, and W. A. Noyes, "Some aspects of benzene vapor phase photochemistry," *The Journal of Chemical Physics*, vol. 65, no. 7, pp. 2805–2811, 1976.

[47] N. J. Turro, C. A. Renner, and T. J. Katz, "Kinetics and thermochemistry of the rearrangement of benzvalene to benzene. An energy sufficient but non-chemiluminescent reaction," *Tetrahedron Letters*, vol. 46, pp. 4133–4136, 1976.

[48] S. Vanni, M. Garavelli, and M. A. Robb, "A new formulation of the phase change approach in the theory of conical intersections," *Chemical Physics*, vol. 347, no. 1–3, pp. 46–56, 2008.

[49] J. J. Palmer, I. N. Ragazos, F. Bemardi, M. Olivucci, and M. A. Robb, "An MC-SCF study of the S_1 and S_2 photochemical reactions of benzene," *Journal of the American Chemical Society*, vol. 115, no. 2, pp. 673–682, 1993.

[50] H. F. Bettinger, P. R. Schreiner, H. F. Schaefer III, and P. v. R. Schleyer, "Rearrangements on the C_6H_6 potential energy surface and the topomerization of benzene," *Journal of the American Chemical Society*, vol. 120, pp. 5741–5750, 1998.

[51] S. Cogan, A. Kahan, S. Zilberg, and Y. Haas, "Photophysics of (1-Butyl-4-(1H-inden-1-ylidene)-1,4-dihydropyridine (BIDP): an experimental test for conical intersections," *The Journal of Physical Chemistry A*, vol. 112, pp. 5604–5612, 2008.

[52] S. Cogan and Y. Haas, "Self-sensitized photo-oxidation of *para*-indenylidene—dihydropyridine derivatives," *Journal of Photochemistry and Photobiology A: Chemistry*, vol. 193, no. 1, pp. 25–32, 2008.

[53] S. Alfalah, O. Deeb, S. Zilberg, and Y. Haas, "Solvent effect on the conical intersection of 4-cyclopentadienylidene-1,4-dihydropyridine (CPDHP)," *Chemical Physics Letters*, vol. 459, no. 1–6, pp. 100–104, 2008.

[54] A. Kahan, A. Wand, S. Ruhman, S. Zilberg, and Y. Haas, "Solvent tuning of a conical intersection: direct experimental verification of a theoretical prediction," *The Journal of Physical Chemistry A*, vol. 115, no. 40, pp. 10854–10861, 2011.

[55] J.-M. L. Pecourt, J. Peon, and B. Kohler, "Ultrafast internal conversion of electronically excited RNA and DNA nucleosides in water," *Journal of the American Chemical Society*, vol. 122, no. 38, pp. 9348–9349, 2000.

[56] P. C. Ke, C. Naumann, and P. Dubin, "New developments in polymer analytics II. Advances in polymer science," *Journal of the American Chemical Society*, vol. 123, pp. 5166–5166, 2001.

[57] J.-M. L. Pecourt, J. Peon, and B. Kohler, "DNA excited-state dynamics: ultrafast internal conversion and vibrational cooling in a series of nucleosides," *Journal of the American Chemical Society*, vol. 123, no. 42, pp. 10370–10378, 2001.

[58] N. Ismail, L. Blancafort, M. Olivucci, B. Kohler, and M. Robb, "Ultrafast decay of electronically excited singlet cytosine via a π,π^* to n_O,π^* state switch," *Journal of the American Chemical Society*, vol. 124, pp. 6818–6819, 2002.

[59] D. Tuna, A. L. Sobolewski, and W. Domcke, "Mechanisms of ultrafast excited-state deactivation in adenosine," *The Journal of Physical Chemistry A*, vol. 118, no. 1, pp. 122–127, 2014.

[60] L. Blancafort and M. A. Robb, "Key role of a threefold state crossing in the ultrafast decay of electronically excited cytosine," *The Journal of Physical Chemistry A*, vol. 108, pp. 10609–10614, 2004.

[61] G. Steinhauser and T. M. Klapötke, ""Green" pyrotechnics: a chemist's challenge," *Angewandte Chemie International Edition*, vol. 47, no. 18, pp. 3330–3347, 2008.

[62] J. J. Sabatini, A. V. Nagori, G. Chen, P. Chu, R. Damavarapu, and T. M. Klapötke, "High-nitrogen-based pyrotechnics: longer- and brighter-burning, perchlorate-free, red-light illuminants for military and civilian applications," *Chemistry*, vol. 18, no. 2, pp. 628–631, 2012.

[63] A. Vij, J. G. Pavlovich, W. W. Wilson, V. Vij, and K. O. Christe, "Experimental detection of the pentaazacyclopentadienide (pentazolate) anion, *cyclo*-N_5," *Angewandte Chemie International Edition*, vol. 41, pp. 3051–3054, 2002.

[64] H. Östmark, S. Wallin, T. Brinck et al., "Detection of pentazolate anion (cyclo-N_5^-) from laser ionization and decomposition of solid *p*-dimethylaminophenylpentazole," *Chemical Physics Letters*, vol. 379, no. 5-6, pp. 539–546, 2003.

[65] L. Belau, Y. Haas, and S. Zilberg, "Formation of the *cyclo*-pentazolate N_5^- anion by high-energy dissociation of phenyl-pentazole anions," *The Journal of Physical Chemistry A*, vol. 108, no. 52, pp. 11715–11720, 2004.

[66] A. Cembran, F. Bernardi, M. Garavelli, L. Gagliardi, and G. Orlandi, "On the mechanism of the cis-trans isomerization in the lowest electronic states of azobenzene: S-0, S-1, and T-1," *Journal of the American Chemical Society*, vol. 126, pp. 3234–3243, 2004.

[67] W. R. Browne and B. L. Feringa, "Light switching of molecules on surfaces," *Annual Review of Physical Chemistry*, vol. 60, pp. 407–428, 2009.

[68] B. C. Arruda and R. J. Sension, "Ultrafast polyene dynamics: the ring opening of 1, 3-cyclohexadiene derivatives," *Physical Chemistry Chemical Physics*, vol. 16, pp. 4439–4455, 2014.

[69] T. Cusati, G. Granucci, E. Martínez -Núñez, F. Martini, M. Persico, and S. Vázquez, "Semiempirical hamiltonian for simulation of azobenzene photochemistry," *The Journal of Physical Chemistry A*, vol. 116, no. 1, pp. 98–110, 2012.

[70] G. Tiberio, L. Muccioli, R. Berardi, and C. Zannoni, "How does the trans-cis photoisomerization of azobenzene take place in organic solvents?" *ChemPhysChem*, vol. 11, pp. 1018–1028, 2010.

[71] J. J. Blavins, D. L. Cooper, and P. B. Karadakov, "Modern valence bond description of the electronic mechanisms of S_N2 identity reactions," *The Journal of Physical Chemistry*, vol. 108, pp. 914–920, 2004.

[72] S. Shaik and P. C. Hiberty, "Myth and reality in the attitude toward valence-bond (VB) theory: are its "Failures" real?" *Helvetica Chimica Acta*, vol. 86, no. 4, pp. 1063–1084, 2003.

[73] G. A. Worth and L. S. Cederbaum, "Mediation of ultrafast electron transfer in biological systems by conical intersection," *Chemical Physics Letters*, vol. 338, no. 4–6, pp. 219–223, 2001.

[74] D. R. Yarkony, "On the role of conical intersections of two potential energy surfaces of the same symmetry in photodissociation. I. $CH_3SH \rightarrow CH_3S+H$ and CH_3+SH," *The Journal of Chemical Physics*, vol. 100, no. 5, pp. 3639–3644, 1994.

21

Anticancer Activities of Mononuclear Ruthenium(II) Coordination Complexes

William M. Motswainyana[1,2] and Peter A. Ajibade[1]

[1]*Department of Chemistry, Faculty of Science and Agriculture, University of Fort Hare, Private Bag X1314, Alice 5700, South Africa*
[2]*Botswana Institute for Technology Research and Innovation, Private Bag 0082, Gaborone, Botswana*

Correspondence should be addressed to William M. Motswainyana; mtswil004@gmail.com

Academic Editor: M. Paula Robalo

Ruthenium compounds are highly regarded as potential drug candidates. The compounds offer the potential of reduced toxicity and can be tolerated *in vivo*. The various oxidation states, different mechanism of action, and the ligand substitution kinetics of ruthenium compounds give them advantages over platinum-based complexes, thereby making them suitable for use in biological applications. Several studies have focused attention on the interaction between active ruthenium complexes and their possible biological targets. In this paper, we review several ruthenium compounds which reportedly possess promising cytotoxic profiles: from the discovery of highly active compounds imidazolium [*trans*-tetrachloro(dmso)(imidazole)ruthenate(III)] (NAMI-A), indazolium [*trans*-tetrachlorobis(1H-indazole)ruthenate(III)](KP1019), and sodium *trans*-[tetrachloridobis(1H-indazole)ruthenate(III)] (NKP-1339) to the recent work based on both inorganic and organometallic ruthenium(II) compounds. Half-sandwich organometallic ruthenium complexes offer the opportunity of derivatization at the arene moiety, while the three remaining coordination sites on the metal centre can be functionalised with various coordination groups of various monoligands. It is clear from the review that these mononuclear ruthenium(II) compounds represent a strongly emerging field of research that will soon culminate into several ruthenium based antitumor agents.

1. Introduction

The compound *cis*-diamminedichloroplatinum(II) (cisplatin) has been an established antineoplastic agent in the treatment of various cancer cells since its accidental discovery by Rosenberg in the 1960s [1, 2]. Despite its success in the treatment of tumours, the clinical effectiveness of cisplatin has been greatly limited by drug resistance and significant side effects [3, 4]. These drug dynamics have in some cases prompted patients to refuse further treatment. The documented cisplatin limitations have triggered the development and screening of a large number of platinum-based complexes as potential antitumor agents, with a view to develop less toxic, but equally effective cisplatin analogues [3–5]. To date, two more platinum compounds are currently in clinical trial, but initial signs show that these complexes fall short in addressing many of the drawbacks associated with cisplatin [6]. Consequently, these limitations have provided an incentive for further research into other transition metal complexes in an attempt to develop novel drugs that would overcome the disadvantages associated with cisplatin therapy.

Ruthenium compounds are well known for their high relevance as drug candidates, though they have very little in common with the already existing platinum-based drugs. Antitumor potential of these compounds was established over two decades ago, but the interest to explore their cytotoxic profile was very low, possibly because they do not mimic cisplatin in their mode of action [7]. The compounds also offer the potential of reduced toxicity and can be better tolerated *in vivo*. This phenomenon is attributed to the ability of ruthenium to mimic the binding of iron to serum transferrin, which solubilises and transports iron in plasma, thereby exploiting the body's mechanism of nontoxic delivery of iron [8–10]. Furthermore, the various oxidation states, different mechanism of action, and kinetics of ruthenium compounds appear to give them several advantages over platinum-based complexes. For instance, ruthenium is known to be stable in oxidation states II, III, and IV at physiological conditions. In addition, ruthenium(III) complexes have an octahedral coordination sphere in contrast to square-planar platinum(II) complexes [11, 12]. The ligand substitution kinetics of ruthenium(II) and ruthenium(III) are also similar to platinum(II) compounds but tunable due to the strong influence of the coordinated ligands. This makes ruthenium compounds suitable for use in biological applications due to their slow ligand exchange rates, which are close to those of cellular processes [13]. It is well known that ligand exchange is an important determinant of biological activity, as very few metal drugs reach their biological targets without being chemically modified. Under physiological conditions, metal interaction with nucleic acids, proteins, sulfur, or oxygen containing compounds and water could occur in the cells, and such interactions are important for inducing therapeutic effect of a drug [13].

In recent years, many research groups have explored ruthenium compounds for their ability to inhibit tumour growth. The major focus has been on the interaction between active ruthenium complexes and their possible biological targets such DNA, RNA, transferrin, albumin, and cytochrome c [14–17]. It is generally accepted that cytotoxicity of ruthenium(III) and ruthenium(II) complexes is related to their ability to bind DNA [17], although some exceptions have been reported [18]. Studies have also revealed that some ruthenium compounds could inhibit DNA replication, produce mutagenic effects, induce SOS repair, bind to DNA, and reduce RNA synthesis, thereby suggesting a DNA interaction [7]. A major research breakthrough in anti-tumor potency of ruthenium compounds is reflected in the discovery of three most promising ruthenium(III) complexes, imidazolium [trans-tetrachloro(dmso) (imidazole)ruthenate(III)] (NAMI-A) (1) [19–21], indazolium [trans-tetrachlorobis(1H-indazole)ruthenate(III)](KP1019) (2a) [22, 23], and sodium trans-[tetrachloridobis(1H-indazole)ruthenate(III)] (NKP-1339) (2b) [24] (Figure 1). NKP-1339 is a water-soluble sodium salt of KP1019. These complexes adopt octahedral coordination geometry with respect to the ruthenium(III) metal centre, but they exhibit different biological activities despite their structural and chemical similarities. The complexes have since entered clinical trials. NKP-1339 was evaluated in a clinical phase I trial for its safety, tolerability, pharmacokinetics, and pharmacodynamics [25]. In the case of KP1019, an open-label flat-dose escalation trial was performed in patients with advanced solid tumours without further therapeutic options [24, 25].

NAMI-A interferes with the interactions of tumour cells with the extracellular matrix, including an increase of actin-dependent cell adhesion and reduction of cell invasiveness and migration [26, 27]. This mechanism of action by NAMI-A clearly shows that its effects are directed against the process of metastasis, with little mentioned on its inhibitory potential against established tumours [26, 27], possibly due to its low capacity of binding to DNA [28]. However, it remained to be seen if the compound would successfully make it due to its negligible cytotoxicity. Previous clinical results showed that effective compounds for treating metastasis should also have capacity to exert their effect on primary tumours [7].

The redox activities of KP1019 and NPK-1339 conform to the proposed mode of action of ruthenium compounds, which involves disturbance of the cellular redox balance, followed in succession by induction of G2/M cell cycle arrest, blockage of DNA synthesis, and induction of apoptosis via the mitochondrial pathway [25]. Furthermore, their high tumour targeting potential is probably based on delivery to tumour sites by serum proteins albumin and transferrin, as well as on the activation of the compounds in the reductive tumour milieu [25]. KP1019 and NKP-1339 are administered intravenously, and their interaction with serum proteins has significant relevance. Several research findings have demonstrated a strong affinity of KP1019 and NKP-1339 to proteins in the bloodstream, in particular to albumin and transferrin [29]. A detailed investigation of the binding of the ruthenium(III) drugs to transferrin, in particular KP1019, in vitro was proven by X-ray crystal structure analysis. X-ray crystallography experiments revealed that, after binding to histidine-253, the two indazole ligands still remain bound to the ruthenium centre [30]. It was also shown that the reaction of transferrin with KP1019 or NKP-1339 is very fast and gets completed within several minutes [31]. It is therefore evident from the literature reports that ruthenium(III) compounds have yielded significant success in recent days, and several reports based on their cytotoxicity continue to emerge. In this review, we focus particular attention on ruthenium(II) compounds which have been reported to possess promising cytotoxic profiles.

2. Ruthenium(II) Compounds

2.1. Organometallic Ruthenium(II) Compounds. According to the literature reports, ruthenium(III) complexes are activated by reduction to ruthenium(II), and these reports have focused current research on anticancer potential of "half-sandwich" ruthenium(II) arene compounds of the type [(η^6-arene)Ru(YZ)(X)] (3) [32]. The compounds are also referred to as "piano-stool," where YZ is a bidentate chelating ligand and X is a leaving group (Figure 2) [32, 33]. These half-sandwich complexes offer the opportunity of derivatization at the arene moiety, while the three remaining coordination sites on the metal centre (X, Y, and Z) can be functionalised with various coordination groups of various monoligands or bidentate ligands [32–35]. These ligands can introduce additional functionalities to the coordination sphere around the metal centre, resulting in a metal-ligand drug synergism [32–35]. Therefore, it becomes apparent that half-sandwich arene ruthenium complexes will allow introduction of numerous biologically active groups. Among the first ruthenium(II) arene complexes reported is [Ru(η^6-C$_6$H$_6$)(DMSO)Cl$_2$] (4a) (Figure 3). Investigations into its antitumor potential revealed that the complex strongly inhibits the function of topoisomerase II, which is important for structural organization of the mitotic chromosomal scaffold during cell replication process [36].

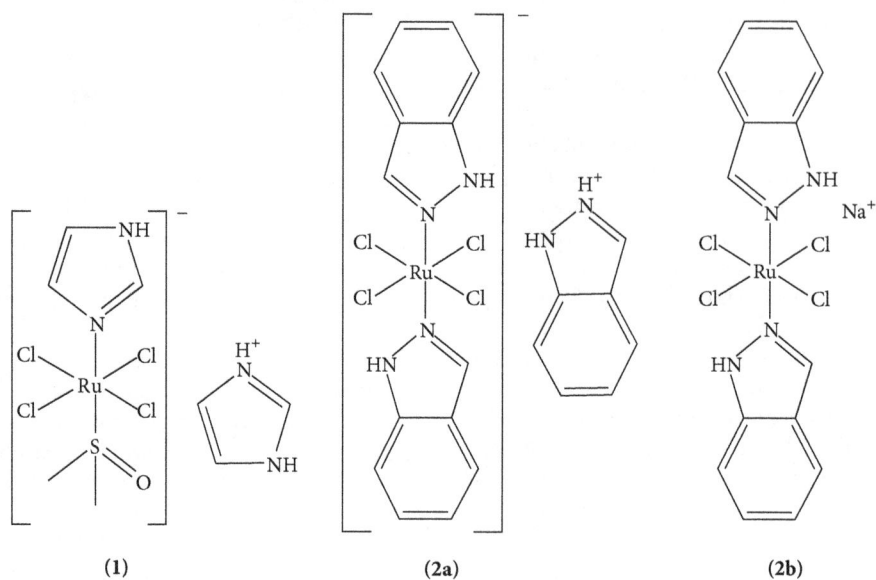

FIGURE 1: Structures of Ru(III) complexes: NAMI-A (**1**), KP1019 (**2a**), and NKP-1339 (**2b**).

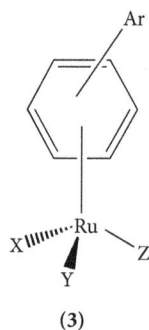

FIGURE 2: General structure of Ru(II) complexes with "piano-stool" configuration.

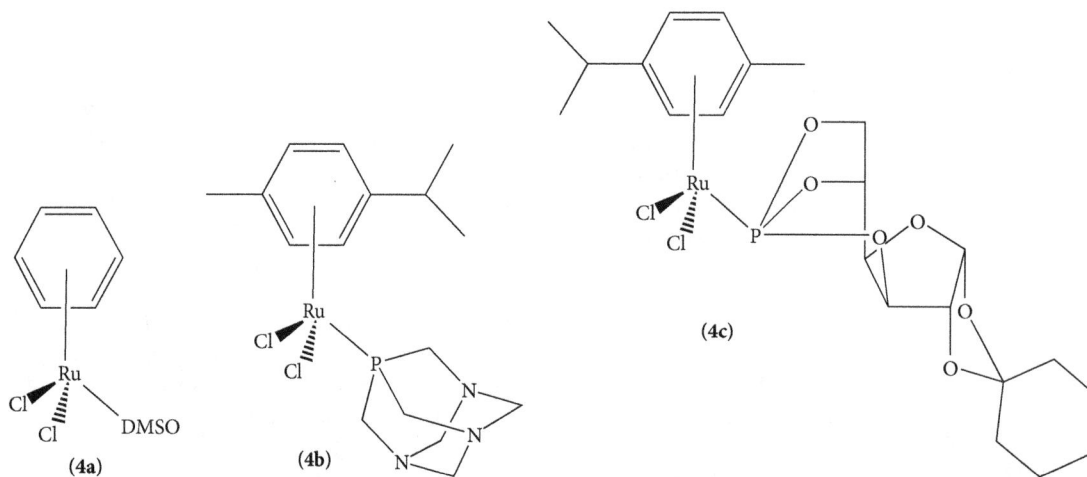

FIGURE 3: Structures of Ru(II)arene complexes [Ru(C_6H_6)(DMSO)Cl_2] (**a**), RAPTA-C (**b**), and KP1558 (**c**).

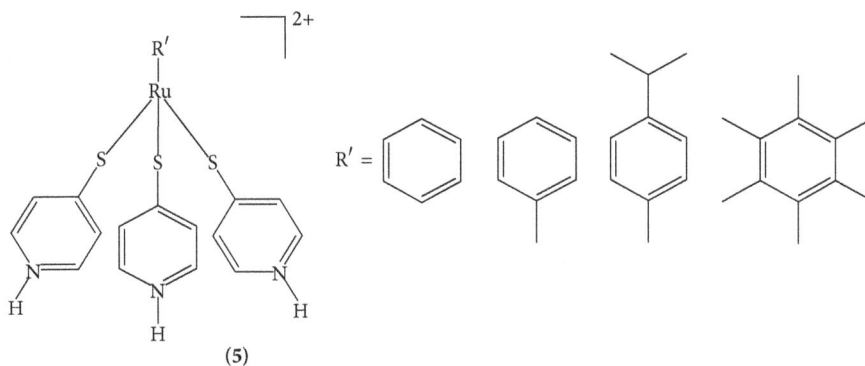

FIGURE 4: Ruthenium(II) complexes containing pyridinethiolato ligands.

Scolaro et al. has shown that when DMSO molecule in compound **4a** was replaced by PTA (1,3,5-triaza-7-phosphatricyclo[3.3.1.1]decane) this ligand exchange increased aqueous solubility of the new complex RAPTA-C (**4b**), thereby giving rise to promising antimetastatic activities, high selectivity, and generally low toxicity which makes them attractive as potential therapeutic agents [37]. It is therefore strongly believed that PTA ligand is responsible for the observed selectivity. Furthermore, methylation of one PTA nitrogen leads to a further increase in toxicity [37, 38]. The compound was found to be noncytotoxic in the metastasizing mouse cell lines TS/A or human HBL100 (breast carcinoma) cell line but showed pronounced antimetastatic activity. Crystallographic experiments of **4b** with the nucleosome core showed that proteins are the primary target of the compound [37, 38].

Another compound KP1558 (**4c**) (Figure 3) and its analogues are prepared by replacing PTA with 3,5,6-bicyclophosphite-α-D-glucofuranoside ligands to obtain highly active compounds [39, 40]. The lipophilicity of the complexes can be modulated by modifying the carbohydrate moiety to yield coordination compounds with high aqueous solubility ideally suited for intravenous administration or hydrophobic species that facilitate cellular uptake. The hydrolytic behaviour, the affinity to proteins and model nucleobases, and the *in vitro* antineoplastic activity of the complexes were investigated against human SW480 (colon adenocarcinoma), CH1 (ovarian cancer), cisplatin-resistant A2780 (ovarian carcinoma), A549 (lung carcinoma), Me300 (melanoma), LNZ308 (glioblastoma), and HCEC (endothelial) cell lines [39, 40]. In water, the phosphite ligand is hydrolyzed, and this process is activated by prior hydrolysis of the chlorido ligand at the ruthenium centre. The complex resulting from hydrolysis of the Ru-Cl bonds and P-O bonds of the ligand showed affinity towards albumin and transferrin, and it was also capable of binding to 9-ethylguanine (9EtG) with formation of 1 : 1 adducts. The complexes showed remarkably high activity in all tested cell lines, and cytotoxic activity of the compounds is believed to be enhanced by the alkyl moiety of the phosphite which makes the compound more lipophilic [39, 40].

Arene ruthenium(II) complexes containing pyridinethiolato ligands (**5**) have been subject of investigation due to their remarkable *in vitro* cytotoxicity profile (Figure 4) [41]. The work was motivated by previous reports on platinum complexes containing pyridinium-thiolato ligands, which demonstrated excellent anticancer activity [42]. The mononuclear ruthenium complexes $[(\eta^6\text{-arene})\text{-Ru}(SC_5H_4NH)_3]^{2+}$ are prepared by reaction of dinuclear arene ruthenium complexes $[(\eta^6\text{-arene})_2\text{Ru}_2(\mu_2\text{-Cl})_2\text{Cl}_2]$ (arene = C_6H_6, C_6H_5Me, $p\text{-}^iPrC_6H_4Me$, C_6Me_6) with 4-pyridinethiol. The reaction formally involves the transfer of the hydrogen atom in 4-pyridinethiol from sulfur to nitrogen to form a 4-pyridinium–thiolato ligand [41]. The redox data of the complexes shows that the complexes generally undergo two consecutive irreversible oxidations. However, there is no clear correlation between the IC$_{50}$ values against human A2780 (ovarian cancer) cells and the E_{pa} data, since the benzene derivative is more cytotoxic compared to the alkyl-substituted arene derivatives. The hexamethylbenzene derivative is the most cytotoxic, and it is worth noting that the observed cytotoxic values are superior to those of cisplatin and other platinum-based agents [41].

In an attempt to overcome multidrug resistance, arene ruthenium complexes of the general formula $[\text{Ru}(\eta^6\text{-}p\text{-cymene})\text{Cl}_2(L)]$ (**6–8**) have been explored (L = anthracene-based multidrug resistance modulator ligands) (Figure 5) [43]. The general toxicity test of complexes **6–8** shows that **8** is the most active while **6** is the least cytotoxic substance. Compound **8** provides a relatively homogeneous cytotoxicity profile across all the examined cell lines, showing similar activity in all tumorigenic cell lines and the nontumorigenic human cell line HBL-100 (breast carcinoma), which could indicate a generalized mechanism of action. Further studies on **8** show that it induces cell death *via* inhibition of DNA synthesis. Since the complex is fluorescent, its uptake in cells was studied, and relative to the free anthracene-based ligand, uptake of the complex is accelerated and accumulation of the complex in the cell nucleus is observed. Interestingly, **6** exhibits highly cytotoxic properties in human HT-29 (colon adenocarcinoma) cells only, but its selectivity towards a certain type of cancer makes it a promising compound [43].

A recent report based on excellent *in vitro* anticancer activity of long-chain isonicotinic ester ruthenium(0) nanoparticles [44] motivated the group of Khan et al. to

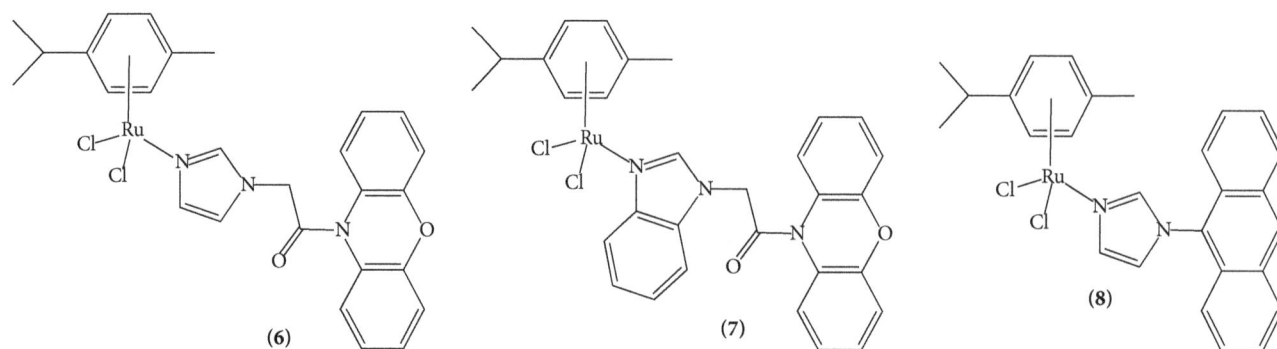

FIGURE 5: Ruthenium(II) complexes of the general formula $[Ru(\eta^6\text{-}p\text{-cymene})Cl_2(L)]$.

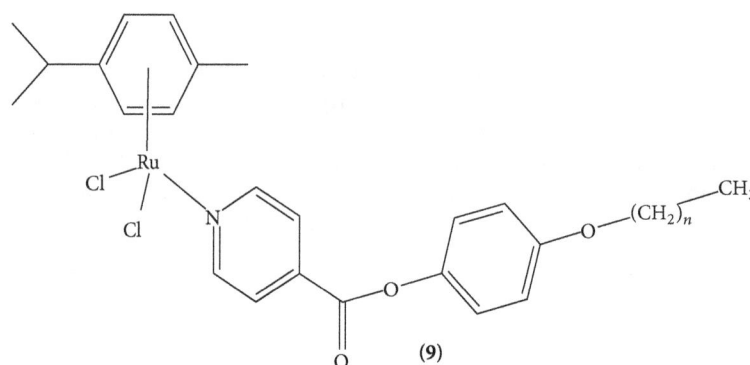

FIGURE 6: Arene ruthenium dichloro complexes containing isonicotinic ester ligands.

adopt long-chain isonicotinic esters in their ligand backbone as lipophilic components, with a view to increase anticancer activity of their arene ruthenium complexes [45]. The work eventually led to the development of a series of p-cymene ruthenium dichloro complexes containing isonicotinic ester ligands, $[(\eta^6\text{-arene}) RuCl_2NC_5H_4\text{-}4\text{-COO-}C_6H_4\text{-}p\text{-O-}(CH_2)_n\text{-}CH_3]$ (n = 1, 3, 5, 7, 9, 11, 13, 15) (9) from p-cymene ruthenium dichloro dimer, and the corresponding isonicotinic ester ligand (Figure 6). The molecules adopt a pseudotetrahedral piano-stool geometry, while the isonicotinic ester ligand is coordinated through the nitrogen atom. Isonicotinic acid (pyridine-4-carboxylic acid) is widely used for the synthesis of antibiotics and antituberculosis preparations [46]. This encouraging pharmacological profile of isonicotinic acid derivatives coupled with amphiphilic arene ruthenium moieties makes this combination desirable for drug design. The p-cymene ruthenium dichloro complexes were evaluated for anticancer activity towards human A2780 (ovarian carcinoma) and A2780cisR (ovarian carcinoma, cisplatin resistant) cells, and high activities were observed only for n = 9 (presenting a chain with ten carbon atoms) [45].

There is remarkable prospective among ruthenium half-sandwich structured complexes with cyclopentadienyl ligands [47–49]. The aromatic ligand $\eta^5\text{-}C_5H_5$ occupies three coordination positions, thereby offering stability to the metal centre [47]. As expected for potentially active ruthenium half-sandwich compounds, the remaining coordination sites can be occupied by a ligand that could impart antitumor activity together with ligands that can control electronic properties at the ruthenium centre [47]. Molecular structures of many of these compounds were successfully solved by X-ray crystallography. Garcia et al. first reported biological studies of compounds containing the "RuCp" fragment (Cp = η^5-cyclopentadienyl) with heteroaromatic ligands (10) (Figure 7) [49]. A dose response of each compound was performed in human LoVo (colon adenocarcinoma) and Mia-PaCa (pancreatic) cell lines, and both compounds caused a significant effect in cell viability in the nanomolar range. The complex with pyridazine (10b) produced a decrease of cell survival of more than 90% at 500 nM in both cell lines, whereas the complex with triazine (10a) was less effective [49]. Atomic force microscopy (AFM) images suggest different mechanisms of interaction with the plasmid pBR322 DNA, while the mode of binding of the compounds could be intercalation between base pairs of DNA (10a) or covalent bond formation with N from the purine base (10b) [49]. The IC_{50} values derived from these compounds are within the lowest IC_{50} values in half-sandwich stool ruthenium complexes reported elsewhere [50, 51].

Moreno et al. further explored cytotoxic properties of compounds of general formula $[Ru(\eta^5\text{-}C_5H_5)(PP)L]^+$ (11 and 12), where L is a N-heteroaromatic σ-bonded ligand, chosen within planar molecules, in order to potentiate intercalation of compounds in DNA, besides their possible covalent binding to N7 guanine residues (Figure 7) [48]. They observed

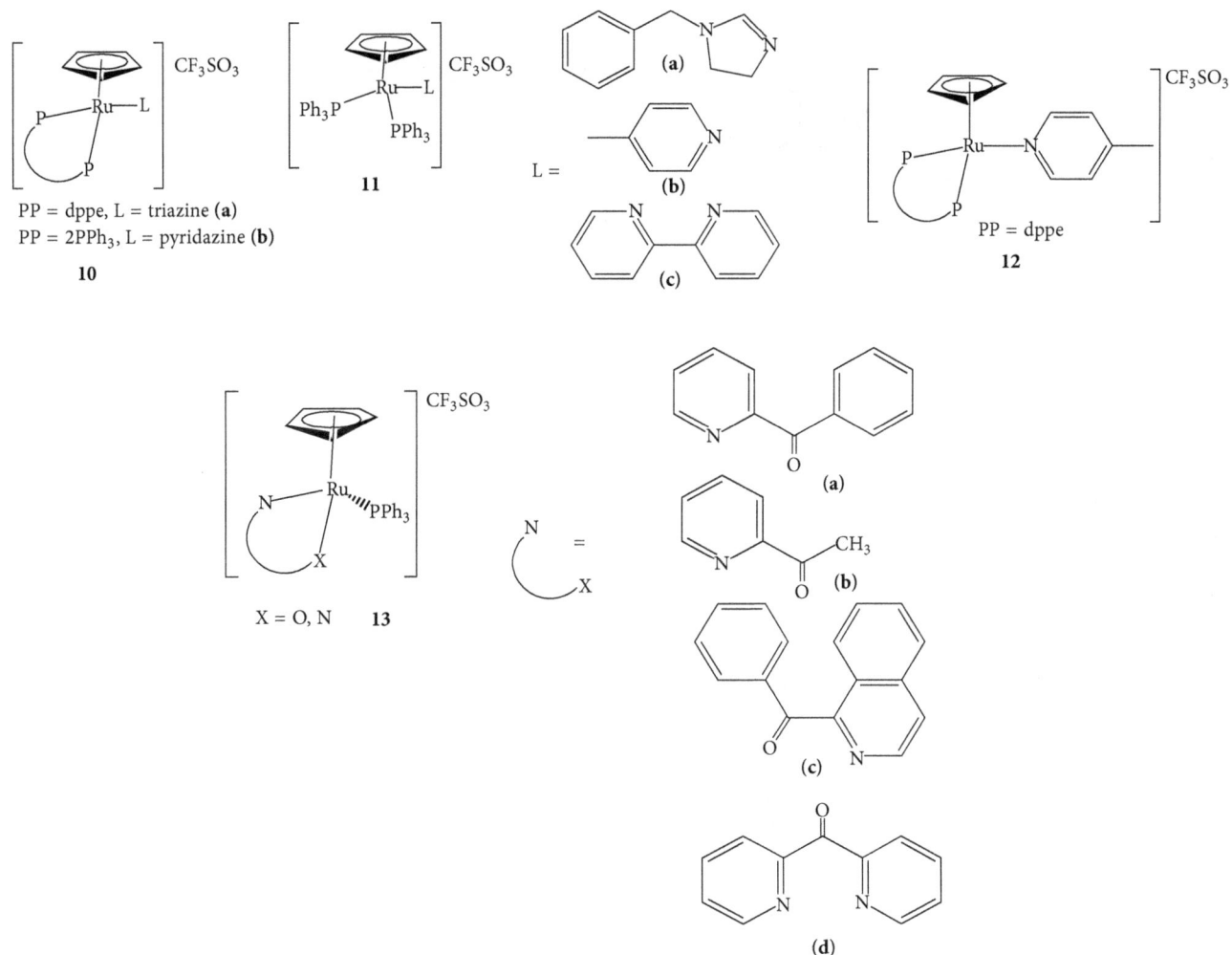

FIGURE 7: Ruthenium-cyclopentadienyl complexes with heteroatomic ligands.

interesting properties of stability in addition to the excellent cytotoxicity in Ru(η^5-C$_5$H$_5$) systems coordinated to N-heteroatomic molecules, thereby suggesting the effectiveness of a chelating ligand [48]. Their investigation mainly focused on different hapticity of L, which can be monodentate, such as 1-benzylimidazole (11a), 4-methylpyridine (11b), or 2,2′-bipyridyl (11c), occupying two coordination sites, leading to a different mechanism of action of the complexes [48]. Compound 12 had a bidentate phosphine ligand to check if there would be significant difference in cytotoxicity profile of the complex compared to 11b. Interaction of these new compounds with the plasmid pBR322 DNA was studied by AFM, electrophoretical mobility, and viscosity measurements, and their cytotoxicity was examined on human HL-60 (leukemia cancer) cells [48]. Generally, cytotoxicity of the compounds was much higher than that found in cisplatin against human HL-60 (leukemia cancer) cells, thereby indicating that, for these type of ruthenium compounds, DNA is possibly one of the targets of their action inside the cells [48]. Further biological studies were performed by incubating the compounds with free plasmid pBR322 DNA and pBR322 DNA, and the resulting images obtained by

AFM showed several supercoiled forms of plasmid DNA strongly modified, which could suggest interaction with plasmid DNA [48]. The interaction was further confirmed by viscosity measurements of solutions of free DNA and DNA incubated with different concentrations of the compounds. The complexes contain 1-benzylimidazole ligand or phenyl groups on the two phosphine ligands which are capable of intercalation between base pairs of DNA [48].

This architectural friendliness observed in half-sandwich arene ruthenium complexes offers an opportunity to improve their cytotoxic profiles [47]. The complexes were prepared by σ coordination of ligands L, presenting two heteroatoms, 2-benzoylpyridine (13a), 2-acetylpyridine (13b), 1-isoquinolinyl phenyl ketone (13c), and di(2-pyridyl)ketone (13d). The compounds presented coordination sites other than nitrogen, and they proved to be very stable when exposed to air and moisture [47]. Experimental data (^{1}H, ^{13}C NMR, UV-vis, and DFT theoretical calculations) of these compounds revealed that metal-ligand binding was strengthened by an electronic flow from the metal centre to the σ-bonded ligand [47]. Cytotoxic studies in human A2780 (ovarian carcinoma), A2780cisR (ovarian carcinoma, cisplatin

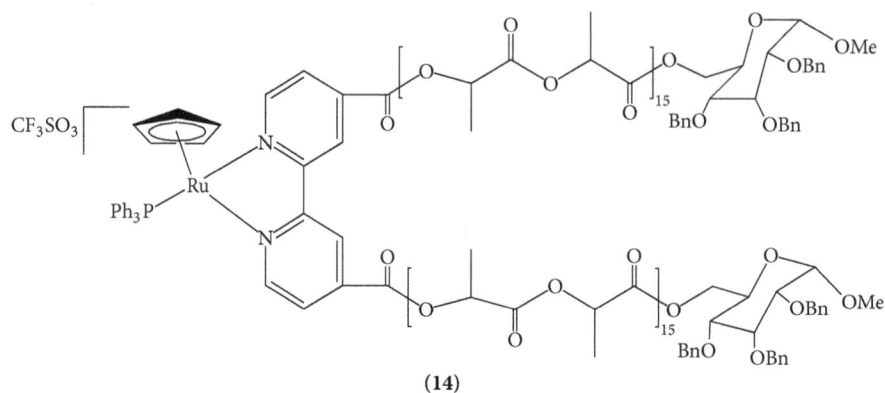

FIGURE 8: D-glucose end capped polylactide ruthenium-cyclopentadienyl.

resistant), MCF-7 (breast adenocarcinoma-hormone dependent), MDAMB231 (hormone-independent breast adenocarcinoma), HT-29 (colon adenocarcinoma), and PC3 (prostate cancer) cell lines revealed an exceptional activity of the complexes, with IC_{50} values in the nanomolar range [47]. However, the isoquinolinyl moiety in complex **13c** seemed to reduce its cytotoxic activity when compared with **13a**, **13b**, and **13d**. Despite compound **13c** ranking as the least active complex, it still presented higher cytotoxic activity than the reference drug cisplatin even for the more chemioresistant cancer cells PC3 [47]. Further biological studies are yet to be carried out to highlight the potential of these compounds as future drugs considering that cytotoxicity is not exclusive for determining suitability of metallodrugs.

Recently, Valente et al. reported on the first polymer based ruthenium-cyclopentadienyl complex [CpRu(P)(bpyPLA)]$^+$ (**14**) (Cp = cyclopentadienyl, P = triphenylphosphine, and bpyPLA = 2,2'-bipyridine-4,4'-D-glucose end-capped polylactide), which demonstrated potential as anticancer agent (Figure 8) [52]. The work was motivated by earlier findings which revealed how polymeric controlled drug delivery could prove to be a promising alternative to the conventional drug delivery approaches in cancer therapy [53]. A degradation study of the complex in HEPES (4-(2-hydroxyethyl)-1-piperazineethanesulfonic acid) buffer was performed to investigate the polymer hydrolysis at physiological and tumour cell pH. Results showed that the complex is stable over some days in an aqueous environment at physiologic pH [52]. Cytotoxicity of compound **14** was assayed in human MCF-7 (breast adenocarcinoma-hormone dependent) and MDAMB231 (hormone-independent breast adenocarcinoma) and A2780 (ovarian adenocarcinoma) cell lines, and the compound proved to be 6-fold more cytotoxic compared to cisplatin especially for the MCF-7 cell lines, thereby supporting the initial suggestion that polymer ruthenium-cyclopentadienyl complexes could soon replace the known anticancer agent [52]. Further tests on cellular distribution of the complex, carried out using inductively coupled plasma mass spectrometry (ICP-MS), showed that higher amount of ruthenium localized in the nucleus than in the membrane, therefore representing high uptake of ruthenium by cells. The results possibly suggest DNA as

the potential target for ruthenium-cyclopentadienyl complexes [52].

Although ruthenium-cyclopentadienyl compounds bearing carbohydrate-derived ligands remain an unexplored area, there is emerging evidence of their application as antitumor agents. For example, new η^5-cyclopentadienyl ruthenium(II) complexes of general formula $[(\eta^5\text{-}C_5H_5)Ru(PP)(L)]^+$, isolated as PF_6^- salts, have been reported [54], where L represents the galactose and fructose carbohydrate derivative ligands, functionalized with nitrile, tetrazole, and 1,3,4-oxadiazole N-coordinating moieties (**15a–c**) (Figure 9) [54]. Electronic density and the stereochemical environment of the metal centre are controlled by using two different phosphanes, PPh$_3$ and dppe, used as coligands. Cytotoxic studies on human HeLa (cervical carcinoma) cancer cell lines revealed very good activities, with IC_{50} values in the low micromolar range, better than cisplatin. However, further work is yet to be carried out on their cellular transport distribution and possible mode of action [54].

The presence of a chelating ligand in the half-sandwich ruthenium(II) complexes appears to offer advantages of structural stability and the opportunity to "fine-tune" the electronics of the ruthenium centre, which is an important parameter for drug development [5, 55–62]. The tuning of the ligand also results in a changed preference of the targeted nucleobases. Based on this background, a group of ruthenium(II) arene complexes containing diamine moieties in the coordination sphere (**16**) has been reported (Figure 10). The complexes contain chelating bidentate ligands such as ethylenediamine [55–60], paullones (**17** and **18**) (Figure 11) [61], and staurosporine (where the π-arene is a cyclopentadienyl, Cp) (**19**) inserted into the ruthenium scaffolds (Figure 12) [5, 55–60, 62]. According to the rules governing structure-activity relationships for an effective platinum anticancer drug, the two nonleaving cis-coordinated amine ligands are crucial for anticancer activity [63]. These rules would also apply to ruthenium compounds.

These ethylenediamine based complexes exhibit high in vitro and in vivo anticancer activities against human ovarian A2780 cell lines. The complexes bind coordinatively to N7 of guanine in DNA, which can be complemented by intercalative binding of an extended arene and specific hydrogen

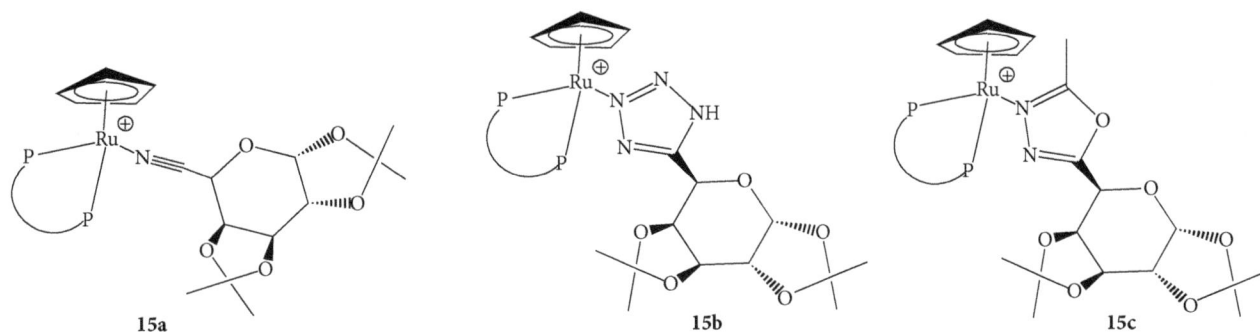

FIGURE 9: Ruthenium-cyclopentadienyl complexes with carbohydrate-derived ligands.

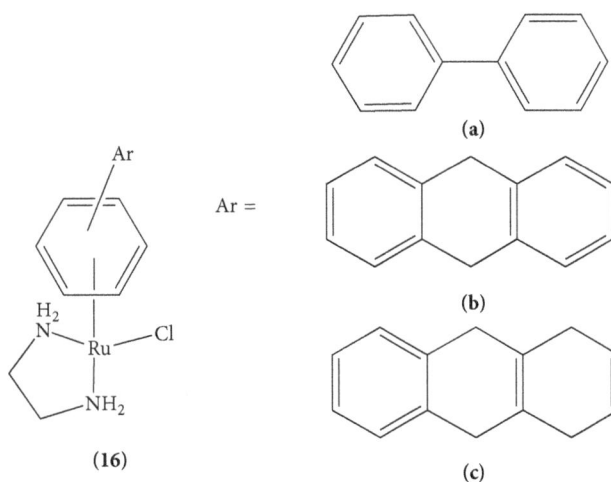

FIGURE 10: Structures of ethylenediamine based Ru(II) complexes with isonicotinic ester ligands benzene derivatives.

bonding interactions of the ethylenediamine NH_2 groups with C6O of guanine [55–60]. These additional interactions result in unique binding modes to duplex DNA and induce different structural distortions in DNA compared to cisplatin, which explains why these complexes are not cross-resistant with cisplatin [64]. Among these ethylenediamine based arene ruthenium complexes, compound 16a has proved to be a lead, by demonstrating p53 and p21/WAF1-dependent early growth arrest in HCT116 (colorectal cancer) cell lines [65]. The paullones, known as cyclin-dependent kinase (CDK) and glycogen synthase kinase-3 inhibitors, contain a seven-member folded azepine ring, which makes the molecule nonplanar [61]. The complexes were investigated for their antiproliferative activity against A529 (non-small lung cancer) cell, CH1 (ovarian carcinoma), and SW480 (colon adenocarcinoma), where they exhibited cytotoxic activities that were in the micromolar concentration range, therefore showing great potential as antitumor agents. This work was immediately followed by the development of arene ruthenium(II) indoloquinoline complexes (20) (Figure 13) with a view to elucidate the structure-activity relationship of these compounds [64]. The complexes have a similar framework to the paullones but have a six-membered ring. Binding of indoloquinolines to a metal-arene scaffold is believed to make the products soluble enough in biological media to allow for assaying their antiproliferative activity [64]. Their cytotoxic profiles were observed to be higher than their paullone counterparts in the three human cancer cell lines. These complexes were later modified by introducing a methyl and halo substituents in position 8 of the molecule backbone (21) (Figure 13) [65]. However, the complexes have a five-membered ring. The effect of substituents in position 8 of the indoloquinoline backbone had been previously investigated on copper(II) complexes, where they returned cytotoxic values in the nanomolar concentration range [66]. The arene ruthenium(II) indoloquinoline compounds with halo substituents in position 8 were found to be more effective cytotoxic agents in vitro than the previously reported species halogenated in position 2 of the indoloquinoline backbone as expected [65].

Staurosporine type complexes (19) were initially explored by Meggers et al. with a view to develop efficient and economical synthetic strategy for the rapid modification of the cyclopentadienyl moiety of ruthenium half-sandwich protein kinase inhibitors [55–60]. They adopted half-sandwich strategy and inserted a ligand derived from the kinase inhibitor to the ruthenium coordination sphere [55–60]. The approach has led to the identification of Pim-1 and GSK-3R inhibitors with improved potencies and selectivities. Before coordination, natural staurosporine has a useful property

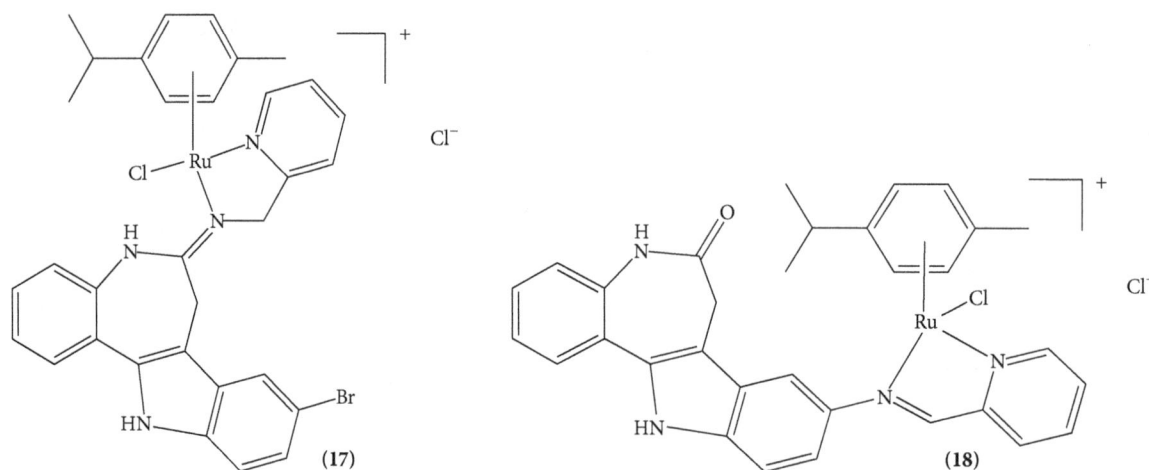

FIGURE 11: Some paullone-type arene ruthenium(II) complexes.

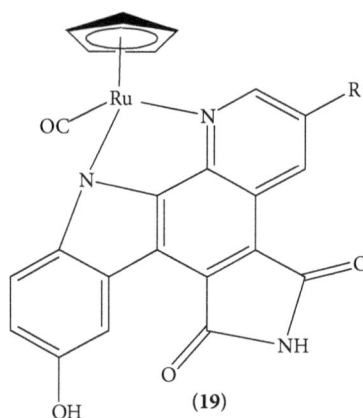

FIGURE 12: Staurosporine-type arene ruthenium complexes.

of inhibiting various kinases, which has led to the development of many anticancer drugs based on staurosporine derivatives. However, recent research has shown that the activity of the ligands becomes more pronounced when they are coordinated to the metal [67]. The strategy of designing metal complexes as protein kinase inhibitors using staurosporine as a lead structure was therefore aimed at mimicking the binding mode of staurosporine [55–60]. The approach was based on the hypothesis that complementing an organic scaffold with a metal center opens new avenues for the design of small molecules with novel three-dimensional structures and thus offers an opportunity to develop compounds with novel biological properties [55–60]. The metal complexes bear a bidentate ligand which retains the structural features of the indolocarbazole heterocycle and the approach targets the metal complexes to the ATP-binding site [55–60]. The remaining ligand sphere of the ruthenium atom gives the opportunity to create interactions with other parts of the ATP-binding site. Meggers et al. have already proved that the ligand sphere around the ruthenium centre substantially influences kinase binding affinities [55–60]. It is therefore not surprising that staurosporine type

arene ruthenium(II) complexes tested against melanoma cell lines gave nanomolar activity [68]. The complexes are potent activators of p53 proteins and induce p53 activated apoptosis *via* mitochondrial pathway [55–60]. At the same time, these new scaffolds have other properties which are clearly distinct from their parent indolocarbazole alkaloids. For example, whereas staurosporine is a nanomolar inhibitor for most protein kinases, some racemic mixtures of ruthenium half-sandwich compounds show a remarkable preference for just a few kinases like GSK-3 and Pim-1, thereby demonstrating improved selectivity [55–60].

Research has demonstrated that arene ruthenium(II) complex $[(\eta^6\text{-arene})Ru^{II}(en)Cl]^+$, which exhibits great inhibitory activity against various human tumor cells, can unwind the double-strand helix of DNA by forming monofunctional adducts with DNA in intercalative mode [69, 70]. Furthermore, cell cycle arrest in G1 phase through p53-dependent and p53-independent mechanism has also been reported for these compounds [71]. However, the little available information on the S-phase cell cycle arrest through DNA damage-mediated p53 phosphorylation induced by arene ruthenium complexes has prompted Wu et al. to study

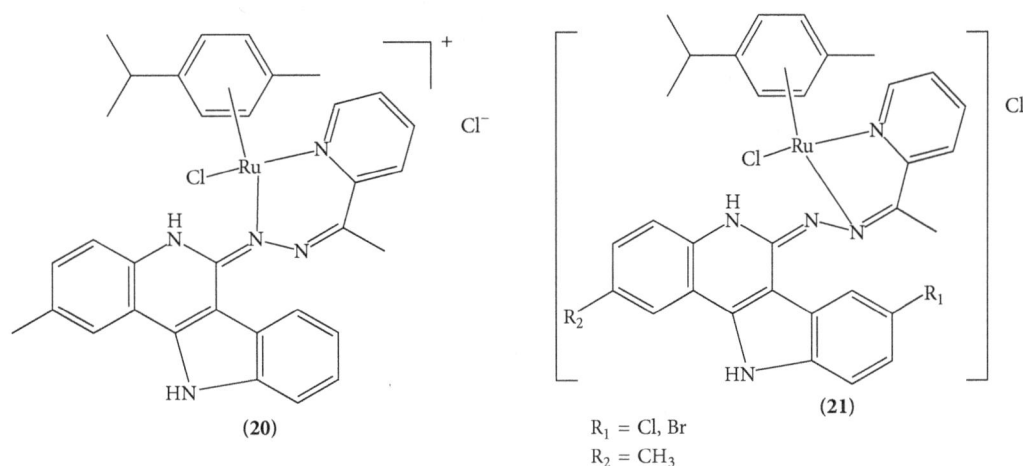

FIGURE 13: Indoloquinoline type arene ruthenium(II) complexes.

R₁ = Cl, Br
R₂ = CH₃

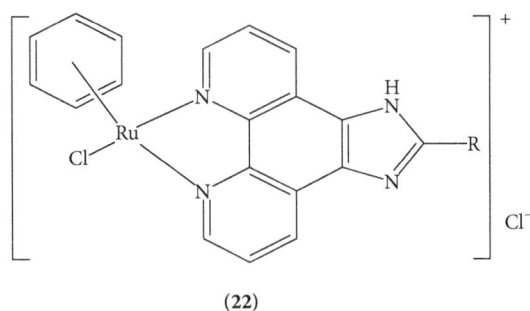

FIGURE 14: Arene ruthenium(II) complexes coordinated by phenanthroimidazole derivatives.

a series of arene ruthenium(II) complexes coordinated by phenanthroimidazole derivatives (**22**) (Figure 14) [72]. The complexes displayed acceptable inhibitory activity against various tumour cells, especially human MG-63 (osteosarcoma) cells. The complexes also bind to DNA molecule by intercalative mode to disturb the biofunction of tumor cells. Further investigation on the mechanisms revealed that this type of arene ruthenium(II) complexes could induce S-phase arrest in tumor cells through DNA damage-mediated p53 phosphorylation [72].

Several other ruthenium based compounds containing chelating pyrone (O,O) and thiopyrone (S,O) (**23**) (Figure 15) [73, 74] and tetrahydroisoquinoline (N,O) (**24**) (Figure 16) moieties [75] have been prepared with a view to study the ruthenium-ligand interaction and its effect on the *in vitro* anticancer activity. The thiopyrones are reported to be more lipophilic than their pyrone analogues. Fernández and coworkers have demonstrated that a change of donor ligand has a profound effect on the electronic properties of the Ru(II) complex [76]. Antiproliferative activities of type **23** compounds were investigated against the colon carcinoma SW480 and ovarian carcinoma CH1 cancer cell lines, and the thiopyrone complexes were more active compared to their pyrone analogues. DFT calculations show that thiopyrones have stronger binding to ruthenium compared to pyrones, and the different stabilities of these compounds may be responsible for the observed influence of the donor atoms on *in vitro* anticancer activity [75]. Ruthenium(II) arene complexes bearing 1,2,3,4-tetrahydroisoquinoline (TIQ) amino alcohol ligands (**24**) have been investigated for their activity against human cancer cell lines MCF-7, A549 (lung adenocarcinoma), MDAMB-231 (hormone-independent breast adenocarcinoma), and normal MDBK cells [75]. Synthetic TIQ derivatives have been found to exhibit interesting biological activities which range from histidine H3 antagonism, antidiabetic activity, and multidrug resistance reversal for some identified cancers [77–79]. These remarkable properties suggest that incorporation of the TIQ moiety as a backbone in metal complexes could be a viable anticancer drug discovery strategy [75]. The complexes are inactive against the A549 and MDAMB-231 cell lines. However, they exhibited moderate activity against MCF-7 cancer cells with IC_{50} values ranging from 34 to 218 μM, a surprising observation because the compounds differ by only one diastereomeric centre. The most fascinating result was the remarkable selectivity displayed for MCF-7 cells in comparison to normal MDBK cells [75].

Anticancer activities of ruthenium(II) compounds based on N-substituted 2-pyridinecarbothioamides (PCAs) (**25**) have been documented in the literature (Figure 17) [80]. The 2-pyridinecarbothioamide ligands have previously shown notable activity as gastric mucosal protectants and low acute toxicity *in vivo* [81]. However, their coordination

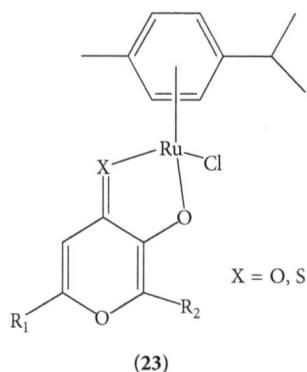

FIGURE 15: Pyrone and thiopyrone ruthenium(II)-cymene complexes.

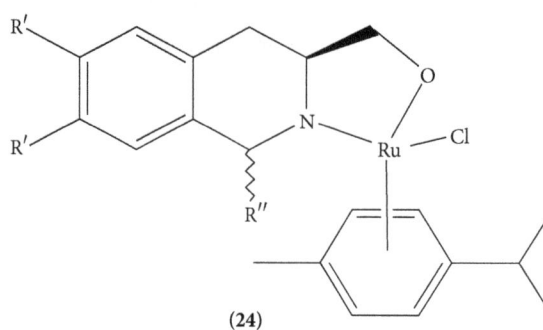

FIGURE 16: Ruthenium(II) arene complexes bearing TIQ amino alcohol ligands.

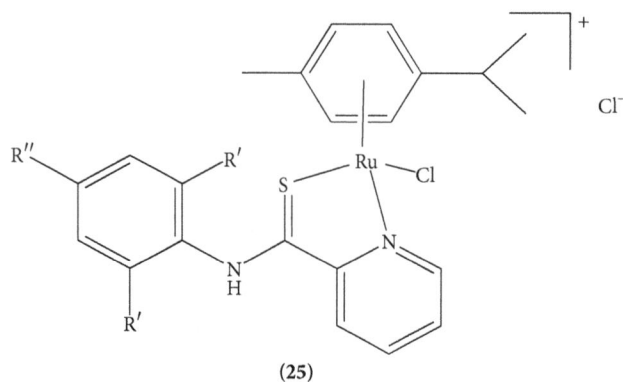

FIGURE 17: Ruthenium(II) complexes of N-substituted 2-pyridinecarbothioamides.

to ruthenium(II) yielded highly cytotoxic agents in the more chemoresistant SW480 (colon adenocarcinoma) and A549 cell lines. The complexes boast exceptional stability in hydrochloric acid, coupled with low reactivity towards biological molecules. Their unexpected aqueous chemistry renders them suitable for oral administration. Crystallographic studies with the nucleosome core revealed that the compounds form exclusively histidine-adducts, while being devoid of DNA binding [80].

2.2. Inorganic Ruthenium(II) Compounds. Ruthenium complexes containing arylazopyridine (azpy) ligands have attracted attention. Three of the possible five isomers of

$[Ru(azpy)_2Cl_2]$ (α(**26a**), β(**26b**) and γ(**26c**)) have been reported, where the α isomer represents *cis,trans,cis*, the β isomer *cis,cis,cis,* and the γ isomer *trans,cis,cis* positions of the chlorides, the pyridine, and azo nitrogens, respectively (Figure 18) [82, 83]. The compounds show promising cytotoxic activity that is structurally dependent (α and γ isomers show higher toxicities than the β isomer). DFT calculations suggest that the ability of the isomers to intercalate to DNA decreases from $\gamma > \alpha > \beta$ isomers on the basis of the geometric and electronic factors, and this supports the observed cytotoxicity [84].

A similar compound to $[Ru(azpy)_2Cl_2]$, the *trans*-$[Ru^{II}$ $(Hpyrimol)_2Cl_2]$ (Hpyrimol = 4-methyl-2-N (2-pyridyl-

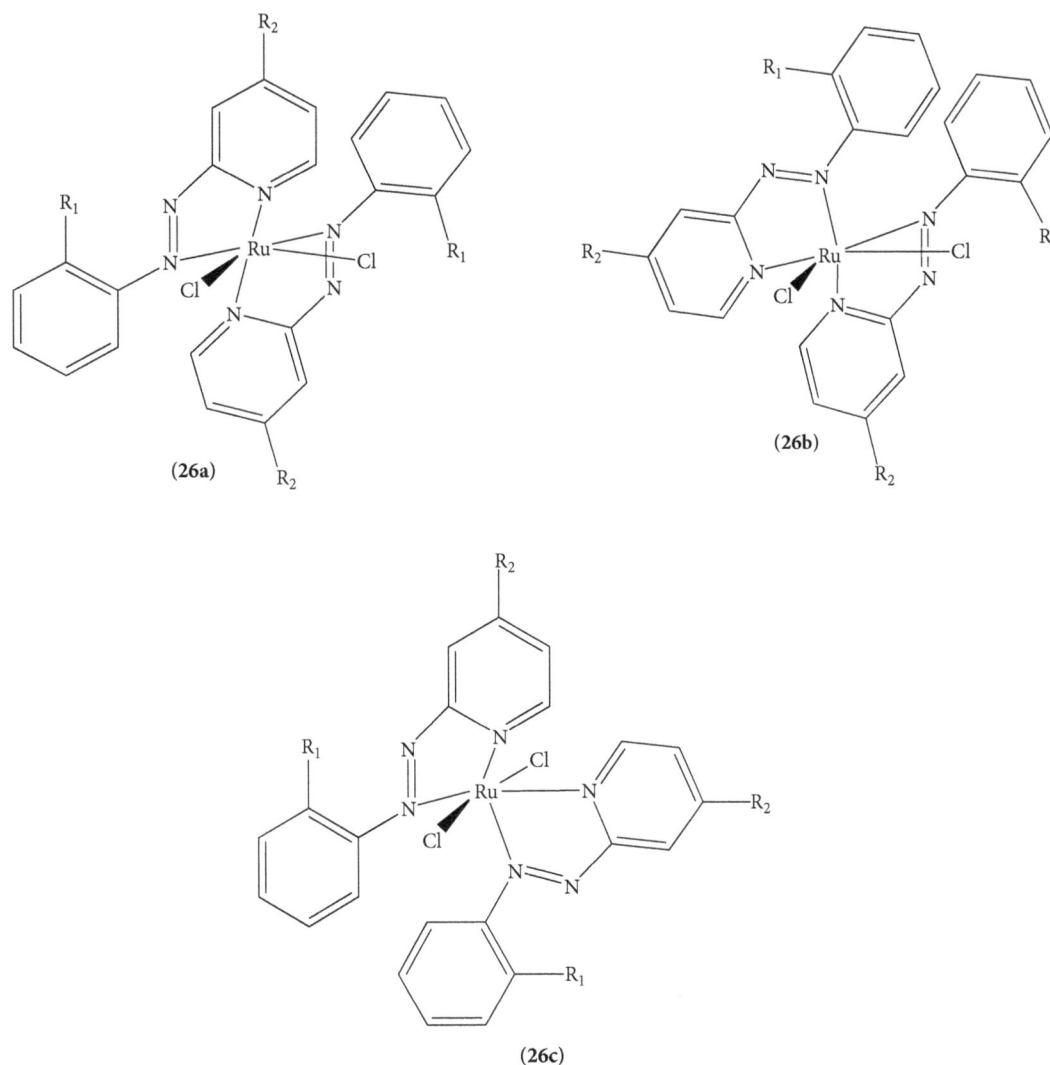

FIGURE 18: Representation of the three isomeric arylazopyridine ruthenium(II) complexes.

methylene) aminophenol) (**27**), has been synthesised and studied in detail to explore its DNA binding and cleaving properties (Figure 19) [85]. The coordinated neutral ligand Hpyrimol is formed through ligand dehydrogenation of the redox-active ligand Hpyramol. Although the ligand is potentially tridentate and anionic, the coordination motif of Hpyrimol is bidentate, as a N,N-donor and a noncoordinated neutral phenol group. Interestingly, when starting from Ru(III) chloride, a Ru(II) compound is formed, in which the ligand Hpyrimol remains neutral and chelates as a bidentate ligand. The neutral phenol groups form stable intramolecular hydrogen bonds with the coordinated chloride anions [85].

Not only does compound **27** bind to DNA, but also it additionally cleaves DNA *via* single-strand cutting. In the DNA-interaction studies, hypochromism is observed with increasing amounts of DNA in the reaction mixture, indicating intercalative binding of the ruthenium compound between the DNA base pairs [86, 87]. Although the compound is not a planar molecule, it may intercalate partially using the ligand surface to slide into the DNA base pairs. An

in vitro cytotoxicity assay shows that the compound is quite active and even comparable to cisplatin. However, its IC_{50} values show that the cytotoxic activity of the compound is significantly lower than the azpy derivatives [85]. Only a few *trans*-chloridoruthenium(II) compounds have been reported in the literature, and therefore this makes it necessary for more Ru(II) compounds with similar ligand architecture to be studied in order to find possible structure-activity relationships.

Ruthenium complexes with pyridyl-based ligands are of special interest due to a combination of easily constructed rigid chiral structures and useful photophysical properties. The large, rigid, and multidentate polypyridyl ligands confer shape and chirality to the ruthenium complexes that could be utilized to achieve customized DNA binding properties [88–90]. Following the literature reports on the pharmacological benefits of attaching carboxylic acid (COOH) group in coordination complexes [91], ruthenium(II) complexes [RuHCl(bpy)(EPh$_3$)(CO)] (bpy = 2,2$'$-bipyridine)

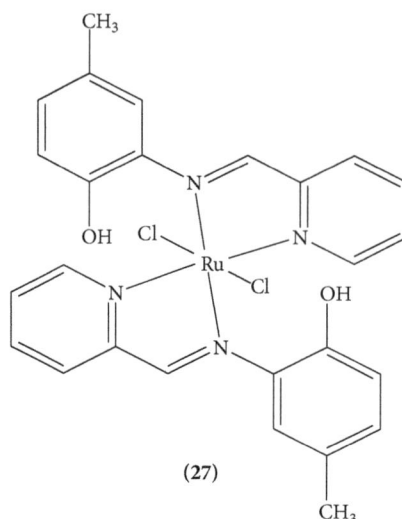

FIGURE 19: Ru(II) compound *trans*-[RuII(Hpyrimol)$_2$Cl$_2$].

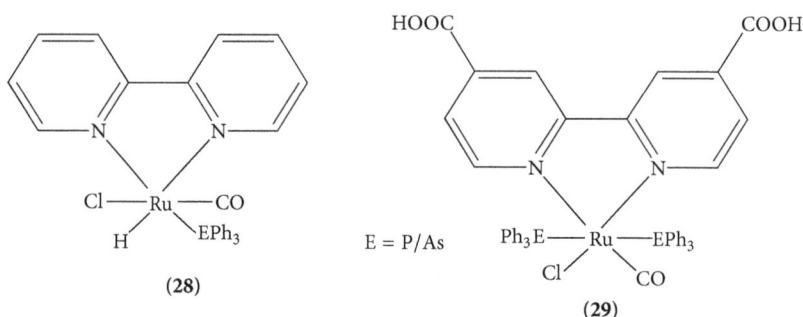

FIGURE 20: Ruthenium(II) bipyridyl complexes.

and [RuCl(HL)(EPh$_3$)$_2$(CO)] (HL = 2,2′-bipyridine-4,4′-dicarboxylic acid, E = P or As) (**28 and 29**) have been prepared and explored for their interaction with DNA (Figure 20) [92]. An appropriate attachment of carboxylic acid could modulate the solubility of the complex, cell transport, and biological activity [91]. The interactions of these complexes with DNA revealed that the complexes could bind to CT-DNA through nonintercalation. The *in vitro* cytotoxic and antioxidant activities of the complexes against a panel of cancer cell lines and free radicals showed that carboxylic acid bearing compounds possess quite high anticancer and antioxidant activities and an apparent dependence of biological activities on incorporation of COOH in bipyridine moiety was noticed [91]. Furthermore, complexes containing bipyridine showed better binding affinity compared to those containing bipyridine dicarboxylic acids, and this could be attributed to the substitution of carboxyl acid groups in bipyridine, leading to differences in space configuration and the electron density distribution affecting the effective binding to DNA. The presence of ancillary ligands may also have an effect on the binding property since all octahedral complexes bind to DNA in three dimensions as reflected in better binding affinity for compounds which contained only

one PPh$_3$ or AsPh$_3$ compared to those which contain two PPh$_3$ or AsPh$_3$ [92].

Cytostatic and cytotoxic properties of ruthenium(II) compounds (RDCs) containing ruthenium(II) atom covalently linked to carbon and nitrogen atoms (**30–32**) have been described (Figure 21) [93, 94]. The compounds induced a G$_1$ cell cycle arrest and DNA fragmentation in mouse RDM4 (lymphoma) cells, human TK6 (B lymphoblastoid) cells, and A172 (glioblastoma) cells as sufficiently as cisplatin. The signalling pathways underlying these effects were also studied and revealed an induction of p53 and p57 protein levels, but with different intensities and kinetics, thereby suggesting p53-dependent and p53-independent modes of action. The results generally demonstrate a structure-function relationship because improved cytotoxicity was realised in ruthenium(II) compounds having a phenanthroline ligand [94].

On the basis of this structure-relationship studies, Vidimar et al. substituted the two acetonitrile ligands of RDC11 by a second phenanthroline, naming the new molecule RDC34 (**33a**) (Figure 22) [95]. An equivalent of RDC34 with another counter-ion (PF$_6^-$) was also synthesized (RDC37) (**33b**) (Figure 22). The redox potential of RDC34 was modified by adding either a NO$_2$ (electron withdrawing) or a NH$_2$ unit

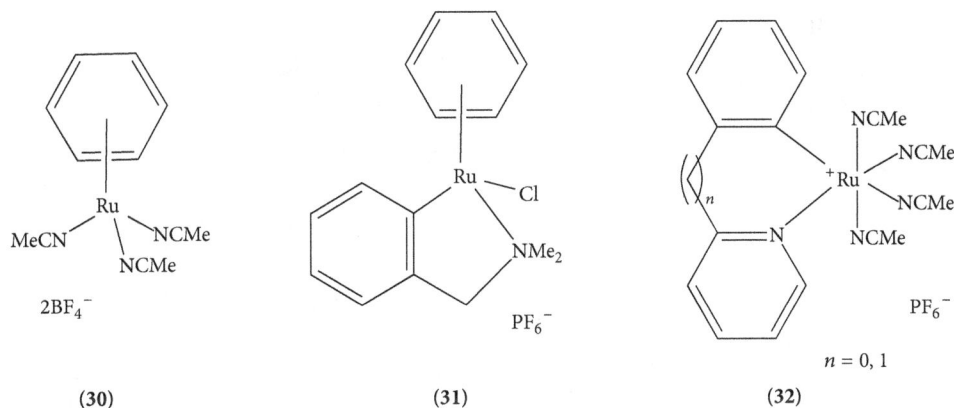

FIGURE 21: Ruthenium(II) complexes with carbon and nitrogen moieties.

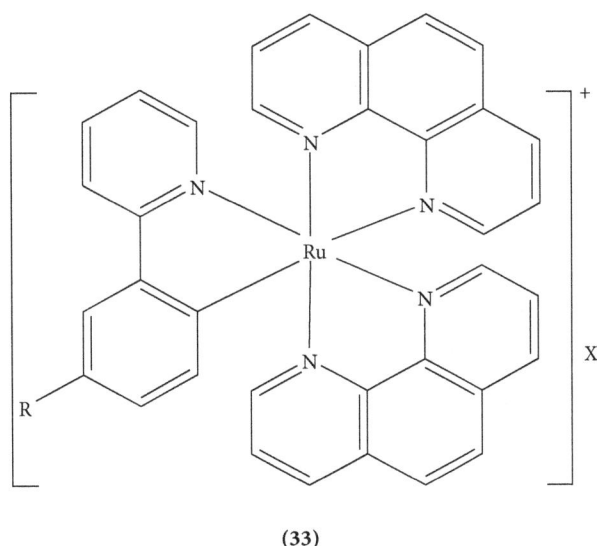

(33)

(a) R = H, X = $CF_3SO_3^-$ (RDC 34)

(b) R = H, X = PF_6^- (RDC 37)

(c) R = NO_2, X = $CF_3SO_3^-$ (RDC 40)

(d) R = NH_2, X = $CF_3SO_3^-$ (RDC 41)

FIGURE 22: Structures of RDC 34, RDC 37, RDC 40, and RDC 41.

(electron releasing) on the phenylpyridine ligand, leading to RDC40 (33c) and RDC41 (33d) (Figure 22), respectively [95]. The water-solubility of RDC34 was finally improved by adding a spermine unit to the phenylpyridine ligand, leading to RDC44 (34) (Figure 23) [96]. Cytotoxicity of the RDCs was first tested on human HCT116 (colon cancer) cells since colon cancers are one of the indications for platinum-derived treatments. Interestingly, compounds 33c containing the NO_2 group showed an increased cytotoxicity (IC_{50} < $2\,\mu M$) compared to RDC11 and cisplatin. However, the addition of the NH_2 group decreased the cytotoxic activity (IC_{50} = 2–$4\,\mu M$, 33d) and that of the spermine moiety had an even worse effect, with the IC_{50} rising to over $16\,\mu M$ [95]. The work demonstrates that, by changing the ligands around

the ruthenium, the ability of the compounds to interact with DNA is modified. Nevertheless, the complexes are more potent inducers of cancer cell death and they trigger the production of reactive oxygen species and the activation of caspase-8. The results also show that blocking reactive oxygen species production or caspase-8 activities reduces the activity of the compounds significantly [95]. Altogether, the data suggest that water-soluble ruthenium(II)-derived compounds represent an interesting class of molecules that can target several proapoptotic signalling pathways leading to reactive oxygen species production and caspase-8 activation [95].

A similar work is seen in ruthenium(II) methylimidazole complexes with the general formula $[Ru(MeIm)_4(N\text{-}N)]^{2+}$

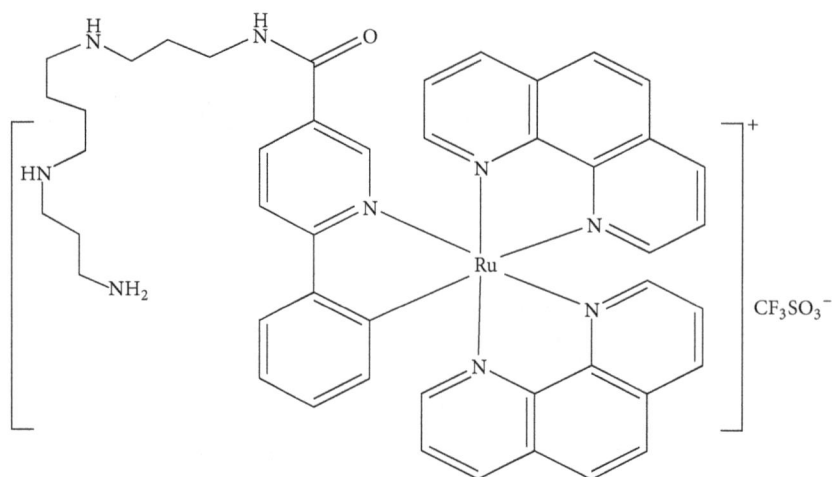

(34)

FIGURE 23: Structures of RDC 44.

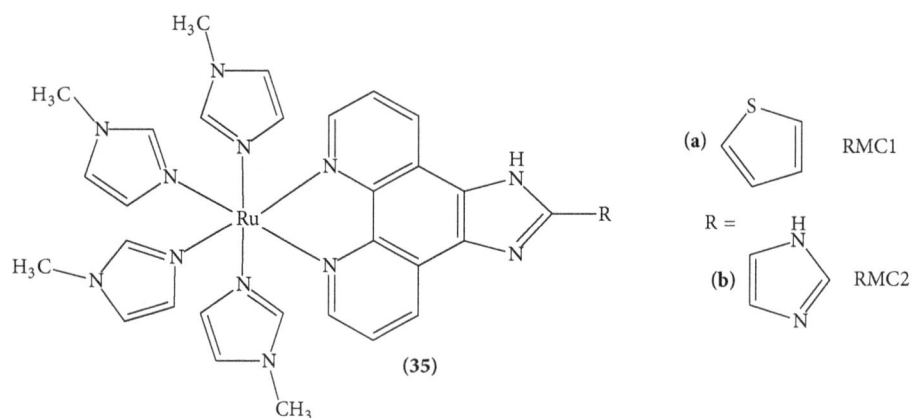

(35)

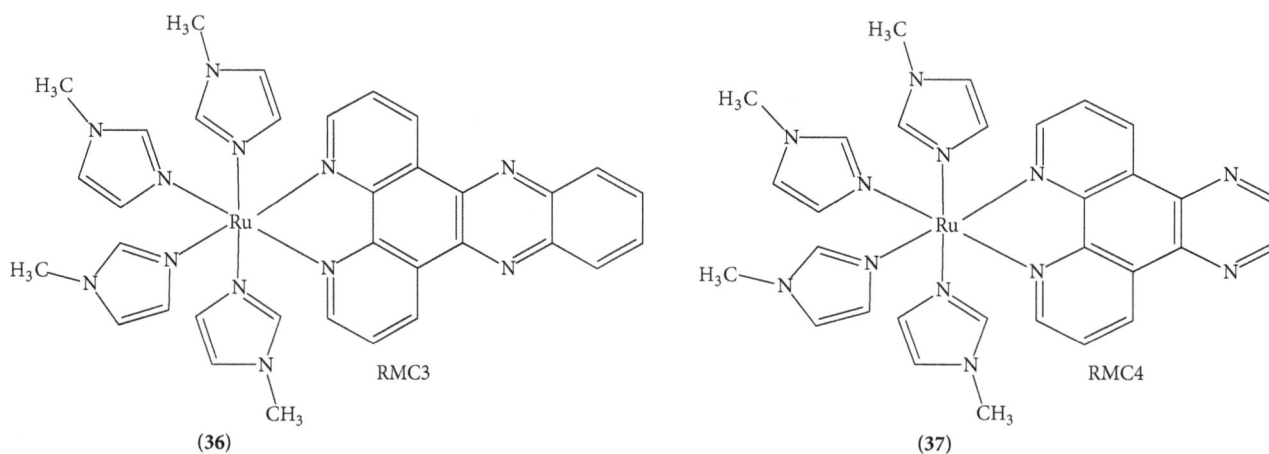

FIGURE 24: Ruthenium(II) methylimidazole complexes of general formula [Ru(MeIm)$_4$(N-N)]$^{2+}$.

(N-N = 2-(thiophene-2-yl)-1H-imidazo [4,5-f] [1, 10]phe-nanthroline (RMC1) (35a), 2-(1H-imidazol-4-yl)-1H-imi-dazo [4,5-f] [1,10]phenanthroline (RMC2) (29b), dipyri-do[3,2-a:2,3′-c]phenazine (RMC3) (36), and pyrazino [2,3-f]

[1,10]phenanthroline) (RMC4) (37), MeIm = 1-methy-limidazole (Figure 24) [97]. The methylimidazole based complexes have been prepared and studied with the hope that the combination of methylimidazole and N-N ligand

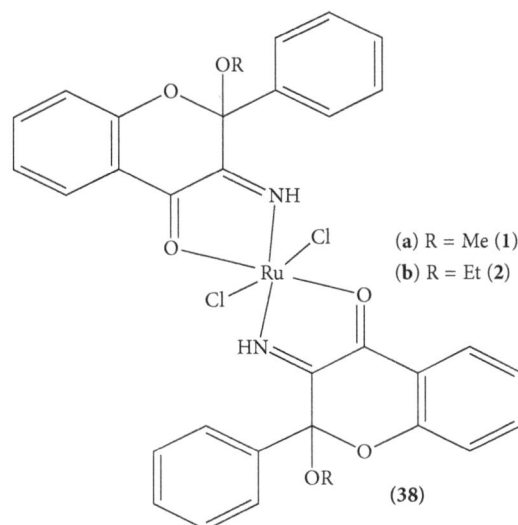

FIGURE 25: Structures of ruthenium(II) compounds: *cis*-dichloridobis(3-imino-2-methoxyflavanone)ruthenium(II)·3H$_2$O (**a**) and *cis*-dichloridobis(3-imino-2-ethoxyflavanone)ruthenium(II)·2H$_2$O (**b**).

could enhance the antitumor activity and water-solubility. However, preliminary screening experiments showed that **36** and **37**, containing methylimidazole ligand, possessed some biological activity [97]. In a successful attempt to improve water-solubility and of course antitumour activity, two new complexes **35a** and **35b**, bearing methylimidazole as the ancillary ligand and the main ligand incorporated with the thiophene and imidazole rings, were prepared. Methylimidazole is mostly used in drug design, as it is small molecule compounds with good water-solubility and a certain degree of biological activity [98, 99]. The most active compound was **35a**, and it inhibited the growth of human A549 (lung adenocarcinoma) cells through induction of apoptotic cell death, as evidenced by the accumulation of cell population in sub-G1 phase. The compound also induced the depletion of mitochondrial membrane potential in A549 cells by regulating the expression of prosurvival and proapoptotic Bcl-2 family members. These results demonstrated that **35a** induced cancer cell death by acting on both mitochondrial and death receptor apoptotic pathways [97].

The N-O moiety has been utilized in the preparation of active compounds *cis*-dichloridobis(3-imino-2-methoxyflavanone) ruthenium(II)·3H$_2$O (**38a**) and *cis*-dichloridobis(3-imino-2-ethoxyflavanone)ruthenium(II)·2H$_2$O (**38b**) (Figure 25) [100]. The compounds are active towards human bladder carcinoma cell line (EJ) and its cisplatin resistant subline (EJcisR). The compounds were found to overcome the resistance to cisplatin of EJcisR cells and are less toxic to healthy human lymphocytes *in vitro* than cisplatin. The possible mechanism of CDDP-resistance of EJcisR cells of the two novel ruthenium compounds was investigated, and it was deduced that CDDP-resistance of EJcisR cells is based on partial loss of apoptotic pathway activating caspase-8 and increased resistance to DNA strand breaks and/or alkali-labile sites. The compounds induced apoptosis so fast and this was attributed in part to

the presence of large, lipophilic flavanone-based ligands that may facilitate their *trans*-membrane transport and redox activity [100].

An investigation of the interaction of ruthenium(II) complex, [Ru(dmp)$_2$PMIP]$^{2+}$ (dmp = 2,9-dimethyl-1,10-phenanthroline, PMIP = 2-(4-methylphenyl)imidazo[4,5-f]1,10-phenanthroline) (**39**), a structural analogue of platinum(II) complex, with calf thymus DNA (CT-DNA), has been carried out on human A549 (lung cancer) cells in order to elucidate its binding mechanism and biological impact of the interactions (Figure 26) [12]. The steric hindrance exerted by the methyl group located on the ligands was expected to produce a moderate intercalative complex. Cell viability experiments indicated that the ruthenium complex showed significant dose-dependent cytotoxicity on human lung cancer cell line, and further experiments on the interaction with DNA indeed showed that the complex and CT-DNA formed a tight 1:1 complex with a binding constant exceeding 10^6 M^{-1} and with a binding mode of intercalation, suggesting that antitumour activity of the ruthenium(II) complex could be related to its interaction with DNA [12].

More interest has developed in ruthenium(II) polypyridyl complexes, and this is reflected in the preparation of ruthenium(II) β-carboline complex [Ru(tpy)(Nh)$_3$]$^{2+}$ (tpy = 2,2′:6′,2″-terpyridine, Nh = Norharman) (**40**) (Figure 27) [101]. The compound was explored as a potentially antiproliferative agent. The β-carboline alkaloids are a class of synthetic and naturally occurring compounds that possess a large spectrum of pharmacological properties, which include sedative, anxiolytic, antiviral, antimicrobial, and antitumor activities [101]. Reports have revealed that some β-carboline alkaloids exert antitumor activities through multiple mechanisms, such as DNA binding and inhibiting topoisomerases I and II [102, 103]. The [Ru(tpy)$_2$]$^{2+}$ complexes are achiral, with the two tpy units perpendicular to each other, for which binding to DNA appears to be sterically possible.

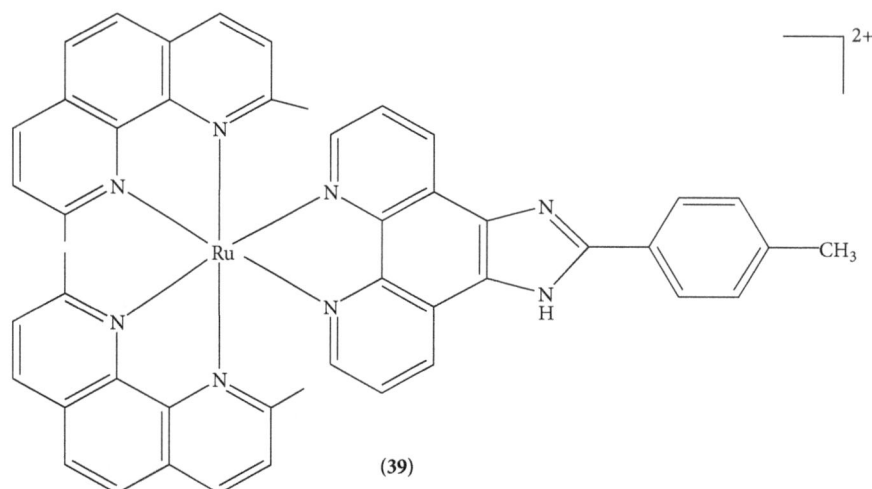

FIGURE 26: Ruthenium(II) polypyridyl complex.

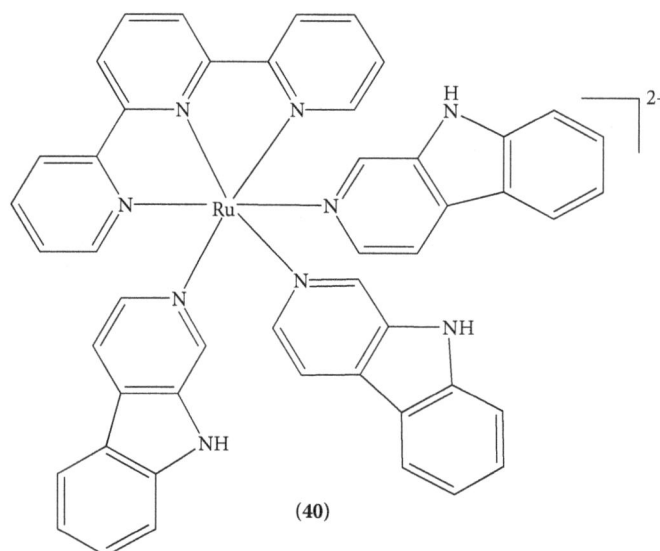

FIGURE 27: Ruthenium(II) β-carboline complex [Ru(tpy)(Nh)$_3$]$^{2+}$.

The compound induced apoptosis against various cancer cell lines and had high selectivity between tumour cells and normal cells. *In vivo* examination indicated that the ruthenium(II) complex decreased mouse MCF-7 and HepG2 (hepatocellular carcinoma) tumour growth. Signalling pathways analysis demonstrated that this complex induced apoptosis *via* the mitochondrial pathway, as evidenced by the loss of mitochondrial membrane potential and the release of cytochrome *c*. When put together, these findings suggest that the compound exhibits high and selective cytotoxicity induced p53-mediated apoptosis [101].

3. Conclusion

After careful review of ruthenium(II) compounds, it becomes apparent that the compounds offer a promising approach to the development of new anticancer agents because they show

remarkable features such as low general toxicity, the ability to mimic iron binding to biomolecules (transferrin, albumin), and stronger affinity for cancer tissues over normal tissues. Some of the compounds interact with DNA at the same initial sites (N7-guanine) as platinum compounds. However, the broad spectrum of anticancer activities displayed by the complexes makes it difficult to deduce their mechanism of action. Generally, cytotoxicity of the compounds is comparable or even better than that of cisplatin against a range of human cancer cells, thereby indicating that, for these types of ruthenium compounds, DNA is one of the targets of their action inside the cells. The use of chelating ligands with stronger binding to ruthenium appears to be desirable because these ligands offer advantages of structural stability in aqueous solution, thereby influencing *in vitro* anticancer activity of the complexes quite significantly. Furthermore, aqueous solubility of these ruthenium complexes gives rise to promising antitumor activities and high selectivity while also rendering them suitable for oral administration. However, instability and the difficult ligand exchange chemistry of inorganic ruthenium complexes present setbacks which can only be overcome with more stable organoruthenium complexes, in order to enhance their potential as drug candidates.

Half-sandwich organometallic ruthenium(II) arene complexes are emerging as promising candidates for cancer treatment. As seen in the review, the aromatic ligand attached to the complex occupies three coordination positions, thereby offering stability to the metal centre, while the remaining coordination sites can be occupied by the ligand that could impart antitumor activity together with ligands that can control electronic properties at the ruthenium centre. This architectural friendlies observed in half-sandwich arene ruthenium complexes offers an opportunity to improve their cytotoxic profile. The compounds possess excellent antitumor activities, with IC_{50} values much lower than those found for cisplatin. This could indicate that DNA is one of the targets of their action. Recent work on polymeric controlled drug delivery using polymer based ruthenium-cyclopentadienyl complexes could prove to be a promising alternative to the conventional drug delivery approaches in cancer therapy. All these findings suggest that further development of ruthenium compounds may contribute to the improvement of future chemotherapeutic protocols.

Abbreviations

AFM: Atomic force microscopy
CT: Calf thymus
DFT: Density functional theory
DMSO: Dimethyl sulfoxide
DNA: Deoxyribonucleic acid
NMR: Nuclear magnetic resonance
RDCs: Ruthenium(II) derived compounds
RMCs: Ruthenium(II) methylimidazole complexes
RNA: Ribonucleic acid.

Conflict of Interests

The authors declare that there is no conflict of interests regarding the publication of this paper.

Acknowledgments

The authors gratefully acknowledge the financial support from Govan Mbeki Research and Development Centre, University of Fort Hare, South Africa.

References

[1] B. Rosenberg, L. van Camp, and T. Krigas, "Inhibition of cell division in *Escherichia coli* by electrolysis products from a platinum electrode," *Nature*, vol. 205, no. 4972, pp. 698–699, 1965.

[2] B. Rosenberg, L. VanCamp, J. E. Trosko, and V. H. Mansour, "Platinum compounds: a new class of potent antitumour agents," *Nature*, vol. 222, pp. 385–386, 1969.

[3] A. Alama, B. Tasso, F. Novelli, and F. Sparatore, "Organometallic compounds in oncology: implications of novel organotins as antitumor agents," *Drug Discovery Today*, vol. 14, no. 9-10, pp. 500–508, 2009.

[4] T. V. Segapelo, I. A. Guzei, L. C. Spencer, W. E. V. Zyl, and J. Darkwa, "(Pyrazolylmethyl)pyridine platinum(II) and gold(III) complexes: synthesis, structures and evaluation as anticancer agents," *Inorganica Chimica Acta*, vol. 362, no. 9, pp. 3314–3324, 2009.

[5] T. W. Hambley, "The influence of structure on the activity and toxicity of Pt anti-cancer drugs," *Coordination Chemistry Reviews*, vol. 166, pp. 181–223, 1997.

[6] S. H. van Rijt and P. J. Sadler, "Current applications and future potential for bioinorganic chemistry in the development of anticancer drugs," *Drug Discovery Today*, vol. 14, no. 23-24, pp. 1089–1097, 2009.

[7] M. A. Jakupec, M. Galanski, V. B. Arion, C. G. Hartinger, and B. K. Keppler, "Antitumour metal compounds: more than theme and variations," *Dalton Transactions*, no. 2, pp. 183–194, 2008.

[8] I. Kostova, "Ruthenium complexes as anticancer agents," *Current Medicinal Chemistry*, vol. 13, no. 9, pp. 1085–1107, 2006.

[9] I. Ott and R. Gust, "Non platinum metal complexes as anticancer drugs," *Archiv der Pharmazie*, vol. 340, no. 3, pp. 117–126, 2007.

[10] D. Griffith, S. Cecco, E. Zangrando, A. Bergamo, G. Sava, and C. J. Marmion, "Ruthenium(III) dimethyl sulfoxide pyridinehydroxamic acid complexes as potential antimetastatic agents: synthesis, characterisation and in vitro pharmacological evaluation," *Journal of Biological Inorganic Chemistry*, vol. 13, no. 4, pp. 511–520, 2008.

[11] R. Margalit, H. B. Gray, M. J. Clarke, and L. Podbielski, "Chemical and biological properties of pentaammineruthenium-bleomycin complexes," *Chemico-Biological Interactions*, vol. 59, no. 3, pp. 231–245, 1986.

[12] P. Zhang, J. Chen, and Y. Liang, "DNA binding, cytotoxicity, and apoptotic-inducing activity of ruthenium(II) polypyridyl complex," *Acta Biochimica et Biophysica Sinica*, vol. 42, no. 7, pp. 440–449, 2010.

[13] J. Reedijk, "Metal-ligand exchange kinetics in platinum and ruthenium complexes," *Platinum Metals Review*, vol. 52, no. 1, pp. 2–11, 2008.

[14] L. Messori, A. Casini, D. Vullo, S. G. Haroutiunian, E. B. Dalian, and P. Orioli, "Effects of two representative antitumor ruthenium(III) complexes on thermal denaturation profiles of DNA," *Inorganica Chimica Acta*, vol. 303, no. 2, pp. 283–286, 2000.

[15] E. Gallori, C. Vettori, E. Alessio et al., "DNA as a possible target for antitumor ruthenium(III) complexes: a spectroscopic and molecular biology study of the interactions of two representative antineoplastic ruthenium(III) complexes with DNA," *Archives of Biochemistry and Biophysics*, vol. 376, no. 1, pp. 156–162, 2000.

[16] L. Messori, P. Orioli, D. Vullo, E. Alessio, and E. Iengo, "A spectroscopic study of the reaction of NAMI, a novel ruthenium(III) anti-neoplastic complex, with bovine serum albumin," *European Journal of Biochemistry*, vol. 267, no. 4, pp. 1206–1213, 2000.

[17] W. H. Ang and P. J. Dyson, "Classical and non-classical ruthenium-based anticancer drugs: towards targeted chemotherapy," *European Journal of Inorganic Chemistry*, vol. 2006, no. 20, pp. 4003–4018, 2006.

[18] C. X. Zhang and S. J. Lippard, "New metal complexes as potential therapeutics," *Current Opinion in Chemical Biology*, vol. 7, no. 4, pp. 481–489, 2003.

[19] A. Bergamo and G. Sava, "Ruthenium complexes can target determinants of tumour malignancy," *Dalton Transactions*, no. 13, pp. 1267–1272, 2007.

[20] P. J. Dyson and G. Sava, "Metal-based antitumour drugs in the post genomic era," *Dalton Transactions*, no. 16, pp. 1929–1933, 2006.

[21] B. K. Keppler, H. Henn, U. M. Juhl, M. R. Berger, R. Niebel, and F. E. Wagner, "New ruthenium complexes for the treatment of cancer," in *Ruthenium and Other Non-Platinum Metal Complexes in Cancer Chemotherapy*, vol. 10 of *Progress in Clinical Biochemistry and Medicine*, pp. 41–69, Springer, Berlin, Germany, 1989.

[22] C. G. Hartinger, M. A. Jakupec, S. Zorbas-Seifried et al., "KP1019, a new redox-active anticancer agent—preclinical development and results of a clinical phase I study in tumor patients," *Chemistry and Biodiversity*, vol. 5, no. 10, pp. 2140–2155, 2008.

[23] P. Heffeter, K. Böck, B. Atil et al., "Intracellular protein binding patterns of the anticancer ruthenium drugs KP1019 and KP1339," *Journal of Biological Inorganic Chemistry*, vol. 15, no. 5, pp. 737–748, 2010.

[24] C. G. Hartinger, M. A. Jakupec, S. Zorbas-Seifried et al., "KP1019, a new redox-active anticancer agent—preclinical development and results of a clinical phase I study in tumor patients," *Chemistry and Biodiversity*, vol. 5, no. 10, pp. 2140–2155, 2008.

[25] R. Trondl, P. Heffeter, C. R. Kowol, M. A. Jakupec, W. Berger, and B. K. Keppler, "NKP-1339, the first ruthenium-based anticancer drug on the edge to clinical application," *Chemical Science*, vol. 5, no. 8, pp. 2925–2932, 2014.

[26] G. Sava, S. Zorzet, C. Turrin et al., "Dual action of NAMI-A in inhibition of solid tumor metastasis: selective targeting of metastatic cells and binding to collagen," *Clinical Cancer Research*, vol. 9, no. 5, pp. 1898–1905, 2003.

[27] G. Sava, F. Frausin, M. Cocchietto et al., "Actin-dependent tumour cell adhesion after short-term exposure to the antimetastasis ruthenium complex NAMI-A," *European Journal of Cancer*, vol. 40, no. 9, pp. 1383–1396, 2004.

[28] D. Pluim, R. C. A. M. van Waardenburg, J. H. Beijnen, and J. H. M. Schellens, "Cytotoxicity of the organic ruthenium anticancer drug Nami-A is correlated with DNA binding in four different human tumor cell lines," *Cancer Chemotherapy and Pharmacology*, vol. 54, no. 1, pp. 71–78, 2004.

[29] F. Kratz, B. K. Keppler, M. Hartmann, L. Messori, and M. R. Berger, "Comparison of the antiproliferative activity of two antitumour ruthenium(III) complexes with their apotransferrin and transferrin-bound forms in a human colon cancer cell line," *Metal-Based Drugs*, vol. 3, no. 1, pp. 15–23, 1996.

[30] C. A. Smith, A. J. Sutherland-Smith, B. K. Keppler, F. Kratz, and E. N. Baker, "Binding of ruthenium(III) anti-tumor drugs to human lactoferrin probed by high resolution X-ray crystallographic structure analyses," *Journal of Biological Inorganic Chemistry*, vol. 1, pp. 424–431, 1996.

[31] A. R. Timerbaev, A. V. Rudnev, O. Semenova, C. G. Hartinger, and B. K. Keppler, "Comparative binding of antitumor indazolium [trans-tetrachlorobis(1*H*- indazole)ruthenate(III)] to serum transport proteins assayed by capillary zone electrophoresis," *Analytical Biochemistry*, vol. 341, no. 2, pp. 326–333, 2005.

[32] B. Therrien, "Functionalised η^6-arene ruthenium complexes," *Coordination Chemistry Reviews*, vol. 253, no. 3-4, pp. 493–519, 2009.

[33] R. E. Aird, J. Cummings, A. A. Ritchie et al., "In vitro and in vivo activity and cross resistance profiles of novel ruthenium (II) organometallic arene complexes in human ovarian cancer," *British Journal of Cancer*, vol. 86, no. 10, pp. 1652–1657, 2002.

[34] G. S. Smith and B. Therrien, "Targeted and multifunctional arene ruthenium chemotherapeutics," *Dalton Transactions*, vol. 40, no. 41, pp. 10793–10800, 2011.

[35] M. J. Clarke, "Ruthenium metallopharmaceuticals," *Coordination Chemistry Reviews*, vol. 236, no. 1-2, pp. 209–233, 2003.

[36] Y. N. V. Gopal, D. Jayaraju, and A. K. Kondapi, "Inhibition of topoisomerase II catalytic activity by two ruthenium compounds: a ligand-dependent mode of action," *Biochemistry*, vol. 38, no. 14, pp. 4382–4388, 1999.

[37] C. Scolaro, A. Bergamo, L. Brescacin et al., "In vitro and in vivo evaluation of ruthenium(II)-arene PTA complexes," *Journal of Medicinal Chemistry*, vol. 48, no. 12, pp. 4161–4171, 2005.

[38] J. R. van Beijnum, A. Casini, A. A. Nazarov et al., "Organometallic ruthenium(II) arene compounds with antiangiogenic activity," *Journal of Medicinal Chemistry*, vol. 54, no. 11, pp. 3895–3902, 2011.

[39] I. Berger, M. Hanif, A. A. Nazarov et al., "In vitro anticancer activity and biologically relevant metabolization of organometallic ruthenium complexes with carbohydrate-based ligands," *Chemistry*, vol. 14, no. 29, pp. 9046–9057, 2008.

[40] E. E. Nifantyev, M. P. Koroteev, A. M. Koroteev et al., "Metal complexes based on monosaccharide bicyclophosphites as new available chiral coordination systems," *Journal of Organometallic Chemistry*, vol. 587, no. 1, pp. 18–27, 1999.

[41] M. Gras, B. Therrien, G. Süss-Fink, P. Š. Renfrew, A. K. Renfrew, and P. J. Dyson, "Water-soluble arene ruthenium complexes containing pyridinethiolato ligands: synthesis, molecular structure, redox properties and anticancer activity of the cations $[(\eta^6\text{-arene})Ru(\textbf{p-}SC_5H_4NH)_3]^{2+}$," *Journal of Organometallic Chemistry*, vol. 693, no. 21-22, pp. 3419–3424, 2008.

[42] K. Becker, C. Herold-Mende, J. J. Park, G. Lowe, and R. Heiner Schirmer, "Human thioredoxin reductase is efficiently inhibited by (2,2′:6′,2′-terpyridine)platinum(II) complexes. Possible implications for a novel antitumor strategy," *Journal of Medicinal Chemistry*, vol. 44, no. 17, pp. 2784–2792, 2001.

[43] C. A. Vock, W. H. Ang, C. Scolaro et al., "Development of ruthenium antitumor drugs that overcome multidrug resistance mechanisms," *Journal of Medicinal Chemistry*, vol. 50, no. 9, pp. 2166–2175, 2007.

[44] G. Süss-Fink, F.-A. Khan, L. Juillerat-Jeanneret, P. J. Dyson, and A. K. Renfrew, "Synthesis and anticancer activity of long-chain isonicotinic ester ligand-containing arene ruthenium complexes and nanoparticles," *Journal of Cluster Science*, vol. 21, no. 3, pp. 313–324, 2010.

[45] F.-A. Khan, B. Therrien, G. Süss-Fink, O. Zava, and P. J. Dyson, "Arene ruthenium dichloro complexes containing isonicotinic ester ligands: synthesis, molecular structure and cytotoxicity," *Journal of Organometallic Chemistry*, vol. 730, pp. 49–56, 2013.

[46] G. V. Tsarichenko, V. I. Bobrov, and M. V. Smarkov, "Toxicity of isonicotinic acid," *Pharmaceutical Chemistry Journal*, vol. 11, no. 4, pp. 481–483, 1977.

[47] T. S. Morais, T. J. L. Silva, F. Marques et al., "Synthesis of organometallic ruthenium(II) complexes with strong activity against several human cancer cell lines," *Journal of Inorganic Biochemistry*, vol. 114, pp. 65–74, 2012.

[48] V. Moreno, M. Font-Bardia, T. Calvet et al., "DNA interaction and cytotoxicity studies of new ruthenium(II) cyclopentadienyl derivative complexes containing heteroaromatic ligands," *Journal of Inorganic Biochemistry*, vol. 105, no. 2, pp. 241–249, 2011.

[49] M. H. Garcia, T. S. Morais, P. Florindo et al., "Inhibition of cancer cell growth by ruthenium(II) cyclopentadienyl derivative complexes with heteroaromatic ligands," *Journal of Inorganic Biochemistry*, vol. 103, no. 3, pp. 354–361, 2009.

[50] C. Scolaro, A. Bergamo, L. Brescacin et al., "In vitro and in vivo evaluation of ruthenium(II)-arene PTA complexes," *Journal of Medicinal Chemistry*, vol. 48, no. 12, pp. 4161–4171, 2005.

[51] B. Serli, E. Zangrando, T. Gianferrara et al., "Is the aromatic fragment of piano-stool ruthenium compounds an essential feature for anticancer activity? The development of new Ru^{II}-[9]aneS$_3$ analogues," *European Journal of Inorganic Chemistry*, vol. 2005, no. 17, pp. 3423–3434, 2005.

[52] A. Valente, M. H. Garcia, F. Marques, Y. Miao, C. Rousseau, and P. Zinck, "First polymer 'ruthenium-cyclopentadienyl' complex as potential anticancer agent," *Journal of Inorganic Biochemistry*, vol. 127, pp. 79–81, 2013.

[53] H. Maeda, G. Y. Bharate, and J. Daruwalla, "Polymeric drugs for efficient tumor-targeted drug delivery based on EPR-effect," *European Journal of Pharmaceutics and Biopharmaceutics*, vol. 71, no. 3, pp. 409–419, 2009.

[54] P. Florindo, I. J. Marques, C. D. Nunes, and A. C. Fernandes, "Synthesis, characterization and cytotoxicity of cyclopentadienyl ruthenium(II) complexes containing carbohydrate-derived ligands," *Journal of Organometallic Chemistry*, vol. 760, pp. 240–247, 2014.

[55] D. S. Williams, G. E. Atilla, H. Bregman, A. Arzoumanian, P. S. Klein, and E. Meggers, "Switching on a signaling pathway with an organoruthenium complex," *Angewandte Chemie—International Edition*, vol. 44, no. 13, pp. 1984–1987, 2005.

[56] E. Meggers, G. E. Atilla-Gokcumen, K. Gründler, C. Frias, and A. Prokop, "Inert ruthenium half-sandwich complexes with anticancer activity," *Dalton Transactions*, no. 48, pp. 10882–10888, 2009.

[57] E. Meggers, G. E. Atilla-Gokcumen, H. Bregman et al., "Exploring chemical space with organometallics: ruthenium complexes as protein kinase inhibitors," *Synlett*, no. 8, pp. 1177–1189, 2007.

[58] J. É. Debreczeni, A. N. Bullock, G. E. Atilla et al., "Ruthenium half-sandwich complexes bound to protein kinase Pim-1," *Angewandte Chemie International Edition*, vol. 45, no. 10, pp. 1580–1585, 2006.

[59] D. S. Williams, G. E. Atilla, H. Bregman, A. Arzoumanian, P. S. Klein, and E. Meggers, "Switching on a signaling pathway with an organoruthenium complex," *Angewandte Chemie—International Edition*, vol. 44, no. 13, pp. 1984–1987, 2005.

[60] H. Bregman and E. Meggers, "Ruthenium half-sandwich complexes as protein kinase inhibitors: an N-succinimidyl ester for rapid derivatizations of the cyclopentadienyl moiety," *Organic Letters*, vol. 8, no. 24, pp. 5465–5468, 2006.

[61] K. S. M. Smalley, R. Contractor, N. K. Haass et al., "An organometallic protein kinase inhibitor pharmacologically activates p53 and induces apoptosis in human melanoma cells," *Cancer Research*, vol. 67, no. 1, pp. 209–217, 2007.

[62] H.-K. Liu, S. J. Berners-Price, F. Y. Wang et al., "Diversity in guanine-selective DNA binding modes for an organometallic ruthenium arene complex," *Angewandte Chemie International Edition*, vol. 45, no. 48, pp. 8153–8156, 2006.

[63] R. L. Hayward, Q. C. Schornagel, R. Tente et al., "Investigation of the role of Bax, p21/Waf1 and p53 as determinants of cellular

responses in HCT116 colorectal cancer cells exposed to the novel cytotoxic ruthenium(II) organometallic agent, RM175," *Cancer Chemotherapy and Pharmacology*, vol. 55, no. 6, pp. 577–583, 2005.

[64] L. K. Filak, G. Mühlgassner, F. Bacher et al., "Ruthenium- and osmium-arene complexes of 2-substituted indolo[3,2-c]quinolines: synthesis, structure, spectroscopic properties, and antiproliferative activity," *Organometallics*, vol. 30, no. 2, pp. 273–283, 2011.

[65] L. K. Filak, S. Göschl, S. Hackl, M. A. Jakupec, and V. B. Arion, "Ruthenium- and osmium-arene complexes of 8-substituted indolo[3,2-c]quinolines: synthesis, X-ray diffraction structures, spectroscopic properties, and antiproliferative activity," *Inorganica Chimica Acta*, vol. 393, pp. 252–260, 2012.

[66] M. F. Primik, S. Göschl, M. A. Jakupec, A. Roller, B. K. Keppler, and V. B. Arion, "Structure-activity relationships of highly cytotoxic copper(II) complexes with modified indolo[3,2-c]quinoline ligands," *Inorganic Chemistry*, vol. 49, no. 23, pp. 11084–11095, 2010.

[67] K. Sharma, R. V. Singh, and N. Fahmi, "Palladium(II) and platinum(II) derivatives of benzothiazoline ligands: synthesis, characterization, antimicrobial and antispermatogenic activity," *Spectrochimica Acta—Part A: Molecular and Biomolecular Spectroscopy*, vol. 78, no. 1, pp. 80–87, 2011.

[68] A. C. G. Hotze, S. E. Caspers, D. de Vos et al., "Structure-dependent in vitro cytotoxicity of the isomeric complexes [Ru(L)$_2$Cl$_2$] (L=o-tolylazopyridine and 4-methyl-2-phenylazopyridine) in comparison to [Ru(azpy)$_2$Cl$_2$]," *Journal of Biological Inorganic Chemistry*, vol. 9, no. 3, pp. 354–364, 2004.

[69] R. E. Morris, R. E. Aird, P. del Socorro Murdoch et al., "Inhibition of cancer cell growth by ruthenium(II) arene complexes," *Journal of Medicinal Chemistry*, vol. 44, no. 22, pp. 3616–3621, 2001.

[70] S. W. Magennis, A. Habtemariam, O. Novakova et al., "Dual triggering of DNA binding and fluorescence via photoactivation of a dinuclear ruthenium(II) arene complex," *Inorganic Chemistry*, vol. 46, no. 12, pp. 5059–5068, 2007.

[71] C. Gaiddon, P. Jeannequin, P. Bischoff, M. Pfeffer, C. Sirlin, and J. P. Loeffler, "Ruthenium (II)-derived organometallic compounds induce cytostatic and cytotoxic effects on mammalian cancer cell lines through p53-dependent and p53-independent mechanisms," *Journal of Pharmacology and Experimental Therapeutics*, vol. 315, no. 3, pp. 1403–1411, 2005.

[72] Q. Wu, C. Fan, T. Chen et al., "Microwave-assisted synthesis of arene ruthenium(II) complexes that induce S-phase arrest in cancer cells by DNA damage-mediated p53 phosphorylation," *European Journal of Medicinal Chemistry*, vol. 63, pp. 57–63, 2013.

[73] W. Kandioller, C. G. Hartinger, A. A. Nazarov et al., "Tuning the anticancer activity of maltol-derived ruthenium complexes by derivatization of the 3-hydroxy-4-pyrone moiety," *Journal of Organometallic Chemistry*, vol. 694, no. 6, pp. 922–929, 2009.

[74] W. Kandioller, C. G. Hartinger, A. A. Nazarov et al., "From pyrone to thiopyrone ligands-rendering maltol-derived ruthenium(II)-arene complexes that are anticancer active in vitro," *Organometallics*, vol. 28, no. 15, pp. 4249–4251, 2009.

[75] M. P. Chelopo, S. A. Pawar, M. K. Sokhela, T. Govender, H. G. Kruger, and G. E. M. Maguire, "Anticancer activity of ruthenium(II) arene complexes bearing 1,2,3,4-tetrahydroisoquinoline amino alcohol ligands," *European Journal of Medicinal Chemistry*, vol. 66, pp. 407–414, 2013.

[76] R. Fernández, M. Melchart, A. Habtemariam, S. Parsons, and P. J. Sadler, "Use of chelating ligands to tune the reactive site of half-sandwich ruthenium(II)-arene anticancer complexes," *Chemistry*, vol. 10, no. 20, pp. 5173–5179, 2004.

[77] A. S. Capilla, M. Romero, M. D. Pujol, D. H. Caignard, and P. Renard, "Synthesis of isoquinolines and tetrahydroisoquinolines as potential antitumour agents," *Tetrahedron*, vol. 57, no. 39, pp. 8297–8303, 2001.

[78] M. A. Letavic, J. M. Keith, J. A. Jablonowski et al., "Novel tetrahydroisoquinolines are histamine H$_3$ antagonists and serotonin reuptake inhibitors," *Bioorganic and Medicinal Chemistry Letters*, vol. 17, no. 4, pp. 1047–1051, 2007.

[79] Y. Li, H. B. Zhang, W. L. Huang, X. Zhen, and Y. M. Li, "Synthesis and biological evaluation of tetrahydroisoquinoline derivatives as potential multidrug resistance reversal agents in cancer," *Chinese Chemical Letters*, vol. 19, no. 2, pp. 169–171, 2008.

[80] S. M. Meier, M. Hanif, Z. Adhireksan et al., "Novel metal(II) arene 2-pyridinecarbothioamides: a rationale to orally active organometallic anticancer agents," *Chemical Science*, vol. 4, no. 4, pp. 1837–1846, 2013.

[81] W. A. Kinney, N. E. Lee, R. M. Blank et al., "N-Phenyl-2-pyridinecarbothioamides as gastric mucosal protectants," *Journal of Medicinal Chemistry*, vol. 33, no. 1, pp. 327–336, 1990.

[82] A. H. Velders, H. Kooijman, A. L. Spek, J. G. Haasnoot, D. de Vos, and J. Reedijk, "Strong differences in the in vitro cytotoxicity of three isomeric dichlorobis(2-phenylazopyridine)ruthenium(II) complexes," *Inorganic Chemistry*, vol. 39, no. 14, pp. 2966–2967, 2000.

[83] A. H. Velders, K. van der Schilden, A. C. G. Hotze, J. Reedijk, H. Kooijman, and A. L. Spek, "Dichlorobis(2-phenylazopyridine)ruthenium(II) complexes: characterisation, spectroscopic and structural properties of four isomers," *Dalton Transactions*, no. 3, pp. 448–455, 2004.

[84] J. C. Chen, J. Li, L. Qian, and K. C. Zheng, "Electronic structures and SARs of the isomeric complexes α-, β-, γ-[Ru(mazpy)$_2$Cl$_2$] with different antitumor activities," *Journal of Molecular Structure: THEOCHEM*, vol. 728, no. 1–3, pp. 93–101, 2005.

[85] S. Roy, P. U. Maheswari, A. Golobic, B. Kozlevc, and J. Reedijk, "Synthesis, crystal structure and biological studies of the highly anticancer active compound *trans*-dichloridobis(4-methyl-2-N-(2-pyridylmethylene)-aminophenol)ruthenium(II)," *Inorganica Chimica Acta*, vol. 393, pp. 239–245, 2012.

[86] L. M. Chen, J. Liu, J. C. Chen et al., "Experimental and theoretical studies on the DNA-binding and spectral properties of water-soluble complex [Ru(MeIm)$_4$(dpq)]$^{2+}$," *Journal of Molecular Structure*, vol. 881, no. 1–3, pp. 156–166, 2008.

[87] L. M. Chen, J. Liu, J. C. Chen et al., "Synthesis, characterization, DNA-binding and spectral properties of complexes [Ru(L)$_4$(dppz)]$^{2+}$ (L = Im and MeIm)," *Journal of Inorganic Biochemistry*, vol. 102, pp. 330–341, 2008.

[88] C. W. Jiang, H. Chao, X. L. Hong, H. Li, W. J. Mei, and L. N. Ji, "Enantiopreferential DNA-binding of a novel dinuclear complex [(bpy)$_2$Ru(bdptb)Ru(bpy)$_2$]$^{4+}$," *Inorganic Chemistry Communications*, vol. 6, no. 6, pp. 773–775, 2003.

[89] J. G. Liu, B. H. Ye, Q. L. Zhang et al., "Enantiomeric ruthenium(II) complexes binding to DNA: binding modes and enantioselectivity," *Journal of Biological Inorganic Chemistry*, vol. 5, pp. 119–128, 2000.

[90] H.-L. Huang, Z.-Z. Li, Z.-H. Liang, and Y.-J. Liu, "Cell cycle arrest, cytotoxicity, apoptosis, DNA-binding, photocleavage, and antioxidant activity of octahedral ruthenium(II) complexes," *European Journal of Inorganic Chemistry*, vol. 2011, no. 36, pp. 5538–5547, 2011.

[91] M. Galanski and B. K. Keppler, "Carboxylation of dihydroxoplatinum(IV) complexes via a new synthetic pathway," *Inorganic Chemistry*, vol. 35, no. 6, pp. 1709–1711, 1996.

[92] T. Sathiya Kamatchi, N. Chitrapriya, S. K. Kim, F. R. Fronczek, and K. Natarajan, "Influence of carboxylic acid functionalities in ruthenium (II) polypyridyl complexes on DNA binding, cytotoxicity and antioxidant activity: synthesis, structure and in vitro anticancer activity," *European Journal of Medicinal Chemistry*, vol. 59, pp. 253–264, 2013.

[93] S. Fernandez, M. Pfeffer, V. Ritleng, and C. Sirlin, "An effective route to cycloruthenated n-ligands under mild conditions," *Organometallics*, vol. 18, no. 12, pp. 2390–2394, 1999.

[94] C. Gaiddon, P. Jeannequin, P. Bischoff, M. Pfeffer, C. Sirlin, and J. P. Loeffler, "Ruthenium (II)-derived organometallic compounds induce cytostatic and cytotoxic effects on mammalian cancer cell lines through p53-dependent and p53-independent mechanisms," *The Journal of Pharmacology and Experimental Therapeutics*, vol. 315, no. 3, pp. 1403–1411, 2005.

[95] V. Vidimar, X. Meng, M. Klajner et al., "Induction of caspase 8 and reactive oxygen species by ruthenium-derived anticancer compounds with improved water solubility and cytotoxicity," *Biochemical Pharmacology*, vol. 84, no. 11, pp. 1428–1436, 2012.

[96] L. Fetzer, B. Boff, M. Ali et al., "Library of second-generation cycloruthenated compounds and evaluation of their biological properties as potential anticancer drugs: passing the nanomolar barrier," *Dalton Transactions*, vol. 40, no. 35, pp. 8869–8878, 2011.

[97] X. Yang, L. Chen, Y. Liu et al., "Ruthenium methylimidazole complexes induced apoptosis in lung cancer A549 cells through intrinsic mitochondrial pathway," *Biochimie*, vol. 94, no. 2, pp. 345–353, 2012.

[98] F. J. Huo, C. X. Yin, and P. Yang, "The crystal structure, self-assembly, DNA-binding and cleavage studies of the [2]pseudorotaxane composed of cucurbit[6]uril," *Bioorganic & Medicinal Chemistry Letters*, vol. 17, no. 4, pp. 932–936, 2007.

[99] R. S. Myers, R. E. Amaro, Z. A. Luthey-Schulten, and V. J. Davisson, "Reaction coupling through interdomain contacts in imidazole glycerol phosphate synthase," *Biochemistry*, vol. 44, no. 36, pp. 11974–11985, 2005.

[100] M. M. Kasprzak, L. Szmigiero, E. Zyner, and J. Ochocki, "Proapoptotic activity *in vitro* of two novel ruthenium(II) complexes with flavanone-based ligands that overcome cisplatin resistance in human bladder carcinoma cells," *Journal of Inorganic Biochemistry*, vol. 105, no. 4, pp. 518–524, 2011.

[101] Y. Chen, M.-Y. Qin, L. Wang, H. Chao, L.-N. Ji, and A.-L. Xu, "A ruthenium(II) β-carboline complex induced p53-mediated apoptosis in cancer cells," *Biochimie*, vol. 95, no. 11, pp. 2050–2059, 2013.

[102] S. L. Xiao, W. Lin, C. Wang, and M. Yang, "Synthesis and biological evaluation of DNA targeting flexible side-chain substituted β-carboline derivatives," *Bioorganic and Medicinal Chemistry Letters*, vol. 11, no. 4, pp. 437–441, 2001.

[103] Y. Li, F. S. Liang, W. Jiang et al., "DH334, a β-carboline anticancer drug, inhibits the CDK activity of budding yeast," *Cancer Biology & Therapy*, vol. 6, pp. 1193–1199, 2007.

22

Synthesis, Characterization, and Antihypertensive Evaluation of Some Novel 2,2,8,8-Tetramethyl-2,3,7,8-tetrahydro-4,6-diamino-3,7-dihydroxy-6,7-epoxy-benzo-[1,2-*b*:5,4-*b'*]dipyran Derivatives

Pankaj Dwivedi,[1] Kuldipsinh P. Barot,[2] Shailesh Jain,[2] and Manjunath D. Ghate[2]

[1]*Department of Pharmaceutical Chemistry, Krupanidhi College of Pharmacy, Bangalore 35, India*
[2]*Department of Pharmaceutical Chemistry, Institute of Pharmacy, Nirma University, Ahmedabad, Gujarat 382481, India*

Correspondence should be addressed to Manjunath D. Ghate; manjunath.ghate@nirmauni.ac.in

Academic Editor: Georgia Melagraki

A series of 2,2,8,8-tetramethyl-2,3,7,8-tetrahydro-4,6-diamino-3,7-dihydroxy-6,7-epoxy-benzo-[1,2-*b*:5,4-*b'*]dipyran derivatives **7a–e** and **8a–e** were synthesized from resorcinol. All the synthesized compounds were characterized by FTIR, mass spectra, and ^{1}H NMR. These compounds were evaluated for antihypertensive activity using Wister Albino Rat model. Direct antihypertensive activity was performed using the instrument BIOPAC System MP-36 Santa Barbara, California, for recording blood pressure response. Among the title compounds, compounds **7b**, **7c**, and **7d** showed potent antihypertensive activity and other compounds were also found to exert low and moderate antihypertensive activity. The relaxant potency in rat aorta and trachea was used for biological characterization of the benzopyrans. Structure-activity relationships study was investigated around position-4 of the benzopyran nucleus.

1. Introduction

Potassium specific channels are assorted group of ion channels and play a fundamental role in the modulation of cell excitability [1, 2]. Potassium channel classifications and their pharmacological activities have been reviewed extensively [3]. The term "potassium channel openers (KCOs)" was introduced to designate a group of novel synthetic molecules which are specified by cromakalim. It led to a new direction in the pharmacology of ion channels by reporting that cromakalim evoked smooth muscle relaxant effects by the opening of K[+] channels in cell membranes [4]. It has initiated major research efforts in the search for other such molecules and in the determination of the specific channel(s) involved [5]. KCO properties are demonstrated in a diverse range of synthetic chemical structures and endogenous substances [6].

Cromakalim evoked a contractile response in rabbit aorta bathed in a Ca[2+] free solution which is related to the effects on intracellular Ca[2+] stores [7]. These findings support those obtained from vascular smooth muscle where contractile responses to noradrenaline depend on intracellular calcium stores which are attenuated by cromakalim [8]. In contrast, the effect of cromakalim on rat pulmonary artery did not appear to involve an action on Ca[2+] release from internal stores [9]. A variety of compounds having a benzopyran such as levocromakalim, bimakalim, and Y-27152 generally exhibit potent antihypertensive activity. Benzodipyrans have structural and chemical similarity with the cromakalim [10]. The ATP-sensitive potassium channel (K_{ATP}) openers (e.g., chromakalim) were originally developed for the treatment of hypertension due to their potent peripheral vasodilating properties [11]. To find more potent vasodilators, various benzopyran derivatives modified at position-4 were synthesized and structure-activity relationship was examined by evaluation of the extent and duration of the increase in coronary blood flow in anesthetized dogs [12]. Compounds having a 1, 6-dihydro-6-oxopyridazin-3-yl amino group at

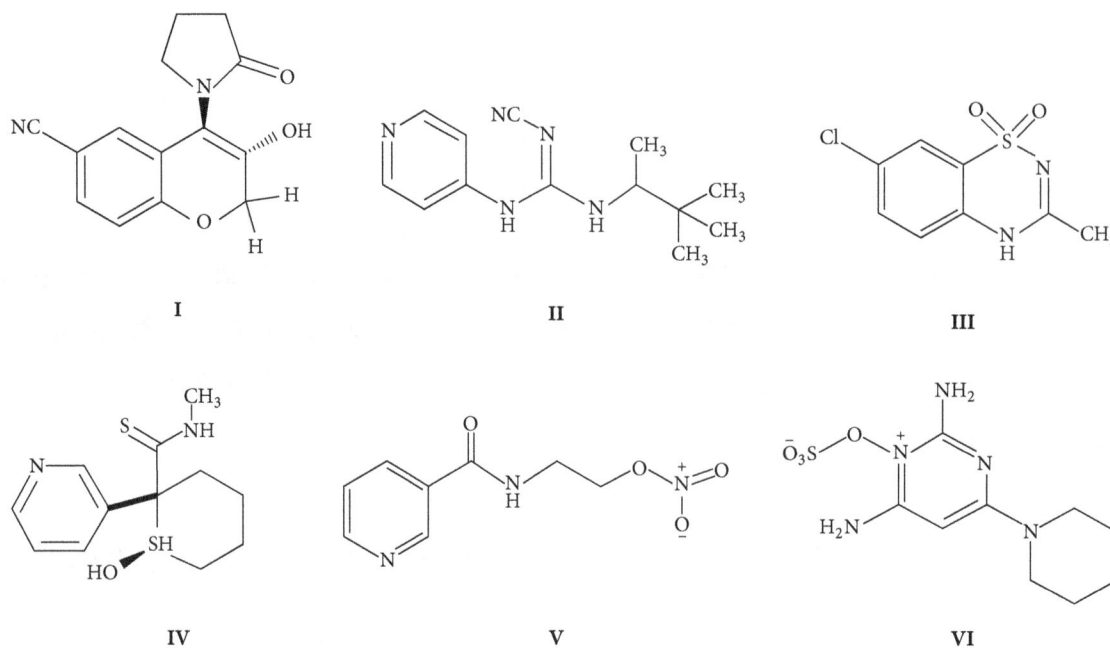

FIGURE 1: First generation potassium channel openers (KCOs) as antihypertensive agents.

position-4, in addition to the two methoxymethyl groups at position-2, were found to be more potent and have an improved duration of action [13].

Myocardial preconditioning as KCOs is of great interest as myocardial protecting agents [14]. The first generation (K_{ATP}) openers I–VI (Figure 1) are potent peripheral vasodilators, but the use of these compounds for the treatment of acute myocardial ischemia is limited due to the possibility of hemodynamic alterations upon systemic administration which can result in under perfusion of the area that is already at risk [15]. It was presumed that clinical utility of these agents for the treatment of hypertension is due to their peripheral vasodilating properties, as they are widely known to open potassium channels in several tissue types. But relevant studies have shown that K_{ATP} openers have direct cardioprotective properties independent of their vasodilator effect. Therefore, tissue selective K_{ATP} openers are clearly required to explore the potential of these agents [16].

2. Experimental

2.1. General. All the reagents were purchased from Sigma-Aldrich Chemicals (Bangalore, India) and were used without further purification. All solvents were distilled and dried using dry sieves as the usual manner. TLC analysis was carried out on aluminum foil precoated with silica gel 60 F254 (Sigma-Aldrich, Bangalore dealer). Melting points were determined on a Thomas micro-hot stage apparatus and are uncorrected. FTIR spectra were determined as KBr solid discs on a Shimadzu model 470 spectrophotometer. ^{1}H NMR spectra were recorded using a Jeol Eclipse 400 MHz spectrometer using CDCl$_3$ as NMR solvent and are reported in ppm down field from the residual CDCl$_3$. ^{1}H NMR

spectrum exhibited different signals at different ppm which were assigned to the different types of protons. The synthetic route leading to the title compounds is summarized in Scheme 1.

2.2. Synthesis

2.2.1. 2,4-Diacetyl Resorcinol (2). Dry resorcinol 1 (1.0 g, 9.09 mmol) was added to a mixture of zinc chloride (2.467 g, 18.18 mmol) in dried acetic anhydride (1.89 mL, 18.18 mmol) in a round bottom flask quickly with stirring. The reaction mixture was slowly heated on wire gauze and kept at 145–150°C for 15 min. The resulting viscous reaction mixture was allowed to cool at room temperature and ice cold aqueous hydrochloride solution was added to it with constant stirring. An orange-red crystalline compound separated out, which was purified by column chromatography to obtain white color solid compound. Yield 70.58%. mp 175–177°C. IR (KBr) ν (cm^{-1}): 3414.3, 3079.8, 2926.1, 1658.6, 1588.6, 1256.7. ^{1}H NMR (CDCl$_3$, δ ppm): 12.93 (s, Ar, 2H), 8.19 (s, 1H), 6.39 (s, 1H), 2.62 (s, 6H). MS (EIMS, FAB, *m/z*): 195.4 (M + 1).

2.2.2. 2,2,8,8-Tetramethyl-2,3,7,8-tetrahydro-4H,6H-benzo-[1, 2-b:5,4-b']-dipyran-4,6-dione (3). 2,4-Diacetyl resorcinol 2 (1.0 g, 5.15 mmol) was mixed with piperidine (0.875 g, 10.3 mmol) and acetone (3.0 mL) in toluene in a round bottom flask which was fixed with Dean Stark apparatus. The resulting reaction mixture was slowly heated at 120–125°C for 24 hr. After completion of reaction, the reaction mixture was distilled off to remove the solvents. It was quenched with ice cold water and extracted with chloroform. Organic layer was separated and dried over sodium sulfate and solvent

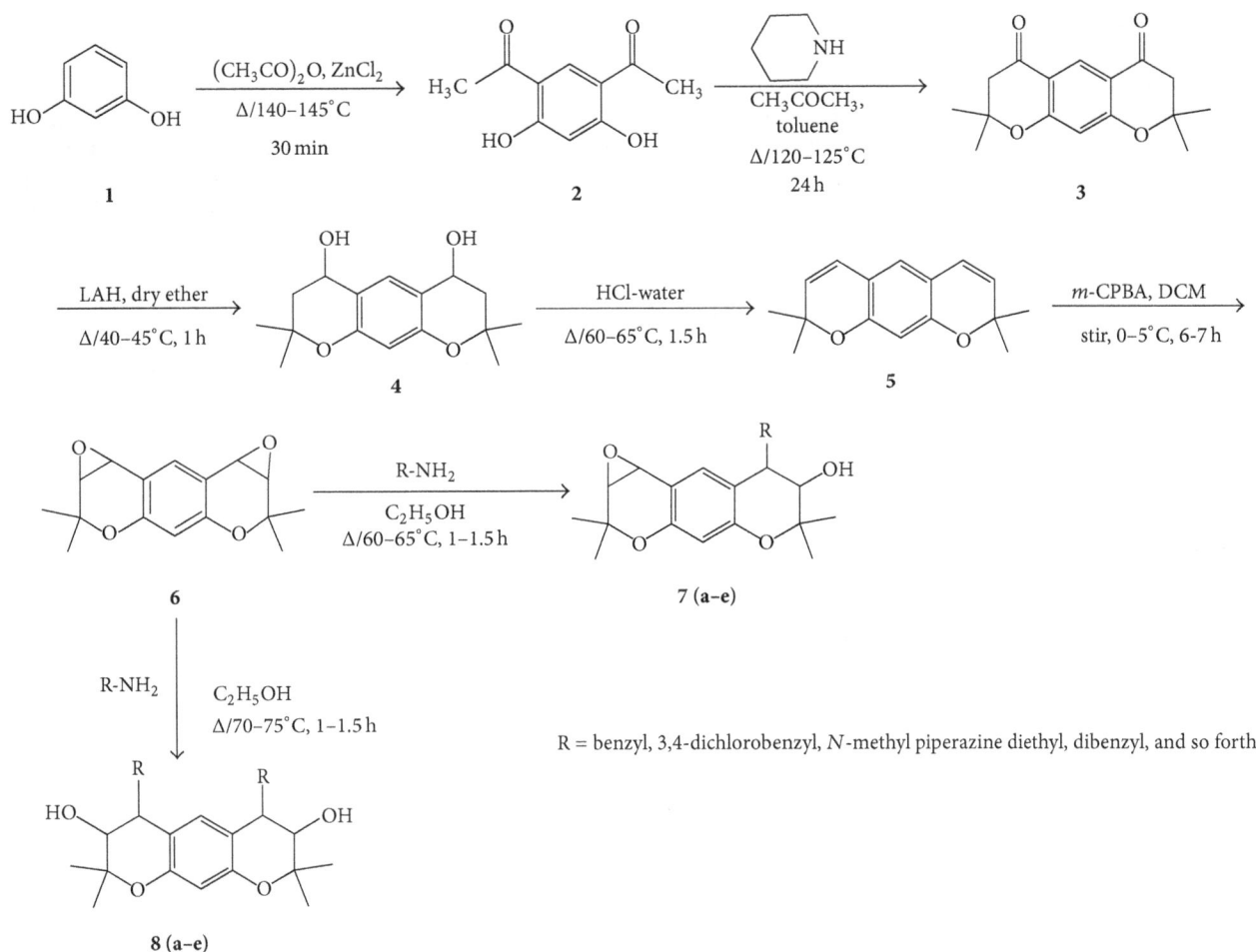

SCHEME 1: Synthesis of 2,2,8,8-tetramethyl-2,3,7,8-tetrahydro-4,6-diamino-3,7-dihydroxy-6,7-epoxy-benzo-[1,2-b:5,4-b']dipyran derivatives **7a–e** and **8a–e** from resorcinol for antihypertensive activity.

was evaporated off. The compound was purified by column chromatography over silica gel to get a white color solid product. Yield 51.85%. mp 184–186°C. IR (KBr) ν (cm^{-1}): 2973.4, 2929.7, 1703.1, 1600.9, 1234.2. ^{1}H NMR (CDCl$_3$, δ ppm): 8.46 (s, 1H), 6.38 (s, 1H), 2.69 (s, 4H), 1.45 (s, 12H). MS (EIMS, FAB, m/z): 275.1 (M + 1).

2.2.3. 2,2,8,8-Tetramethyl-2,3,7,8-tetrahydro-4H,6H-benzo-[1, 2-b:5,4-b']dipyran-4,6-dihydroxy (4). To a solution containing 500 mg (0.013 mol) of lithium aluminum hydride (LAH) in 25 mL ether, the corresponding chromanone **3** (1.37 g, 0.005 mole) in 30 mL of ether was added drop wise with stirring in a round bottom flask. The resulting reaction mixture was heated to reflux for an hour, allowed to cool, and then filtered. Acetone (20 mL) was added to the resulting filtrate to decompose the excess of lithium aluminum hydride and the reaction was monitored by TLC. Yield 65.43%. mp 183–185°C. IR (KBr) ν (cm^{-1}): 3281.7, 2972.2, 2361.8, 1630.1, 1254. ^{1}H NMR (CDCl$_3$, δ ppm): 8.33 (d, 3H), 6.46 (s, 2H), 2.92 (s, 2H), 1.52 (s, 4H). MS (EIMS, FAB, m/z): 279.7 (M + 1).

2.2.4. 2,2,8,8-Tetramethyl-2H,8H-benzo[1,2-b:5,4-b']dipyran (5). Compound **4** was refluxed with 6 M HCl (10 mL) for 10 min. Then, 50 mL water was added to it and the reaction mixture was further refluxed for 1.0 h, allowed to cool, solvent was evaporated off, and aqueous phase was extracted with methylene chloride. Organic layer was dried over sodium sulfate, concentrated, and purified by column chromatography over silica gel to get a white color solid product. Yield 83.19%. mp 195–197°C. IR (KBr) ν (cm^{-1}): 3042.7, 2977.5, 1562.7, 1212.7. ^{1}H NMR (CDCl$_3$, δ ppm): 6.61 (s, 1H), 6.27 (s, 2H), 6.25 (d, 1H, J = 9.9 Hz), 5.47 (d, 1H, J = 9.9 Hz), 1.42 (s, 12H). MS (EIMS, FAB, m/z): 243.5 (M + 1).

2.2.5. 2,2,8,8-Tetramethyl-2H,8H-benzo-[1,2-b:5,4-b']dipyran Oxide (6). Compound **5** (100 mg, 0.413 mmol) was dissolved in dichloromethane. *m*-Chloroperbenzoic acid (*m*-CPBA) (213 mg, 1.239 mmol) was added to resulting reaction mixture. It was allowed to stir at 0°C for 1h. Solvent was evaporated at low temperature and excess of *m*-CPBA was decomposed by NaHCO$_3$ solution. The aqueous solution was extracted using dichloromethane and organic layer was separated. It was concentrated and purified by column chromatography over silica gel to get a white color solid product. Yield 37.09%. mp 186–188°C. IR (KBr) ν (cm^{-1}): 3074.3, 2934.5, 1553.5, 1243.3, 1043.4. ^{1}H NMR (CDCl$_3$, δppm): 7.54 (d, 1H, J = 9.6 Hz), 6.32 (d, 1H, J = 7.8 Hz), 6.27 (s, 1H), 2.46 (s, 3H). MS (EIMS, FAB, m/z): 275.1 (M + 1).

2.2.6. 2,2,8,8-Tetramethyl-2,3,7,8-tetrahydro-4,6-diamino-3,7-dihydroxy-6,7-epoxy-benzo-[1,2-b:5,4-b′]dipyran Derivatives (7a–e) and 2,2,8,8-Tetramethyl-2,3,7,8-tetrahydro-4,6-diamino-3,7-dihydroxy-benzo-[1,2-b:5,4-b′]dipyran Derivatives (8a–e).

Compound 6 (100 mg, 0.364 mmol) was dissolved in 30 mL ethanol in a 100 mL round bottom flask. Benzylamine (0.78 mL, 729 mmol) was added to the above reaction mixture and was allowed to reflux at 80°C for 3 h and reaction was monitored by TLC. After completion of reaction, it was distilled off to remove solvent from the reaction mixture. Then, it was quenched with ice cold water and extracted with dichloromethane. Organic layer was distilled off to get crude product. Further purification was accomplished by column chromatography. Mobile phase: ethyl acetate: hexane = 6:4.

2.3. Antihypertensive Activity

2.3.1. Toxicity Studies to Fix Up LD$_{50}$.
Toxicity studies were carried out according to the OECD guidelines numbers 420 and 421 in order to fix up the dose to carry out the antihypertensive activity [17, 18]. Wister Albino Rats weighing 200–250 g were chosen and oral route is selected for the drug administration. Six groups of animals each containing three animals were initially selected as per the guidelines numbers 420 and 421. Given dose of 70 mg/kg body weight was monitored in the animal for the toxic symptoms as well as mortality [19]. The animals showed high toxicity symptoms such as increased intestinal motility, diarrhoea, tail erection, and irritation to nose, and all the animals were dead after 3.0 h. Hence, we decreased the dose to 50 mg/kg body weight and administered to the next group of animals, monitored for toxic symptoms and mortality. In this dose, animals were safe but showed fewer toxic symptoms and only few were died. Toxicity symptoms were diarrhoea, tail erection, and irritation to the nose. Once again, we decreased the dose and it was fixed to a dose of 20 mg/kg body weight to the next set of animals and observed for the toxic symptoms and mortality. All the animals were safe and no toxic symptoms were seen at this specific dose. Hence, it was concluded that 20 mg/kg body weight dose was safe and recommended dose for further antihypertensive activity [20].

2.3.2. Direct Antihypertensive Activity.
Direct antihypertensive activity was carried out using the instrument BIOPAC System MP-36 Santa Barbara California for recording the blood pressure response [21]. The instrument was calibrated before carrying out the experiment and process was thoroughly practiced and understood including handling and surgically cannulating artery for monitoring blood pressure and a vein for drug administration [22, 23].

2.3.3. Preparation of Model.
Male albino rats weighing 200–250 g were used for the antihypertensive activity. Rats were anesthetized using urethane hydrochloride (1.25 g/kg). Rats were prepared by shaving the neck and inguinal region using animal hair clippers. Jugular vein was surgically cannulated for the drug administration. Left carotid artery was isolated and exposed by dissection for blood pressure recording using PE-50 tubing [24]. By means of a three-way plastic stop cock and a stainless steel needle at the end of the PE tubing, arterial cannula and venous cannula were attached to a blood-pressure transducer and syringe, respectively [25, 26]. Fluid was filled in the both cannulae with heparinised saline before cannulation. Arterial cannula was connected via the BSL pressure transducer (SS13L) to the BIOPAC Systems, Inc. Criterion for antihypertensive activity was the reduction of systolic arterial pressure by about 10–20 mmHg [27].

2.3.4. Experimental Procedure.
Adrenaline (5.0 μg/kg, i.v.) was administered intravenously for the sympathetic system activation to induce hypertension [28, 29]. Venous cannula was flushed with 0.2 mL of normal saline and allowed to return to preinjection level. Test compound 20 mg/kg solution was injected intravenously and allowed to equilibration in the system. Adrenaline (5.0 μg/kg, i.v.) was repeated as described previously. Blood-pressure response was observed and recorded to each procedure [30–32]. Antihypertensive activity of the benzodipyran derivatives 7a–e and 8a–c was summarized in Table 1.

3. Result and Discussion

3.1. Chemistry. 4, 6-Diacetyl resorcinol 2 was synthesized using a mixture of zinc chloride in dried acetic anhydride from dry resorcinol 1 by constant stirring. It was kept at high temperature for 30 min and was purified through column chromatography [33, 34]. Compound 3 (2,2,8,8-tetramethyl-2,3,7,8-tetrahydro-4H,6H-benzo[1,2-b:5,4-b′]dipyran-4,6-dione) was synthesized from 4,6-diacetyl resorcinol using acetone and piperidine in a solution in a Dean Stark apparatus using toluene as solvent. The reaction mixture was slowly heated at 120–125°C for 24 h to get the compound 3 [35, 36]. Compound 4 (2,2,8,8-tetramethyl-2,3,7,8-tetrahydro-4H,6H-benzo[1,2-b:5,4-b′]dipyran-4,6-dihydroxy) was synthesized using lithium aluminum hydride (LAH) in ether; the corresponding chromanone in ether and both these solutions were added drop wise with stirring. The resulting reaction mixture was heated to reflux for 1 h [37]. This reaction mixture was taken as such for the synthesis of 2,2,8,8-tetramethyl-2H,8H-benzo[1,2-b:5,4-b′]dipyran 5 by adding 6 M HCl and allowed to reflux for 1.5 h. After completion of the reaction, solvent was evaporated and the remaining aqueous phase was extracted with methylene chloride [38, 39]. 2,2,8,8-Tetramethyl-2H,8H-benzo[1,2-b:5,4-b′]dipyran oxide 6 was synthesized from compound 5 by dissolving in dichloromethane and required quantity of m-chloroperbenzoic acid (m-CPBA) was added to it. The resulting reaction mixture was allowed to stir at 0–5°C for 6–7 h [40]. Different benzodipyran derivatives 7a–e and 8a–e

TABLE 1: Antihypertensive activity of the benzodipyran derivatives **7a–e** and **8a–c**.

Comp. code	Parameter	Baseline	With adrenaline	With test alone	With test + adrenaline	Inference
7a	SBP	101.2 (±0.693)	147.8 (±1.308)	140.6 (±1.121)	141.4 (±2.040)	Minimal activity
	DBP	57.9 (±0.769)	79.2 (±0.861)	83.2 (±1.240)	73.9 (±1.167)	
	MABP	77.5 (±0.480)	143.9 (±1.202)	107.2 (±1.387)	109.3 (±0.894)	
	HR	196.5 (±3.201)	291.9 (±3.158)	248.1 (±3.118)	260.2 (±3.201)	
7b	SBP	106.4 (±0.482)	160.6 (±1.198)	124.6 (±1.389)	126.0 (±1.406)	Antihypertensive
	DBP	47.8 (±0.927)	75.1 (±3.054)	53.0 (±0.107)	59.3 (±1.541)	
	MABP	72.97 (±0.890)	122.1 (±1.883)	83.44 (±0.488)	109.2 (±0.931)	
	HR	299.9 (±3.106)	326.0 (±5.431)	321.0 (±0.557)	302.4 (±3.658)	
7c	SBP	105.7 (±1.203)	148.2 (±1.055)	127.1 (±1.031)	132.4 (±2.049)	Antihypertensive
	DBP	63.1 (±1.202)	86.5 (±1.218)	62.1 (±3.983)	84.2 (±1.160)	
	MABP	68.1 (±0.399)	109.1 (±0.963)	119.8 (±1.991)	101.3 (±0.896)	
	HR	278.8 (±2.697)	302.1 (±2.616)	298.7 (±2.065)	296.4 (±3.120)	
7d	SBP	120.8 (±1.061)	162.7 (±1.211)	143.6 (±0.892)	121.7 (±0.231)	Antihypertensive
	DBP	64.2 (±0.381)	86.8 (±0.665)	84.6 (±0.820)	71.1 (±1.858)	
	MABP	98.2 (±0.485)	120.7 (±1.390)	111.7 (±1.057)	99.2 (±1.524)	
	HR	209.5 (±1.161)	267.2 (±1.108)	248.9 (±2.868)	266.8 (±2.830)	
7e	SBP	108.9 (±0.519)	135.4 (±1.328)	160.8 (±0.940)	167.0 (±3.027)	Moderately active
	DBP	61.0 (±1.291)	87.1 (±1.966)	68.2 (±1.332)	91.3 (±1.857)	
	MABP	88.9 (±1.135)	102.1 (±1.503)	106.2 (±1.887)	119.3 (±0.197)	
	HR	232.2 (±2.097)	241.0 (±3.088)	209.2 (±3.210)	216.1 (±2.118)	
8a	SBP	128.8 (±3.62)	182.7 (±2.63)	168.8 (±0.85)	198.0 (±3.43)	Minimal activity
	DBP	70.89 (±2.33)	108.4 (±2.02)	134.3 (±0.95)	143.1 (±3.93)	
	MABP	103.7 (±3.28)	139.3 (±1.532)	149.9 (±1.152)	163.9 (±0.852)	
	HR	286.4 (±5.128)	304.6 (±3.210)	338.6 (±3.308)	341.9 (±1.409)	
8b	SBP	106.8 (±2.367)	178.6 (±1.349)	179.3 (±2.101)	189.3 (±0.239)	No activity
	DBP	84.0 (±2.105)	126.2 (±0.968)	128.7 (±1.151)	148.6 (±1.059)	
	MABP	112.9 (±2.250)	146.7 (±2.152)	182.3 (±1.788)	195.6 (±1.272)	
	HR	306.3 (±3.977)	331.8 (±3.300)	34.6 (±2.312)	361.8 (±2.008)	

TABLE 1: Continued.

Comp. code	Parameter	Baseline	With adrenaline	With test alone	With test + adrenaline	Inference
8c	SBP	89.82 ($\pm$1.363)	121.1 ($\pm$2.015)	143.3 ($\pm$1.38)	147.5 ($\pm$1.843)	Minimal activity
	DBP	67.1 ($\pm$0.379)	89.5 ($\pm$0.895)	102.8 ($\pm$3.516)	123.3 ($\pm$2.331)	
	MABP	88.0 ($\pm$3.837)	135.5 ($\pm$1.293)	142.8 ($\pm$2.870)	150.2 ($\pm$0.922)	
	HR	269.8 ($\pm$3.897)	302.1 ($\pm$4.430)	348.5 ($\pm$3.772)	356.9 ($\pm$1.573)	

SBP: systolic blood pressure, DBP: diastolic blood pressure, MABP: mean arterial blood pressure, and HR: heart rate. Values are expressed in mean $\pm$ S.E.M. Number of readings: 03.

TABLE 2: Physical and spectral analysis of 2,2,8,8-tetramethyl-2,3,7,8-tetrahydro-4,6-diamino-3,7-dihydroxy-6,7-epoxy-benzo-[1,2-b:5,4-b'] dipyran derivatives 7a–e.

Comp. code	R	Molecular formula	R_f	FTIR (cm^{-1})	^{1}H NMR (ppm)	Mass (M + 1)
7a	(benzyl-NH$_2$)	$C_{27}H_{23}O_4N$	3.1	3352.7 (O–H str), 2934.2 (Ali C–H str), 1138.8 (C–N str), 1615 (C=C str)	δ7.16 (m, 12H), δ2.15 (s, 4H), δ1.30 (s, 12H)	382.5
7b	H$_3$C–N NH	$C_{21}H_{30}O_4N_2$	3.3	3317.8 (O–H str), 2924.4 (Ali C–H str), 1094.4 (C–N str), 1229.8 (C–O str), 1611.1 (C=C str)	δ5.37 (d, 1H), δ2.15 (s, 6H), δ1.26 (s, 12H)	374.0
7c	Cl Cl NH$_2$	$C_{23}H_{25}Cl_2O_4N$	2.7	3395.5 (O–H str), 2927.9 (Ali C–H str), 1256.6 (C–O str), 1655.1 (C=C str), 1094.6 (C–N)	δ7.41 (s, 1H), δ6.37 (s, 1H), δ4.03 (d, 4H), δ1.50 (s, 4H), δ1.43 (s, 12H)	451.1
7d	(diphenylamine)	$C_{28}H_{29}O_4N$	3.3	3266.4 (O–H str), 2940.8 (Ali C–H str), 1251.6 (C–O str), 1650.8 (C=C str), 1029.1 (C–N)	δ7.43 (m, 22H), δ2.29 (s, 6H), δ1.20 (s, 12H)	444.0
7e	(diethylamine)	$C_{20}H_{29}O_4N$	4.2	3427 (O–H str), 3020.7 (Ali C–H str), 1216.4 (C–O str), 1562.7 (C=C str), 1041.1 (C–N)	δ3.47 (m, 4H), δ2.29 (s, 4H), δ1.20 (s, 12H)	348.2

TABLE 3: Physical and spectral analysis of 2,2,8,8-tetramethyl-2,3,7,8-tetrahydro-4,6-diamino-3,7-dihydroxy-benzo[1,2-b:5,4-b'] dipyran 8a–e.

Comp. code	R	Molecular formula	R_f	FTIR (cm^{-1})	^{1}H NMR (ppm)	Mass (M + 1)
8a	(benzyl-NH$_2$)	$C_{30}H_{36}O_4N_2$	2.8	3352.7 (O–H str), 2934.2 (Ali C–H str), 1138.8 (C–N str), 1615 (C=C str)	δ7.16 (m, 12H), δ2.15 (s, 4H), δ1.30 (s, 12H)	489.1
8b	H$_3$C–N NH	$C_{26}H_{42}O_4N_4$	2.7	3317.8 (O–H str), 2924.4 (Ali C–H str), 1094.4 (C–N str), 1229.8 (C–O str), 1611.1 (C=C str)	δ5.37 (d, 1H), δ2.15 (s, 6H), δ1.26 (s, 12H)	475.4
8c	Cl Cl NH$_2$	$C_{30}H_{32}Cl_4O_4N_2$	2.3	3395.5 (O–H str), 2927.9 (Ali C–H str), 1256.6 (C–O str), 1655.1 (C=C str), 1094.6 (C–N)	δ7.41 (s, 1H), δ6.37 (s, 1H), δ4.03 (d, 4H), δ1.50 (s, 4H), δ1.43 (s, 12H)	625.0
8d	(diphenylamine)	$C_{40}H_{40}O_4N_2$	2.7	3266.4 (O–H str), 2940.8 (Ali C–H str), 1251.6 (C–O str), 1650.8 (C=C str), 1029.1 (C–N)	δ7.43 (m, 22H), δ2.29 (s, 6H), δ1.20 (s, 12H)	613.6
8e	(diethylamine)	$C_{23}H_{37}O_4N_2$	3.6	3427 (O–H str), 3020.7 (Ali C–H str), 1216.4 (C–O str), 1562.7 (C=C str), 1041.1 (C–N)	δ3.47 (m, 4H), δ2.29 (s, 4H), δ1.20 (s, 12H)	421.1

were synthesized from different amines such as diethylamine, 3,4-dichlorobenzylamine, dibenzylamine, benzylamine, and N-methyl piperazine. These derivatives were synthesized by the ring opening of epoxide and were identified by different spectroscopic techniques [41]. The synthesized compounds were screened for antihypertensive activity and some of these compounds showed significant antihypertensive activity. Physical and spectral analysis of 2,2,8,8-tetramethyl-2,3,7,8-tetrahydro-4,6-diamino-3,7-dihydroxy-6,7-epoxy-benzo-[1,2-b:5,4-b']dipyran derivatives **7a–e** are summarized in Table 2. Physical and spectral analysis of 2,2,8,8-tetramethyl-2,3,7,8-tetrahydro-4,6-diamino-3,7-dihydroxy-6,7-epoxy-benzo-[1,2-b:5,4-b']dipyran derivatives **8a–e** are summarized in Table 3.

4. Conclusion

The present study describes the synthesis and evaluation of the antihypertensive activity of novel 2,2,8,8-tetramethyl-2,3,7,8-tetrahydro-4,6-diamino-3,7-dihydroxy-6,7-epoxy-benzo-[1,2-b:5,4-b']dipyran derivatives **7a–e** and **8a–e**. Compounds **7b**, **7c**, and **7d** showed potent antihypertensive activity and can constitute lead compounds. Compounds **7a**, **8a**, and **8c** showed minimal antihypertensive activity while other compounds showed moderate antihypertensive activity.

Conflict of Interests

The authors declare that there is no conflict of interests regarding the publication of this paper.

Acknowledgments

The authors would like to thank Institute of Pharmacy, Nirma University, Ahmedabad, Gujarat, India, for providing continuous support and Krupanidhi College of Pharmacy, Bangalore, for providing necessary facilities for experimental work. The authors are also thankful to SAIF, CDRI, Lucknow, for spectral analysis of the synthesized compounds. The authors are also thankful to Council of Scientific and Industrial Research (CSIR), India, and Department of Science & Technology (DST), India, for providing for financial support.

References

[1] B. Rudy, "Diversity and ubiquity of K$^+$ channels," *Neuroscience*, vol. 25, pp. 729–735, 1988.

[2] L. Aguilar-Bryan, C. G. Nichols, S. W. Wechsler et al., "Cloning of the β cell high-affinity sulfonylurea receptor: a regulator of insulin secretion," *Science*, vol. 268, no. 5209, pp. 423–426, 1995.

[3] K. S. Atwal, G. J. Grover, S. Z. Ahmed et al., "Cardioselective anti-ischemic ATP-sensitive potassium channel openers," *Journal of Medicinal Chemistry*, vol. 36, no. 24, pp. 3971–3974, 1993.

[4] F. Dreyer, "Peptide toxins and potassium channels," *Reviews of Physiology Biochemistry and Pharmacology*, vol. 115, pp. 93–136, 1990.

[5] I. Baczkó, I. Leprán, and J. G. Papp, "K$_{ATP}$ channel modulators increase survival rate during coronary occlusion-reperfusion in anaesthetized rats," *European Journal of Pharmacology*, vol. 324, no. 1, pp. 77–83, 1997.

[6] K. S. Atwal, P. Wang, W. L. Rogers et al., "Small molecule mitochondrial F$_1$F$_0$ ATPase hydrolase inhibitors as cardioprotective agents. Identification of 4-(N-arylimidazole)-substituted benzopyran derivatives as selective hydrolase inhibitors," *Journal of Medicinal Chemistry*, vol. 47, no. 5, pp. 1081–1084, 2004.

[7] A. Noma, "ATP-regulated K$^+$ channels in cardiac muscle," *Nature*, vol. 305, pp. 147–148, 1983.

[8] A. D. Wickenden, "K$^+$ channels as therapeutic drug targets," *Pharmacology & Therapeutics*, vol. 94, no. 1-2, pp. 157–182, 2002.

[9] S. J. H. Ashcroft and F. M. Ashcroft, "Properties and functions of ATP-sensitive K-channels," *Cellular Signalling*, vol. 2, no. 3, pp. 197–214, 1990.

[10] G. Edwards and A. H. Weston, "The pharmacology of ATP-sensitive potassium channels," *Annual Review of Pharmacology and Toxicology*, vol. 33, pp. 597–637, 1993.

[11] J. Anabuki, M. Hori, H. Ozaki, I. Kato, and H. Karaki, "Mechanisms of pinacidil-induced vasodilatation," *European Journal of Pharmacology*, vol. 190, no. 3, pp. 373–379, 1990.

[12] R. H. Grimm Jr., "Antihypertensive therapy: taking lipids into consideration," *The American Heart Journal*, vol. 122, no. 3, pp. 910–918, 1991.

[13] S. L. Archer, J. Huang, T. Henry, D. Peterson, and E. K. Weir, "A redox-based O$_2$ sensor in rat pulmonary vasculature," *Circulation Research*, vol. 73, no. 6, pp. 1100–1112, 1993.

[14] J. Bellemin-Baurreau, A. Poizot, P. E. Hicks, L. Rochette, and J. Michael Armstrong, "Effects of ATP-dependent K$^+$ channel modulators on an ischemia-reperfusion rabbit isolated heart model with programmed electrical stimulation," *European Journal of Pharmacology*, vol. 256, no. 2, pp. 115–124, 1994.

[15] V. M. Bolotina, "Calcium-activated potassium channels in cultured human endothelial cells are not directly modulated by nitric oxide," *Nature*, vol. 368, pp. 850–854, 1994.

[16] G. Edwards and A. H. Weston, "The pharmacology of ATP-sensitive potassium channels," *Annual Review of Pharmacology and Toxicology*, vol. 33, pp. 597–637, 1993.

[17] R. Hong, J. Feng, R. Hoen, and G.-Q. Lin, "Synthesis of (±)-3,3′-bis(4-hydroxy-2H-benzopyran): a literature correction," *Tetrahedron*, vol. 57, no. 41, pp. 8685–8689, 2001.

[18] K. Tanaka, H. Kawasaki, K. Kurata, Y. Aikawa, Y. Tsukamoto, and T. Inaba, "T-614, a novel antirheumatic drug, inhibits both the activity and induction of cyclooxygenase-2 (COX-2) in cultured fibroblasts," *Japanese Journal of Pharmacology*, vol. 67, no. 4, pp. 305–314, 1995.

[19] E. Tyrrell, K. H. Tesfa, I. Greenwood, and A. Mann, "The synthesis and biological evaluation of a range of novel functionalised benzopyrans as potential potassium channel activators," *Bioorganic and Medicinal Chemistry Letters*, vol. 18, no. 3, pp. 1237–1240, 2008.

[20] J. M. Evans, C. S. Fake, T. C. Hamilton, R. H. Poyser, and G. A. Showell, "Synthesis and antihypertensive activity of 6,7-disubstituted trans-4-amino-3,4-dihydro-2,2-dimethyl-2H-1-benzopyran-3-ols," *Journal of Medicinal Chemistry*, vol. 27, no. 9, pp. 1127–1131, 1984.

[21] G. C. Rovnyak, S. Z. Ahmed, C. Z. Ding et al., "Cardioselective antiischemic ATP-sensitive potassium channel (K$_{ATP}$) openers. 5. Identification of 4-(N-aryl)-substituted benzopyran derivatives with high selectivity," *Journal of Medicinal Chemistry*, vol. 40, no. 1, pp. 24–34, 1997.

[22] H. H. Herman, S. H. Pollock, L. C. Fowler, and S. W. May, "Demonstration of the potent antihypertensive activity of phenyl-2-aminoethyl sulfides," *Journal of Cardiovascular Pharmacology*, vol. 11, no. 5, pp. 201–210, 1988.

[23] J. L. Wang, K. Aston, D. Limburg et al., "The novel benzopyran class of selective cyclooxygenase-2 inhibitors. Part III: the three microdose candidates," *Bioorganic & Medicinal Chemistry Letters*, vol. 20, no. 23, pp. 7164–7168, 2010.

[24] Y.-M. Lee, M.-H. Yen, Y.-Y. Peng et al., "The antihypertensive and cardioprotective effects of (−)-MJ-451, an ATP-sensitive K^+ channel opener," *European Journal of Pharmacology*, vol. 397, no. 1, pp. 151–160, 2000.

[25] N. Kaur, A. Kaur, Y. Bansal, D. I. Shah, G. Bansal, and M. Singh, "Design, synthesis, and evaluation of 5-sulfamoyl benzimidazole derivatives as novel angiotensin II receptor antagonists," *Bioorganic and Medicinal Chemistry*, vol. 16, no. 24, pp. 10210–10215, 2008.

[26] S. Prasanna, E. Manivannan, and S. C. Chaturvedi, "Quantitative structure-activity relationship analysis of a series of 2,3-diaryl benzopyran analogues as novel selective cyclooxygenase-2 inhibitors," *Bioorganic and Medicinal Chemistry Letters*, vol. 14, no. 15, pp. 4005–4011, 2004.

[27] A. Bali, Y. Bansal, M. Sugumaran et al., "Design, synthesis, and evaluation of novelly substituted benzimidazole compounds as angiotensin II receptor antagonists," *Bioorganic and Medicinal Chemistry Letters*, vol. 15, no. 17, pp. 3962–3965, 2005.

[28] L. Xing, B. C. Hamper, T. R. Fletcher et al., "Structure-based parallel medicinal chemistry approach to improve metabolic stability of benzopyran COX-2 inhibitors," *Bioorganic and Medicinal Chemistry Letters*, vol. 21, no. 3, pp. 993–996, 2011.

[29] J. Gierse, M. Nickols, K. Leahy et al., "Evaluation of COX-1/COX-2 selectivity and potency of a new class of COX-2 inhibitors," *European Journal of Pharmacology*, vol. 588, no. 1, pp. 93–98, 2008.

[30] H. J. Finlay, J. Lloyd, M. Nyman et al., "Pyrano-[2,3*b*]-pyridines as potassium channel antagonists," *Bioorganic and Medicinal Chemistry Letters*, vol. 18, no. 8, pp. 2714–2718, 2008.

[31] B. Becker, M.-H. Antoine, Q.-A. Nguyen et al., "Synthesis and characterization of a quinolinonic compound activating ATP-sensitive K^+ channels in endocrine and smooth muscle tissues," *British Journal of Pharmacology*, vol. 134, no. 2, pp. 375–385, 2001.

[32] Y. H. Joo, J. K. Kim, S.-H. Kang et al., "2,3-Diarylbenzopyran derivatives as a novel class of selective cyclooxygenase-2 inhibitors," *Bioorganic and Medicinal Chemistry Letters*, vol. 13, no. 3, pp. 413–417, 2003.

[33] R. Hong, J. Feng, R. Hoen, and G.-Q. Lin, "Synthesis of (±)-3,3′-bis(4-hydroxy-2*H*-benzopyran): a literature correction," *Tetrahedron*, vol. 57, no. 41, pp. 8685–8689, 2001.

[34] J. Lim, I.-H. Kim, H. H. Kim, K.-S. Ahn, and H. Han, "Enantioselective syntheses of decursinol angelate and decursin," *Tetrahedron Letters*, vol. 42, no. 24, pp. 4001–4003, 2001.

[35] M. C. Breschi, V. Calderone, A. Martelli et al., "New benzopyran-based openers of the mitochondrial ATP-sensitive potassium channel with potent anti-ischemic properties," *Journal of Medicinal Chemistry*, vol. 49, no. 26, pp. 7600–7602, 2006.

[36] T. Takahashi, H. Koga, H. Sato, T. Ishizawa, N. Taka, and J.-I. Imagawa, "Synthesis and vasorelaxant activity of 2-fluoromethylbenzopyran potassium channel openers," *Bioorganic and Medicinal Chemistry*, vol. 6, no. 3, pp. 323–337, 1998.

[37] H. Cho, S. Katoh, S. Sayama et al., "Synthesis and selective coronary vasodilatory activity of 3,4-dihydro-2,2-bis(methoxymethyl)-2H-1-benzopyran-3-ol derivatives: novel potassium channel openers," *Journal of Medicinal Chemistry*, vol. 39, no. 19, pp. 3797–3805, 1996.

[38] N. Taka, H. Koga, H. Sato, T. Ishizawa, T. Takahashi, and J.-I. Imagawa, "6-Substituted 2,2-bis(fluoromethyl)-benzopyran-4-carboxamide K^+ channel openers," *Bioorganic and Medicinal Chemistry*, vol. 8, no. 6, pp. 1393–1405, 2000.

[39] R. Thompson, S. Doggrell, and J. O. Hoberg, "Potassium channel activators based on the benzopyran substructure: synthesis and activity of the C-8 substituent," *Bioorganic & Medicinal Chemistry*, vol. 11, no. 8, pp. 1663–1668, 2003.

[40] R. Mannhold, G. Cruciani, H. Weber et al., "6-Substituted benzopyrans as potassium channel activators: synthesis, vasodilator properties, and multivariate analysis," *Journal of Medicinal Chemistry*, vol. 42, no. 6, pp. 981–991, 1999.

[41] J. Renaud, S. F. Bischoff, T. Buhl et al., "Estrogen receptor modulators: identification and structure—activity relationships of potent ERα-selective tetrahydroisoquinoline ligands," *Journal of Medicinal Chemistry*, vol. 46, no. 14, pp. 2945–2951, 2003.

23

Telomerization of Vinyl Chloride with Chloroform Initiated by Ferrous Chloride-Dimethylacetamide under Ultrasonic Conditions

Hua Qian[1] and Hengdao Quan[2]

[1]School of Chemical Engineering, Nanjing University of Science and Technology, Xiaolingwei 200, Nanjing, Jiangsu 210094, China
[2]National Institute of Advanced Industrial Science and Technology (AIST), Tsukuba, Ibaraki 305-8565, Japan

Correspondence should be addressed to Hua Qian; qianhua@njust.edu.cn

Academic Editor: Oliver Höfft

Telomerization of vinyl chloride with chloroform was investigated using ferrous chloride-dimethylacetamide system, and 42.1% yield, more than four times the one reported before, was achieved. The addition of ultrasound further improved the reaction and yield was raised to 51.9% with trace byproducts at highly reduced reaction time and temperature. Ferrous chloride-dimethylacetamide under ultrasonic irradiation acts as a very efficient catalyst system for the 1 : 1 telomerization.

1. Introduction

1,1,3,3-tetrachloropropane is an important intermediate for organic synthesis and is also the last one of twenty-nine possible chloropropanes that was not characterized until 1951 [1]. In the literature, 1,1,3,3-tetrachloropropane can be prepared by the chlorination of 1,1,3-trichloropropane [1] or the reaction of vinyl chloride with chloroform [2–4]. The first method is uncompetitive, for the preparation of 1,1,3-trichloropropane is also difficult and not in favor of environment. The latter is a classic telomerization. Compared with the telomerization of olefin and tetrachloromethane [5–8], chloroform is much less reactive, and the highest yield that has been reported is less than 10% [2].

During the past few years, researchers all confirmed that the effect of solvents was not remarkable [9, 10]; they focused on effective catalyst system [11–16]. Catalysts of telomerization must be able to produce free radicals, such as organic initiator, UV light, electron beam, and oxyreductive initiator, in which amine-metallic salts are widely regarded as an effective initiating system to synthesize halogenated alkanes.

The key to enhance yield is to increase the contacting opportunity of reactants and stabilize the corresponding intermediates. So, different from other researchers' efforts

to search effective initiators, solvent effects and reaction conditions were investigated in our research and a high yield of more than 50% was achieved.

2. Experiments

Vinyl chloride (purity > 99%) was directly used from Sumitomo Seika chemicals Co. Ltd. Chloroform was purified by distillation after water wash, the purity being examined by gas chromatography. Nitrogen was produced in our institute and its purity was more than 99.9%. Other reagents were obtained from Wako Company without further purification.

GC analysis was performed on Shimadzu 17A instrument with an HP CP-PoraBond Q column (25 m × 0.32 mm). Mass spectrum was measured on Shimadzu QP2010 mass spectrometers. ^{1}H NMR and ^{13}C NMR spectra were recorded on JEOL NMR 300 instruments, using TMS as internal standard and $CDCl_3$ as solvent. UV-visible absorption spectral measurements were carried out using a recording double beam spectrophotometer (Shimadzu, UV-2600, Japan). Sonication was performed in Bransonic 5510J-MT ultrasonic cleaner with the frequency of 42 Hz and an output power of 135 W.

The reaction was carried out in a 500 mL stainless steel autoclave with a magnetic stirrer. Chloroform, initiators, and

TABLE 1: Effects of catalyst systems[a].

Entry	Initiator	Solvents	Conversion[b] (%)	Yield[b] (%)
1	$FeCl_2 \cdot 4H_2O$, $(HOCH_2CH_2)_3N$	$MeOH^c$	73.4	5.7
2	$CuCl \cdot 2H_2O$, $(HOCH_2CH_2)_3N$	CH_3CN	11.4	8.7
3	$FeCl_2 \cdot 4H_2O$, $(HOCH_2CH_2)_3N$	CH_3CN	22.1	18.3
4	$FeCl_2 \cdot 4H_2O$, $(HOCH_2CH_2)_3N$	DMF	35.9	24.6
5	$FeCl_2 \cdot 4H_2O$, $(HOCH_2CH_2)_3N$	DMSO	40.5	28.0
6	$FeCl_2 \cdot 4H_2O$, $(HOCH_2CH_2)_3N$	DAMC	67.7	40.5
7	$FeCl_2 \cdot 4H_2O$	DAMC	64.7	42.1

[a]The molar ratio of vinyl chloride, chloroform, catalysts, and solvent is 1 : 4 : 0.02 : 0.2, respectively; reaction time 30 h; temperature 130°C; initial pressure 0.3 MPa.
[b]It is calculated by vinyl chloride. [c]Methanol is the same molar amount as chloroform.

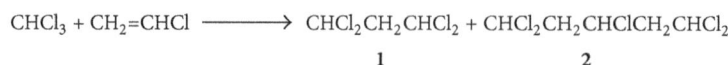

$$CHCl_3 + CH_2=CHCl \longrightarrow CHCl_2CH_2CHCl_2 + CHCl_2CH_2CHClCH_2CHCl_2$$
$$1 \qquad\qquad\qquad 2$$

SCHEME 1

solvents were placed in the autoclave. Then autoclave was cooled down by liquid nitrogen and vacuumed to 20 Pa below. Quantitative vinyl chloride was inhaled to the autoclave through vacuum line, and then nitrogen was charged to maintain 0.03 MPa. The autoclave was heated to a prescribed temperature. After the reaction, autoclave was cooled to room temperature and mixtures were washed by water. The organic layer was dried by anhydrous Na_2SO_4, then filtered, and subjected to GC and MS analysis. Unconverted vinyl chloride and excess chloroform were recovered by distillation at atmosphere pressure. Distillation at 20 kPa and 53°C gave the product, transparency liquid with the boiling point of 162°C:

^{1}H NMR (300 MHz, $CDCl_3$): δ 3.08–3.12 (t, J = 5.0, 2H), 5.89–5.93 (t, 2H);

^{13}C NMR (75.5 MHz, $CDCl_3$): δ 55.29, 68.66;

MS: m/z = 145 (M^+–Cl,7), 109 (100), 97 (8), 83 (29), 61 (8), 49 (7), 39 (8).

As shown in the NMR spectra, the ^{1}H-NMR spectrum of the product in $CDCl_3$ shows two triplets which are assigned as follows: δ 3.08–3.12 (CH_2), δ 5.89–5.93 (CCl_2H). The ^{13}C-NMR spectra of the product show two easily recognized signals at about δ 55.29 (CH_2) and δ 68.66 (CCl_2H), respectively.

3. Results and Discussion

In this study of addition reaction of vinyl chloride with chloroform, amine-metallic salts were used as a model compound. In all cases, 1 : 1 adduct **1** was obtained as the aim product, and the 1 : 2 adduct **2** was the byproduct, with the structures presented in Scheme 1.

Several catalyst systems with different solvents were examined and the results were summarized in Table 1.

Entry 1 was examined using methanol (MeOH) as solvent according to the patent [15]. The conversion is high, while little aim product was observed, for most vinyl chloride was polymerized to form white spherical solids. Low yield can be explained that methanol is a protonic solvent with the ability of absorbing free radicals, causing hindrance of

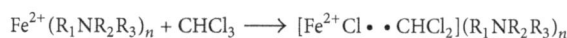

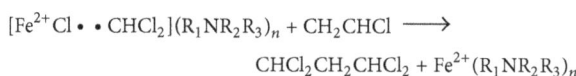

$$Fe^{2+} + nR_1NR_2R_3 \longrightarrow Fe^{2+}(R_1NR_2R_3)_n$$

$$Fe^{2+}(R_1NR_2R_3)_n + CHCl_3 \longrightarrow [Fe^{2+}Cl \bullet \bullet CHCl_2](R_1NR_2R_3)_n$$

$$[Fe^{2+}Cl \bullet \bullet CHCl_2](R_1NR_2R_3)_n + CH_2CHCl \longrightarrow$$
$$CHCl_2CH_2CHCl_2 + Fe^{2+}(R_1NR_2R_3)_n$$

SCHEME 2: Catalytic mechanism of ferrous iron.

the telomerization. Compared with entries 2 and 3, ferrous chloride is much more effective than cuprous chloride, so ferrous chloride-diethanolamine was used to evaluate the effect of solvents.

In the previous work, methanol, propanol, heptane, dioxane, and benzene were all tried, and the effects were not remarkable [9, 15, 17, 18]. In our experiment, non-protonic solvents with better solubility of both metallic salt and chloroform, such as acetonitrile (CH_3CN), N,N-dimethylformamide (DMF), dimethyl sulfoxide (DMSO), and N,N-dimethylacetamide (DAMC), were tried (entries 3–7), and the yield was enhanced to 42.1%.

According to the effective chlorine-transfer mechanism [19], amine can form the complex with ferrous ion, which is the actual initiator. Assuming the existence of such complexes, the reaction scheme shown in Scheme 2 would follow.

Organic amine was used as a stabilizer shown in Scheme 2. The complex as a reaction intermediate would be formed through the interaction of chlorine radical with ferrous ion and the stabilization by coordinated amines. The attack of vinyl chloride on this complex affords trichloropropyl radical, and the resulting radical instantly abstracts chlorine radical from the complex. The presence of such complexes greatly decreases activation energy for the insertion of vinyl chloride into the Cl-$CHCl_2$ bond. DMAC, as both an effective complexing agent and organic amine, can highly catalyze the telomerization, which was also verified by entry 7.

In order to further confirm our thoughts, UV-visible spectra of solutions of DMAC, $FeCl_2 \cdot 4H_2O$, and their mixture in C_2H_5OH were measured as shown in Figure 1, and

TABLE 2: Effects of reaction conditions[a].

Entry	Condition	Temperature (°C)	Initial pressure (MPa)	Time (h)	Conversion[b] (%)	Yield[b] (%)	
						1	2
1	Silent	110	0.3	30	16.9	14.0	1.4
2	Silent	130	0.3	30	64.7	42.1	15.6
3	Silent	130	0.1	30	24.6	21.0	2.2
4	Silent	130	0.3	20	46.0	27.8	3.7
5	Silent	130	0.3	40	69.2	43.6	18.9
6	)))[c]	130	0.3	30	73.2	54.1	18.2
7	)))	100	0.3	8	52.3	50.1	0.7

[a]The molar ratio of vinyl chloride, chloroform, $FeCl_2 \cdot 4H_2O$, and DMAC is $1:4:0.02:0.2$. [b]It is calculated by vinyl chloride. [c]Ultrasonic irradiation.

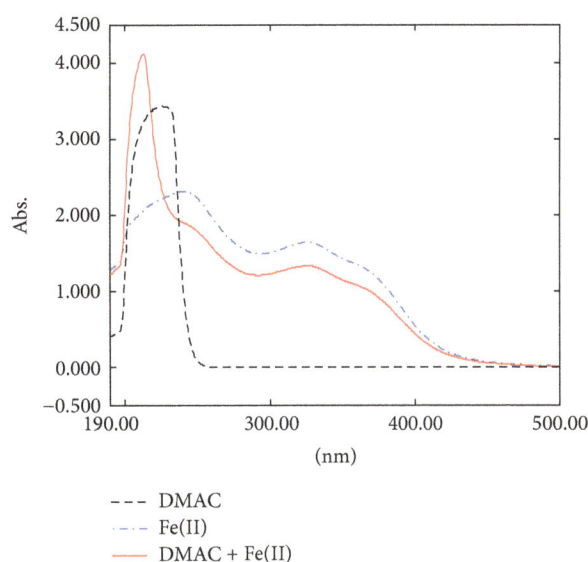

--- DMAC
·−·− Fe(II)
—— DMAC + Fe(II)

FIGURE 1: UV-visible spectra of reactants.

spectra were recorded against solvent blank from 190 to 500 nm. It was confirmed that the absorption maxima of DMAC, $FeCl_2 \cdot 4H_2O$, and their mixture was, respectively, 224 nm, 241 nm, and 213 nm, which indicated that a new complex was formed. It is the complex that actually catalyzes the telomerization.

As shown in Table 1, the difference between conversion and yield is mainly due to the formation of the 1 : 2 adduct **2**. To improve the selectivity, we extended the reaction conditions and the results were presented in Table 2.

Reaction temperature offered the energy for the telomerization and the lower temperature sharply decreased the yield (entries 1 and 2). High pressure is needful during the formation of telomeres for raw materials were all in the gaseous state. The yield of telomeres increased with the elongation of reaction time from 20 h to 40 h.

As shown in entries 1–5, high yield usually accompanies low selectivity, which is mainly caused by high reaction temperature and long reaction time. In recent years, ultrasound has been employed in various chemical transformations with considerable enhancement in rate and yield and, in several cases, facilitates organic transformations at ambient conditions [20, 21]. So, ultrasound was introduced to continue our research.

The pronounced effect of ultrasound can be easily demonstrated from the results shown by the entries 6 and 7. It is expected that these radical generation processes are facilitated by ultrasonic irradiation. For the mild reaction, the ultrasound mediated telomerization has led to high yields in a relatively short period with high selectivity.

4. Conclusion

A breakthrough was made in the telomerization of vinyl chloride with chloroform initiated by ferrous chloride-dimethylacetamide under ultrasonic irradiation. With trace byproducts, 1,1,3,3-tetrachloropropane was gained in 51.9% yield, more than four times the one reported before. The important achievements of our work are (1) DMAC selected as both solvent and cocatalyst; (2) reduction of the reaction time and temperature by ultrasound introduction; (3) control over generation of byproduct; (4) remarkable increase in the yield of products. With the use of this new catalyst system, the telomerization of less reactive materials, such as chloroform, on a large scale may become possible.

Conflict of Interests

The authors declare that there is no conflict of interests regarding the publication of this paper.

Acknowledgments

The authors would like to express their thanks to Professor J. Mizukado for useful discussions and Dr. Y. Suzuki for assistance in the measurements.

References

[1] A. M. Whaley and H. W. Davis, "Preparation of 1,1,3,3-tetrachloropropane," *Journal of the American chemical society*, vol. 73, no. 3, pp. 1382–1383, 1951.

[2] A. Teruzo, S. Manabu, and C. C. Wu, "Telomerization of olefin with chloroform initiated by amine and metal salt," *Indian Journal of Chemistry*, vol. 72, no. 8, pp. 1818–1822, 1969 (Japanese).

[3] B. Boutevin, Y. Pietrasanta, M. Taha, and T. El Sarraf, "Synthesis of macromers of vinylchloride and vinylidene chloride from telomers," *Polymer Bulletin*, vol. 10, no. 3-4, pp. 157–161, 1983.

[4] B. Boutevin, Y. Pietrasanta, and M. Taha, "Télomères mono-fonctionnels du chlorure de vinyle—I: synthèse et caractérisation d'étalons de télomères du chlorure de vinyle," *European Polymer Journal*, vol. 18, no. 8, pp. 675–678, 1982.

[5] M. Kotora and M. Hájek, "Selective additions of polyhalogenated compounds to chloro substituted ethenes catalyzed by a copper complex," *Reaction Kinetics & Catalysis Letters*, vol. 44, no. 2, pp. 415–419, 1991.

[6] N. P. Zhiryukhina, A. A. Kamyshova, E. T. Chukovskaya, and R. K. Freidlina, "Synthesis of polychloroalkanes with several different chlorine-containing groups," *Bulletin of the Academy of Sciences of the USSR*, vol. 32, no. 1, pp. 129–133, 1983.

[7] V. Mathieu and J. P. Schoebrechts, "Process for preparing halogenated hydrocarbons," U.S. Patent no. 6,500,993, 2002.

[8] A. Lambert and V. Mathieu, "Process for preparing halohydrocarbons in the presence of a co-catalyst," U.S. Patent 6,452,057, 2002.

[9] A. Teruzo, S. Manabu, and C. C. Wu, "Telomerization of vinyl chloride with carbon tetrachloride using aming-cupric chloride. Effects of amines and solvent," *Bulletin of the Chemical Society of Japan*, vol. 43, no. 4, pp. 1127–1131, 1970.

[10] M. Asscher and D. Vofsi, "Chlorine-activation by redox-transfer. Part III. The "Abnormal" addition of chloroform to olefins," *Journal of the Chemical Society*, pp. 3921–3927, 1963.

[11] A. Teruzo, S. Manabu, and O. Noritaka, "Telomerization of ethylene with chloroform initiated by N-chloroalkylamines," *Bulletin of the Chemical Society of Japan*, vol. 47, no. 8, pp. 2007–2010, 1974.

[12] A. Teruzo, S. Manabu, and O. Noritaka, "Reactions of ethylene with chloroform initiated by amine-ferrous chloride system," *Bulletin of the Chemical Society of Japan*, vol. 47, no. 12, pp. 3142–3145, 1974.

[13] J. R. Briggs, H. Hagen, S. Julka, and J. T. Patton, "Palladium-catalyzed 1,3-butadiene telomerization with methanol. Improved catalyst performance using bis-o-methoxy substituted triarylphosphines," *Journal of Organometallic Chemistry*, vol. 696, no. 8, pp. 1677–1686, 2011.

[14] I. Chung, "Monte Carlo simulation of free radical telomerization," *Polymer*, vol. 41, no. 15, pp. 5643–5651, 2000.

[15] I. Wlassics and V. Tortelli, "Tetrafluoroethylene telomerization using dibromohaloethanes as telogens," *Journal of Fluorine Chemistry*, vol. 127, no. 2, pp. 240–248, 2006.

[16] N. A. Porter, R. Breyer, E. Swann et al., "Free-radical telomerization of chiral acrylamides: control of stereochemistry in additions and halogen-atom transfer," *Journal of the American Chemical Society*, vol. 113, no. 18, pp. 7002–7010, 1991.

[17] J. P. Schoebrechts, V. Mathieu, and F. Janssens, "Method for the preparation of 1,1,1,3,3-pentachlorobutane," US Patent 6,399,840, 2002.

[18] X. Cheng, C. P. Zhang, and L. Jian, "Advances in the catalysts for synthesisi of halohydrocarbons by liquid-phase catalytic telomerization," *Industrial Catalysis*, vol. 15, no. 11, pp. 11–14, 2007.

[19] M. Asscher and D. Vofsi, "Chlorine activation by redox-transfer, part I: the reaction between aliphatic amines and carbon tetrachloride," *Journal of the Chemical Society*, vol. 439, pp. 2261–2264, 1961.

[20] Q. Hua, Y. Zhiwen, and L. Chunxu, "Ultrasonically promoted nitrolysis of DAPT to HMX in ionic liquid," *Ultrasonics Sonochemistry*, vol. 15, no. 4, pp. 326–329, 2008.

[21] Q. Hua, L. Dabin, and L. Chunxu, "Ultrasonically-promoted synthesis of mandelic acid by phase transfer catalysis in an ionic liquid," *Ultrasonics Sonochemistry*, vol. 18, no. 5, pp. 1035–1037, 2011.

24

Synthesis and Structural Studies of a New Complex of Di[hexabromobismuthate (III)] 2,5-Propylaminepyrazinium [$C_{10}H_{28}N_4$]Bi_2Br_{10}

Mohamed El Mehdi Touati and Habib Boughzala

Laboratoire de Matériaux et Cristallochimie, Faculté des Sciences, Université de Tunis El Manar, 2092 Tunis, Tunisia

Correspondence should be addressed to Habib Boughzala; boughzala@yahoo.com

Academic Editor: Jolanta N. Latosinska

A new organic-inorganic hybrid material, [$C_{10}H_{28}N_4$]Bi_2Br_{10}, has been synthesized and characterized. The compound crystallizes in monoclinic P2_1/c space group with $a = 11.410(4)$ Å, $b = 11.284(4)$ Å, $c = 12.599(3)$ Å, $\beta = 115.93(2)°$, and $V = 1458.8(8)$ Å^3. The structure consists of discrete dinuclear [Bi_2Br_{10}]$^{4-}$ anions and [$C_{10}H_{28}N_4$]$^{4+}$ cations. It consists of a 0-D anion built up of edge-sharing bioctahedron. The crystal net contains N–H$\cdots$Br hydrogen bonds. The differential scanning calorimetry (DSC) reveals an irreversible phase transition at $-17°$C. The frontier molecular orbital and the energy gap between the highest occupied molecular orbital (HOMO) and the lowest unoccupied molecular orbital (LUMO) calculation allow the classification of the material as an insulator.

1. Introduction

The results of systematic structural investigations of halobismuthate (III) compounds reveal a great variety of different anionic frameworks. Most of these compounds are described by a general formula R^{+a} ($M_b X_{3b+a}$)$^{-a}$ (where R is organic cations, M is Bi, and X is Cl, Br, and I) and have a tendency to constitute bi- or polynuclear anions in the crystalline state.

Generally, in these compounds, the coordination sphere of bismuth appears to be dominated by the tendency towards hexacoordination with polybismuthate species arising from corner, edges, or faces sharing BiX_6 distorted octahedra.

The formation of the anionic sublattice is clearly determined by the counteractions, but the effects of their most evident properties such as charge, size, and shape are almost not predictable. The organic moiety can be used as physical and electronic barrier, contributing to original electrical and optical behaviour. In addition, since, in the crystal state, important contribution to the lattice stabilization is due to hydrogen bonding interactions, it should be possible to influence the bismuth coordination geometry acting on the number and orientation of the hydrogen bond donor sites of the cations [1–4].

2. Synthesis Experimental Protocol

The title compound was synthesized by dissolving stoichiometric amounts of bismuth (III) bromide in piperazine in a mixture of water and HBr. The resulting solution was stirred well and then kept at room temperature. Few weeks later, transparent crystals, as bright-yellow prism, were grown by slow evaporation. The purity of synthesized compound was improved by successive recrystallization process.

3. Results and Discussion

3.1. X-Ray Data Collection. The X-ray diffraction intensities from a single crystal of about (0.3 × 0.3 × 0.1) mm^3 were collected with a CAD-4 (Enraf Nonius) diffractometer using the Mo Kα radiation ($\lambda = 0.71073$ Å). The crystal structure was solved by direct methods using SHELXS-97 [5]. Full-matrix F^2 least-squares refinement and subsequent Fourier synthesis procedures were performed using SHELXL-97 [6]. Molecular graphics were prepared using Diamond 3 [7]. CCDC-1008226 contains the supplementary crystallographic data for this compound. These data can be obtained free of charge at http://www.ccdc.cam.ac.uk/conts/retrieving.html or from

TABLE 1: Experimental data for $C_{10}H_{28}N_4Bi_2Br_{10}$.

Crystal data		Data collection	
Empirical formula	$C_{10}H_{28}N_4Bi_2Br_{10}$	Diffractometer	Enraf-Nonius CAD-4
Formula weight	1417.42 (g·mol^{-1})	Wavelength: λ	0.71073 Å
Crystal system	Monoclinic	θ range	$2.55° \leq \theta \leq 26.97°$
Space group	P2$_1$/c	Temperature	298 (2) K
Unit cell dimensions	$a = 11.410$ (4) Å $b = 11.284$ (4) Å $c = 12.599$ (3) Å $\beta = 115.93$ (2)°	Limiting indices h, k, and l	$-14 \leq h \leq 13$ $0 \leq k \leq 14$ $0 \leq l \leq 16$
Crystal habit	Yellow prism	Absorption correction	Psi scan
Volume	$V = 1458.8$ (8) Å^3	Standard reflection	2 every 120 min
Z	$Z = 4$	Measured reflections	3697
Absorption coefficient	$\mu = 25.75$ mm^{-1}	Independent reflections	3177
$F(000)$	1264	Observed reflections $I > 2\sigma(I)$	2352
Crystal size (mm^3)	$(0.3 \times 0.3 \times 0.1)$	R_{int}	0.062
Density$_{calc}$	$D_{calc} = 3.236$	T_{min}/T_{max}	0.001/0.074
Refinement			
Refinement	Least square on F^2	$\Delta\rho_{max}$	4.95 eÅ^{-3}
R$[F^2 > 2\sigma(F^2)]$	0.066	$\Delta\rho_{min}$	$-7,09$ eÅ^{-3}
S	1.06	wR	0.169

the Cambridge Crystallographic Data Centre (CCDC), 12 Union Road, Cambridge CB2 1EZ, UK.

Hydrogen atoms were located at their idealized positions using appropriate HFIX instructions in SHELXL-97 and included in subsequent least-squares refinement cycles in riding-motion approximation. Crystal data and parameters of the final refinement are reported in Table 1.

3.2. Crystal Morphology. Crystal morphology is a key element in many industrial processes and has an enormous impact in the materials processing stages. Thus, rationalization of the relationships between crystal morphology and the arrangement of atoms in the bulk crystal lattice is of great interest in many areas of science.

The crystal morphology prediction was obtained by BFDH (Bravais-Fridel and Donnay-Harker) [8–10] algorithm calculation using Mercury (CSD 3.0.1) [11].

The program uses the crystal lattice parameters and the symmetry space group to generate a list of possible growth faces and their relative growth rates.

The qualitative analysis result obtained by energy dispersive X-ray spectroscopy (EDX) is presented in Figure 1. It reveals the presence of the chemical elements identified by the single crystal X-ray diffraction.

The scanning electron microscopy with energy dispersive X-ray spectroscopy (SEM/EDX) high resolution images of the surface topography produces an image of the focused crystal. The view of the observed and calculated crystal morphologies reveals a similarity between the two shapes (Figure 2). This result allows identifying the crystallographic axis and shows the absence of preferential orientation of crystallites.

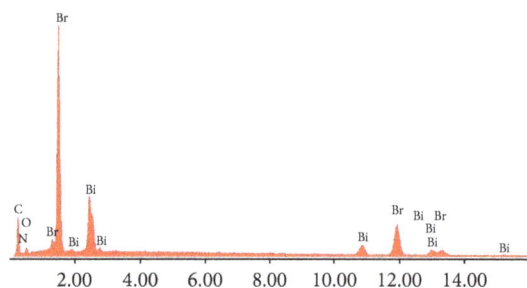

FIGURE 1: Qualitative analysis by EDX of $C_{10}H_{28}N_4Bi_2Br_{10}$.

3.3. Structure Description. At room temperature, the present compound crystallizes in the monoclinic P2$_1$/c space group. The asymmetric unit contains two bromobismuthate $[Bi_2Br_{10}]^{4-}$ anions and $[C_{10}H_{28}N_4]^{4+}$ cation, as shown in Figure 3.

The organic and inorganic moieties are linked by N–H$\cdots$Br hydrogen bonds ensuring the structure cohesion.

The 1 : 5 stoichiometry of anionic part $[Bi_2Br_{10}]^{4-}$ can be realized by different types of anionic sublattices.

In this compound, two BiBr$_6$ octahedra share two bridging Br atoms and consequently form dinuclear $[Bi_2Br_{10}]^{4-}$ anion. One bismuth (III) ion is surrounded by six bromine anions; however, Bi–Br distances fall into two ranges: 2.793(2) to 2.894(2) Å for terminal Br and 2.981(4) to 3.079(2) Å for the bridging ones (Table 2). Lower than the sum of Van der Waals radii (4.35 Å according Pauling [12]), we deduce that the bismuth-bromine bonds have a dominant covalent character.

(a)   (b)

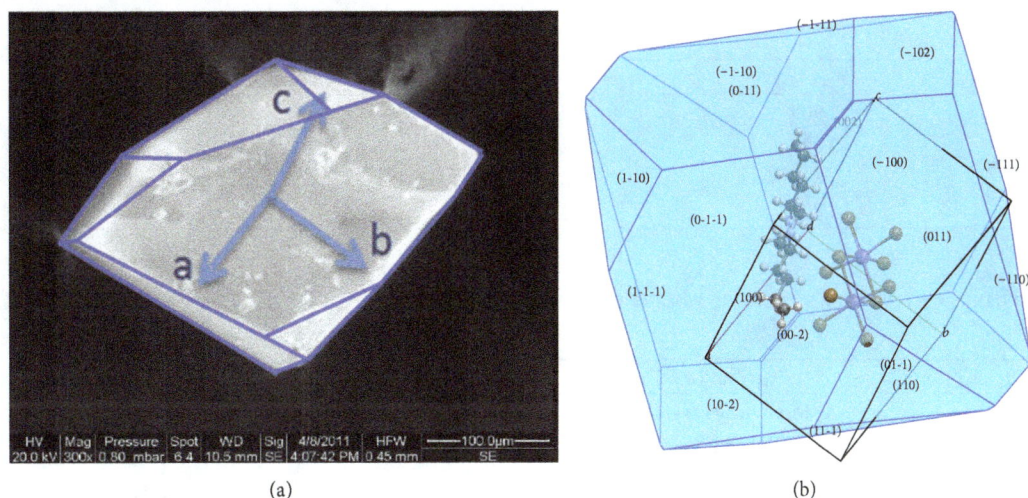

FIGURE 2: Observed and calculated morphologies of $C_{10}H_{28}N_4Bi_2Br_{10}$ crystals.

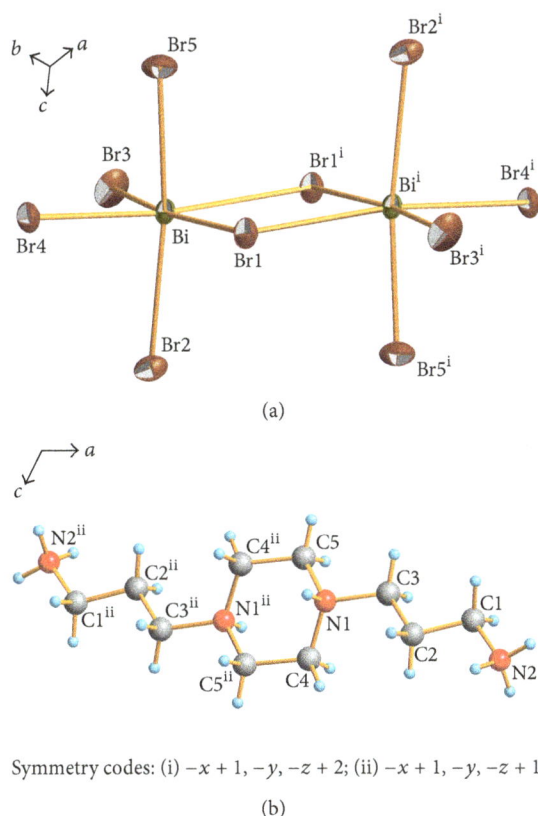

(a)

Symmetry codes: (i) $-x+1, -y, -z+2$; (ii) $-x+1, -y, -z+1$

(b)

FIGURE 3: ORTEP of the inorganic part $[Bi_2Br_{10}]^{4-}$ in (a) and the organic moiety $[C_{10}H_{28}N_4]^{4+}$ in (b) with 50% of probability level.

TABLE 2: Selected interatomic distances (Å) and angles (°) in the structure of $C_{10}H_{28}N_4Bi_2Br_{10}$.

(a) Octahedra BiBr$_6$

Bi–Br4	2.744 (2)	Br5–Bi–Br2	169.15 (6)
Bi–Br3	2.748 (3)	Br4–Bi–Br1	98.30 (6)
Bi–Br5	2.793 (2)	Br3–Bi–Br1	170.84 (6)
Bi–Br2	2.894 (2)	Br5–Bi–Br1	84.48 (7)
Bi–Br1	2.981 (4)	Br2–Bi–Br1	91.96 (6)
Bi–Br1^i	3.079 (2)	Br4–Bi–Br1^i	170.99 (5)
Br1–Bi^i	3.079 (2)	Br3–Bi–Br1^i	82.72 (6)
Br4–Bi–Br3	90.61 (7)	Br5–Bi–Br1^i	90.14 (6)
Br4–Bi–Br5	84.30 (6)	Br2–Bi–Br1^i	100.00 (6)
Br3–Bi–Br5	94.36 (7)	Br1–Bi–Br1^i	88.19 (5)
Br4–Bi–Br2	86.09 (6)	Bi–Br1–Bi^i	91.81 (5)
Br3–Bi–Br2	90.75 (7)		

(b) Organic group

C1–N2	1.490 (2)	N2–C1–C2	111.30 (1)
C1–C2	1.500 (2)	C1–C2–C3	109.00 (1)
C2–C3	1.500 (2)	C2–C3–N1	113.30 (1)
C3–N1	1.501 (2)	N1–C4–C5^{ii}	113.60 (1)
C4–N1	1.454 (2)	N1–C5–C4^{ii}	112.50 (1)
C4–C5^{ii}	1.520 (2)	C4–N1–C3	115.10 (1)
C5–N1	1.510 (2)	C4–N1–C5	110.00 (1)
C5–C4^{ii}	1.520 (2)	C3–N1–C5	109.50 (1)

Symmetry codes: (i) $-x+1, -y, -z+2$; (ii) $-x+1, -y, z+1$.

The BiBr$_6$ octahedra are somewhat distorted. As described by Shannon [13], the distortion index of this polyhedra ($ID_{(Bi-Br)} = 1.88 \cdot 10^{-3}$) indicates a significant dissymmetry in the dinuclear entity $[Bi_2Br_{10}]^{4-}$

With ID $(Bi-Br) = \dfrac{1}{6}\sum_{i=1}^{6}\left(\dfrac{(BiBr)_i - (BiBr)_m}{(BiBr)_m}\right)^2$,   (1)

where $(BiBr)_m$ is the average value of the Bi–Br bond length. The bond angles values (see Table 3) confirm the octahedral distortion since they are, in some cases, 10° less than the ideal values.

The intermolecular N–H···Br hydrogen bonds can be also the reason of geometrical distortion of $[Bi_2Br_{10}]^{4-}$ anion, due to possibility of shifting of the halogen atoms in the direction of the positive charge located on the cations.

TABLE 3: Hydrogen bonds for $C_{10}H_{28}N_4Bi_2Br_{10}$ (D: donor; A: acceptor).

D–H$\cdots$A	D–H (Å)	H$\cdots$A (Å)	D$\cdots$A (Å)	D–H$\cdots$A (°)
N1–H1$\cdots$Br5	0.909	2.779 (3)	3.534 (1)	141.27 (7)
N2–HA$\cdots$Br3	0.890	2.703 (2)	3.537 (2)	156.38 (1)
N2–HB$\cdots$Br2	0.890	2.792 (2)	3.521 (2)	149.03 (1)
N2–HB$\cdots$Br4	0.890	2.829 (2)	3.487 (2)	131.92 (1)

TABLE 4: Infrared bands observed and assigned to vibration modes for $C_{10}H_{28}N_4Bi_2Br_{10}$.

Observed frequencies (cm^{-1})	Attributions
3470	Stretching (N–H)
2925	Stretching (C–H)
1640	Bending (N–H)
1389	Bending (C–H) and stretching (C–C)
1114	Stretching (C–N)

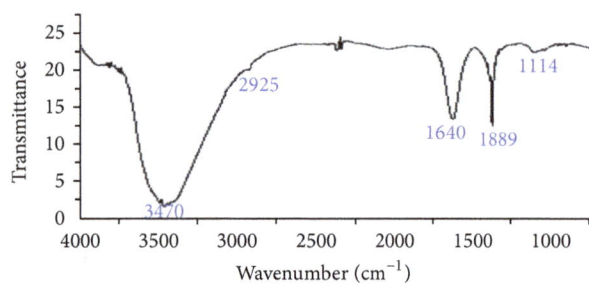

FIGURE 4: Observed IR spectrum of the compound $C_{10}H_{28}N_4Bi_2Br_{10}$.

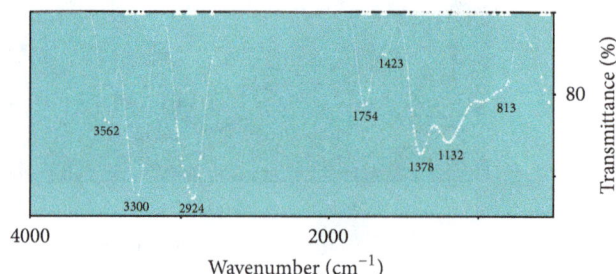

FIGURE 5: Calculated IR spectrum of the compound $C_{10}H_{28}N_4Bi_2Br_{10}$.

The protonation of $[C_{10}H_{28}N_4]$ leads to $[C_{10}H_{28}N_4]^{4+}$. These cations are stacked between the planes containing the bioctahedrons $[Bi_2Br_{10}]^{4-}$.

Each bioctahedron is sandwiched between two sheets of organic cations, setting out its 6 vertices Br$^-$ where three links to the higher plane and the others with the lower one by hydrogen bonds involve the terminal amino group of the organic cations.

Hydrogen bonds are linking the organic and inorganic moieties (Table 3). This fact can explain the observed fragility of the crystals. The hydrogen bonds ensure the crystal cohesion by connecting the alternating organic-inorganic layers and building a three-dimensional framework.

3.4. Infrared Spectroscopy.
The IR spectrum of $C_{10}H_{28}N_4Bi_2Br_{10}$ (Figure 4 and Table 4) was recorded at room temperature in the range of 400 to 4000 cm^{-1} using the VERTEX 80/80 v FT-IR research spectrometer, by dispersing 2% of the studied compound in KBr discs. We have calculated the vibrational spectrum by using semiempirical PM3 geometry optimization by "CAChe" program [14].

After an optimization of the molecular configuration, the calculated spectrum, presented in Figure 5, is very helpful for the attribution of the observed spectroscopic bands. On the other hand, the observed bands assignment becomes easier by comparing the observed frequencies and those calculated.

Based on the previous literature results and the theoretical simulation of the IR spectrum, the large band around 3470 cm^{-1} is attributed to the stretching modes of (N–H) in the amine group. The out of plane bending mode of this group is probably responsible for the band located at 1640 cm^{-1}. The (C–H) stretching of methylene group is centered on 2925 cm^{-1}.

The band around 1389 cm^{-1} is probably the result of the bending vibration of (C–H) and the stretching of (C–C). The stretching modes of (C–N) are probably observed around 1114 cm^{-1}.

3.5. Thermal Properties.
The differential scanning calorimetry (DSC) thermogram, shown in Figure 6 was performed with a DSC EVO131 instrument under nitrogen atmosphere. Firstly, no phase transition was observed when the sample was under cooling until $-45°$C; then, during heating from -45 to $20°$C, we observed an energetic effect that reveals a phase transition in the temperature range of $(-17/-12)°$C accompanied by a significant enthalpy transition (ΔH) evaluated at 2.732 Jg^{-1}. Comparing the heating and the cooling curves, the observed phenomenon seems to be irreversible. Further treatments are scheduled to know the structural behaviour of the resulting compound after the DSC experiment.

3.6. X-Ray Powder Diffraction.
The X-ray powder diffraction technique (XRD) was used to control the crystalline phase purity. The diffraction pattern was obtained on a D8 ADVANCE Bruker diffractometer with a Lynxeye accelerator using Cu (Kα_1/α_2 = 1.54060/1.54439 Å) wavelength. The measurement was performed in spinning mode (60 tr·mn^{-1}) in order to minimize the preferential orientation effect of the crystallites with step-scanning ($\Delta 2\theta = 0.02°$) constant time

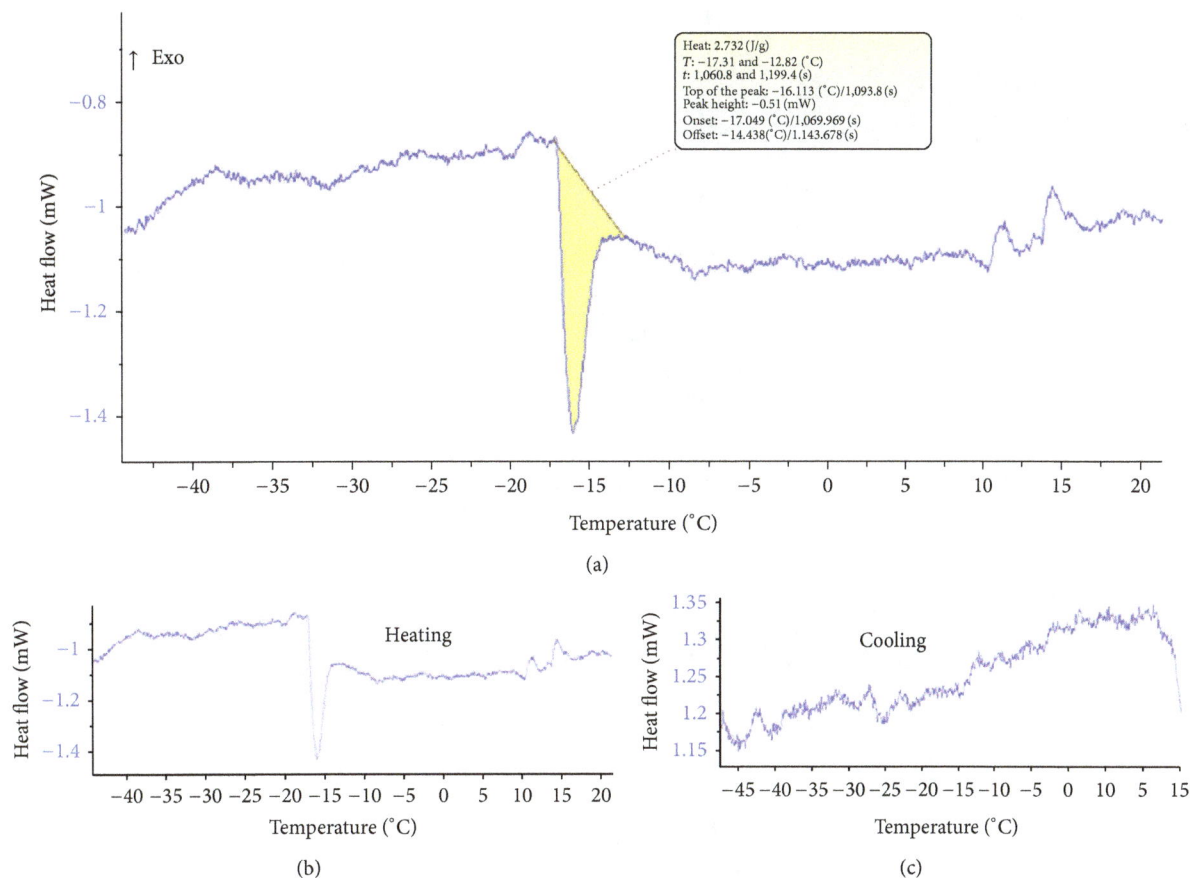

FIGURE 6: Differential scanning calorimetry curves of $C_{10}H_{28}N_4Bi_2Br_{10}$.

interval of 0.1 s. The quantitative criteria of goodness of fits are the following agreement R factors:

$$Rp = \frac{\sum |Y_i \cdot (\text{obs}) - (1/c)\,Y_i\,(\text{cal})|}{\sum Y_i \cdot (\text{obs})},$$

$$Rwp = \sqrt{\frac{\sum w_i\,[Y_i \cdot (\text{obs}) - (1/c)\,Y_i \cdot (\text{cal})]^2}{w_i\,[Y_i \cdot (\text{obs})]^2}},$$ (2)

$$GOF = \sqrt{\frac{\sum w_i\,[Y_i \cdot (\text{obs}) - (1/c)\,Y_i \cdot (\text{cal})]^2}{N - p}},$$

where $Y_i(\text{obs})$ and $Y_i(\text{cal})$ are the observed and calculated intensities at the ith step in the pattern, respectively. w_i is the reciprocal of the variance of each observation; the summation is carried out over all the observations and "c" is a scale factor.

The refinements were carried out using TOPAS program [15].

The basic structural model for the $C_{10}H_{28}N_4Bi_2Br_{10}$ was taken from Topa et al. output [16]. Details of the refinement are given in Table 5.

Figure 7 shows good agreement between the observed and calculated XRD patterns which confirms the crystalline purity of the prepared compound with an experimental error of 3% of mass. Furthermore, the XRPD raw diffraction is

TABLE 5: Unit cell parameters and details of Rietveld refinement of $C_{10}H_{28}N_4Bi_2Br_{10}$.

Formula	$C_{10}H_{28}N_4Bi_2Br_{10}$
System	Monoclinic
Space group	$P2_1/c$
Z	4
Unit cell	
$a/Å$	11.387 (1)
$b/Å$	11.260 (1)
$c/Å$	12.577 (1)
$\beta/(°)$	115.925 (3)
Volume/Å^3	1450.45 (2)
Density	3.245 (5)
Zero point $2\theta/(°)$	0.0182 (8)
Reliability factors (%)	
Rp	3.85
Rwp	5.89
R_B	4.917
R_F	2.22

marked with the presence of a large hump centered around $2\theta = 13°$, signature of an amorphous part. The TOPAS degree of crystallinity calculation gives about 86%.

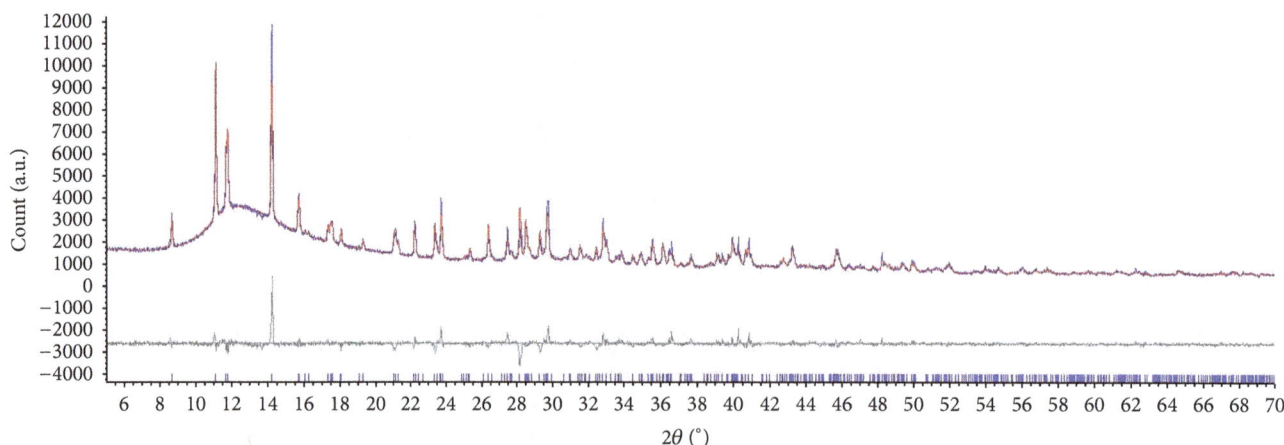

FIGURE 7: Experimental and calculated X-ray diffraction patterns and their difference for $C_{10}H_{28}N_4Bi_2Br_{10}$.

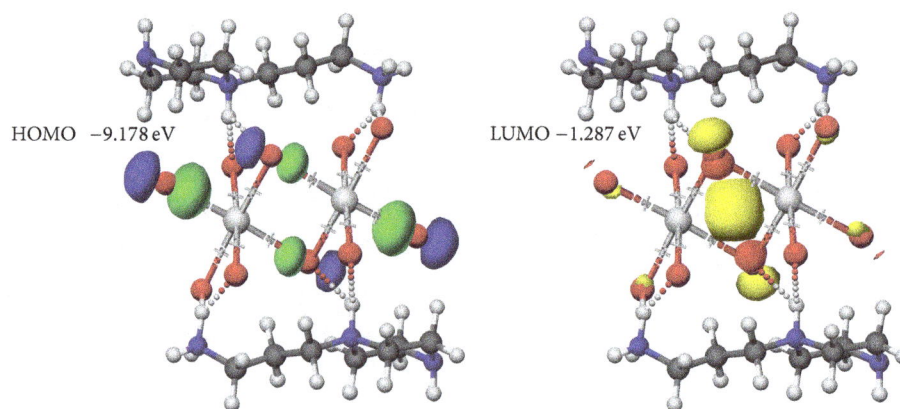

FIGURE 8: Calculated frontier molecular orbital of the title compound.

3.7. The HOMO-LUMO Gap. Crystalline materials can be classified according to their band gap. The $C_{10}H_{28}N_4Bi_2Br_{10}$ exhibits absorption bands around $(-1,287) - (-9,179) = 7,891$ eV calculated by the program "CAChe" using the semiempirical PM3 method and corresponding to the electronic transition from the highest occupied molecular orbital (HOMO) to the lowest unoccupied one (LUMO). The atomic orbital compositions of the frontier molecular orbital are sketched in Figure 8.

We can note the HOMO contribution coming from the halogen. In the LUMO, the percentage of contribution from the bromine atoms is dominant. These calculations show that the electronic properties are roughly imposed by the inorganic part. The Bi coordination geometry is the dominant structural factor influencing the electronic structure of studied compound [17].

4. Conclusion

The present paper has shown that the new organic-inorganic hybrid $C_{10}H_{28}N_4Bi_2Br_{10}$ was synthesized by slow evaporation. Its structure is built up by dibutylpyrazinium dications and discrete (0-D) bromobismuthate anions. Several experimental techniques have been used to characterize the new compound.

The crystal structure was solved by single crystal X-ray diffraction. The vibrational properties were studied by Raman scattering and infrared spectroscopy; the crystal morphology was carries out using the Bravais-Freidel and Donnay-Harker model, and the X-ray powder diffraction measurement was carried out to check the title compound's purity.

The crystal structure of $C_{10}H_{28}N_4Bi_2Br_{10}$ consists of discrete $[Bi_2Br_{10}]^{4-}$ anions with dinuclear geometry of two $BiBr_6$ octahedra sharing two bridging Br atoms and $[C_{10}H_{28}N_4]^{4+}$. The cohesion is assumed by hydrogen bonds. The title compound undergoes one low-temperature phase transitions at $-17°C$ identified by differential scanning calorimetry.

Conflict of Interests

The authors declare that there is no conflict of interests regarding the publication of this paper.

References

[1] A. Piecha, V. Kinzhybalo, R. Jakubas, J. Baran, and W. Medycki, "Structural characterization, molecular dynamics, dielectric and spectroscopic properties of tetrakis(pyrazolium) bis(μ_2-bromo-tetrabromobismuthate(III)) dihydrate, $[C_3N_2H_5]_4 \cdot [Bi_2Br_{10}] \cdot 2H_2O$," *Journal of Coordination Chemistry*, vol. 9, pp. 1036–1048, 2007.

[2] G. A. Bowmaker, P. C. Junk, M. Lee Aaron, B. W. Skelton, and A. H. White, "Synthetic, structural and vibrational spectroscopic studies in bismuth(III) halide/*N,N'*-Aromatic bidentate base systems. I large-cation (2,2'-bipyridinium and 1,10-phenanthrolinium) salts of polyhalobismuthate (III) ions," *Australian Journal of Chemistry*, vol. 51, no. 4, pp. 293–309, 1998.

[3] F. Benetollo, G. Bombieri, G. Alonzo, N. Bertazzi, and G. A. Casella, "Synthesis and crystal structure of [(PhenH)(PhenH$_2$)]. [BiCl$_6$]·2H$_2$O with different o-phenanthroline protonations," *Journal of Chemical Crystallography*, vol. 28, no. 11, pp. 791–796, 1998.

[4] G. Alonzo, F. Benetollo, N. Bertazzi, and G. Bombieri, "Synthesis and crystal structures of 4,4'-bipyridinium and 2,2'-bipyridinium pentachloro complexes containing the polynuclear anions [Bi2Cl10]4- and [Bi4Cl20]8-," *Journal of Chemical Crystallography*, vol. 29, no. 8, pp. 913–919, 1999.

[5] G. M. Sheldrick, *SHELXS-97, Program for Crystal Structure Solution*, University of Göttingen, Göttingen, Germany, 1997.

[6] G. Sheldrick M, *SHELXL-97: Program for the Refinement of Crystal Structures*, University of Göttingen, Göttingen, Germany, 1997.

[7] K. Brandenburg, *Diamond*, Crystal Impact GbR, Bonn, Germany, 2008.

[8] A. Bravais, *Etudes Cristallographiques*, Gauthier-Villars, Paris, France, 1913.

[9] G. Fridel, "Crystal habits of minerals," *Bulletin de la Société Chimique de France*, pp. 326–455, 1907.

[10] J. D. Donnay and H. Harker D, *Springer Handbook of Crystal Growth*, American Mineralogist, 1937.

[11] (CCDC), *Mercury CSD 3. 0. 1 (Build RC6)*, Cambridge Crystallographic Data Centre (CCDC), 2011.

[12] L. Pauling, *The Nature of Chemical Bond*, Cornell University Press, Ithaca, NY, USA, 1960.

[13] R. D. Shannon, "Revised effective ionic radii and systematic studies of interatomic distances in halides and chalcogenides," *Acta Crystallographica A*, vol. 32, part 5, pp. 751–767, 1976.

[14] *Cache: Worksystem 7.5.0.85*, Fujitsu Limited, Oxford, UK, 2000–2006.

[15] A. A. Coelho, *TOPAS, Version 4.2 (Computer Software)*, Coelho Software, Brisbane, Australia, 2009.

[16] D. Topa, E. Makovicky, H. J. Schimper et al., "The crystal structure of SnSb4S7, A new member of the meneghinite homologous series," *Canadian Mineralogist*, vol. 48, no. 5, pp. 1119–1126, 2010.

[17] J. E. Monat, J. H. Rodriguez, and J. K. McCusker, "Ground- and excited-state electronic structures of the solar cell sensitizer bis(4,4'-dicarboxylato-2,2'-bipyridine)bis(isothiocyanato)ruthenium(II)," *Journal of Physical Chemistry A*, vol. 106, no. 32, pp. 7399–7406, 2002.

25

Zr(IV), La(III), and Ce(IV) Chelates with 2-[(4-[(Z)-1-(2-Hydroxyphenyl)ethylidene]aminobutyl)-ethanimidoyl]phenol: Synthesis, Spectroscopic Characterization, and Antimicrobial Studies

M. M. El-ajaily,[1] H. A. Abdullah,[2] Ahmed Al-janga,[3] E. E. Saad,[2] and A. A. Maihub[4]

[1]Department of Chemistry, Faculty of Science, Benghazi University, Benghazi, Libya
[2]Department of Chemistry, Faculty of Science, Sebha University, Sebha, Libya
[3]Department of Zoology, Faculty of Science, Sebha University, Sebha, Libya
[4]Department of Chemistry, Faculty of Science, Tripoli University, Tripoli, Libya

Correspondence should be addressed to M. M. El-ajaily; melajaily@yahoo.com

Academic Editor: Mahmut Ulusoy

La(III), Zr(IV), and Ce(IV) chelates of 2-[(4-[(Z)-1-(2-hydroxyphenyl)ethylidene]aminobutyl)-ethanimidoyl]phenol were synthesized and characterized by using several physical techniques. The Schiff base was obtained by refluxing of o-hydroxyacetophenone with 1,4-butanediamine in 2:1 molar ratio. The CHN elemental analysis results showed the formation of the Schiff base and the chelates has been found to be in 1:1 [M:L] ratio. The molar conductance measurements revealed that all the chelates are nonelectrolytes. Structural elucidations of the ligand and its chelates were based on compatible analytical and spectroscopic evidences. The infrared spectral data revealed that the Schiff base coordinates to the metal ions through active sites which are –OH and –C=N groups. According to the electronic spectral data, an octahedral geometry was proposed for the chelates. The synthesized ligand and its metal chelates were screened for their antimicrobial activity against two Gram negative (*Escherichia coli*, *Salmonella kentucky*) and two Gram positive (*Lactobacillus fermentum*, *Streptococcus faecalis*) bacterial strains, unicellular fungi (*Fusarium solani*), and filamentous fungi (*Aspergillus niger*). The activity data showed that the metal chelates have antibacterial and antifungal activity more than the parent Schiff base ligand against one or more bacterial or fungi species. The results also indicated that the metal chelates are higher sensitive antimicrobial agents as compared to the Schiff base ligand.

1. Introduction

Schiff bases are most widely used as chelating ligands in coordination chemistry and have been investigated extensively for the past several decades leading to new synthetic routes of structure, biological, and industrial applications [1]. They are also useful in catalyst chemistry and in medicine (pharmacology) as antibiotic, antiallergic, and antitumor agents [2]. The complexes of the type MLXn, where M=VO(IV), Mn(II), Fe(III), Co(II), Ni(II), Zn(II) and Cd(II), X=H_2O/Cl⁻, and L is the Schiff base ligand derived from 2,4-dihydroxy-5-acetylacetophenone and 1,4-diaminobutane have been synthesized and isolated in solid state; they are stable in air. The physicochemical data suggested a square pyramidal structure for VO(IV), pseudo octahedral structure for Cu(II), and an octahedral structure for Mn(II), Fe(III), Co(II), Ni(II), and Cd(II) complexes. The ligand field parameters have been calculated and related to the electronic environment. The Schiff base and its complexes were screened for their antimicrobial activities against various bacteria and fungi [3]. The lanthanide(III) complexes of the chloro, hydroxo substituted 14-membered tetraazamacrocyclic solid complexes of La(III), Ce(III), and Pr(III) have been synthesized and characterized. From the microanalytical data, the stoichiometry of the complexes has been found to be 1:1 (metal:ligand). The TGA-DSC data suggested that all lanthanide(III) complexes have one ionic nitrate, two coordinated nitrate ions, two water molecules for Ce(III), and four water molecules for La(III) and Pr(III). The X-ray diffraction data suggested an orthorombic crystal system for La(III) and monoclinic crystal system for Ce(III) and Pr(III) complexes [4].

In the present study we have reviewed the synthesis and characterization of Schiff base and its chelates with Zr(IV), La(III), and Ce(IV) ions. The antimicrobial activity of the Schiff base and its chelates also was screened against some pathogenic bacteria and fungi.

2. Experimental

2.1. Materials and Methods. The chemicals used in this investigation are of pure grade (Merck or Aldrich), including 2-hydroxyacetophenone, 1,4-diaminobutane, $ZrOCl_2 \cdot 8H_2O$, $La(NO_3)_3 \cdot 6H_2O$, and $Ce(SO_4)_2 \cdot 4H_2O$, C_2H_5OH, $CHCl_3$, DMF, DMSO, and NH_4OH solution. The synthesized Schiff base and its chelates were subjected to CHN elemental analyses using Perkin-Elmer 2400 elemental analyzer, infrared spectra were obtained by KBr disc technique by using IFS-25DPUSR\IR spectrometer (Bruker) in the range of 4000–400 cm^{-1}, and proton nuclear magnetic resonance spectrum of the Schiff base was recorded on Varian Gemini 200-200 MHz spectrometer using TMS as internal standard and D^6 DMSO as a solvent (Figure 11). The electronic spectra of the Schiff base and its Zr(IV), La(III), and Ce(IV) chelates were measured in DMSO solvent using a Perkin-Elmer-Lambda β-spectrophotometer. The mass spectra were carried out by using Shimadzu QP-2010 Plus. The molar conductivity of the chelates was measured in DMSO solvent using digital conductivity meter CMD 650, at Chemistry Department, Sebha University, Sebha, Libya. All the mentioned analyses were done at Micro Analytical Center, Cairo University, Giza, Egypt.

2.2. Preparation of the Schiff Base (SB). A Schiff base (**SB**) of 2-[(4-[(Z)-1-(2-hydroxyphenyl)ethylidene]aminobutyl)-ethanimidoyl]phenol (**1**) was synthesized by mixing an ethanolic solution of o-hydroxyacetophenone (2.40 g, 0.02 mole) and 1,4-diaminobutane (0.88 g, 0.01 mole). The reaction mixture was left under refluxing for two hours. The formed solid product was separated by filtration, purified by crystallization from ethanol, and dried under vacuum over anhydrous calcium chloride [5]. The pale lemon-yellow product was produced in 85.61% yield and its melting point is in the range of 192–194°C.

2.3. Synthesis of Schiff Base Chelates. The Schiff base metal chelates were synthesized by dropwise addition of an ethanolic solution of the metal salts (0.01 mole; 3.22, 4.33 and 4.04 g) of $ZrOCl_2 \cdot 8H_2O$, $La(NO_3)_3 \cdot 6H_2O$, and $Ce(SO_4)_2 \cdot 4H_2O$ to the same ratio of an ethanolic solution of the Schiff base in a 1 : 1 [M : L] ratio. If the chelates were not isolated, a few drops of ammonia solution were added to adjust the pH = 7-8. The reaction mixtures were refluxed for four hours, then collected, and washed several times with hot ethanol until the filtrates become clear. The chelates were dried in desiccators

over anhydrous calcium chloride. The yields of the chelates were in the range of 63.76–86.42% and their melting points are above 300°C.

2.4. Antimicrobial Assays. The *in vitro* biological sensitivity of different compounds conjugated with Schiff base and its chelates was studied by Bauer et al. [6] paper disc diffusion method; they were tested against the bacterial species and fungal species.

2.5. Antibacterial Assay. The antibacterial activity of the Schiff base and its chelates was studied against two Gram positive bacteria (*Lactobacillus fermentum and Streptococcus faecalis*) and Gram negative bacteria (*Escherichia coli, Salmonella kentucky*) [7]. A stock solution of the metal chelate dissolved in DMSO solvent was prepared at a concentration of 1 mg/mL. Whatman filter papers number 1 were cut and sterilized in autoclave. These paper discs were soaked in 10 μL of different concentrations of the ligand/chelates (5, 50, and 500 μg/mL) solutions in DMSO solvent as negative control and were then placed aseptically in the Petri dishes containing Mueller Hinton Agar (MHA, Oxoid) inoculated with the above mentioned two bacteria separately. The Petri dishes were incubated at 37 ± 1°C and the inhibition zones were recorded after 20 h of incubation. The obtained results were compared with known antibiotic, ciprofloxacin. Three replicates were taken and average values are given in Tables 3 and 4.

2.6. Antifungal Assay. The compounds were screened for their antifungal activity against fungi, namely, *Aspergillus niger* and *Fusarium solani* [8]. These fungal species were isolated from the infected parts of the host plants, that is, Saburaud Dextrose Agar (SDA). The compounds were tested at the concentrations of the ligand/chelates (5, 50, and 500 μg/mL) in DMSO solvent prepared from a solution at a concentration of 1 mg/mL and compared with a known antibiotic as control. Miconazole was prepared for testing against spore germination of each fungus. The culture of fungi was purified by single spore isolation technique. Filter paper discs of 6 mm size, prepared by using Whatman filter paper number 1, were cut and sterilized in an autoclave. These paper discs were soaked with 10 μL of the compounds dissolved in DMSO solvent as negative control. The fungal culture plates were inoculated and incubated at 25 ± 2°C for 40 h are presented in Table 5. Since all the tested compounds and standard drugs were prepared in freshly distilled DMSO solvent, its zone of inhibition was found to be very negligible and taken as zero mm. Activity was determined by measuring the diameter of the zone showing complete inhibition (mm). Growth inhibition was compared with standard drugs. The plates were then observed and the diameters of the inhibition zones (in mm) were measured and tabulated and the activity index was also calculated by using the following equation (see [8, 9]):

$$*\% \text{ Activity index} = \frac{C}{T} \times 100, \quad (1)$$

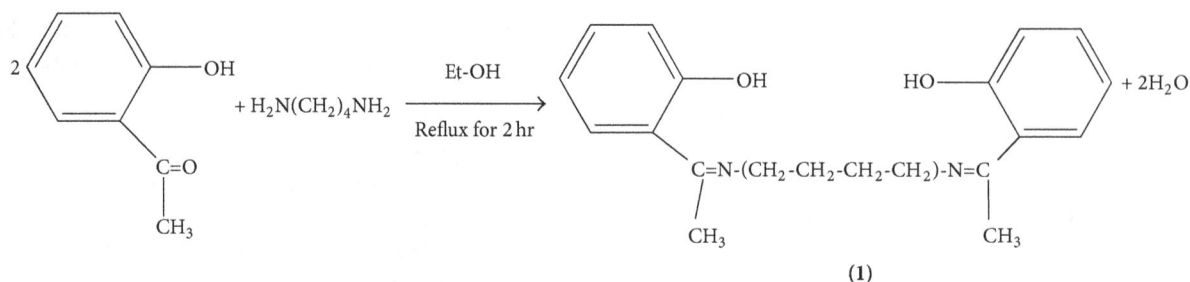

FIGURE 1: Synthetic scheme for construction of 2-[(4-[(Z)-1-(2-hydroxyphenyl)ethylidene]aminobutyl)-ethanimidoyl]phenol.

TABLE 1: CHN elemental analysis data and some physical properties of Schiff base ligand and its chelates.

Schiff base/chelate	Mol.Wt	Colour	m.p. (°C)	Yield %	Elemental analysis% Cal (Obs)			Λ^*
					C%	H%	N%	
SB	324.43	Lemon-yellow	194	85.61	74.20 (74.00)	7.38 (7.46)	8.69 (8.64)	—
[Zr(L)(OH$_2$)$_2$]·4H$_2$O	537.84	Shinny white	>300	63.76	38.14 (38.24)	7.37 (7.45)	4.45 (4.55)	5.28
[La(L)(OH$_2$)$_2$]·5H$_2$O	587.55	Yellow canary	>300	73.91	47.99 (48.52)	7.51 (7.39)	3.66 (3.62)	4.31
[Ce(L)(OH)$_2$]·3H$_2$O	550.56	Canary	>300	67.70	34.43 (35.13)	7.08 (7.09)	4.01 (4.33)	2.49

$\Lambda^* = \Omega^{-1} \, cm^2 \, mol^{-1}$.

where AI% means activity index, C diameter of zone inhibition of microorganisms in check, and T diameter of the disc (the zone of inhibition was measured after 18–20 hrs). Ciprofloxacin (5 μg disc^{-1}) and miconazole (25 μg disc^{-1}) were used as positive standard.

3. Results and Discussion

Schiff base ligand (1) (Figure 1) was obtained by the reaction of o-hydroxyacetophenone and 1,4-diaminobutane in (2 : 1) ratio. Only one type of Schiff base compound (1) was formed.

3.1. CHN Elemental Analyses and Molar Conductivity. It is found that the found data are in good agreement with those theoretical ones (Table 1). The newly synthesized Schiff base chelates are very stable in air and generally soluble in DMF and DMSO solvents. The CHN elemental analytical data of the chelates reveal that the chelates are formed in 1 : 1 [M : L] ratio. The molar conductance values of the chelates were determined by using 10^{-3} M concentration in DMSO solvent, and their values (Table 1) suggest nonelectrolyte nature [10, 11].

3.2. Thermogravimetric Analysis of Ce(IV) Chelate. The thermogravimetric analysis was performed to assist in predicting the molecular structures of the chelates and the weight losses were measured from the ambient temperature up to 1000°C using a heating rate of 10°C/min [12]. The TG curve of Ce(IV) chelate of the formula [Ce(L)(OH)$_2$]·3H$_2$O exhibits three steps of decomposition. The first step of decomposition involves the elimination of three hydrated water molecules

at weight loss of 10.00% (calcd. 9.81%) at temperature of 50–315°C. In the second step the decomposition occurs at temperature of 315–762°C indicating the loss of the Schiff base as carbonate or oxalate ion [13], whereas, in the final step, stable state was observed above 762°C indicating the presence of thermally stable residual metal oxide. The weight percentage of residual metal oxide (CeO$_2$) was found to be 50.00% which is very close to the theoretical value 48.15% [14] (Figure 2). Thermal stability places the chelate in the following order:

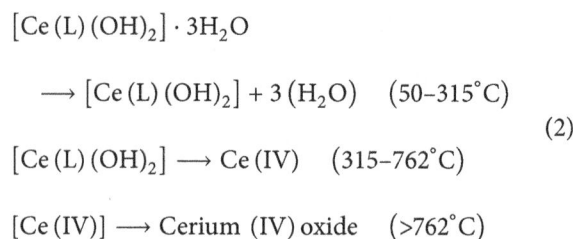

$$[Ce(L)(OH)_2] \cdot 3H_2O$$
$$\longrightarrow [Ce(L)(OH)_2] + 3(H_2O) \quad (50\text{–}315°C)$$
$$[Ce(L)(OH)_2] \longrightarrow Ce(IV) \quad (315\text{–}762°C)$$
$$[Ce(IV)] \longrightarrow Cerium(IV) \, oxide \quad (>762°C)$$

(2)

3.3. Infrared Spectra. The IR spectra of the chelates were compared with those of the free ligand in order to determine the involvement of coordination sites in chelation. Characteristic peaks in the spectra of the ligand and chelates were considered and compared. The significant IR bands for the Schiff base ligand as well as for its metal chelates and their tentative assignments are compiled and represented in Table 2 and Figures 3–6. In the IR spectrum 2-[(4-[(Z)-1-(2-hydroxyphenyl)ethylidene]aminobutyl)-ethanimidoyl]phenol exhibits a band of the azomethine (C=N) at 1612 cm^{-1}.

TABLE 2: Infrared band assignments (cm^{-1}) and electronic spectral data (nm, cm^{-1}) of Schiff base (SB) and its chelates.

Comp.	IR spectral data, cm^{-1}							Electronic spectral data, **nm** (cm^{-1})
	ν(OH)	ν(C=N)	ν(CH$_3$)	ν(CH)	ν(CH$_2$)	ν(M–O)	ν(M–N)	
SB	3416	1612	1308	755	2932	—	—	**269** (37175), **329** (30395)
[Zr(L)(OH$_2$)$_2$]·4H$_2$O	3413	1617	1330	863	2921	527	444	**249** (40160), **322** (31055)
[La(L)(OH$_2$)$_2$]·5H$_2$O	3545	1642	1339	819	2926	565	424	—
[Ce(L)(OH)$_2$]·3H$_2$O	3402	1633	1428	748	3079	617	482	—

TABLE 3: Effect of used salts and Schiff base and its chelates on bacteria growth.

		Diameter of inhibition zone (6 mm); concentration in μg/mL											
		Gram negative organisms											
S.N	Comp.	Escherichia coli						Salmonella kentucky					
		% Activity index						% Activity index					
		5 μg/mL	50 μg/mL	500 μg/mL	5 μg/mL	50 μg/mL	500 μg/mL	5 μg/mL	50 μg/mL	500 μg/mL	5 μg/mL	50 μg/mL	500 μg/mL
1B	HAPH	0	0	0	0	0	0	5	8	10	13	20	25
2B	BD	0	10	14	0	25	35	7	11	19	28	35	45
3B	ZrOCl$_2$·8H$_2$O	0	0	0	0	0	0	0	0	16	0	0	40
4B	La(NO$_3$)$_3$·6H$_2$O	8	13	18	20	33	45	8	11	16	23	31	46
5B	Ce(SO$_4$)$_2$·4H$_2$O	0	0	8	0	0	20	8	10	13	23	29	37
1C	SB	0	0	0	0	0	0	0	0	0	0	0	0
2C	La(III) chelate	0	0	0	0	0	0	0	0	0	0	0	0
3C	Zr(IV) chelate	0	8	10	0	20	25	0	0	0	0	0	0
4C	Ce(IV) chelate	0	0	8	0	0	20	0	0	8	0	0	23
C	DMSO (Control)	0	0	0	0	0	0	0	0	0	0	0	0
CIP	Ciprofloxacin (Standard)	40	40	40	100	100	100	35	35	35	100	100	100

HAPH (o-hydroxyacetophenone), BD (1,4-butanediamine), *microgram per mL.
$I\% = C - T/C \times 100$.

TABLE 4: Effect of used salts and Schiff base and its chelates on bacteria growth.

		Diameter of inhibition zone (6 mm); concentration in μg/mL											
		Gram positive organisms											
S.N	Comp.	Lactobacillus fermentum						Streptococcus faecalis					
		% Activity index						% Activity index					
		5 μg/mL	50 μg/mL	500 μg/mL	5 μg/mL	50 μg/mL	500 μg/mL	5 μg/mL	50 μg/mL	500 μg/mL	5 μg/mL	50 μg/mL	500 μg/mL
1B	HAPH	5	8	10	13	20	25	0	1	2	0	3	6
2B	BD	11	14	18	28	35	45	1	3	11	3	9	31
3B	ZrOCl$_2$·8H$_2$O	0	0	16	0	0	40	0	0	0	0	0	0
4B	La(NO$_3$)$_3$·6H$_2$O	0	0	16	0	0	40	0	7	15	0	20	43
5B	Ce(SO$_4$)$_2$·4H$_2$O	0	9	15	0	23	38	8	13	16	23	37	46
1C	SB	0	0	0	0	0	0	0	3	6	0	9	17
2C	La(III) chelate	0	0	0	0	0	0	0	2	16	0	6	46
3C	Zr(IV) chelate	0	0	0	0	0	0	0	0	9	0	0	26
4C	Ce(IV) chelate	0	0	0	0	0	0	0	0	0	0	0	0
C	DMSO (Control)	0	0	0	0	0	0	0	0	0	0	0	0
CIP	Ciprofloxacin (Standard)	40	40	40	100	100	100	35	35	35	100	100	100

HAPH (o-hydroxyacetophenone), BD (1,4-butanediamine), *microgram per mL.
$I\% = C - T/C \times 100$.

TABLE 5: Effect of used salts and Schiff base and its chelates on fungi growth.

		Diameter of inhibition zone (6 mm); concentration in μg/mL										
		Fungi										
S.N	Comp.	Aspergillus niger						Fusarium solani				
					% Activity index						% Activity index		
		5 μg/mL	50 μg/mL	500 μg/mL	5 μg/mL	50 μg/mL	500 μg/mL	5 μg/mL	50 μg/mL	500 μg/mL	5 μg/mL	50 μg/mL	500 μg/mL
1B	HAPH	0	0	19	0	0	90	0	0	0	0	0	0
2B	BD	11	17	23	52	81	109	0	0	0	0	0	0
3B	ZrOCl$_2$·8H$_2$O	0	0	4	0	0	19	0	0	0	0	0	0
4B	La(NO$_3$)$_3$·6H$_2$O	0	0	3	0	0	14	0	0	0	0	0	0
5B	Ce(SO$_4$)$_2$·4H$_2$O	0	1	8	0	5	38	0	0	0	0	0	0
1C	SB	0	0	0	0	0	0	0	0	0	0	0	0
2C	La(III) chelate	0	0	0	0	0	0	0	0	0	0	0	0
3C	Zr(IV) chelate	0	0	0	0	0	0	0	0	0	0	0	0
4C	Ce(IV) chelate	0	0	0	0	0	0	0	0	0	0	0	0
C	DMSO (Control)	0	0	0	0	0	0	0	0	0	0	0	0
CIP	Ciprofloxacin (Standard)	21	21	21	100	100	100	13	13	13	100	100	100

HAPH (o-hydroxyacetophenone), BD (1,4-butanediamine), *microgram per mL.
I% = C − T/C × 100.

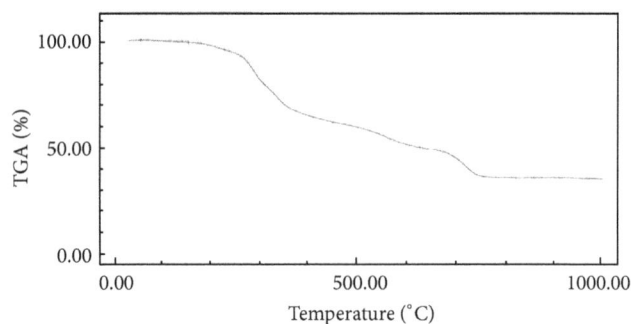

FIGURE 2: Thermogravimetric analysis of Ce(IV) chelate.

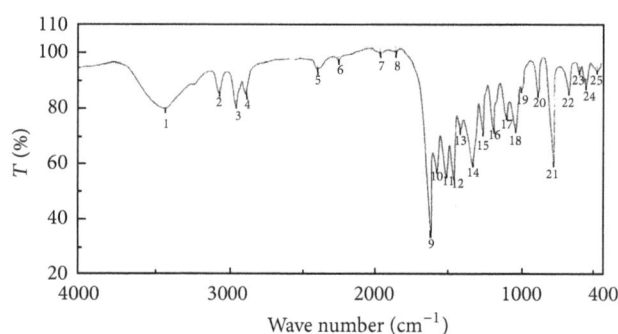

FIGURE 3: Infrared spectrum of the 2-[(4-[(Z)-1-(2-hydroxyphenyl) ethylidene]aminobutyl)-ethanimidoyl]phenol.

The shifting of this band during the chelate formation (Table 2) indicates its involvement in coordination with the metal ions through nitrogen atom of the azomethine group [15, 16]. The appearance of broad bands in the IR spectra of the Schiff base ligand and its chelates in the range of 3402–3545 cm^{-1} is due to the existence of water molecules [17]. The spectrum of Zr(IV) chelate shows a band at 1030 cm^{-1} which could be due to the presence of Zr=O group [18]. New bands in the range of 527–617 cm^{-1} and 424–482 cm^{-1} which are not present in the free Schiff base are due to v(M-O) and v(M-N) vibration [19], and the appearance of these vibrations supports the involvement of nitrogen and oxygen atoms of the azomethine and C-OH groups in chelation with the metal ions.

3.4. Electronic Spectra and Magnetic Moments of the Chelates. The electronic spectral data of 2-[(4-[(Z)-1-(2-hydroxy-phenyl)ethylidene]aminobutyl)-ethanimidoyl]phenol and its

Zr(IV), La(III), and Ce(IV) chelates were recorded in DMSO solvent and shown in Table 2 and Figures 7–10. In the spectrum of the Schiff base it exhibits two bands at 269 nm (37175 cm^{-1}) and 329 nm (30395 cm^{-1}) assigned to $\pi \rightarrow \pi^*$ and $n \rightarrow \pi^*$ transitions [20]. In the metal chelates, the octahedral chelates that contain a metal ion of d^0 electronic configuration are diamagnetic. The electronic spectrum of Zr(IV) chelate displays two absorption bands at 249 nm (40161 cm^{-1}) and 322 nm (31056 cm^{-1}) due to charge transfer transition [21]. The f-f transitions of the complexes are characteristic of the lanthanide and are not influenced by the ligand. Intensity of the peaks also varies according to the metal ion. The f-f bands are sharp and line-like. This is because of the effective shielding of the 4f orbital by the 5s, 5p octet and consequently minimum ligand field perturbation

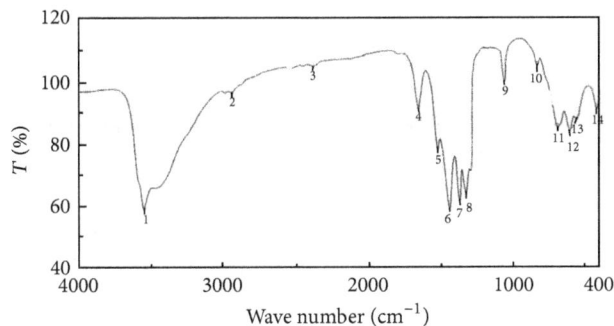

FIGURE 4: Infrared spectrum of the Zr(IV) chelate.

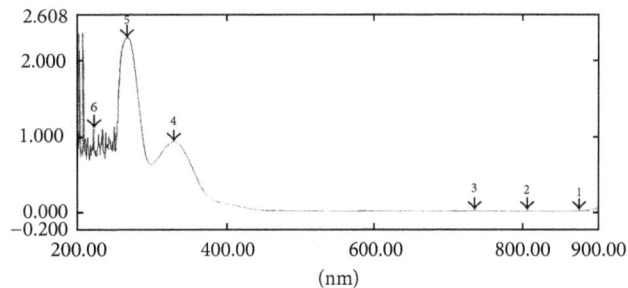

FIGURE 7: Ultraviolet spectrum of the 2-[(4-[(Z)-1-(2-hydroxy-phenyl)ethylidene]aminobutyl)-ethanimidoyl]phenol.

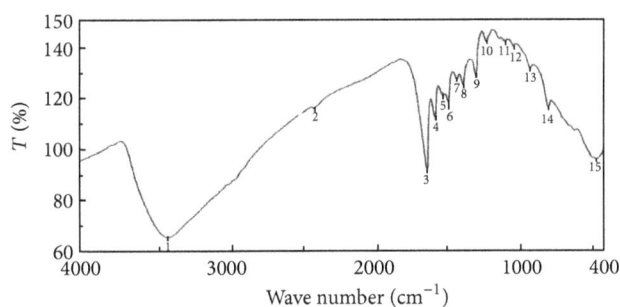

FIGURE 5: Infrared spectrum of the La(III) chelate.

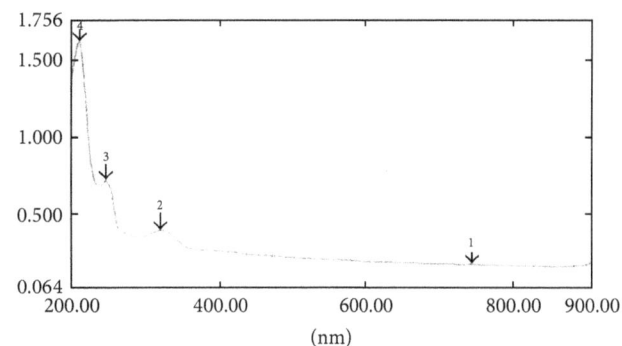

FIGURE 8: Electronic spectrum of the Zr(IV) chelate.

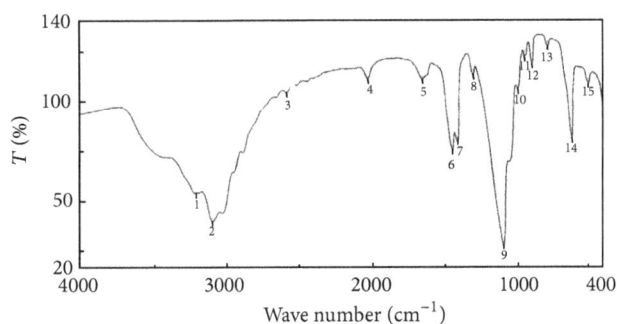

FIGURE 6: Infrared spectrum of Ce(IV) chelate.

of the electronic energy levels in lanthanides. La(III) has no observable visible spectra and Ce(IV) has no transition in this region [22]. The observed f-f transitions in 500–700 nm region and the tentative assignments are given in Table 2.

3.5. ^1HNMR Spectra of the Zr(IV) Chelate.
^{1}HNMR spectra of Schiff base and its Zr(IV) chelate were recorded in d^6-DMSO solvent. The ^{1}HNMR spectrum of the Schiff base 2-[(4-[(Z)-1-(2-hydroxyphenyl)ethylidene]aminobutyl)-ethanimidoyl]phenol in DMSO solvent shows signals of methyl protons at 2.368 ppm (s, 3H) and –CH at 6.712–7.323 (s, H). In addition, peak appeared at 2.25 ppm (s, 3H) which is assigned to methyl proton [23]. The spectrum shows sharp peak at

16.625 (S, 1H) due to OH of 2-hydroxyacetophenone moiety indicating the formation of Schiff base ligand system, but in the case of Zr(IV) chelate, this group disappeared indicating the involvement of phenolic oxygen in the coordination via deprotonation. The appearance of a new singlet peak at δ 3.316 ppm is assigned to the presence of methyl group. The multiplet signals observed in the region at 6.987–7.556 ppm are assigned to the aromatic protons [24, 25].

3.6. Mass Spectra of Schiff Base and Schiff Base Chelates.
The electron impact mass spectra of Schiff base ligand and chelates are recorded and investigated at 70 eV of electron energy. The mass spectra of the studied Schiff base and its chelates are shown in Schemes 1–4, and Figures 12–15. Fragment at m/z = 325 is due to the original molecular weight of the free Schiff base (SB). The fragment of Schiff base at m/z = 191 corresponds to $C_{12}H_{17}NO$ ion. Whereas the fragment at m/z 136 is analogous to $C_8H_{10}NO$ ion, the other molecular ion fragments that appeared in the mass spectrum are attributed to two fragmentations; first fragment at m/z 121 is corresponding to the loss of methyl group, and the second fragment at m/z 112 is analogous to the loss of C_2H_3N from the compound. The last fragment displays two fragments at m/z 80 and m/z 57 attributed to loss of different atoms which are shown in Scheme 1. The last fragment at m/z 55 is attributed to $C_4H_7^-$ ion. For $[Zr(L)(OH_2)_2]\cdot4H_2O$ chelate, the spectrum exhibits a fragment at m/z = 464

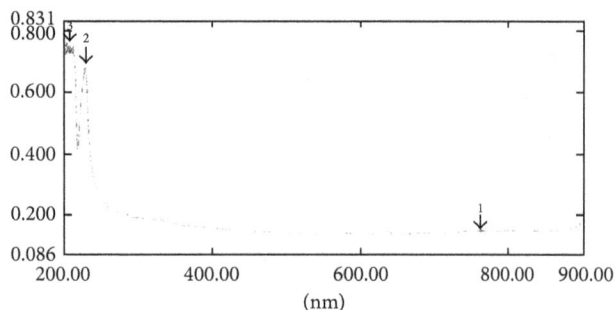

FIGURE 9: Electronic spectrum of the La(III) chelate.

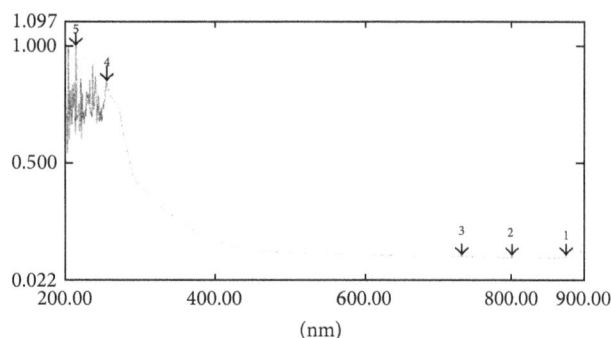

FIGURE 10: Electronic spectrum of the Ce(IV) chelate.

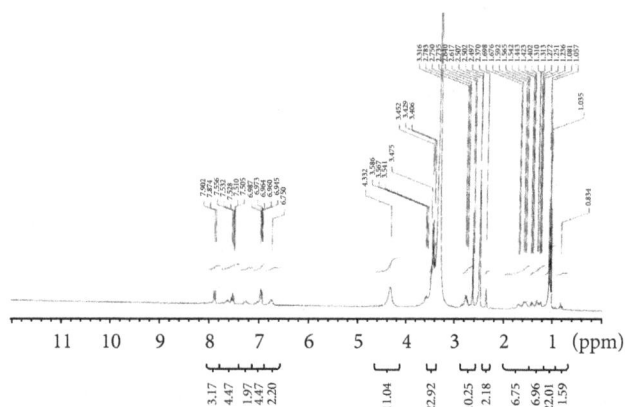

FIGURE 11: Proton magnetic resonance spectrum of the Zr(IV) chelate.

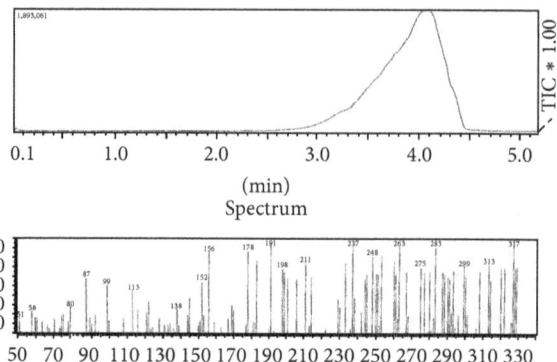

FIGURE 12: Mass spectrum of the 2-[(4-[(Z)-1-(2-hydroxyphenyl) ethylidene]aminobutyl)-ethanimidoyl]phenol.

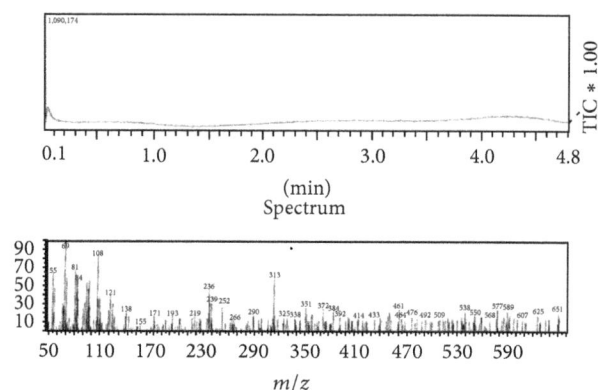

FIGURE 13: Mass spectrum of the Zr(IV) chelate.

which is analogous to the loss of four water molecules and the fragments at $m/z = 239$ and 135 assigned to the loss of zirconium oxide. Meanwhile, the fragments have been observed at m/z values, 121, 108, and 81, suggesting different fragmentations (Scheme 2). Another important fragment appears at $m/z = 69$ which is due to the loss of C_5H_9 and the fragment at $m/z = 55$ is attributed to C_4H_7. The mass spectrum of $[La(L)(OH_2)_2]\cdot5H_2O$ chelate has been recorded. The spectrum of the La(III) chelate exhibited the molecular ion (M^+) peak at m/z 588 suggesting the monomeric nature of the chelate. The fragments at m/z values 496, 460, and 329 are analogous to the loss of $C_8H_{21}NO_8$ molecule from chemical formula of the chelate and fragments at m/z 232

may be due to $C_6H_{11}N$ ion. The same spectrum of chelate shows fragment at m/z 96 is corresponding to the loss of lanthanum(III) ion and the other important fragment at m/z 55 is assigned to the loss of C_4H_7 and C_4H_7. For Ce(IV) chelate, the spectrum exhibits a fragment at m/z 550 due to the original molecular weight of the chelate. The mass spectrum of cerium chelate exhibits fragments at m/z values 461, 329, 260, 247, and 232 suggesting different fragmentations and the other fragment at m/z 93 is corresponding to the loss of cerium(IV) ion from chemical formula of the chelate. The final fragment at m/z 64 is analogous to the appearance of C_5H_4; these fragments are attributed to loss of different atoms (see Scheme 3). The above fragmentations illustrate the formation of the Schiff base and the formation of the chelates in 1 : 1 [M : L] ratio.

3.7. Biological Assay of the Schiff Base and Its Chelates. The *in vitro* biological activity of the synthesized Schiff base and its chelates were screened for their activity against *Escherichia coli* and *Salmonella kentucky* (Gram negative), *Lactobacillus fermentum* and *Streptococcus faecalis* (Gram positive), and antifungal activity against *Aspergillus niger* and *Fusarium solani* were carried out. The bactericidal and fungicidal investigation data of the compounds are summarized in the following results.

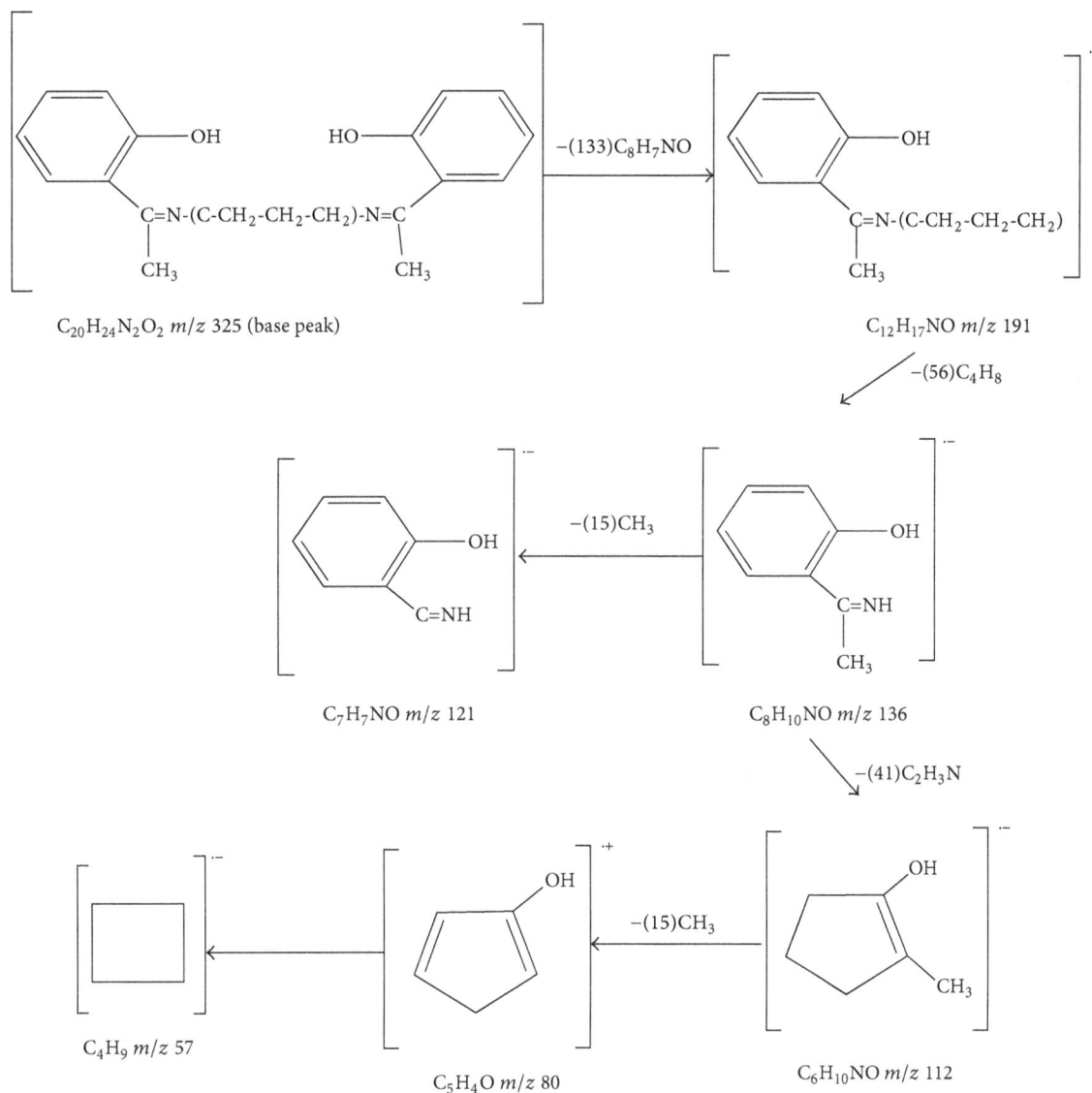

SCHEME 1: Mass spectral fragmentations of the 2-[(4-[(Z)-1-(2-hydroxyphenyl)ethylidene]aminobutyl)-ethanimidoyl]phenol.

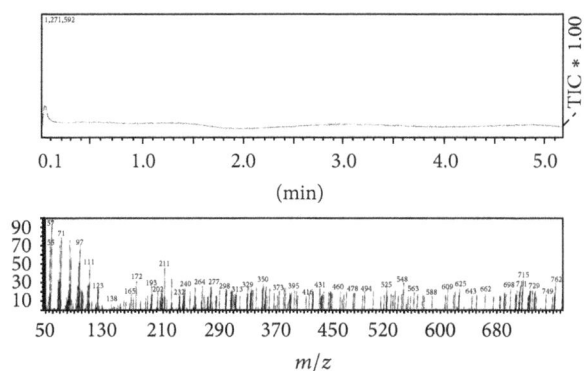

FIGURE 14: Mass spectrum of the La(III) chelate.

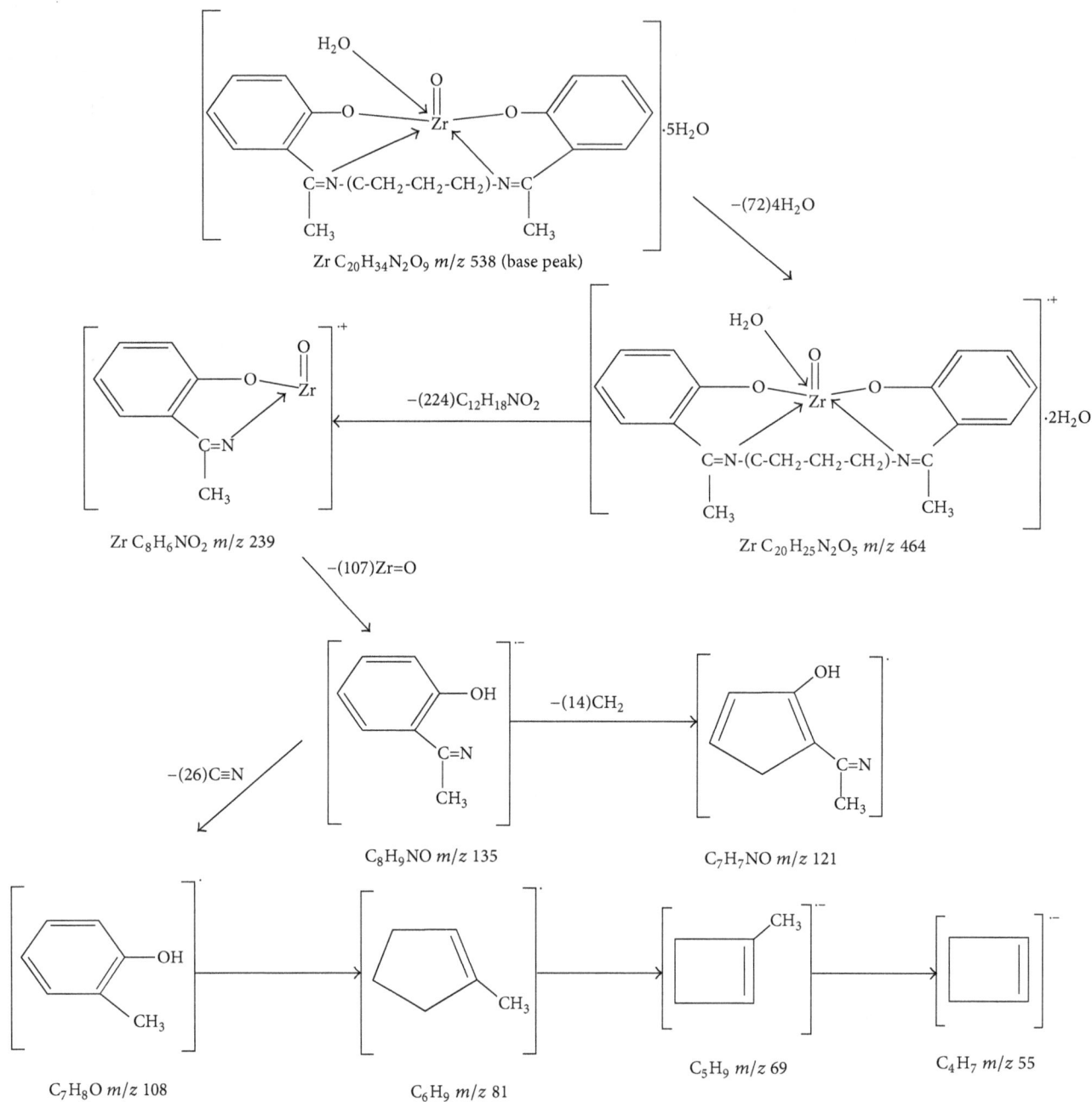

SCHEME 2: Mass spectral fragmentations of the Zr(IV) chelate.

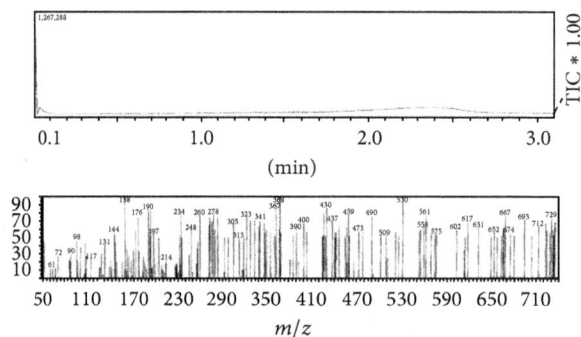

FIGURE 15: Mass spectrum of the Ce(IV) chelate.

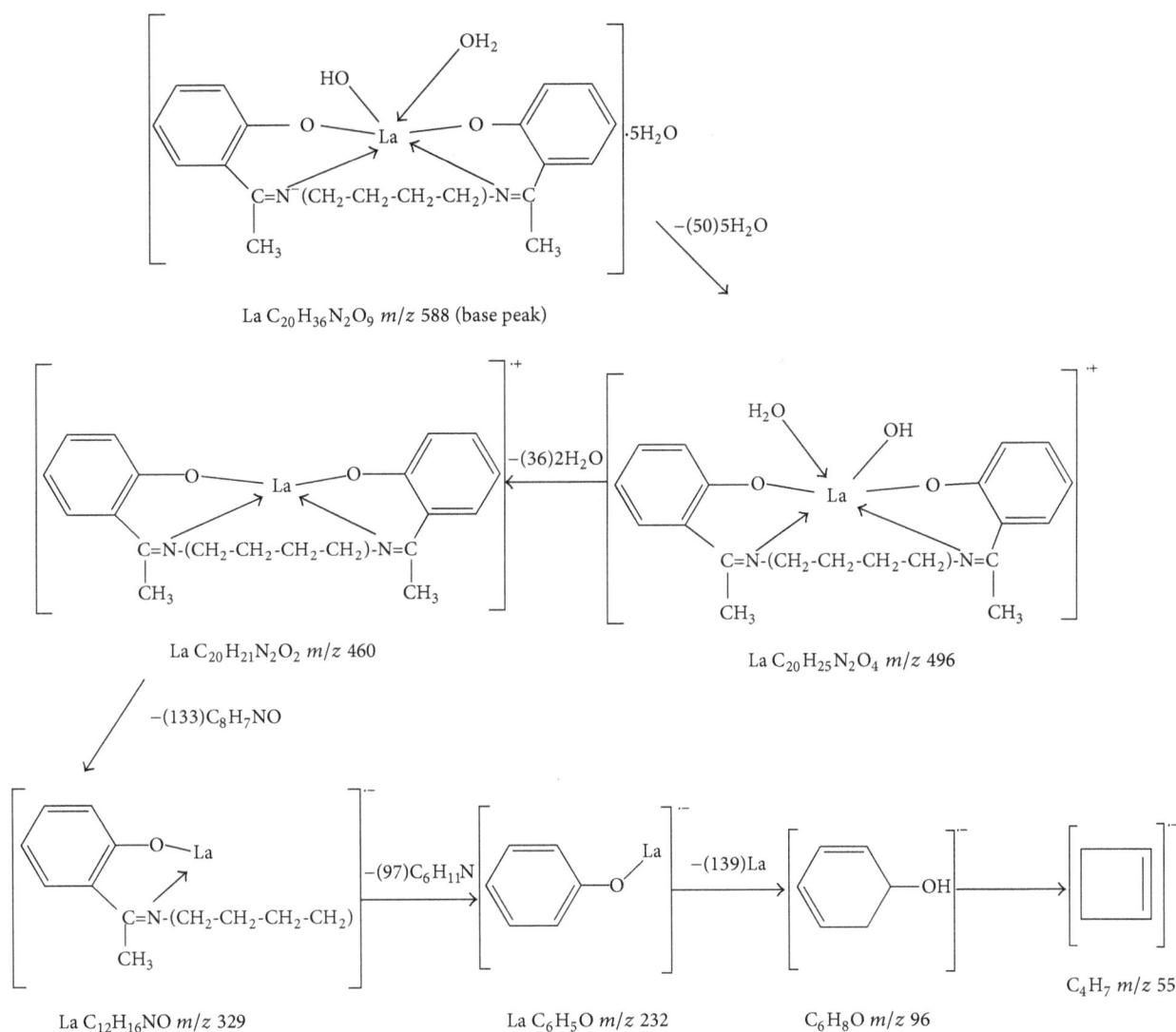

SCHEME 3: Mass spectral fragmentations of the La(III) chelate.

3.7.1. Antibacterial Study. The results of antibacterial activity are presented in Tables 3 and 4 and Figures 16–19 and 22–25. The samples for the reagents to synthesize Schiff base and its corresponding chelates (1B, 2B, 3B, 4B, and 5B) showed good to moderate activity against the tested bacteria strains with a wide activity index range that varied between 0 and 46%. The sample (1B) for o-hydroxyacetophenone was ineffective against the proliferation of Gram negative bacteria *Escherichia coli* and Gram positive bacteria *Streptococcus faecalis* but slightly effective against *Salmonella kentucky* and *Lactobacillus fermentum* with activity index within 3–13%. The sample (2B) for 1,4-butanediamine was found to be active against all organisms at all tested concentrations with the activity index within 0–45%. The sample (3B) was found to be ineffective against *Escherichia coli* and *Streptococcus faecalis* at all the concentrations tested; however, it showed moderate activity against *Salmonella kentucky* and *Lactobacillus fermentum* at 500 μg/mL concentration with an activity index

of 40%. The samples (4B) and (5B) for La(NO₃)₃·6H₂O and Ce(SO₄)₂·4H₂O exhibited moderate activity against all the tested bacteria in the range of 0–46% activity index. Schiff base (1C) and La(III) chelate (2C) showed higher activity against *Streptococcus faecalis* with activity index of 17% and 46%, respectively, at 500 μg/mL concentration compared to the other bacterial strains. Zr(IV) chelate (3C) displayed antibacterial activity against *Escherichia coli* and *Streptococcus faecalis* with activity index of 25 and 26%, respectively, while it was ineffective against other bacteria. The sample 4C (Ce(IV) chelate) showed no growth inhibition against Gram positive bacteria and slight activity against Gram negative bacteria with activity index of 20–23%. To sum up, the best activity was recorded by Zr(IV) chelate and Ce(IV) chelate against both Gram negative bacteria tested. La(III) chelate showed the highest activity against Gram positive bacterium *Streptococcus faecalis*. However, all of the compounds were ineffective at growth inhibition of Gram negative bacterium

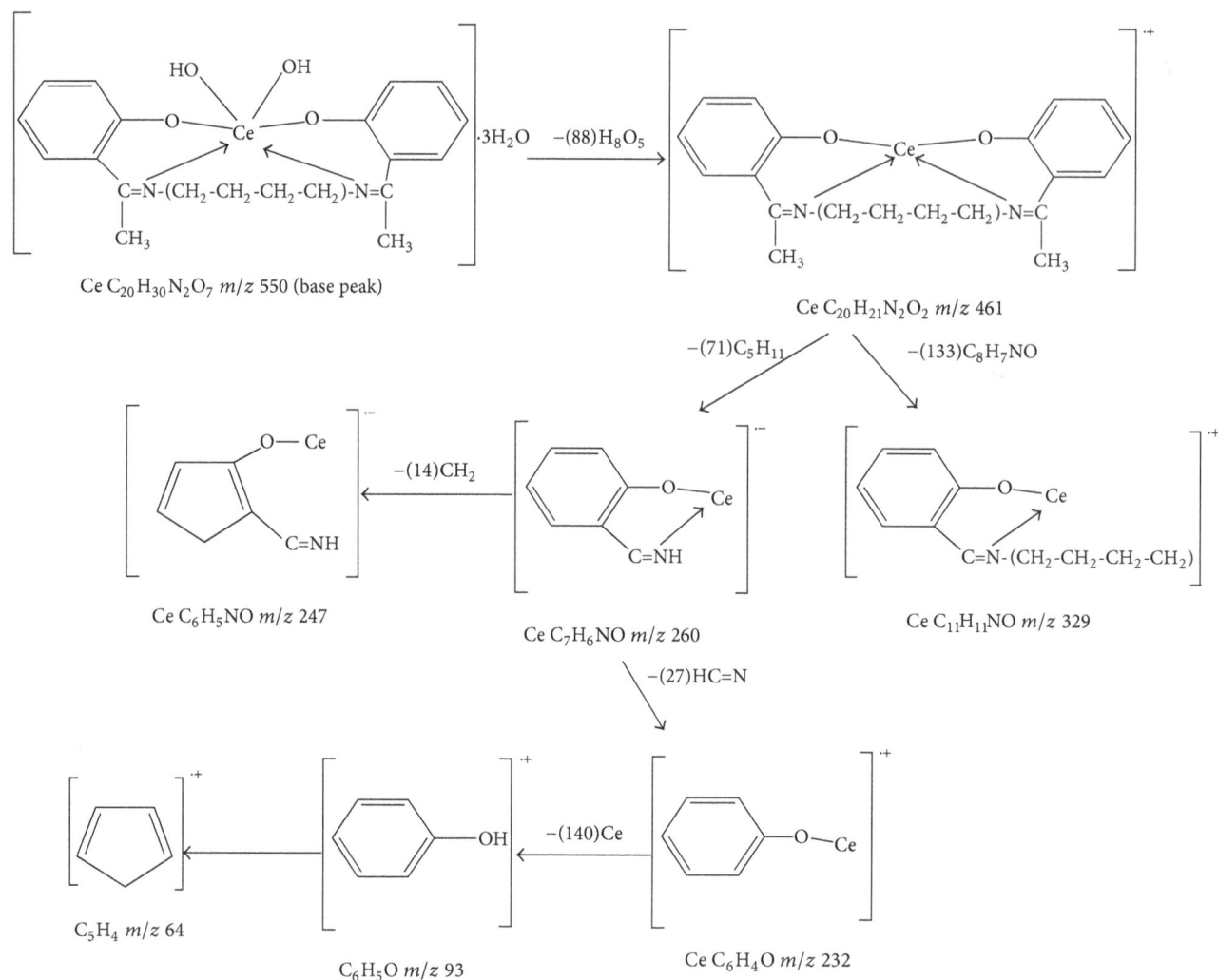

SCHEME 4: Mass spectral fragmentations of the Ce(IV) chelate.

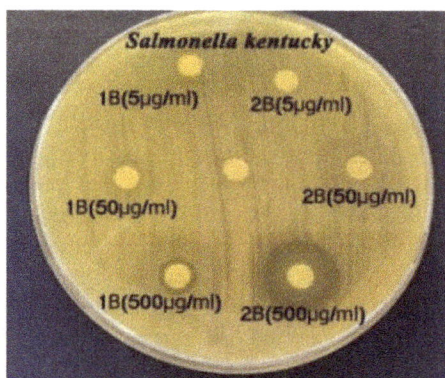

FIGURE 16: Effect of *o*-hydroxyacetophenone and 1,4-butanediamine on *Escherichia coli*.

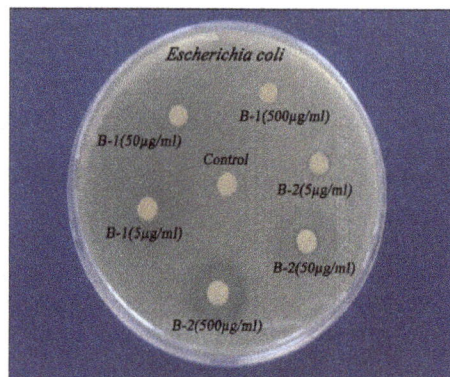

FIGURE 17: Effect of *o*-hydroxyacetophenone and 1,4-butanediamine on *Salmonella Kentucky*.

Lactobacillus fermentum. In general, judging from the results it seems that chelation improved the antibacterial activity compared to the Schiff base.

3.7.2. Antifungal Study. The antifungal activity results are listed in Table 5 and Figures 20, 21, 26, and 27. The sample 2B was most effective against *Aspergillus niger* at 19 and 90% at concentration of 500 μg/mL and ineffective against *Fusarium*

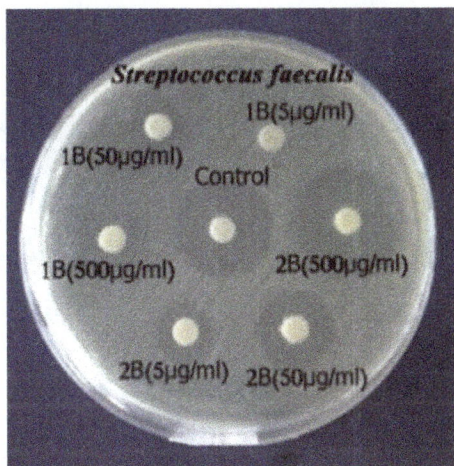

FIGURE 18: Effect of *o*-hydroxyacetophenone and 1,4-Butanediamine on *Streptococcus faecalis*.

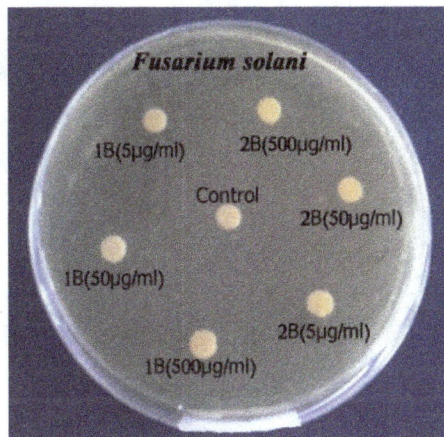

FIGURE 19: Effect of *o*-hydroxyacetophenone and 1,4-butanediamine on *Lactobacillus fermentum*.

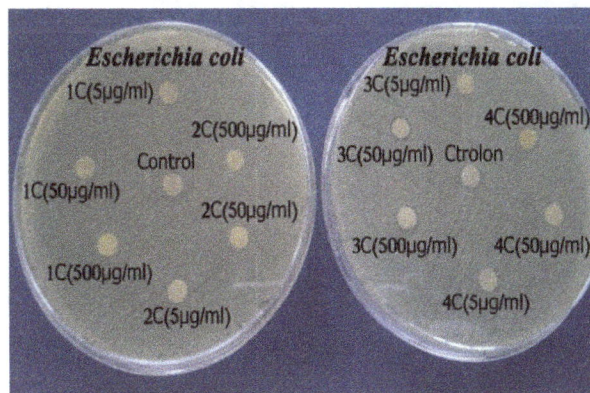

FIGURE 20: Effect of ketone and 1,4-butanediamine on *Aspergillus niger*.

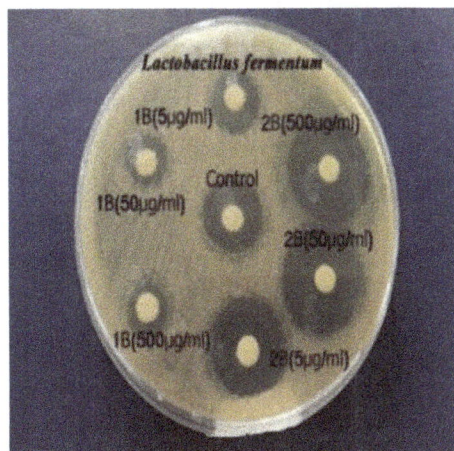

FIGURE 21: Effect of ketone and 1,4-butanediamine on *Fusarium solani*.

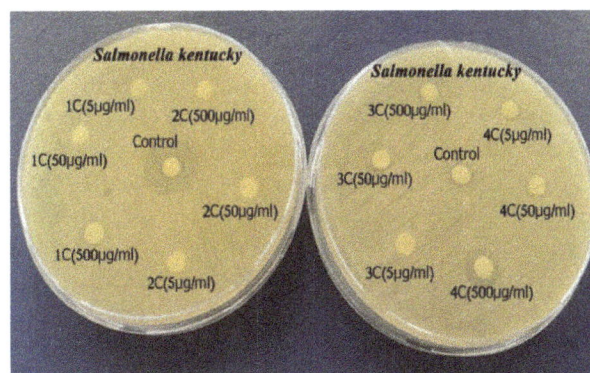

FIGURE 22: Effect of SB1 and Zr(IV), La(III), and Ce(IV) chelates on *Escherichia coli*.

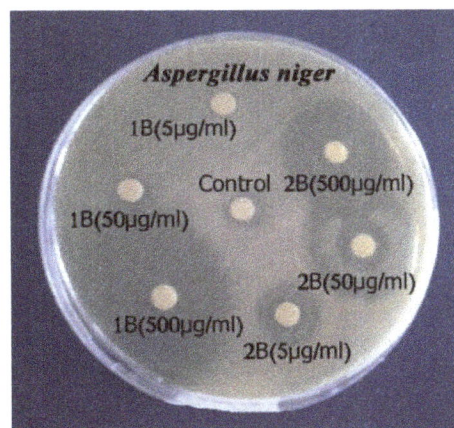

FIGURE 23: Effect of SB1 and Zr(IV), La(III), and Ce(IV) chelates on *Salmonella kentucky*.

solani. Meanwhile, the CIP exhibited higher activity against fungus in the range of 13–100% activity index. The samples 1B, 3B, 4B, and 5B exhibited slight to moderate activity with the activity index of 0–38%, when compared to the standard drug miconazole with growth inhibition against *Aspergillus niger* expect *Fusarium solani* which showed ineffective. Therefore,

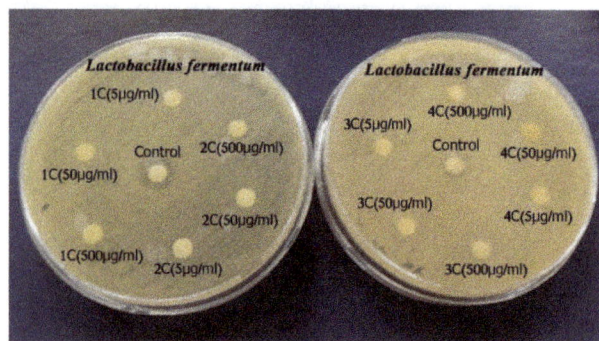

FIGURE 24: Effect of SB and Zr(IV), La(III), and Ce(IV) chelates on *Lactobacillus fermentum*.

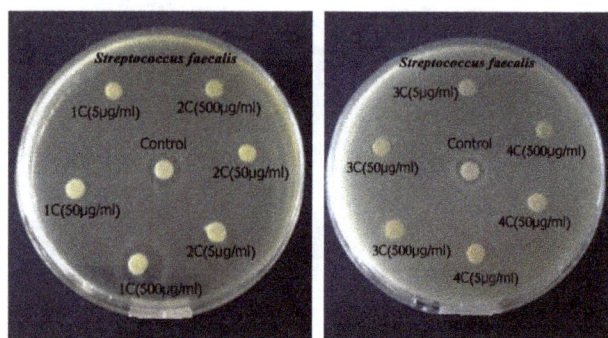

FIGURE 25: Effect of SB and Zr(IV), La(III), and Ce(IV) chelates on *Streptococcus faecalis*.

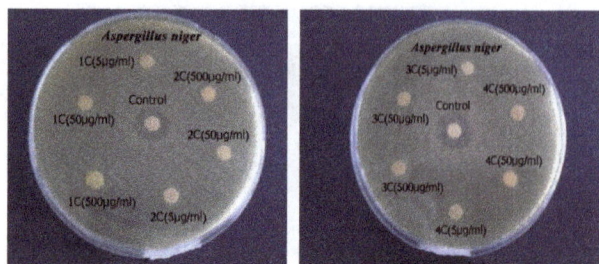

FIGURE 26: Effect of SB and Zr(IV), La(III), and Ce(IV) chelates on *Aspergillus niger*.

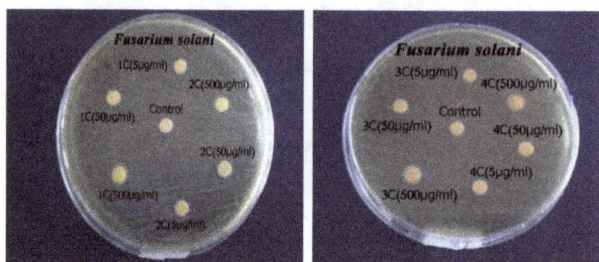

FIGURE 27: Effect of SB and Zr(IV), La(III), and Ce(IV) chelates on *Fusarium solani*.

Schiff base (1C), La(III) chelate (2C), Zr(IV) chelate (3C), and Ce(IV) chelate (4C) were found to be ineffective against the tested fungus represented by *Aspergillus niger* and *Fusarium solani*. The Schiff base and its chelates could enhance the antimicrobial effect on both strains probably by the azomethine nitrogen groups. The activities of all the tested chelates may be explained on the basis of chelation theory; chelation reduces the polarity of the metal atom mainly because of partial sharing of its positive charge with the donor groups and possible π-electron delocalization within the whole chelate ring. Also, chelation increases the lipophilic nature of the central atom which subsequently favors its permeation through the lipid layer of the cell membrane [26–29].

4. Conclusion

In this paper, a Schiff base ligand 2-[(4-[(Z)-1-(2-hydroxyphenyl)ethylidene]aminobutyl)-ethanimidoyl]phenol is derived from the condensation of 2-hydroxyacetophene and 1,4-diaminobutane. The synthesized ligand is used in making some chelates with Zr(IV), La(III), and Ce(IV) ions. Based on the data of the elemental analysis, molar conductivity,

thermogravimetric analysis, IR, ^{1}HNMR, and electronic and mass spectra, an octahedral structure was suggested for all the chelates as shown in Figure 28.

The biological activity results showed that all chelates have been found to be moderately potent compared to their ligand because the process of chelation dominantly affects the overall biological behavior of the chelates. Some of the chelates have higher antibacterial and antifungal activities than the ligand. However, biological activities are less than standards. All these observations put together lead us to propose six coordinated octahedral structures to Zr(IV), La(III), and Ce(IV) chelates. These observations showed that the majority of the salts are more active than their respective Schiff bases. In some cases, Schiff base and its chelates have similar activity against bacteria and resistive fungi. Chelation may enhance or suppress the biochemical potential of bioactive organic species. The higher activity of the chelates may be owing to the effect of metal ions on the normal cell membrane. Metal chelates bear polar properties together; this makes them suitable for permeation to the cells and tissues. Changing hydrophilicity and lipophilicity probably leads to bringing down the solubility and permeability barriers of cell, which in turn enhances the bioavailability of chemotherapeutics on one hand and potentiality on another [30]. The biological activities of the Schiff base ligand under investigation and its chelates against bacterial and fungal organisms are promising which need further and deep studies on animals and humans.

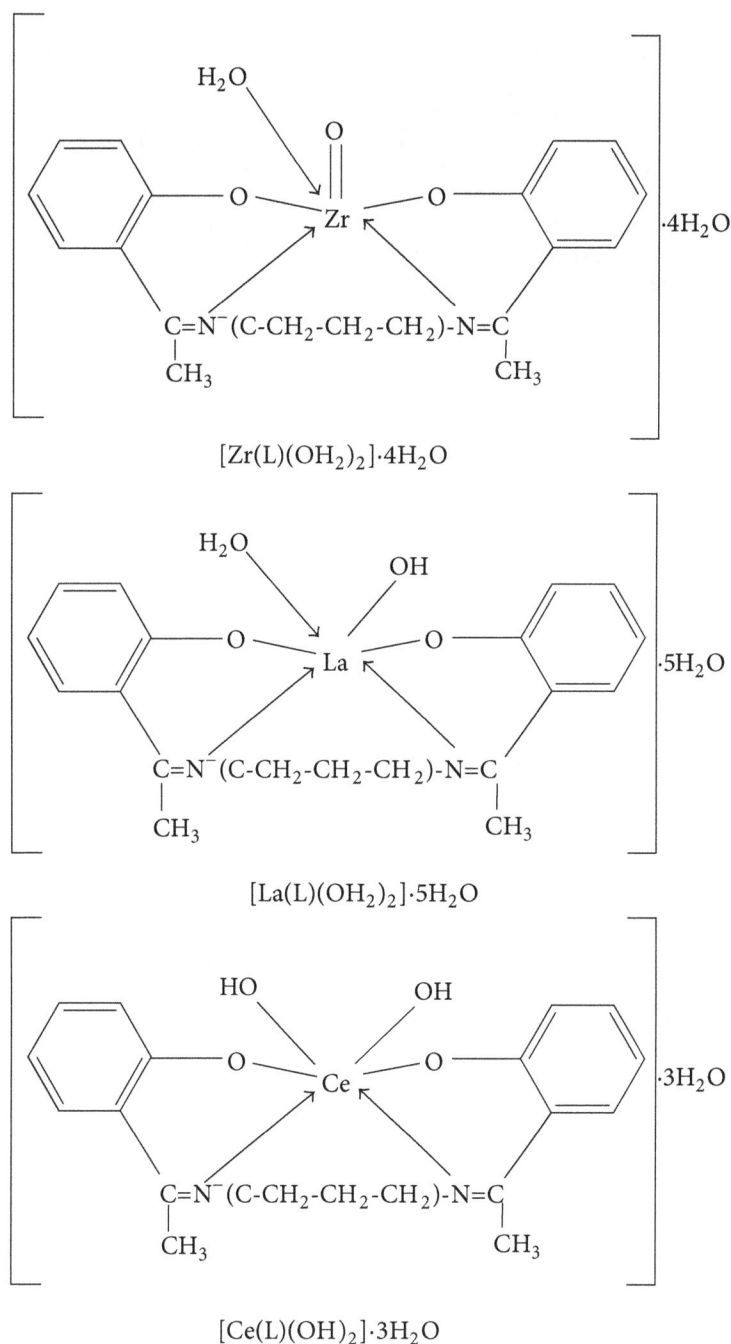

$[Zr(L)(OH_2)_2]\cdot 4H_2O$

$[La(L)(OH_2)_2]\cdot 5H_2O$

$[Ce(L)(OH)_2]\cdot 3H_2O$

FIGURE 28

Conflict of Interests

The authors declare that there is no conflict of interests regarding the publication of this paper.

References

[1] B. Sindhukumari, G. Rijuulal, and K. Mohanan, "Microwave assisted synthesis, spectroscopic, thermal and biological studies of some lanthanide(III) chloride complexes with a heterocyclic Schiff base," *Synthesis and Reactivity in Inorganic, Metal-Organic, and Nano-Metal Chemistry*, vol. 39, no. 1, pp. 24–30, 2009.

[2] K. Singh, M. S. Barwa, and P. Tyagi, "Synthesis, characterization and biological studies of Co(II), Ni(II), Cu(II) and Zn(II) complexes with bidentate Schiff bases derived by heterocyclic ketone," *European Journal of Medicinal Chemistry*, vol. 41, no. 1, pp. 147–153, 2006.

[3] P. Liu, M. Huang, W. Pan, Y. Zhang, J. Hu, and W. Deng, "Synthesis and luminescence properties of europium and terbium complexes with pyridine- or bipyridine-linked oligothiophene ligand," *Journal of Luminescence*, vol. 121, no. 1, pp. 109–112, 2006.

[4] M. A. Sakhare, S. L. Khillare, M. K. Lande, and B. R. Arbad, "Synthesis, characterisation and antimicrobial studies

on La(III), Ce(III) and Pr(III) complexes with a tetraaza macrocyclic Ligand," *Advances in Applied Science Research*, vol. 4, no. 1, pp. 94–100, 2013.

[5] H. Naeimi, F. Salimi, and K. Rabiei, "Convenient, mild and rapid synthesis and characterization of some Schiff-base ligands and their complexes with uranyl(II) ion," *Journal of Coordination Chemistry*, vol. 61, no. 22, pp. 3659–3665, 2008.

[6] A. W. Bauer, W. M. Kirby, J. C. Sherris, and M. Turck, "Antibiotic susceptibility testing by a standardized single disk method," *The American Journal of Clinical Pathology*, vol. 45, no. 4, pp. 493–496, 1966.

[7] A. S. Munde, V. A Shelke, S. M. Jadhav et al., "Synthesis, characterization and antimicrobial activities of some transition metal complexes of biologically active asymmetrical tetradentate ligands," *Advances in Applied Science Research*, vol. 3, no. 1, pp. 175–182, 2012.

[8] P. S. Mane, S. G. Shirodkar, B. R. Arbad, and T. K. Chondhekar, *Indian Journal of Chemistry*, vol. 40, p. 648, 2001.

[9] G. Monica et al., *International Journalof Pharmaceutical and Biomedical Research*, vol. 2, no. 2, p. 110, 2001.

[10] S. M. E Khalil, S. L. Stefan, and K. A. Bashir, "The synthesis and Co(II), Ni(II), Cu(II) and Uo₂(Vi) complexes of 1,2-O-Benzal-4-Aza-7-Aminoheptane," *Synthesis and Reactivity in Inorganic and Metal-Organic Chemistry*, vol. 29, no. 10, p. 1685, 1999.

[11] M. Pekerci and E. Tap, "The synthesis and characterization of 1,2-O-cyclohexylidene-4-aza-8-aminooctane and some of its transition metal complexes," *Heteroatom Chemistry*, vol. 11, pp. 254–260, 2000.

[12] O. E. Sherif and H. M. Abd El-Fattah, "Characterization of transition metal ion chelates with 8-(arylazo) chromones using thermal and spectral techniques," *Journal of Thermal Analysis and Calorimetry*, vol. 74, no. 1, pp. 181–200, 2003.

[13] D. C. Dash, A. Mahapatra, U. K. Mishra, and R. K. Mohapatra, "Synthesis and characterization of transition metal complexes with benzimidazolyl-2-hydrazones of salicylidene acetone and salicylidene acetophenone," *Journal of Chemical and Pharmaceutical Research*, vol. 4, no. 1, pp. 279–285, 2012.

[14] T. Nishide, M. Sato, and H. Hara, "Crystal structure and optical property of TiO_2 gels and films prepared from Ti-edta complexes as titania precursors," *Journal of Materials Science*, vol. 35, no. 2, pp. 465–469, 2000.

[15] H. A. Abdullah, M. M. El-ajaily, E. E. Saad, A. A. Janga, A. A. Maihub, and J. Pharm, "Preparation, spectroscopic investigation and biological evaluation of schiff's base La(III), Zr(IV)and Ce(IV) chelates," *International Journal of Pharmaceutical and Chemical Sciences*, vol. 3, no. 2, p. 463, 2014.

[16] F. M. Morad, M. M. El-Ajaily, and A. A. Maihub, *Egyptian Journal of Analytical Chemistry*, vol. 15, p. 98, 2006.

[17] K. R. Krishnapriya and M. Kandaswamy, "Coordination properties of a dicompartmental ligand with tetra- and hexadentate coordination sites towards copper (II) and nickel (II) ions," *Polyhedron*, vol. 24, no. 1, pp. 113–120, 2005.

[18] S. M. Ben-Saber, A. A. Maihub, S. S. Hudere, and M. M. El-Ajaily, "Complexation behavior of Schiff base toward transition metal ions," *The American Microchemical Journal*, vol. 81, no. 2, pp. 191–194, 2005.

[19] A. A. A. Emara, "Structural, spectral and biological studies of binuclear tetradentate metal complexes of N₃O Schiff base ligand synthesized from 4,6-diacetylresorcinol and diethylenetriamine," *Spectrochimica Acta Part A: Molecular and Biomolecular Spectroscopy*, vol. 77, pp. 117–125, 2010.

[20] S. B. Kalia, K. Lumba, G. Kaushal, and M. Sharma, "Magnetic and spectral studies on cobalt(II) chelates of a dithiocarbazate derived from isoniazid," *Indian Journal of Chemistry Section A*, vol. 46, no. 8, pp. 1233–1239, 2007.

[21] M. El-Ajaily, A. A. Maihub, F. I. Moshety, and H. A. Boshalla, "A complex formation of TiO (IV), Cr (III) and Pb (II) ions using 1,3-bis(2-hydroxybenzylidene)thiourea as ligand," *Egyptian Journal of Analytical Chemistry*, vol. 20, no. 16, pp. 18–25, 2011.

[22] D. Suresh Kumar and V. Alexander, "Synthesis of lanthanide(III) complexes of chloro- and bromo substituted 18-membered tetraaza macrocycles," *Polyhedron*, vol. 18, no. 11, pp. 1561–1568, 1999.

[23] M. Usharani, E. Akila, P. Jayaseelan, and R. Rajavel, "Structural elucidation of newly synthesized potentially active binuclear Schiff base Cu(II), Ni(II), Co(II) and Mn(II) complexes using physicochemical methods," *International Journal of Scientific & Engineering Research*, vol. 4, no. 7, pp. 1055–1064, 2013.

[24] R. M. Silverstein and F. X. Webster, *Spectrometric Identification of Organic Compounds*, John Wiley & Sons, New Delhi, India, 1st edition, 2007.

[25] M. M. El-Ajaily, H. F. Bomorawiaha, and A. A. Maihub, *Egyptian Journal of Analytical Chemistry*, vol. 16, p. 16, 2007.

[26] M. Tümer, H. Köksal, M. Kasim Sener, and S. Serin, "Antimicrobial activity studies of the binuclear metal complexes derived from tridentate Schiff base ligands," *Transition Metal Chemistry*, vol. 24, no. 4, pp. 414–420, 1999.

[27] M. Imran, J. Iqbal, S. Iqbal, and N. Ijaz, "In vitro antibacterial studies of ciprofloxacin-imines and their complexes with Cu(II),Ni(II),Co(II), and Zn(II)," *Turkish Journal of Biology*, vol. 31, no. 2, pp. 67–72, 2007.

[28] F. Azam, S. Singh, S. L. Khokhra, and O. Prakash, "Synthesis of Schiff bases of naphtha[1,2-*d*]thiazol-2-amine and metal complexes of 2-(2′-hydroxy)benzylideneaminonaphthothiazole as potential antimicrobial agents," *Journal of Zhejiang University. Science B*, vol. 8, no. 6, pp. 446–452, 2007.

[29] S. M. Pradeepa, H. S. B. Naik, B. V. Kumar, K. I. Priyadarsini, A. Barik, and T. R. R. Naik, "Cobalt(II), Nickel(II) and Copper(II) complexes of a tetradentate Schiff base as photosensitizers: quantum yield of $^1O_2$ generation and its promising role in anti-tumor activity," *Spectrochimica Acta Part A: Molecular and Biomolecular Spectroscopy*, vol. 101, pp. 132–139, 2013.

[30] D. Rehder, "Biological and medicinal aspects of vanadium," *Inorganic Chemistry Communications*, vol. 6, no. 5, pp. 604–617, 2003.

Permissions

List of Contributors

Namdeo R. Jadhav, Ramesh S. Kambar and Sameer J. Nadaf
Department of Pharmaceutics, Bharati Vidyapeeth College of Pharmacy, Kolhapur, Maharashtra 416013, India

Sadjia Bennour and Fatma Louzri
Laboratory of Polymer Materials, Faculty of Chemistry, University of Sciences and Technology Houari Boumediene, BP 32, El Alia,16111 Algiers, Algeria

Douglas L. Strout
Department of Physical Sciences, Alabama State University, Montgomery, AL 36101, USA

Saikat S. Mallick and Vidya V. Dighe
Chemistry Department, Ramnarain Ruia College, Matunga East, Mumbai, Maharashtra 400 019, India

Rakesh Kumar
Department of Chemistry, Bio-organic Laboratory, Kirori Mal College, University of Delhi, Delhi 110007, India

Sonam Ruhil, Neetu Phougat and Anil K. Chhillar
Centre for Biotechnology, Maharshi Dayanand University, Rohtak 124 001, India

Jyoti Arora
Department of Chemistry, Bio-organic Laboratory, Kirori Mal College, University of Delhi, Delhi 110007, India
Department of Chemistry, Bio-organic Laboratory, University of Delhi, Delhi 110 007, India

Ashok K. Prasad
Department of Chemistry, Bio-organic Laboratory, University of Delhi, Delhi 110 007, India

Hassan H. Hammud
Chemistry Department, Faculty of Science, King Faisal University, Al-Ahsa 31982, Saudi Arabia
Chemistry Department, Faculty of Science, Beirut Arab University, Beirut, Lebanon

Ali El-Shaar
Chemistry Department, Faculty of Science, Beirut Arab University, Beirut, Lebanon

Essam Khamis
City of Scientific Research & Technological Applications, Borg Al Arab, Alexandria, Egypt

El-SayedMansour
Faculty of Science, Alexandria University, Alexandria, Egypt

Isam Eldin Hussein Elgailani and Christina Yacoub Ishak
Department of Chemistry, Faculty of Science, University of Khartoum, P.O Box 321, 11115 Khartoum, Sudan

Mathieu Bosch and Muwei Zhang
Department of Chemistry, Texas A&M University, College Station, TX 77842, USA

Hong-Cai Zhou
Department of Chemistry, Texas A&M University, College Station, TX 77842, USA
Department of Materials Science and Engineering, Texas A&M University, College Station, TX 77842, USA

Mohammad Reza Nazarifar
Young Researchers and Elites Club, Shiraz Branch, Islamic Azad University, Shiraz, Iran

Abdelrahman A. Badawy
Physical Chemistry Department, Center of Excellence for Advanced Science, Renewable Energy Group, National Research Center, Dokki, Cairo 12622, Egypt

Shaymaa E. El-Shafey and Gamil A. El-Shobaky
Physical Chemistry Department, National Research Center, Dokki, Cairo 12622, Egypt

Suzan Abd El All
Radiation Physics Department, National Center for Radiation Research and Technology (NCRRT), Nasr City, Cairo 11762, Egypt

C. Díaz
Departamento de Química Módulo 13, Universidad Autónoma de Madrid, 28049 Madrid, Spain

Siafu Ibahati Sempeho and Askwar Hilonga
Department of Materials Science and Engineering, Nelson Mandela African Institution of Science and Technology, P.O. Box 447, Arusha, Tanzania

Hee Taik Kim
Department of Chemical Engineering, Hanyang University, 1271 Sa 3-dong, Sangnok-gu, Ansan-si, Gyeonggi-do 426-791, Republic of Korea

Egid Mubofu
Chemistry Department, University of Dar es Salaam, P.O. Box 35091, Dar es Salaam, Tanzania

Aditi Anand, Navjeet Kaur, and Dharma Kishore
Department of Chemistry, Banasthali University, Banasthali, Raj 304022, India

Anil Kumar Koneti
Department of Physics, Sri Vani School of Engineering, Chevuturu, Andhra Pradesh 521 229, India

Srinivasu Chintalapati
Department of Physics, Andhra Loyola College, Vijayawada, Andhra Pradesh 520 008, India

Laura Micheli
Department of Chemical Sciences and Technologies, University of Rome "Tor Vergata", Via della Ricerca Scientifica, 00133 Rome, Italy
Consorzio Interuniversitario Biostrutture e Biosistemi "INBB", Viale Medaglie d'Oro 305, 00136 Rome, Italy

ClaudiaMazzuca, Eleonora Cervelli and Antonio Palleschi
Department of Chemical Sciences and Technologies, University of Rome "Tor Vergata", Via della Ricerca Scientifica, 00133 Rome, Italy

Fahim Nawaz, Rashid Ahmad, Ejaz AhmadWaraich and Rana Nauman Shabbir
Department of Crop Physiology, University of Agriculture, Faisalabad 38040, Pakistan

Muhammad Yasin Ashraf
Nuclear Institute for Agriculture and Biology (NIAB), P.O. Box No. 128, Faisalabad, Pakistan

Elaine Rose Maia and Daniela Regina Bazuchi Magalhães
Laboratório de Estudos Estruturais Moleculares (LEEM), Instituto de Química, Universidade de Brasília (UnB), CP 4478, 70904-970 Brasilia, DF, Brazil

Dan A. Lerner and Dorothée Berthomieu
Institut Charles Gerhardt, UMR 5253 CNRS-UM2-ENSCM-UM1, Matériaux Avancés pour la Catalyse et la Santé, ENSCM, 8 rue de l'Ecole Normale, 34296 Montpellier Cedex 5, France

Jean-Marie Bernassau
Sanofi Montpellier, 264 rue du Professeur Blayac, 34080 Montpellier, France

Tatiana Kovalchuk-Kogan, Valery Bulatov, and Israel Schechter
Schulich Department of Chemistry, Technion, Israel Institute of Technology, 32000 Haifa, Israel

Fahad Mohammed Al-Jasass
King Abdulaziz City for Science & Technology, Life Science and Environment Research Institute,National Center for Agricultural Technology, P.O. Box 6086, Riyadh 11442, Saudi Arabia

Muhammad Siddiq
Food Science Consulting,Windsor, ON, Canada N9H 2M4

Dalbir S. Sogi
Department of Food Science and Technology, Guru Nanak Dev University, Amritsar, India

Yehuda Haas
Institute of Chemistry, the Hebrew University of Jerusalem, 91904 Jerusalem, Israel

William M. Motswainyana
Department of Chemistry, Faculty of Science and Agriculture, University of Fort Hare, Private Bag X1314, Alice 5700, South Africa
Botswana Institute for Technology Research and Innovation, Private Bag 0082, Gaborone, Botswana

Peter A. Ajibade
Department of Chemistry, Faculty of Science and Agriculture, University of Fort Hare, Private Bag X1314, Alice 5700, South Africa

Pankaj Dwivedi
Department of Pharmaceutical Chemistry, Krupanidhi College of Pharmacy, Bangalore 35, India

Kuldipsinh P. Barot, Shailesh Jain and Manjunath D. Ghate
Department of Pharmaceutical Chemistry, Institute of Pharmacy, Nirma University, Ahmedabad, Gujarat 382481, India

Hua Qian
School of Chemical Engineering, Nanjing University of Science and Technology, Xiaolingwei 200, Nanjing, Jiangsu 210094, China

Hengdao Quan
National Institute of Advanced Industrial Science and Technology (AIST), Tsukuba, Ibaraki 305-8565, Japan

Mohamed ElMehdi Touati and Habib Boughzala
Laboratoire de Matériaux et Cristallochimie, Faculté des Sciences, Université de Tunis El Manar, 2092 Tunis, Tunisia

M. M. El-ajaily
Department of Chemistry, Faculty of Science, Benghazi University, Benghazi, Libya

H. A. Abdullah and E. E. Saad
Department of Chemistry, Faculty of Science, Sebha University, Sebha, Libya

Ahmed Al-janga
Department of Zoology, Faculty of Science, Sebha University, Sebha, Libya

A. A. Maihub
Department of Chemistry, Faculty of Science, Tripoli University, Tripoli, Libya